Cognitive Radio and Interference Management:

Technology and Strategy

Meng-Lin Ku
National Central University, Taiwan, R.O.C.

Jia-Chin Lin
National Central University, Taiwan, R.O.C.

Information Science
REFERENCE

Managing Director:	Lindsay Johnston
Senior Editorial Director:	Heather A. Probst
Book Production Manager:	Jennifer Romanchak
Publishing Systems Analyst:	Adrienne Freeland
Managing Editor:	Joel Gamon
Development Editor:	Myla Merkel
Assistant Acquisitions Editor:	Kayla Wolfe
Typesetter:	Henry Ulrich
Cover Design:	Nick Newcomer

Published in the United States of America by
Information Science Reference (an imprint of IGI Global)
701 E. Chocolate Avenue
Hershey PA 17033
Tel: 717-533-8845
Fax: 717-533-8661
E-mail: cust@igi-global.com
Web site: http://www.igi-global.com

Library of Congress Cataloging-in-Publication Data

Cognitive radio and interference management: technology and strategy / Meng-Lin Ku and Jia-Chin Lin, editors.
p. cm.
Includes bibliographical references and index.
Summary: “This book brings together state-of-the-art research results on cognitive radio and interference management from both theoretical and practical perspectives, serving as a bridge between people who are working to develop theoretical and practical research in cognitive radio and interference management, and therefore facilitate the future development of cognitive radio and its applications”-- Provided by publisher.
ISBN 978-1-4666-2005-6 (hardcover) -- ISBN 978-1-4666-2006-3 (ebook) -- ISBN 978-1-4666-2007-0 (print & perpetual access) 1. Cognitive radio networks. 2. Radio--Interference--Prevention. I. Ku, Meng-Lin, 1980- II. Lin, Jia-Chin, 1972-
TK5103.4815.C46 2013
621.382’1--dc23

2012012203

British Cataloguing in Publication Data
A Cataloguing in Publication record for this book is available from the British Library.

List of Reviewers

Han-Kui Chang, *National Central University, Taiwan*
Yu-Ting Sun, *National Central University, Taiwan*
Chi-Sheng Lin, *National Central University, Taiwan*
Kao-Peng Chou, *National Central University, Taiwan*
Keshav Singh, *National Central University, Taiwan*

Table of Contents

Detailed Table of Contents

Nowadays, cognitive radio is one of the most promising paradigms in the arena of wireless communications, as it aims at the proficient use of radio resources. Proper utilization of the radio spectrum requires dynamic spectrum accessing. To this end, spectrum sensing is undoubtedly necessary. In this chapter, various approaches for dynamic spectrum access scheme are presented, together with a survey of spectrum sensing methodologies for cognitive radio. Moreover, the challenges are analyzed that are associated with spectrum sensing and dynamic spectrum access techniques. Sensing beacon transmitted from different cognitive terminals creates significant interference to the primary users if proper precautions have not been taken into consideration. Consequently, cognitive radio transmitter power control are finally addressed to analyze energy efficiency aspects.

The cognitive radio has been widely investigated to support modern wireless applications. To exploit the spectrum vacancies in cognitive radios, the chapter considers the collaborative spectrum sensing by multiple sensor nodes in the likelihood ratio test (LRT) frameworks. In this chapter, the functions of sensors can be served through the cooperative regular nodes in the cognitive radio, or the specifically deployed sensor nodes for spectrum sensing. In the LRT, the sensors make individual decisions. These individual decisions are then transmitted to the fusion center to make the final decision, which provides better detection accuracy than the individual sensor decisions. The author provides the lowered-bounded probability of detection (LBPD) criterion as an alternative criterion to the conventional Neyman-Pearson (NP) criterion. In the LBPD criterion, the detector pursues the minimization of the probability of false alarm while maintaining the probability of detection above the pre-defined value. In cognitive radios, the LBPD criterion limits the probabilities of channel conflicts to the primary users. Under the NP and LBPD criteria, the chapter provides explicit algorithms to solve the LRT fusion rules, the probability of false alarm, and the probability of detection for the fusion center. The fusion rules generated by the algorithms are optimal under the specified criteria. In the spectrum sensing, the fading channels influence the detection accuracies. The chapter investigates the single-sensor detection and collaborative detections of multiple sensors under various fading channels and derives testing statistics of the LRT with known fading statistics.

Harmonic coexistence of cognitive radio systems and licensed systems requires the secondary users to have the capability of sensing and keeping track of primary users' transmissions. While existing spectrum sensing methods usually assume known distributions of the primary signals, such an assumption is often not true in practice. As a result, applying existing sensing methods will often lead to unreliable detection performance in practical networks. In this chapter, the authors try to investigate the sensing performance under the distribution uncertainty of primary signals. They first investigate the performance bounds for single user detection with unknown distribution, and provide an analytical expression for the lower bound of detection probability. Moreover, they bring the distribution uncertainty into multi-user cooperative sensing. The authors formulate the optimal sensing design as a robust optimization problem, and propose an iterative algorithm to determine the optimal decision threshold for each user. Extensive simulations demonstrate the effectiveness of the proposed algorithm.

In cognitive wireless networks, secondary nodes are required to sense the spectrum and to be cognizant of their interference on the primary receiving nodes. In this chapter, the authors show that the throughputs in the primary and the secondary networks have an entangled relationship and need to be considered jointly for optimization. Assuming that the primary and secondary networks use ALOHA as their medium access control protocol, the chapter shows that the throughputs in primary and secondary networks are influenced by the choices of transmission attempt probabilities in both of these networks. Authors show that the interference constraint limits the feasible space of transmission probabilities over which these networks can optimize their throughputs. They propose a progressive transmission probability optimizations algorithm and show through simulation that the algorithm converges to the optimum values of throughput in few time slots. Future research directions and the application of the constrained game theory to the throughput optimization problem are also discussed.

Interference is a performance limiting factor in dense cellular networks with aggressive frequency reuse. Cooperation among base stations (BSs) is a promising approach for improving data rates by eliminating or mitigating interference. For the downlink, the highest spectral efficiency gains are achieved through precoding with full coordination, which requires complete channel state information (CSI) and data sharing among BSs at the cost of significant utilization of the backhaul. In this chapter, the authors introduce distributed precoding techniques for the multicell MIMO downlink. Each BS designs its own precoder without requiring data or downlink CSI of links from other BSs. The authors also study the effect of imperfect CSI on these precoders and introduce a robust precoder in the presence of CSI uncertainty. Simulations show that these methods enjoy a rate increase with SNR similar to multicell joint dirty paper coding in the high SNR regime due to effective interference mitigation.

In this chapter, the authors present the application of fuzzy logic for power control in spectrum sharing cognitive radio network. The investigated network consists of a pair of primary users (PUs) and a pair of cognitive radio users (CRs). In order to compute an optimal power scale for the CR link, a fuzzy-based optimal power control strategy is investigated based on the Mamdani fuzzy control model using two input variables: the PU's signal-to-noise ratio (SNR) and PU's interference channel gain (channel gain from the cognitive radio users to primary users). Applying the fuzzy-based optimal power control strategy, the desired quality of service (QoS) is assured on the PU link, and the efficiency of spectrum usage is also maximized on the CR link in the spectrum sharing network. In addition, the bit error rate (BER) is also improved as compared with the spectrum sharing network without power control.

Demands on high data rate applications are increasing and consequently demands on spectral resources are increasing as well. Although electromagnetic spectrum is recently said to be in a scarcity situation, several studies have shown that this scarcity is mainly due to the legacy command-and-control regulation rather than due to physical scarcity of spectrum. For this reason, researchers have started investigating techniques to better manage the usage of spectrum. Among these techniques there exist the methods allowing the primary/secondary usage of spectrum, or secondary market. Secondary market techniques mainly manage sensing, accessing, and aborting the spectrum usage by the secondary users. Techniques developed for secondary market context are also referred to as algorithms for Cognitive Radio (CR) networks. Regulators worldwide took measures to promote the deployment of primary/secondary context. In this chapter, the authors give an illustrative discussion on CR and on the application of game theory to overcome the spectrum scarcity problem. Game theory is a field of applied mathematics that describes and analyzes scenarios with interactive decisions. In recent years, there has been considerable interest in adopting game theoretic approaches to model many communications and networking problems such as radio resource management and routing. Nowadays, game theory is also used to model interactive situations for CR terminals.

This chapter investigates resource allocation in a Universal Mobile Telecommunication System (UMTS) Long Term Evolution (LTE) network. Users form coalitions, each exploiting resources in a particular femto-cell, while occupying optimal resources in the sense that total network throughput is maximized. Users in each cell collaborate to increase network throughput and simultaneously attempt to increase their own payoffs subject to a fairness criterion. Payoffs to the users are defined as the monetary equivalent of the individual users' achievable throughput in the specified coalition structure. A distributed game-theoretic resource allocation mechanism is studied whereby users autonomously decide which sub-channel in which coalition to join. It is proved that if each user operates according to the proposed algorithm, the sum throughput of all links converges almost surely to its maximum feasible value.

Thanks to the advances of multimedia application, mobile computing platform, and wireless communication technology, the research area has attracted serious attention in order to seamlessly provide interactive and ubiquitous user experience. To make it happen, the pursuit of higher system capacity in resource limited wireless networks is never-ending. Cognitive radio (CR) represents an exciting new communication paradigm with advantages on spectrum management so as to heighten channel utilization and capacity. The bandwidth demanding multimedia applications are excellent candidates to fully exploit the potential of CR. However, the research effort has been focused mainly on spectrum access while the application specific performance has been much less touched. The research considering both spectrum access and application data scheduling is emerging for maximal user experience. In this chapter, the authors first discuss advances in opportunistic spectrum access (OSA) strategies as well as multimedia QoS scheduling schemes, and then introduce the research trend on joint access and scheduling frameworks.

In recent years, vertical handover (VHO) has been identified as the primary vehicle to provision seamless mobility and quality of service (QoS) transparency for end-user in composite network. This allows end-user to enjoy ubiquitous connectivity in the most efficient way, irrespective of time and place, commonly known as always best connected. In this chapter, the authors introduce the notion of cognitive cooperation as a means to provide optimized VHO opportunistically in order to exploit the inherent heterogeneity that exists within such composite network to improve radio resource usage. Through the cognitive cooperation, the chapter proposes a distributed load adaptation strategy (LAS) framework which exploits the benefits of joint optimization, particularly between link adaptation and load adaptation on-demand. The authors advocate that such synergetic interactions between the physical layer (PHY) and medium access control (MAC) layer have advantages over the PHY approach based only on link adaptation. Comprehensive performance analyses show that the LAS framework arbitrates a QoS-balanced system in which statistical QoS guarantee for multimedia traffic can be provisioned and overall system capacity can be maximized.

This chapter addresses the issues in air interface designs for Cognitive Radios. Fuzzy logic system is used as one of the soft computing techniques to learn to sub optimality and vagueness. Many good, simple, and quick fuzzy based solutions have been developed since few decades in many diverse domains. Through this chapter, the authors first discuss the significance and need of soft computing techniques in designing such solutions and then present fuzzy based solutions for spectrum access, mobility, and management. Hierarchical fuzzy systems have been used to get over to the problem of curse of dimensionality. The proposed solutions consider an architecture, similar to the one proposed by IEEE 802.22 working group, for spectrum sharing and management. Models have been designed using fuzzy logic toolbox in MATLAB, and the system performance is checked using SIMULINK.

Nowadays, due to the tremendous growth of wireless communications technologies and multimedia applications, the radio spectrum is starting to be crowded and scarce to meet the continuous growth of frequency requirements. Additionally, interference management is one of the key issues in wireless networks. Therefore, network solutions and evolutions have crucial challenges to overcome the inefficiency in configuring and managing network resources. To optimize wireless network operations and spectrum scarcity, a new networking paradigm, known as cognitive radio networks (CRNs), has been introduced. Due to the limited capabilities of the conventional layered protocol, CRNs adjust layer parameters adaptively according to the spectrum environment and Quality of Service (QoS) requirements. Hence, cross-layer design (CLD) solutions were necessary to allow for improving and optimizing CRNs performance. This chapter provides an extensive and exclusive overview of cognitive networks, CLD methodologies and properties, and cross-layer optimization (CLO) schemes among different layers. Moreover, it presents possible research solutions for cognitive networking. Finally, indispensable highlights of future work research directions are provided.

Most of the frequency spectrum bands have already been licensed, and the licensed spectrum is not being utilized efficiently. Cognitive Radio Networks (CRNs) are the kind of full duplex radio that automatically altered its transmission or reception parameters, in such a way that the entire wireless communication network of which it is a node communicates efficiently, while avoiding interference with primary or secondary users. In this chapter, the authors introduce the concept of security threats that may pose a serious attack in CRN. Due to the unique characteristics of CRN, such network is highly vulnerable to security attacks compared to wireless network or infrastructure-based wireless network. The main objective of this chapter is to assist CR designers and the CR application engineers to consider the security factors in the early development stage of CR techniques. Challenges and various security issues are explored with respect to OSI (Open Systems Interconnection) reference model. Various possible and attacks are discussed broadly and respective solutions are also proposed by this chapter. Different architectures and models are also explained, and compared with the existing models.

Next generation (xG) wireless networks will contain a number of radio access networks (RANs), which mainly depend on various radio access technologies (RATs). Meanwhile, the subscribers in the current networks will be equipped with multi-mode, reconfigurable, and cognitive equipments, with or without multi-homing abilities. This chapter looks at the definition, principle, architecture, optimization objective, specific approach, and optimization of resource for such a composite network /hybrid network context under IEEE1900.4 framework. First, the challenges of spectrum scarcity, and the requirements of a novel resource management framework in the limited resource scenario are summarized. Second, the authors focus on the techniques of resource management and optimization architecture. After a simple description of the basics of IEEE P1900.4, a novel double cognition cycle (DCC) design in the fashion of utility function theory is proposed in this chapter. The DCC contains two cognition cycles corresponding to the terminal and network side, respectively, and is uniformed in the frameworks of network utility maximization (NUM). Third, the authors capture the challenges of resource management protocols of the current developing heterogeneous networks and investigate the dynamic churning behaviors, where the subscribers are equipped with intelligent terminals with the aid of cognitive radios.

Spectrum sensing is the cornerstone of cognitive radio, which detects the availability of a spectrum band for the current time. In theory, the result of spectrum sensing reflects the current channel state, which is the ideal case. However, according to the author's measurements, hardware platforms can introduce a non-negligible time delay on the signal path, which undermines the accuracy of spectrum sensing. To reduce the negative impact of the hardware platform time delay, channel state prediction in cognitive radio is proposed and presented in this chapter. As examples, channel state prediction algorithms based on a modified hidden Markov model (HMM) are given and tested using recorded real-world data. Moreover, as a second stage, cooperative channel state prediction is also proposed and experimentally evaluated. The experimental results approve that channel state prediction in cognitive radio indeed helps improve the accuracy of spectrum sensing in practical cases.

This chapter addresses a few challenges and issues in developing Cognitive Radio Networks (CRNs), and provides unique solutions to enhance security at physical layer and to boost CRN computing power in a distributed manner. In this age of vast connectivity, network security becomes more and more prominent. In addition to the security means added at the upper layers, security can be further enhanced at physical layer. In particular, location based wideband channel characteristic as a unique signature can

Preface

The explosive growths of wireless multimedia service are continuous to promote the waves of developing many potential wireless communication standards and systems in the foreseeable future. The demands of ample licensed bands can therefore be anticipated to support numerous wireless applications in the next decade. The precious limited radio resource, however, obstructs this exciting progress in wireless technologies due to the fact that the static spectrum allocation policy is widely used. Recently, some research reports have indicated that most of licensed spectrum resource is underutilized, and Federal Communications Commission is taking initiative in revising the spectrum usage policies to improve the spectrum utilization. For this purpose, cognitive radio has emerged as a new paradigm shift, which allows different systems to dynamically access and opportunistically exploit the same frequency band in an efficient way.

The advances in software defined radio technology nowadays also tremendously push this prospective trend from practical implementation aspects. The successful realization of cognitive radio, however, faces many challenges since the interference among many coexisting systems is becoming an important factor degrading the system performance. Many research activities and even standardization frameworks such as IEEE wireless regional area network are boosted to handle this critical interference problem. Three essential paradigms, interweaving, underlying, and overlaying, then drive a wide range of research topics and directions such as spectrum sensing, spectrum mobility, multi-antenna interference mitigation, resource allocation and scheduling, dirty paper coding, et cetera. The rapid developments in this area also inspire a burst of research avenues from networking viewpoints, such as cooperative, cognitive, competitive strategies, to effectively manage the interference and dramatically improve the spectrum utilization. It can be expected that all relevant issues and design concepts will play key roles in providing technology roadmap for the next-generation wireless communications in the near future.

In recent years, a lot of research activities and remarkable advances in cognitive radio and interference management have been emerging. Relevant topics in this area still gather significant attention, and researchers from both academia and industry continue making contribution to this emerging area. This book examines the issues, challenges, and immediate state-of-art research results on the evolutionary communication technology: cognitive radio and interference management. Several topics are addressed, including spectrum sensing methodologies, interference mitigation schemes, cognitive radio network analyses, resource allocation and scheduling strategies, spectrum access and mobility management, cross-layer designs, cognitive radio network security, real applications and experiments, and so on. The book is organized in a manner from the communication lower-layer to the communication upper-layer and covers both academic theories and practical applications. This book should enable a wide audience of readers to immediately keep pace to the evolution of their concerned topics. This book intends to serve as a useful bridge between people who are working on the theoretical and practical research in cognitive radio and interference management. Moreover, this book presents several open research problems which could serve as future research topics for interested readers.

This book is composed of sixteen chapters as follows. Chapter 1 gives an introduction to cognitive radios, as well as a comprehensive survey to a wide variety of spectrum sensing approaches. It also addresses several challenges to design cognitive radio networks. Chapter 2 designs collaborative spectrum sensing approaches and fusion center rules for multi-node cognitive radios. The problem is formulated through the likelihood ratio test framework over various fading channels. Chapter 3 focuses on the single-user and multi-user spectrum sensing with unknown distributions of primary users. In addition, a robust spectrum sensing optimization problem is solved by analyzing the lower bound performance of the spectrum detection probability. Chapter 4 discovers an interesting relationship that the throughputs of secondary and primary networks are mutually influenced by the transmission attempt probabilities of both networks, and a progressive algorithm is proposed to jointly optimize the throughputs and transmission probabilities. Chapter 5 presents distributed multicell precoders for efficiently mitigating the interference among cellular networks, studies the effect of imperfect channel state information on precoders, and proposes robust precoders in the presence of channel information uncertainty.

Chapter 6 applies the fuzzy logic for designing power control strategies in spectrum sharing cognitive radio networks. Chapter 7 is focused on the application of game theory to overcome spectrum scarcity problem in cognitive radio networks. Several design examples regarding to game theory are provided in this chapter. Chapter 8 investigates resource allocation strategies for cognitive femtocells and the total throughput subject to the fairness criterion is maximized by using game theory. Chapter 9 is devoted to jointly considering opportunistic spectrum access and multimedia data scheduling for cognitive radio networks. Chapter 10 studies a vertical handover through cognitive cooperation in a composite network so as to improve the radio resource usage. Chapter 11 provides hierarchical fuzzy-based solutions for spectrum access and mobility management in cognitive radio networks. Chapter 12 comprehensively covers the cross-layer design issues and optimization across different layers for cognitive radio networks. Chapter 13 is dedicated to the security issues for cognitive radio networks, and a complete study of numerous challenges and attack types, along with the respective solutions, is discussed. Chapter 14 gives an overview of resource management architectures of IEEE P1900.4 which is being standardized to improve the capacity and quality-of-service in a composite/hybrid network, and resource management protocols are investigated under the IEEE P1900.4 framework.

Chapter 15 discusses the influence of the time delay, caused by cognitive radio hardware platforms, on the spectrum sensing performance, and channel state prediction algorithms with or without cooperation are proposed to enhance the sensing performance. Chapter 16 provides solutions to enhance security from the physical layer perspective in cognitive radio and addresses an idea of distributed computing to save the computing power.

It is our hope that this book will contribute to a better understanding of the cognitive radio and interference management and may motivate further investigation into this new technology.

Meng-Lin Ku
National Central University, Taiwan, R.O.C.

Jia-Chin Lin
National Central University, Taiwan, R.O.C.

Acknowledgment

We would like to acknowledge the authors of each chapter in this book, who have helped us making this book possible and exerted their best effort to present their research works. We also appreciate IGI Global for giving us the opportunity to put these excellent research works together and in particular, Mrs. Myla R. Merkel for her support and guidance during the whole process. Finally, we thanks to our families for their constant love, patience, understanding, and support. This book must be dedicated to them.

Chapter 1
Opportunistic Spectrum Sensing and Transmissions

S. S. Alam
University of Genoa, Italy

L. Marcenaro
University of Genoa, Italy

C. S. Regazzoni
University of Genoa, Italy

ABSTRACT

Nowadays, cognitive radio is one of the most promising paradigms in the arena of wireless communications, as it aims at the proficient use of radio resources. Proper utilization of the radio spectrum requires dynamic spectrum accessing. To this end, spectrum sensing is undoubtedly necessary. In this chapter, various approaches for dynamic spectrum access scheme are presented, together with a survey of spectrum sensing methodologies for cognitive radio. Moreover, the challenges are analyzed that are associated with spectrum sensing and dynamic spectrum access techniques. Sensing beacon transmitted from different cognitive terminals creates significant interference to the primary users if proper precautions have not been taken into consideration. Consequently, cognitive radio transmitter power control are finally addressed to analyze energy efficiency aspects.

1.1 INTRODUCTION

The demand for wireless communications tremendously increased in time and to cope up with this demand, cognitive radio (CR) is a solution of huge prospect. CR can be described as an intelligent and dynamically reconfigurable radio that can adaptively regulate its internal parameters (or similar) in response to changes in the surrounding environment. CR has been made feasible by recent advances such as software-defined radio (SDR), machine learning techniques and smart antennas, etc (e.g., see Bixio et al, 2011). The use of CR technology allows in principle flexible and agile access to the spectrum as well as improving spectrum efficiency substantially. In this sense, CR represents a possible solution to the problem of

DOI: 10.4018/978-1-4666-2005-6.ch001

spectrum scarcity, due to the variety of bandwidth demanding newly developed wireless communication techniques. In particular, spectrum scarcity occurs due to the use of traditional static frequency allocation planning for different communication protocols. Federal Communications Commission, FCC (2002) disclosed that the utilization of the continuously assigned spectrum still only ranges between 15% and 85%. This means that the primary users (also known as licensed or legacy users) do not often occupy the allocated radio resources (code, temporal and spatial domain) incessantly and this leads to their underutilization. Allocated but not used spectrum bands are usually known as spectrum holes or white spaces (WSs) in literature (Zhao & Sadler, 2007). WSs can opportunistically be brought into play in CR network by introducing efficient techniques expressed as dynamic spectrum access (DSA). DSA make it possible to a CR user to sense the vacant spectrum before using it temporarily, thus making it more likely to occur an improved efficient spectrum employment (Zhao & Sadler, 2007).

When a CR user aims at using WSs, interferences with the PU licensees may occur. Therefore, one of the main challenges in CR network is related to the management of the available radio resources among the PUs and cognitive users for satisfying the respective quality-of-service (QoS) requirements while limiting the interference to the PU licensees (Hsien-Po & Schaar, 2009). The growing interest of DSA in CR is specially related to the fact that it is considered as a possible solution of the static spectrum assignment. DSA techniques for the cognitive radio can be classified as *dynamic exclusive-use, shared-use or hierarchical access and open sharing or spectrum common models* as suggested by several researchers as E. Hossain et al (2009) and Couturier & Scheers (2009). In those papers DSA models were described; in the dynamic exclusive-use model, a licensed user can grant to a cognitive user the right to have exclusive access to the spectrum. In the hierarchical access model, an unlicensed user accesses the spectrum opportunistically without interrupting a licensed user when a legacy user is not interested to utilize that portion of the frequency spectrum. In an open sharing model, an unlicensed user can access the spectrum freely.

In order to dig up the benefit from dynamic spectrum access, knowledge about the spectrum WSs is necessary and this spectrum WSs can be discovered by *spectrum sensing,* one of the most important as well as challenging task, a CR has to perform. In particular, a CR should extract the information regarding the existence of active PUs in a specific frequency band and geographical location which is helpful and used to minimize the interference to licensed communications (Arshad et al, 2010). By sensing the spectrum, the terminal collects fundamental data from external environment, in particular through the sensing channel, and it can carry out the typical adaptation of CR (Gandetto & Regazzoni, 2007). The techniques of spectrum sensing in CR can generally be classified into stand alone or single transmitter based detection and distributed or cooperative fashion (Bixio et al, 2008). In their paper Bixio et al (2008) categorize different sensing techniques such as energy detection, matched filtering, radio identification, feature detection based and waveform based sensing as the different techniques which belong to transmitter based spectrum sensing; distributed detection with or without data-fusion (Cattoni et al, 2006) are addressed as sub-classes of feature based sensing techniques when cooperative spectrum sensing is performed. In CR environment cooperative sensing is a potential solution of the type hidden terminal problem is similar to other WLAN (IEEE 802.11) schemes. In spite of the enormous advances made on stand-alone techniques, spectrum sensing remains a complex task in real cognitive radio environment due to the multipath fading, frequency selectivity, time varying channels and noise (Yücek & Arslan, 2010).

DSA techniques, despite enhancing overall spectrum utilization, can carry to the decision of radio signal transmission by cognitive nodes that can produce harmful interference to the PUs. This aspect must be taken into account: to this end, two important design criteria for CR networks have been fixed, i.e., joint maximization of radio resource utilization and minimization of the interference caused to the primary users (Hoang & Liang, 2008). Consequently, power control techniques assumed great importance in CR communication to deal with the minimization of the probable cross-interference originated by multiple users, either primary or secondary, within CR networks. Generally, power control is employed in mobile networks for improving link performance (e.g., see DS-CDMA systems); similar concepts can be applied to CR networks as well. Power control is a crucial task also in CR networks and it is one of the basic modules of radio resource management (Qin & Su, 2009). In a wireless network where users are prioritized (i.e. both PUs and CRs are present), if a CR gets the access to exploit the spectrum, it must take into account the power transmission itself to assure an interference-free transmission for the surrounding users (PUs and CRs both). There are different techniques of this kind. For example, spectrum underlay is one of the existing techniques which permit cognitive users to share the whole spectrum with the PUs as long as they do not exceed the interference threshold at PUs (Pareek et al, 2010). Shi and Hou (2007) have been demonstrated an excellent survey on power control of cognitive radio and this is followed by the known "*protocol model*" for interference modeling. Since power control directly affects the transmission range and interference range (received power at the destination node and interference power at other nodes) of a cognitive terminal, it has profound impact on scheduling feasibility, bandwidth efficiency, and problem complexity (Shi & Hou, 2008). Furthermore, power control is an important issue for the energy-efficient spectrum sensing, the future trend of green cognitive radios.

The remainder of this chapter is organized as follows: Section 2 is engaged with the problems related to the traditional static spectrum allocation, in Section 3, a comprehensive overview of the DSA techniques in CR networks, focus will be on the state of the art in DSA, pros and cons of each DSA sub-classes are drawn, then spectrum sensing is discussed in Section 4 and then optimal, energy efficient power allocation, which will be helpful to mitigate the interference temperature, will be discussed. At last, some conclusions will be drawn.

1.2 SPECTRUM SCARCITY PROBLEMS AND WHY DSA COMES OUT

Nowadays, the availability of frequency spectrums are running out due to the use of traditional static spectrum allocation (SSA) strategy and a growing number of wireless applications and equipments. The frequency spectrum is assigned worldwide by the world radio communications conference (WRC) under the presence of all the member nations as they demand the frequency and of different service operators that require spectrum allocations for mobile, fixed, and satellite technologies. The government authority of each country is responsible to assign the portion of spectrums for different communication vendors, operators or organizations for different purposes. The long-term spectrum allocation is granted in space and time invariant and any changes to it happen under strict control of the regulatory authority. The licensee is responsible for the fixed spectrum whether the spectrum is used or not.

The SSA scheme assigns a specific spectrum for a communication system to be employed for a certain geographic location. This allocation is fixed and cannot be changed over time and space. A large part of the radio spectrum is allocated but barely used in most of the locations and time. The portions of the spectrum which are temporarily unused by the licensee, are called the spectrum holes or spectrum white spaces (WSs) or vacant spaces and that WSs present in temporal, frequency and spatial domains. Several radio bands allocated for military, government and public safety use experience negligible utilization. FCC (2002) disclosed the static partitioning of spectrum has significant operational implications which have been recently brought to light by extensive spectrum utilization measurements in the USA and Europe. The available spectrum can be divided into licensed and license-exempt frequency bands. Usually, licensed frequency spectrum provides for some degree of interference protection which is called guard band because each new licensee must demonstrate compliance with certain standards for limiting interference to other existing nearby licensed systems. The radio parameters are defined as those parameters (e.g., radiated transmitter power level, frequency, SNR, modulation techniques, etc) that may have influence to each licensee. License-exempt bands do not require individual transmitters to be licensed in order to operate, but there are still radiated power restrictions that usually keep power at low levels as a de facto way of limiting interference.

SSA has several shortcomings because of being time and space invariant. In SSA, parts of the radio spectrum have been statically allocated for various wireless networking services for the military, government, commercial, private and public safety systems. As noticed by FCC (2002), the main problem of static spectrum allocation is the underutilization of the radio spectrum, as revealed by extensive measurements of actual spectrum usage; this fact caused several activities in the engineering, economics, and regulation communities in searching for better spectrum management policies. Often, the spectrum reserved by the government authority for a particular purpose could be not in use. Another major problem with the static spectrum allocation and the legacy static radio technology is the lack of interoperability with existing technologies. Interoperability is hampered by the use of multiple frequency bands, incompatible radio equipment, and a lack of standardization. Therefore, first responders (or emergency responders i.e., police, fire or emergency medical services) from different jurisdictions and agencies often cannot communicate during emergencies (Wang et al, 2011). There have been examples of failure in communication between different organizations at world trade center (WTC) on 9/11/2001, which is mentioned in several open literatures. For example, some of the police warnings regarding to the evacuation of people were not received in the area of the second building of WTC. Unfortunately, the fire fighters did not receive alarm messages thus leading to considerable loss of life and other disaster areas due to lack of interoperability among the primary/legacy radios (Dilmaghani & Rao, 2006). The other problem of static frequency allocation is that it cannot change the assigned spectrum properly even if the transmission channel is noisy although there are other options to choose the available spectrum (for less noisy transmission) which is not being used at that instant. Generally, the spectrum usage pattern for military intentions and commercial systems are quite different. It may certainly happen that the usage of spectrum in certain networks is lower than anticipated, while in other location users survive for the spectrum because of high demand in that geographic location. Static allocation of frequency spectrum fails to deal with this kind of issues of spectrum sharing. Static spectrum allocation additionally faces difficulties due to the modification of old technologies. In order to overcome the problems related to SSA, one should move towards the regime of DSA, where flexibility, efficient utilization of spectrum is concerned. DSA and its different possibilities are elaborately expressed in the following sections.

1.3 DYNAMIC SPECTRUM ACCESS (DSA) IN CR NETWORKS

In order to meet the massive demand of frequency spectrum, the CR network has opened up a new way of sensing and utilizing properly wireless radio resources (spectrum both in temporal and spatial domain). Cognitive radio is a dynamically reconfigurable radio which can adjust its radio parameters in response to the surrounding environment. This has been made feasible by recent advances such as software-defined radio (SDR) and smart antennas (Bixio et al, 2010). By using such CR devices enables flexible and agile access to the wireless spectrum, which can in turn, improve efficiency in spectrum utilization significantly. In this section, the state of art DSA schemes will be discussed and many researches are summarized that have been carried out using those techniques in the context of distributed CR networks. Nowadays, wireless communication is suffered from spectrum scarcity due to newly developed communication techniques always required an additional exclusive spectrum access. Latest experiment on spectrum management (Akyildiz et al, 2006) have shown that the licensed frequency bands are rigorously underutilized most of the time and a particular geographic location mainly due to the traditional command and control type spectrum regulation (i.e., static spectrum regulation) that has prevailed for decades. In order to use those remaining spectrum holes or white spaces, effort is put on achieving DSA.

Under such a spectrum policy, each spectrum band is assigned to a designated party, which is given an exclusive spectrum usage right for a specific type of service and radio device. CR can manage in order to mitigate the spectrum scarcity problem by enabling DSA scheme, which allows CRs to identify the unemployed portions of licensed band and utilize them opportunistically as long as the CRs do not interfere with the PUs communication. The diversity of the envisioned spectrum reform ideas is manifested in the number of technical terms coined so far: dynamic spectrum access vs. dynamic spectrum allocation, spectrum property rights vs. spectrum commons, opportunistic spectrum access vs. spectrum pooling, spectrum underlay vs. spectrum overlay. A taxonomy of the DSA scheme (Zhao & Swami, 2007) is illustrated in the following figure (Figure 1).

1.3.1 Hierarchical Access Model

In this model, a hierarchical access pattern for the primary and secondary users will be discussed. The

Figure 1. Fundamental classification of dynamic spectrum access

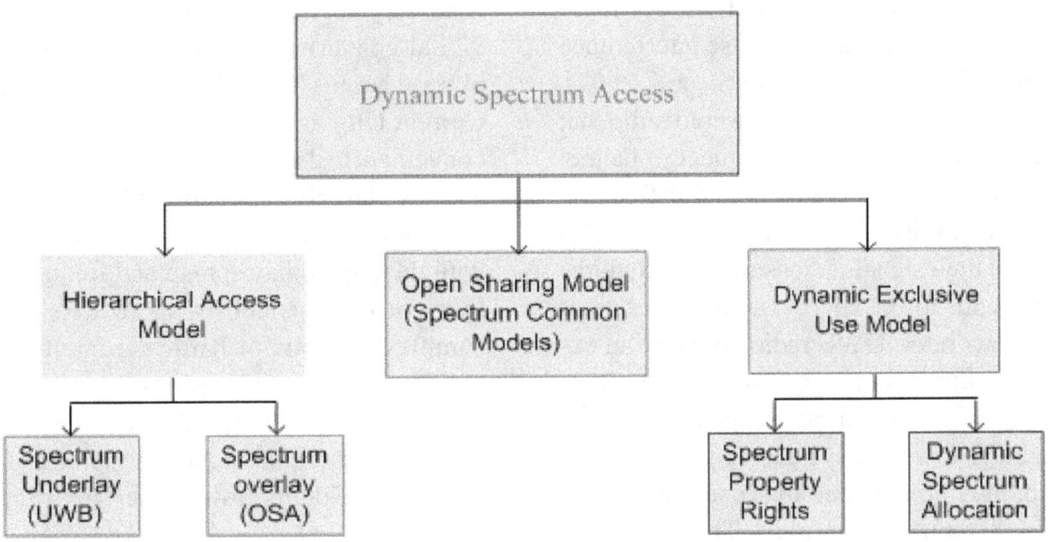

fundamental concept is to open licensed spectrum to cognitive users while limiting the interference perceived by the primary users. This model can be categorized as two different approaches for allocation of the spectrum, i.e., spectrum underlay and spectrum overlay. Both the schemes motivate with the matching goal that CR users need to have an overview about the current spectrum utilization in order to detect and identify PUs. Spectrum underlay exploits the spectrum by using it despite of a PU transmission, but by controlling the interference within a prescribed limit. This can be obtained by using spread spectrum techniques, resulting in a signal with large bandwidth but low spectral power density, which can coexist with primary users. Spectrum overlay intends to use spectrum holes in an opportunistic way (e.g., opportunistic spectrum access, OSA) while without interfering PUs transmission, indicating that the spectrum is periodically monitored by the secondary or cognitive user seeking absence of PUs in order to utilize the vacant spectrum by the CR user.

1.3.1.1 Spectrum Underlay

In an underlay system, regulated spectral masks impose stringent limits on radiated power as a function of frequency, and perhaps location. Radios coexist in the same band with primary licensees, but are regulated to cause interference below prescribed limits (Zhao & Swami, 2007). For simplicity, consider a low-powered radio that could coexist within the same frequency channel with a high-powered broadcast radio. Because of the power limitation, underlay radios (URs) must spread their signals across large bandwidths spreading with lower energy, and/or operate at relatively low rates. UWB radio could be an example of this kind. The power limitation results in a corresponding limit on rate-range capabilities. An advantage of such a system is that radios can be dumb, they do not need to sense the channel in order to defer to primary users. The underlying principle is that the primary users are either sufficiently narrow-band, or sufficiently high-powered, or the URs are sufficiently fast frequency hopping with relatively narrow bandwidth usage in each dwell, so that there is little interference from the URs. An underlay radio spectrum distribution is shown in Figure 2.

In order to spread out the signal over a large bandwidth, underlay radios can use spread spectrum signalling systems, wideband orthogonal frequency division multiplexing (OFDM), or impulse radio. Because of the large front-end bandwidth, URs are susceptible to interference from a sort of co-existing sources, including relatively narrowband signals from primary users. This can cause saturation of the automatic gain control (AGC) circuit leading to signal distortion and loss of dynamic range. It is important to notice that suppressing strong primary signals through front-end notch filters is complex, since there could be many primary signals, and not always at the same frequency locations. Receiver arrays can help notch some primary users by exploiting the spatial degrees of freedom. A second problem is that high-resolution high-rate analog to digital converter (ADC) is extremely challenging due to both the high power consumption of such devices and fundamental limits imposed by the noise level. Consequently, it may be necessary to devise and implement analog or digital converters to achieve high-fidelity sampling at a rate slower than the system bandwidth. Current URs have limited range and rate, not so convenient to deal with the aggregate interference and have largely been confined to indoor applications. Floor URs must also be capable of dealing with the large delay spread and frequency selectivity of the channel. In short, URs tend to be complex in terms of hardware implementation, but relatively dumb in terms of spectrum sensing and access protocols. Challenges exist in hardware implementation, frontend interference suppression, high-fidelity low-power high-rate ADC

Figure 2. Spectrum underlay

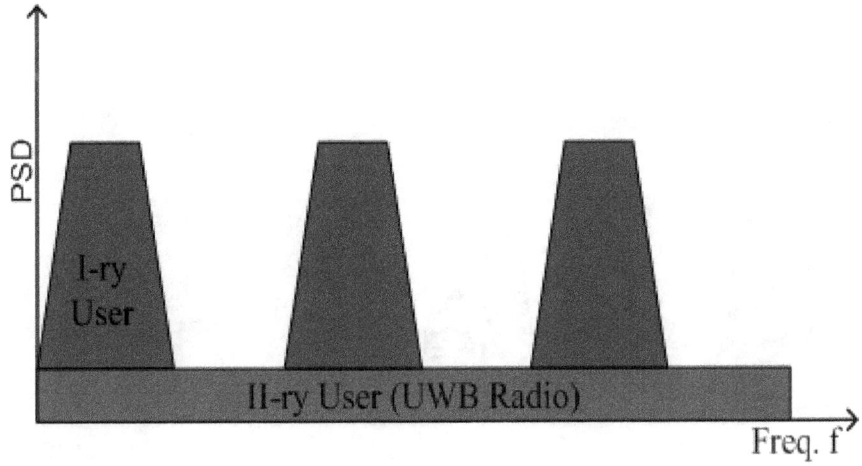

circuit design, and estimation and equalization of long delay-spread channels.

Modeling of aggregate interference from URs and devising algorithms to cope up with the primary receivers have not been adequately addressed. Another aspect of aggregate interference is that spectral masks may have to be adapted to secondary traffic load. An UR could sense the spectrum so as to shape its transmission signal to avoid congested bands. This requires reliable sensing of the spectrum similar to the spectrum overlay systems discussed in the following section.

1.3.1.2 Spectrum Overlay

Spectrum overlay or OSA, can be applied in either temporal or spatial domain. For the first case, secondary users aim to exploit temporal spectrum opportunities resulting from the bursty traffic of primary users and in the latter, cognitive users aim to exploit frequency bands that are not used by primary users in a particular geographic area (Zhao & Swami, 2007). A typical application is the reuse of certain TV white spaces that are not used for TV broadcasting (e.g., digital TV transmission) in a particular geographic location. In the TV broadcasting system, TV-bands assigned to adjacent regions are different to avoid co-site interference. This results in unused frequency bands varying over space. In general, spectrum opportunities vary in both temporal and spatial domains. It is often assumed in the literature that one variation is at a much slower scale than the other. Spectrum overlay mechanism is shown in the following (Figure 3).

OSA is also known as interweaving of frequencies in some papers. Spectrum overlay is therefore defined as doing some pre-coding at the transmitter in order to diminish the interference at the receiver. Therefore extensive knowledge about other signals in the specific spectrum is necessary. This technique is also known as dirty paper coding (Erez & Brink, 2004). The majority of existing work on OSA focuses on the spatial domain where spectrum opportunities are considered static or slowly varying in time. As a consequence, real-time opportunity identification is not as critical a component in this class of applications, and the prevailing approach tackles network design in two separate steps: (i) opportunity identification assuming continuous full spectrum sensing; (ii) opportunity allocation among secondary users assuming perfect knowledge of spectrum opportunities at any location over the entire spectrum.

To find out the spectrum opportunity in the presence of fading and noise there will be some

Figure 3. Spectrum overlay (e.g. spectrum pooling or OSA)

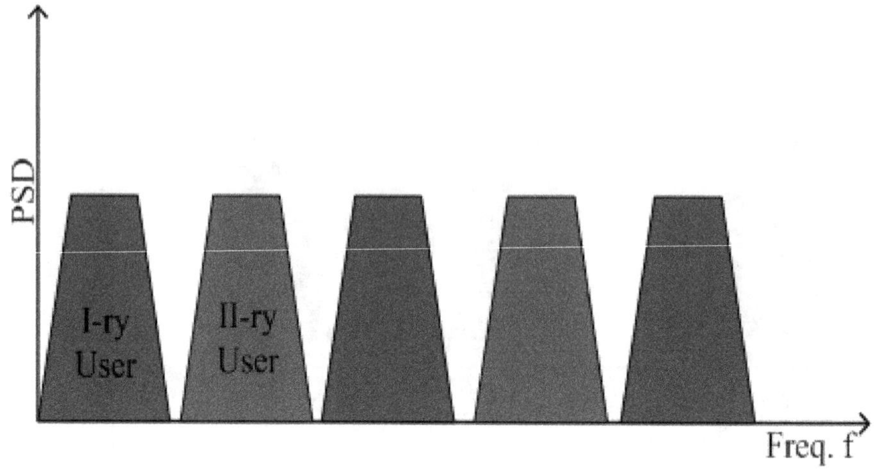

uncertainty and this issue is discussed in some papers and spatial opportunistic allocation is also possible among secondary users. OSA in the time domain needs to design both of the spectrum sensing and access schemes. Tracking the rapidly varying spectrum opportunities would be a crucial issue, and a simple yet sufficiently accurate statistical model of spectrum occupancy is crucial to the efficiency of spectrum opportunity tracking. There are some errors predictable in real-time sensing, and the characteristics of the spectrum sensor should be taken into account in making spectrum access decisions. Initial attempts at addressing the identification and exploitation of temporal spectrum opportunities that also vary in space and for an extended overview of challenges and recent developments in OSA can be investigated by Zhao & Sadler (2007) in their research paper.

Basically, a channel can be considered as an opportunity if it is not currently being used by legacy/primary users. As suggested by the Zhao & Sadler (2007), the spectrum opportunity can be defined as follows.

1. Spectrum opportunity is a local concept expressed as to a particular pair of cognitive users depending not only on the location of the cognitive transmitters but also on the position of secondary receivers. For multicast and broadcast, spectrum opportunity is open for interpretation, and results in networking tradeoffs.
2. Spectrum opportunity is determined by the communication activities of PUs rather than that of cognitive users. Failed communications originated by collisions among SUs do not disqualify a channel from being an opportunity.

1.3.2 Dynamic Exclusive Use Model

In the exclusive-use model for spectrum access, the radio spectrum is licensed to a user/service to be exclusively used under an agreement. This model maintains the basic structure of the current spectrum regulation policy: spectrum bands are licensed to services for exclusive use. The main idea is to introduce flexibility to improve spectrum efficiency. Two schemes have been proposed under this model: spectrum property rights and dynamic spectrum allocation (Couturier & Scheers, 2009).

Generally, PUs do not utilize their spectrum all the time even if they deserve this spectrum to use. Therefore, the PUs can sub-lease those underutilized spectrum to third party thus leading to the spectrum trading. This type of spectrum trading can be given the right to exclusively use those resources without being mandated by a regulation authority. This approach is called Spectrum Property Rights, as the license or the right is based on the three spectrum properties: fixed frequency band, time and a geographic location.

1.3.2.1 Spectrum Property Rights

The detailed explanation of spectrum property rights and challenges of this scheme can be found in (Hatfield & Weiser, 2005). Specifically, L. White (2000) stated in his paper that "the property right would be expressed as the right to transmit over the specified spectrum band, so long as the signals do not exceed a specified strength beyond the specified geographic boundaries during the specified time period". One of the major difficulties in enforcing such spectrum property rights lies in the unpredictability of radio wave propagation in both frequency and space. Spectral and spatial spillover is inevitable, unpredictable, and depending on the characteristics of both transmitters (potential trespassers) and receivers (property right owners).

1.3.2.2 Dynamic Spectrum Allocation

A second approach for the dynamic exclusive use model is dynamic spectrum allocation. For this the temporal and spatial traffic statistics are exploited, which is valuable for sub-leasing long-term applications, such as UMTS or DVB-T. Sub-leasing based on traffic statistics leads to a much more flexible spectrum allocation than in the previous approach. For example, the amount of spectrum allocated to UMTS and DVB-T can vary over region and the-time-of-day. But again, dynamic is limited to the capabilities of the licensee, so it is unlikely that with either of these approaches the spectrum holes can be optimally filled (Couturier & Scheers, 2009).

The literature of dynamic spectrum allocation is extensively discussed by L. Xu et al (2000). This work is related to European Union funded project DRiVE (Dynamic Radio for IP Services in Vehicular Environments), aiming at enabling spectrum-efficient high-quality wireless IP communication in a heterogeneous multi-radio networks considering a common co-ordination channel. DSA opens new possibilities for wireless network operators if a multi-radio infrastructure for optimized interworking of cellular and broadcast network is considered:

- Operators can allocate spectrum for each service access network according to local and temporal needs.
- Users on the move are provided with the benefit of accessing enhanced IP based mobile services on the fly and wherever they are in a cost efficient way.

Multiple networks regulation policy and issues in the context of dynamic spectrum allocation are pointed out in (Leaves et al, 2004). Temporal and Spatial DSA algorithms are discussed there and several implementations of an iterative, real-time, temporal DSA algorithm are possible, but typical steps in the operation of the temporal DSA algorithm include: a) Periodic triggering of DSA algorithm, b) Management of the traffic on the carriers, c) Prediction of the loads on the networks and d) Allocation decision while the goal of spatial DSA is to allocate spectrum to radio access networks (RANs) according to the traffic requirements in each location using DSA scheme. Furthermore, it is necessary to coordinate the spectrum allocation between adjacent DSA areas for avoiding interferences. Especially, the spectrum allocations of different RANs belonging to adjacent DSA areas should not overlap in the same portion of spectrum. In order to avoid this spectrum overlap while still allowing spectrum allocation adaptation to the traffic demand, the guard band needs to be increased to guarantee the coexistence of the different systems. The structure of a usual spatial DSA scheme can be summarized in three main steps: a) calculating the spectrum overlap, b) performing initial assignment and c) optimize the spectrum usage.

Two centralized dynamic spectrum allocation (Recursive Round Robin Algorithm and Greedy-Type Algorithm) protocols that rely on a super base-station and their performance evaluated via simulations are depicted by Aazhang et al (2004). Two simple protocols to redistribute call loading among base stations and facilitate sharing of the spectrum is analyzed in that paper. The protocols proved to be simple yet effective. The protocols also exhibited a tradeoff among fairness, complexity, effectiveness and possibly introducing call handling delays in the cellular network.

1.3.3 Open Sharing Model

The dynamic exclusive use model and the hierarchical access model concern with the primary users having a license to use a certain part of the spectrum, whereas the open sharing model assumes a vacant spectrum with only peer users. Compared to the other two models, many technical issues under this model are perhaps the closest to the conventional medium access control (MAC) problems. Again, two different approaches how to organize interference-free communication are illustrated by Couturier and Scheers (2009). This open sharing model can be categorized as centralized and distributed fashions. Centralized and distributed spectrum sharing strategies have been considered to address the technological challenges under this model.

In a centralized open sharing model, there is only one centralized cognitive manager that controls the whole cognitive radio domain. The cognitive manager can be straightforwardly implemented using an expert system, or the problem can be seen as an optimization problem for which a global optimum has to be found. The centralized approach assumes however that there is a reliable cognitive signalling channel connecting each radio to the centralized manager. It has to be noted that the cognitive manager, in general, will not only influence the proficient spectrum usage, but also other transmission parameters like transmit-power, signal to noise ratio, modulation strategy, etc.

In the distributed open sharing model, decision making is more complicated. Decisions have to be taken locally by all the transmitter-receiver pair, meaning that there must be a cognitive manager in every node. In this case, coordination between pairs or coalitions of pairs can facilitate the spectrum sensing, competent use of radio resources and enhance the quality of the information, on which the pairs can rely to make their decisions.

1.3.3.1 Centralized DSA

In a centralized dynamic spectrum access architecture, a central controller is deployed to collect and process information about the radio environment. With a central controller, the decision of cognitive radio users to access the spectrum can be made such that the desired system-wide objectives are achieved. In a centralized scheme, every CR user always communicates with a central controller to inform their status and in response to that the central controller update to every CR node. The central controller is inside CR network and makes the decisions intelligently as the requirement of CR user. There are two approaches proposed to implement centralized dynamic spectrum access namely, optimization approach and auction-based approach (Hossain et al, p. 274, 2009). With an optimization-based approach, different types of optimization problems can be formulated (e.g., convex optimization, assignment problem, linear programming, and graph theory). Standard methods in optimization theory can then be applied to obtain the optimal solution for dynamic spectrum access. While auction theory driven centralized dynamic spectrum access mainly addresses the spectrum trading in a suitable way. In this approach, CR users submit their bids to the spectrum owner for each vacant spectrum and the highest bidder will then win to utilize the spectrum accordingly.

1.3.3.2 Distributed DSA

In many scenarios such as in ad-hoc cognitive radio networks, deploying a central controller may not be feasible (Hossain et al, 2009). Therefore, distributed dynamic spectrum access would be required in such cognitive radio networks. Due to the absence of any central controller, each unlicensed user has to gather, exchange, and process the information about the wireless environment independently. Also, an unlicensed user has to make decisions autonomously based on the information that available on the radio environment to access the spectrum, so that the unlicensed user can achieve its performance objective under interference constraints. The common behaviors of an unlicensed user in a cognitive radio network without a central controller are as cooperative or non-cooperative behavior, collaborative or non-collaborative behavior and learning ability.

1.3.3.2.1 Cooperative or Non-Cooperative Behavior

Since a central controller which controls a decision of spectrum sharing is not available, a CR user can adopt either cooperative or non-cooperative behavior. A CR with cooperative behavior will make a decision on spectrum access concerning the performance of the overall network (i.e., a group objective), though this decision may not result in the highest individual benefit for each CR user. In contrast, a CR user with non-cooperative behavior will make a decision that is opposite to cooperative behavior i.e. it wants to maximize the individual performance whatever would be the network performance. This feature of the CR user is also called the selfish behavior. Couturier and Scheers (2009) discussed in their paper game theory and iterative water filling approach can be used for the distributed DSA.

To pertain game theory to the process of decision making in a cognitive radio, the decision making process needs to be modeled as a game. First of all, it should be checked whether it is a centralized or a distributed DSA model (i.e., the centralized or the distributed open sharing model). Secondly, it must be decided which performance metric(i.e., the throughput or the latency) is to be optimized. Thirdly, all information about any cognitive radio in the environment of the decision maker needs to be collected (i.e., the possible actions and the preferred strategy).

Finally, a mapping of the elements of a cognitive radio to a game must be carried out, as depicted in Table 1, which shows the number of players can be mapped to the number of transmitter-receiver groups, where a group consists of one transmitter and an arbitrary number of receivers (Couturier & Scheers, 2009).

In general, the CR users are in the same group with the same objective, these users typically choose cooperation among them but if the CRs be independent and have different objectives, non-cooperative behavior would much likely be practiced.

1.3.3.2.2 Collaborative or Non-Collaborative Behavior

Again, since a central controller which can coordinate the gathering and broadcasting of network information is not available, the CR user has to collect network information themselves. In this

Table 1. Mapping of cognitive radio elements to plot a game (Couturier & Scheers, 2009)

Game Theory	Variable	Cognitive Radio
No. of Players	M	No. of nodes to be considered
Entirety of actions	A	Transmission parameters sets
Utility function, Payoff	U	Performance metrics

case, a CR can be collaborative or non-collaborative in nature to exchange network information for distributed dynamic spectrum access. With non-collaborative behavior, all network information is gathered and processed locally by each CR user when there is no interaction among the secondary users. Conversely, with collaborative behavior, the CR user can exchange network information with each other. Again, the choice to become either collaborative or non-collaborative depends on the type of the cognitive radio network and the users. Also, the collaboration of the SUs relates to cooperative and non-cooperative behavior. Typically, collaboration among CR users to exchange network information is required to achieve cooperative behavior. Also, if the CRs are non-collaborative, they are also typically non-cooperative. However, if the CRs are collaborative, they could be either cooperative or non-cooperative. For example, the unlicensed users may agree to reveal some information (e.g., the chosen spectrum access action), but they make a decision to achieve their own objectives (i.e., non-cooperative), rather than a group objective. Therefore, an unlicensed user can be classified into one of the following three categories: collaborative–cooperative, collaborative–non-cooperative, and non-collaborative–non-cooperative. In the case of collaboration among unlicensed users, a protocol will be required for exchanging network information. However, in the case of non-collaborative unlicensed users, the network information has to be observed and learned locally. Therefore, learning ability will be crucial for a cognitive radio which is discussed in the following.

1.3.3.2.3 Learning Ability

The ability to learn and make intelligent decisions is important for distributed dynamic spectrum access. In particular, the CRs have to observe and learn the system state (e.g., the occupancy of the radio spectrum in a certain time). The output of this learning process is the knowledge about the radio environment and the system, which would be useful for a CR user to make a decision on spectrum access. Again, the learning process can be either non-collaborative or collaborative. In the case of non-collaborative learning, the knowledge about the system is produced by each individual unlicensed user without interaction with other users. On the other hand, the unlicensed users can collaborate not only to exchange network information, but also to process and produce the system knowledge. Then, based on this knowledge, an unlicensed user can make the decision whether to achieve the group objective or its individual objective.

1.3.4 Medium Access Control (MAC) for DSA

Basic components of cognitive MAC basic design components of cognitive MAC for the opportunistic spectrum access include a sensing policy for real-time decisions which determines the proper sensing instant and an access guideline that determines how to access the spectrum based on the sensing outcomes (Hossain & Bhargava, p. 272, 2007). The purpose of the sensing policy focus on two aspects: to identify a spectrum opportunity for immediate access and to obtain statistical information on spectrum occupancy for improved future decisions. A trade-off should be considered between these two often conflicting objectives, and the trade-off should adapt to the bursty traffic and energy constraint of the CR user. For example, when there are energy costs associated with sensing, a secondary user may decide to skip sensing when its current estimate of spectrum occupancy indicates that no channels are likely to be idle. Clearly, such decisions should balance the reward in energy savings with the cost in lost spectrum information and potentially missed spectrum opportunities. The objective of the access policy, on the other hand, is to minimize the chance of overlooking an opportunity without violating the constraint of being non-intrusive. Whether the secondary user should adopt an aggressive or a conservative access policy depends on the operating characteristics (probability of false alarm vs. probability of miss detection, and tolerable level of interference) of the spectrum sensor. A joint design of MAC protocols and spectrum sensors at the physical layer is thus necessary to achieve optimality. Energy constraints will further complicate the design of access policies. For energy-constrained OSA in fading environments, the secondary user may avoid transmission when the sensed channel is in a deep fade. Even the residual energy level will play an important role in decision-making. A detailed survey of the essential MAC features for the opportunistic spectrum access is carried out by Pawełczak et al (2008).

1.3.5 Open Issues in DSA

The fundamental concepts of DSA scheme in a cognitive radio network have been discussed in the above section. Different spectrum access models for cognitive radio, namely, the exclusive use, shared-use, and commons models, have been described. Spectrum sensing and spectrum access are the two major functionalities of a cognitive MAC protocol so a brief study on the cognitive MAC protocols has been expressed. DSA poses different sets of challenges in centralized and distributed cognitive radio networks. There are lot of works on sensing have performed even though sensing itself can cause interference to the PUs so the sensing with the learning algorithms are used to build knowledge about both the ambient radio environment and the network based on the observations. This knowledge is then used by the cognitive radio users to make spectrum access decisions. In addition, to support distributed DSA, various signaling protocols are used for exchanging network information between the collaborative unlicensed users.

Above all, there are many issues that have not been standardized yet. The problems associated with the scheduling problems, QoS provisioning, synchronization problems among the CR users etc for the CR users. The scheduling deals with an access protocol for which CR should use the proper channel at the proper time if more than one CR users active in the network. In addition, if a PU is active suddenly in a network, definitely the CR should vacant the channel, so there should be a QoS provisioning for the cognitive terminals. As the radio spectrum is a very expensive resource and it requires efficient allocation to satisfy both licensed and unlicensed users, the economic issues involved in dynamic spectrum sharing are important.

1.4 SPECTRUM SENSING TECHNIQUES FOR CRS

The spectrum has been classified into three types as *black spaces*, *grey spaces* and *white spaces* by estimating the incoming RF stimuli (Wang, 2009). CR user may take the recompenses from grey spaces and white spaces of the spectrum. In order to hit upon those vacant spectrum spaces, the name of the a fundamental contestant is spectrum sensing. There are different methods proposed for identifying the presence of PUs signal transmission. The characterization of the transmitted signal enables not only to choose the signal transmission but also identifying the signal type. Generally, spectrum sensing techniques can be divided into three categories: *transmitter based detection* or *stand alone spectrum sensing, cooperative/distributed detection* and *interference based detection*.

1.4.1 Transmitter Based Detection

The cognitive user is liable to identify the presence of primary user in a certain geographic location and this is figure out in this type of detection. So, this is called transmitter based detection or stand alone detection (Bixio et al, 2009). This is a non-cooperative sensing technique which cannot detect the hidden primary user. This stand alone technique has been studied for military and civilian applications for signal detection, automatic modulation classification, to locate radio source and to perform the jamming activities in communication networks. In this section, some of the most common transmitter based sensing schemes in the CR literature are discussed. This scheme mainly includes *Energy detector based sensing*, *Matched filter*, *Waveform based sensing*, *Feature detection* and *Radio Identification based sensing*.

1.4.2 Energy Detector Based Sensing

The fundamental scheme for spectrum sensing is based on energy detection (also known as radiometry or periodogram) where received signal energy is measured in a specific time epoch and certain frequency intervals. This technique is also prime choice for the engineers due to its low computational and implementation complexities. In addition, it is the most generic (as compared to methods given in this section) idea as receivers do not require any knowledge about the PUs signal (L. Bixio et al. 2009, p.185). The signal is detected by comparing the output of the energy detector with a threshold which depends on the noise floor. Some of the challenges with energy detector based sensing include selection of the threshold for detecting primary users, inability to differentiate interference from primary users and noise, and poor performance under low signal-to-noise ratio (SNR) values (Wang, 2009). Moreover, the dominant problem regarding energy detector as it does not work efficiently for detecting spread spectrum signals. However, this technique is the optimal solution when PUs transmission is not known (Wang, 2009). We want to formulate the energy detection based sensing as illustrated by Yücek and Arslan (2009). Let us consider that the received signal has the following simple form

$$y(n) = s(n) + w(n) \tag{1}$$

where s(n) is the signal to be detected, w(n) is the additive white Gaussian noise (AWGN) sample, and n is the sample index. If s(n) = 0, PU transmission is absent. The decision metric for the energy detector can be written as

$$M = \sum_{n=0}^{N} \left| y(n) \right|^2 \tag{2}$$

where N is the size of the observation vector. The decision on the occupancy of a band can be obtained by comparing the decision metric M against a fixed threshold »$_E$. This is equivalent to distinguishing between the following two hypotheses:

$$\mathcal{H}_0 : y(n) = w(n) \tag{3}$$

$$\mathcal{H}_1 : y(n) = s(n) + w(n) \tag{4}$$

The performance of the detection algorithm can be determined by two probabilities as the probability of detection, P_D and probability of false alarm, P_F. P_D is the probability of detecting a signal on the considered frequency when it truly is present. Thus, a large detection probability should be the desired consideration for the CR system. It can be formulated as

$$P_D = \Pr(M > \lambda_E | \mathcal{H}_1) \tag{5}$$

Alternatively, P_F is the probability that the test incorrectly decides that the considered frequency is occupied when it actually is not, and it can be written as

$$P_F = \Pr(M > \lambda_E | \mathcal{H}_0) \quad (6)$$

P_F should be kept as small as possible in order to prevent underutilization of transmission opportunities. The decision threshold λ_E can be selected for finding an optimum balance between P_D and P_F. However, this requires knowledge of noise and detected signal powers. The noise power can be estimated, but the signal power is difficult to estimate as it changes depending on ongoing transmission characteristics and the distance between the CR and PU (as cited in Yücek & Arslan, 2009). In practice, the threshold is chosen to obtain a certain false alarm rate. Hence, knowledge of noise variance is sufficient for selection of a threshold.

1.4.3 Matched Filter

At a cognitive terminal, in order to maximize the output signal to noise ratio for a certain input signal a matched filter is designed which belongs to the linear filter (Wang, 2009, p. 410). Matched filter detection is applied if a cognitive user has a-priori knowledge of primary user transmitted signal. Therefore, Matched-filtering is known as the optimal strategy for detection of PUs in the presence of stationary Gaussian noise. The main advantage of matched filtering is the short time as it requires only O(1/SNR) samples to meet a given probability of detection constraint as compared to other methods that are discussed in this section. In fact, the required number of samples grows as O(1/SNR) for a target probability of false alarm at low SNRs for matched filtering. However, matched-filtering requires cognitive radio to demodulate received signals. Hence, it requires perfect knowledge of the primary users signaling features such as bandwidth, operating frequency, modulation type and order, pulse shaping, and frame format (Yucek & Arslan, p.123, 2009). Moreover, they discussed that CR needs receiver for all signal types, the implementation complexity of sensing unit is impractically large. In addition, this scheme consumes large power as various receiver algorithms require to be executed for detection. The problem to deal with Matched filter detection, requires a prior knowledge of PUs transmission criteria, if this information is not accurate, the matched filter performance degrades (Wang, 2009).

1.4.4 Feature Detection

Another promising spectrum sensing technique is based on feature detection. Absolute information about the spectrum features can be obtained by identifying the transmission technologies used by primary users. Such identification enables cognitive radio with a higher dimensional knowledge as well as providing higher accuracy (Yucek & Arslan, 2006). A feature captures a specific signature that is inherent characteristics of the primary user signal and the feature is unique for each class of signals. Some of the most common features considered are pilot signal, segment sync, field sync, and also the instantaneous amplitude, phase and frequency (Bixio et al, p. 186, 2009). This property is noticed in many of the signals employed in wireless communication and radar systems (W. Wang, p.411, 2009). There are several features that have been considered in order to detect and classify the primary user signals within a particular radio environment (as cited in Bixio et al, p. 186, 2009). Nowadays, analog to digital conversion has made the use of signal transform practical in order to detect a specific feature. Discrete Fourier transform, Wavelet transform and Wigner-Ville transform are belongs to this kind of ADC to detect the features of PU signal (Bixio et al, p. 186, 2009).

The fundamental and promising feature detection technique is based on the cyclic feature. This technique was introduced by Jondral et al (2007) and Sutton et al (2008) and they proposed the use of the cyclic feature as a reliable sensing technique for CR applications (Bixio et al, p. 186, 2009). Cyclic-feature detection approaches are based on the fact that modulated signal are usually coupled with sinusoidal carriers, hopping sequences, cyclic prefixes, spreading codes, or pulse trains, which result in a built-in periodicity (Bixio et al, 2009). These modulated signals are said to be cyclostationary. Cyclostationary features are originated by the periodicity in the signal in statistical manner like mean and autocorrelation (Gardner, 1991) or they can be intentionally used in order to sustain the spectrum sensing (as cited in Yucek & Arslan, 2009). This periodicity can be used as a feature and can be detected by analyzing a spectral correlation function (SCF), also known as cyclic spectrum (as cited in Bixio et al, 2009). Instead of power spectral density (PSD), cyclic correlation function is used for detecting signals present in a given spectrum. The cyclostationary based detection algorithms can differentiate noise from primary users' signals as the noise is wide-sense stationary (WSS) with no correlation while modulated signals are cyclostationary with spectral correlation due to the redundancy of signal periodicities. The cyclic spectral density (CSD) function of a received signal (1) can be calculated as suggested by (WA. Gardner, 1991).

$$S\left(f,\alpha\right) = \sum_{\tau=-\infty}^{\infty} R_y^{\alpha}(\tau)e^{-j2\pi f\tau} \tag{7}$$

where

$$R_y^{\alpha}\left(\tau\right) = E[y(n+\tau)y^*(n-\tau)e^{j2\pi\alpha n}] \tag{8}$$

is the cyclic autocorrelation function (CAF) and α is the cyclic frequency. The CSD function outputs peak values when the cyclic frequency is equal to the fundamental frequencies of transmitted signal x(n). Cyclic frequencies can be assumed to be known or they can be extracted and used as features for identifying transmitted signals (as cited in Yucek and Arslan, 2009). As a result, cyclostationary feature detector can overcome the energy detector limits in detecting signals in low SNR environments (Akyildiz et al, 2006). In fact, signals with overlapping features in the power spectrum, can have non-overlapping features in the cyclic spectrum (Gardner WA, 1988). Moreover, the cyclic spectrum is a much comfortable domain for signal detection than typical power spectral density. This property provides the flexible use of this technique as a more functional tool (Gardner WA, 1988).

Waveform based sensing is another promising feature detection scheme. Commonly used patterns like preambles, midambles, repeatedly transmitted pilot patterns, spreading sequences, etc, are utilized in wireless systems as a measure of synchronization or used for other purposes. A preamble is a known bit or signal pattern transmitted before each burst and a midamble is transmitted in the middle of a data frame. In the presence of a known pattern, sensing can be performed by correlating the received signal with a known copy of itself (as cited in Yucek & Arslan, p.122, 2009). This method is only applicable to systems with known signal patterns which is considered also a drawback of this kind of sensing, and it is termed as waveform-based sensing or coherent sensing. It is shown that waveform based sensing outperforms energy detector based sensing in reliability and convergence time. Furthermore, it is undoubtedly state that the performance of the sensing algorithm increases if the length of the known signal pattern increases. For analyzing the WLAN channel usage characteristics, packet preambles of IEEE 802.11b signals are exploited.

Another example of feature detection technique is while a PU is identified as a signal of any personal area networks (PANs) which belongs to

Zigbee, Bluetooth, etc. CR can use this kind of information for extracting some useful information in space dimension as the range of signal of this type is in the very short range. Furthermore, CR may want to communicate with the identified communication systems in some applications. In the context of European transparent ubiquitous terminal (TRUST) project, radio identification, feature extraction and classification techniques are used (Farnham et al, 2000). The goal is to identify the presence of some known transmission technologies and achieve communication through them. As suggested by Farnham et al (2000), the two main tasks are initial mode identification (IMI) and alternative mode monitoring (AMM). In IMI, the cognitive device searches for a possible transmission mode following the power on whereas the later scheme monitors the task of other modes while the cognitive device is communicating in a certain mode. In this sensing method, several features are extracted from the received signal and they are used for selecting the most probable PU technology by employing various classification methods. The features obtained by energy detector are used for signal classification. These features contain the energy distribution across the spectrum. Alternatively, channel bandwidth and its shape can be used as reference features. Channel bandwidth is found to be the most discriminating parameter among others. For classification, radial basis function (RBF) neural network is employed (as cited in Yucek & Arslan, 2009). In the research paper of Yucek and Arslan (2009), they illustrated the relevant complexity and accuracy for different types sensing scheme that is shown in Figure 4.

The OFDM waveform is altered before transmission in order to generate system specific signatures or cycle-frequencies at certain frequencies and effective signal categorization is being obtained with those signatures of the signal (as cited in Yucek & Arslan, 2009). Another research work done by Sutton et al (2007), investigated that in order to enhance the robustness against multipath fading, the number of features generated in the signal is increased. However, this comes at the expense of increased overhead and bandwidth loss. Hardware implementation of a cyclostationary feature detector is presented by Tkachenko et al (2007). The main advantage of the feature detection is that the discrimination between the noise energy and the modulated signal energy is very likely prepared. Moreover, cyclostationary feature detection can detect the

Figure 4. Complexities and accuracies of different sensing methods (Yucek & Arslan, 2009)

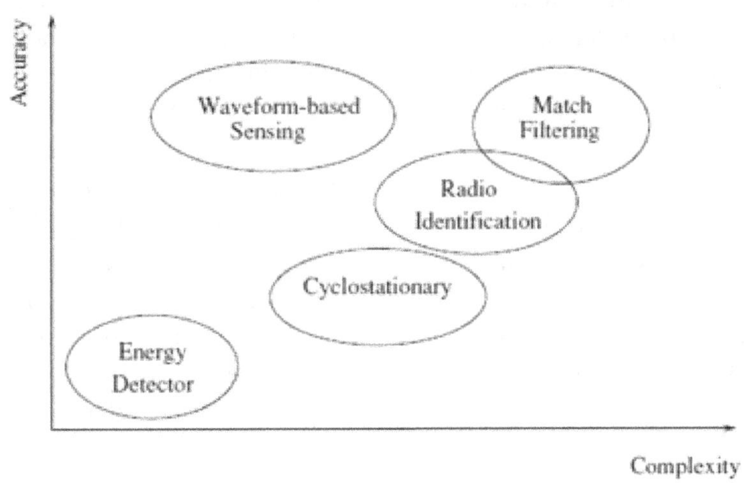

signals with low SNR. In contrast, feature detection requires long observation time and higher computationally complex and also feature detection needs a-prior knowledge of the primary users (W. Wang, 2009).

1.4.5 Cooperative and Distributed Spectrum Sensing

There are several problems like fading, shadowing and low SNR for desired performance associated to sense the signals through the stand alone transmitter and therefore cooperation is proposed as a solution to get rid of it. Cooperation is a sensing scheme which determines the existence of the PUs and the frequency band in use either and cooperation can be among CR users or external sensors can be used to build a common cooperative sensing network. The prime advantage of cooperation is it can solve the hidden terminal problem, when a CR is shadowed or in a deep fade (K. Arshad et al, 2010). In addition, the probabilities of misdetection and false alarm have considerably declined through the cooperative sensing of spectrum. In this sensing nature, several CRs can collaborate with each other in order to make a global decision about the existence of the PU to overcome this problem. Gandetto and Regazzoni (2007) proposed in their paper a distributed cooperative spectrum sensing by the involvement of a pair of cognitive terminals according to a distributed processing strategy. The proposed approach allow spectrum sensing by means of a "distributed cooperative consciousness" and the improved global performances of air interface detection is achieved than the single terminal spectrum approaches by using this method. In addition, it is already proved that collaborative spectrum sensing (CSS) can improve detection performance in the fading channels at the cost of increased computational complexity and bandwidth usage for exchanging information among CRs (as cited in Bixio et al, 2009).

In CSS, every cognitive user (CR) executes its own spectrum sensing tasks and make a local decision regarding the presence of a primary user. All of the CRs forward their decision to a common receiver, often called decision fusion centre or spectrum manager. For this reason, a dedicated feedback channel has to be allocated in order to share collected information which is termed as reporting channel. If the dedicated channel is absent, other methods should be considered which require less overhead (Bixio et al, 2009). The decision can be classified as soft (local measurement) or hard decision. In hard decision, fusion center collects binary decisions from the individual SUs, identifies the available spectrum, and then broadcasts this information to the other CRs. The optimal decision fusion is based on Neyman-Pearson criterion. In contrary, the soft decision, the value of sensing data transmission is higher to the fusion center through the reporting channels (K. Arshad et al, 2010). A fusion center may be an Access Point (AP) in wireless LAN or a CR base station in a cellular system and the fusion centre may be centralized or distributed. In centralized CSS, the fusion center assembles all the decision that comes from all the CRs and then makes a decision that conveys to the CRs through the reporting channel. However, in distributed CSS, each of the CRs may behave as a fusion centre and receive sensing information from the neighboring nodes and take individual decisions.

Challenges of cooperative sensing include developing efficient information sharing algorithms and increased complexity (as cited in Yucek & Arslan, 2009). In cooperative sensing architectures, the control channel (pilot channel) can be implemented using different methodologies. These include a dedicated band, an unlicensed band such as ISM, and an underlay system such as ultra wide band (UWB). Depending on the system requirements, one of these methods can be selected. Control channel can be used for sharing spectrum sensing results among cognitive users as well as for sharing channel allocation information. As far as the networking is concerned, the coordination algorithm should have reduced protocol overhead and it should be robust to changes and failures in the network. Moreover, the coordination algorithm should introduce a minimum amount of delay.

Collaborative spectrum sensing is most efficient when the CR users observe fading or shadowing autonomously. The performance degradation due to correlated shadowing is investigated in terms of missing the opportunities. It is found that it is more advantageous to have the same amount of users collaborating over a large area than over a small area. In order to combat shadowing, beamforming and directional antennas can also be used. Cooperative sensing network can be constructed by the presence of cognitive radios or external sensors may perform sensing tasks together. In the former case, cooperation can be implemented in two fashions: centralized or distributed (as cited in Yucek & Arslan, 2009).

1.4.5.1 Centralized Sensing

Let a set of N cognitive user share the same radio channel. Each CR performs the sensing individually by any of the technique discussed in the earlier section (Bixio et al, 2009). In centralized sensing, a central unit collects those sensing information from cognitive devices, identifies the available spectrum, and broadcasts this information to all the cognitive radios and also the decision of spectrum in use by the CRs. This means it can directly controls the cognitive radio traffic. The sensing results (hard or soft decision) are gathered at a central place which is known as access point (AP) or a cognitive manager. The goal is to mitigate the fading effects of the channel and increase detection performance. Hard and soft information combining methods are investigated for reducing the probability of missed opportunity. In some cases, users send a quantized version of their local decisions to central unit (fusion center). Then, a likelihood ratio test (e.g., Neyman-Pearson test) over the received local likelihood ratios is applied. In contrast, for large number of CR users, the traffic as well as bandwidth required for reporting channel becomes huge. Therefore, local observations of cognitive radios are quantized to one bit (hard/binary decision) to reduce the sharing bandwidth. There are two advantages associated with the hard decision; first it gives relief from the traffic to the reporting channel and later is the energy efficient sensing as for the quantization and transmission of one bit, required energy is lesser than usual. Furthermore, only the cognitive radios with reliable information are allowed to report their decisions to the central unit (Yucek and Arslan, 2009).

1.4.5.2 Distributed Sensing

In the case of distributed sensing, cognitive nodes share information among each other but they make their own decisions as to which part of the spectrum they can use. Distributed sensing is more advantageous than centralized sensing in the motive that there is no need for a backbone infrastructure and it has reduced cost. It is necessary to include in this context, different solutions (distributed detection with or without fusion) can be used depending on the level of cooperation among the CRs. Distributed detection with data sensing scheme provide better performance than the stand alone transmitter based detection. Several problems associated with the distributed detection but centralized data fusion for the CR applications:

- A dedicated reporting channel is needed to share the information which may not be available, if available, the reporting channel can be faded or shadowed.
- High computational capabilities required at the data fusion center.

In order to deal with the above listed issues, distributed detection without fusion spectrum sensing techniques have been developed. As this technique is based on "implicit" cooperation among terminals and it does not require any dedicated channel (Bixio et al, 2009). In these approaches a set of N cooperative cognitive radios share the same radio environment. Each cognitive radio performs spectrum sensing based on its local observation. These local decisions are not fused to obtain a global decision and no sharing of information is required. Yucek and Arslan (2009) introduced an incremental gossiping approach termed as GUESS (gossiping updates for efficient spectrum sensing) in order to perform efficient coordination between CRs in distributed collaborative sensing. They proposed in their algorithm with minimizing complexity as well as reduced protocol overhead.

Sometimes, collaboration for information exchanging is between two CR users. The CR user close to a primary transmitter, which has a better chance of detecting the primary user transmission, cooperates with far away users. An algorithm for pairing secondary users without a centralized mechanism is proposed. A distributed sensing method where CR users share their sensing results among themselves and here only final decisions are shared in order to minimize the network overhead due to collaboration (as cited in Yucek & Arslan, 2009). Features obtained at different radios are shared among cognitive users to improve the detection capability of the system.

1.4.6 External Sensing

Another technique for obtaining spectrum information is external sensing. In external sensing, an external agent (sensor) performs the sensing and broadcasts the channel occupancy information to cognitive radios. External sensing algorithms solve some problems associated with the internal sensing where sensing is performed by the cognitive transceivers internally. Internal sensing is termed as collocated sensing. The main advantages of external sensing are overcoming hidden primary user problem and the uncertainty due to shadowing and fading. Furthermore, as the cognitive radios do not spend time for sensing, spectrum efficiency is increased. The sensing network does not need to be mobile and not necessarily powered by batteries. Hence, the power consumption problem of internal sensing can also be addressed. The presence of passive receivers, viz. television receivers, is detected by measuring the local oscillator power leakage. Once a receiver and the used channel are detected, sensor node notifies cognitive radios in the region of passive primary users via a control channel. A dedicated network composed of only spectrum sensing units is used to sense the spectrum continuously or periodically. The results are communicated to a sink (central) node which further processes the sensing data and shares the information about spectrum occupancy in the sensed area with opportunistic radios. These opportunistic radios use the information obtained from the sensing network for selecting the bands (and time durations) for their data transmission. A pilot channel can be used to distribute the sensing results which are similar to network access and connectivity channel (NACCH). External sensing is one of the methods proposed for identifying primary users in IEEE 802.22 standard as well.

To this end, there is another sensing method named interference based sensing available in the literature. This interference temperature model was introduced by FCC and this can be defined as "the temperature equivalent of the total interference present in RF environment for a particular frequency band and a certain geographic location" (W. Wang, 2009). The CR measures the interference temperature and regulates their transmission in such a way that they avoid raising the interference temperature over the interference temperature limit. Hidden terminal problem can be avoided by implementing this scheme. Instead, difficulties for the measurement of interference temperature put this scheme into a challenge and during detection the CRs cannot differentiate the signals coming from the PU or noise.

1.4.7 Energy Efficient Spectrum Sensing Techniques for the Green CR Network

Nowadays, existing research is going to address the distributed spectrum sensing (DSS) scheme and the challenges of it. In order to perform decision combining, the CR users collect the information from their neighbors and make the decision autonomously through the reporting channel. However, if the reporting channel experiences fading, the sensing performance degrades significantly (Wei & Zhang, 2010). The activated CR users transmit the spectrum sensing results by broadcasting, which is energy-expensive if the secondary users are spread out in a wide area. In 2007, Letaief et al introduced a cluster based DSS method by selecting the secondary user with the largest reporting channel gain as the cluster head (CH) of each cluster though they showed the traditional clustering methods which are not sufficiently efficient in terms of the energy consumption. Wei and Zhang (2010) they proposed a cluster and forward based DSS scheme which is composed by two-tier hierarchical CR network, is discussed in the following sections.

1.4.7.1 Cluster Based Data Fusion

In that scheme, all the secondary users exist in a clustered based geographic location. A pre-defined threshold is chosen to gather the CR users sensing contribution results and the CR user send its spectrum sensing result when its contribution is positive. After each sensing round, all the CR users know the majority sensing decision made by their neighbors. The CH is elected with the largest contribution value among the CR users, which collects and processes the local information sent from its cluster members. The clustering and CH selection at each time step, and the active CHs take turns to be the fusion center is proposed in order to make the CR system energy efficient (Wei & Zhang, 2010).

1.4.7.2 Contribution Based Decision Scheme

Consider that each available CR user gets the decisions from the others. Because of the energy constraint, the CR users exchange the simple message than transferring the complicated information, as proposed by Wei and Zhang, (2010). In this scheme, each CR user cannot send its spectrum sensing result until it provide the positive contribution, and in this way each CR user knows the majority decision of the spectrum sensing at the end of each spectrum sensing round. If the CR user obtains the spectrum sensing result consistent with the consensus, it has the positive contribution; otherwise negative contribution can be considered. A pre-defined threshold for detection is chosen in the scheme. When the CR user's contribution value drops below the threshold, this CR user is considered unreliable and stops it from sending its spectrum sensing results. However, this CR still carrying spectrum sensing and tracking the majority decision. As long as its own spectrum sensing result agrees with the majority decision, its contribution is evaluated. As long as the contribution value exceeds the pre-defined threshold, it starts sending the spectrum sensing results again.

1.4.7.3 Cluster and Forward Scheme

To make the CR network reliable, DSS should send its spectrum sensing decision to its CH. As broadcasting the decisions be an expensive method in the context of energy if the CR users are spread out in a wide area, the forwarding method can be chosen as it requires less energy according to the inverse square-law of power transmission. In the cluster-and-forward scheme, all active CHs and cluster members judged trustworthy as proposed by Wei and Zhang, (2010). Firstly, clustering is done based on geometric locations and the CR users who are close together form a cluster. Then, one of the cluster members with the highest spectrum sensing ability is selected as the cluster head (CH), which processes the spectrum sensing results sent from its cluster members and also collects the individual spectrum sensing decisions. Moreover, due to the energy constraint, the CH is re-selected at each time step as to avoid draining a secondary user's power quickly. After acquiring the spectrum sensing results by the cluster members and also to obtain the spectrum sensing decisions, they transmit not only the results to the selected CH but also the CH adds its own sensing decision and then forwards the all the results of the sensing to the fusion center. In the proposed scheme, the fusion center is dynamically selected from all the active CHs and hence energy saving is achieved by clustering as well as combining data before forwarding. There are some other considerations about the clusters chosen for the sensing scheme, i.e. if the cluster range is small enough so that the local spectrum sensing decision and the local contribution-value collection consume less power while shorter range of clustering increase number of clusters. The more the clusters make the system running in parallel and less delay. However, the more clusters impacts on the data overhead which are undesirable.

1.4.7.4 Compressive Sensing Techniques

Narrow-band spectrum sensing is simpler to implement than the wide-band applications which are nowadays increasingly interested as the deployment of dynamic spectrum access for CR users (Tan & Kong, 2010). However, wide-band spectrum sensing introduces with considerable technical challenges. This wide band sensing can be done in two different ways as the RF front-end can either do narrow-band sensing via a bank of pass-band filters or implement one wide-band RF front-end followed by signal processing blocks to sense the whole wide-band. Nonetheless, the previous associated with the filter design constraints while the later scheme requires high-speed analog-to-digital converter. Recently, potential approach to alleviate the sampling bottleneck in wideband communications is compressive sensing (CS), which asserts that the one can recover sparse signals at sub-Nyquist rates (as cited in S. Hong, 2010). CS relies on this principle of sparsity, so that a concise representation of the signal is possible when expressed in a suitable from. Due to the low percentage of spectrum occupancy by PUs, which originally motivated the development of CRs, wireless communication signals in open-spectrum networks are typically sparse in the frequency domain, allowing us to use compressive sensing to alleviate the sampling bottleneck. While CS is a powerful technique, it does not allow the CR to sample at low rates for free; the resulting increase in computation/complexity is non-trivial, especially in a power constrained mobile cognitive radio. Recent work in CS reveals that a signal having a sparse representation in one basis can be recovered from a small number of projections. Especially, a compressed sampling approach can get the sparse signal at the rates lower than Nyquist sampling; signal reconstruction, which is a solution to a convex optimization problem (Tan & Kong, 2010). As the sampling is done less than the Nyquist rate, so the signals requires less energy for processing at both the transmitter and receiver ends which is supportive to move forward for green cognitive radio.

1.5 POWER CONTROL IN COGNITIVE TRANSCEIVER

The model of DSA techniques, although overall spectrum utilization can be enhanced through exploiting it while radio signal transmission from cognitive nodes can cause harmful interference to the PUs which validates two important design considerations for CR networks, i.e. to maximize the radio resource utilization and to minimize the interference caused to the PUs (Hoang & Liang, 2008). Therefore, power control in CR communication should be necessary regarding the probable interference originated since the presence of multiple users, either primary or secondary, within this network. Generally, power control is employed in mobile networks for improving the link performance that can be applied to CR networks as well, however, power control in CR network is a crucial task and this is the basic module of radio resource management (Qin & Su, 2009). In a wireless network where users are prioritized (i.e. both PUs and CRs are present), if a CR gets the access to exploit the spectrum, it must taking into account the power transmission itself to assure the interference-free transmission for users (PU and CR both) surrounding in it. There are different techniques of this kind, for example, spectrum underlay is one of the existing techniques which permit the CR users to share the whole spectrum simultaneously with the PUs as long as they do not exceed the interference threshold at PUs (Pareek et al, 2010). An excellent survey on power control of CR has been demonstrated and this is followed the so called "protocol model" for interference modeling (Shi & Hou, 2007). Since power control directly affects a node's transmission range and interference range (receive power at the destination node and interference power at other nodes), it has profound impact on scheduling feasibility, bandwidth efficiency, and problem complexity (Shi & Hou, 2008). Furthermore, power control is an important issue for the energy-efficient spectrum sensing, the future trend of green cognitive radio networks.

1.5.1 Power Control for Spectrum Sharing

It can be said that implementation of DSA technique can improve the overall spectrum efficiency though transmission from cognitive devices can cause harmful interference to primary users of the spectrum (Hoang & Liang, 2008). The transmitter power plays an important role in the radio resource management in CR networks as it limits multiuser interference and to maximize the spatial resource reuse (Qin & Su, 2009). Clearly, for the network layer, power control impacts the network topology. For the MAC layer, power control also affects how far apart can two ongoing communication sessions be without interfering with each other while at the physical layer, power control is linked to the signal processing, because the signal processing at the physical layer identifies how stringent the power control requirements to be. All these factors determine the end-to-end performance (Qin & Su, 2009). As a result, there are many cross-layer design methods in the literature have looked at power control in a cross-layer design framework. With the increasing demand for wireless data services, it is necessary to establish power control method for information sources. The power control solution for wireless data solved by the game theoretic approach as a pricing function of the transmit power. Game theory can be so applied to power control in CDMA wireless networks as discussed in (Saraydar et al, 2002).

1.5.2 Spectrum Sharing using Decentralized Scheduling

As proposed by Qin and Su (2009) considering N primary services going on within a CR network of different frequency spectrum and a secondary service is responsible to give service to a set of CR users to share the spectrum with some primary services. The spectrum can be shared among multiple secondary users, where the base station or access point manages the allocated spectrum's radio transmission. In the allocated frequency bands, the CR users' spectrum demand relies on the transmission rate for the adaptive modulation.

In the decentralized scheduling, each node has two spectrum management decisions: a diverse transmission power control with local decentralized scheduling and a decision maker with the acceptance and distribution of frequency bands among participating CRN nodes. As proposed by Qin and Su, (2009) frequency spectrums are submitted by a local user community through the node's decentralized scheduling and flow routing. With the decision maker, it is queued and either started locally or moved to other nodes. The onsite allocation to primary services is eventually done by the local diverse transmission power control. Firstly, the decision maker accepts spectrum from a local user community, it then either allocates frequency bands to the local node's decentralized scheduling and flow routing or forwards them to the decision maker of other nodes.

1.5.3 Design of Distributed Optimization Algorithm

In a wireless network, each node is associated with a certain level of interference range that it can tolerate, which occurs with two pairs of nodes of transmissions that overlap in frequency and time. When a node observes the interference as a low signal to interference and noise ratio (SINR) then it should increase the transmission power to enhance the SINR and through this way each radio should maximize SINR at the anticipated receiver. Undoubtedly, this strategy does not suit well as increasing the transmitter power from each radio than a predefined level results increasing power consumption, decreasing battery life, and raising potential interference to others. To overcome this problem, Qin and Su, (2009) proposed the techniques scheduling-power strategy and global solution with diverse transmission power which select the SINR at a suitable level while maintaining the interference under control. In the earlier strategy, the two-phrase power control with decentralized scheduling (PCDS) algorithm is proposed. At the beginning of each time slot this algorithm checks the interference level developed in some slots. Without interrupting other transmission, the PCDS determines the users set which can transmit in the current slot in security. Basically, it performs two tasks: one is to identify the users set which can try to transmit in a given time slot, the other is to determine the powers set needed to satisfy SINR constraints at respective receivers. The supplementary strategy deals with the process of frequency bands exchange on two different scenarios: the active delegation asks the local decision maker to provide frequency bands to remote nodes in an active way, and the passive delegation asks remote nodes to request frequency bands for the local decision maker. In each scenario, the local decision maker has to publish information of its local waiting queue. The global solution with diverse transmission power builds a connected network, but it does not set all transmission ranges to the same value. Instead, for every node, it tries to find a minimum power level individually.

1.5.4 Power Control in Multiuser CR Networks

Recently, joint beamforming and power control techniques are widely considered for multiuser CR systems. Whitening transform is used for simplifying the capacity optimization, and classic water-filling method is used for beamforming and power control. Wang and Wang (2010) discussed into their paper for the multiuser CR network, they addressed the problem of joint transmit beamforming and power control for CR users when they are allowed to transmit simultaneously with the primary users. By using joint transmit beamforming, sum rate optimization of the network is achieved under the interference constraints of PUs. A transmit beamforming technique is discussed to obtain the achievable rates for the multi-antenna cognitive radio network (Bixio et al, 2011).

1.6 CONCLUDING REMARKS AND POSSIBLE FUTURE SCENARIOS

This chapter overviewed the state-of-the-art techniques for dynamic spectrum management and spectrum sensing for the cognitive radio network as they were proposed so far in literature. The main purpose of this work is to underline the pros and cons of various described techniques. The various proposed schemes have been reviewed and compared having in mind the real-life applications. Dealing with green CR network, several energy-efficient sensing schemes have been proposed. In addition this chapter pointed out how the transmitter power plays an important role in the radio resource management for CR networks; power control optimization was shown as it is needed for limiting multiuser interference as well as for maximizing spatial resource reuse and desired throughput.

Research is still carried out for deploying the dynamic spectrum management: spectrum sensing with single node detection is survived with several problems which could be solved by distributed sensing methods while the computational complexity and hardware constraints push those schemes into challenge. When considering centralized and distributed sensing, the optimization technique should be implemented for both data and decision fusion. There is a possibility for the future wireless network:- flexible radio, which will be increasingly complex and certainly heterogeneous in nature. The concept of flexible radio plays vital role in the future wireless communication (3G-long term evolution: LTE, and LTE-advanced) that must satisfy the adaptability, reconfigurability, modularity, scalability and so on.

REFERENCES

Aazhang, B., Lilleberg, J., & Middleton, G. (2004). Spectrum sharing in a cellular system. *IEEE Eighth International Symposium on Spread Spectrum Techniques and Applications*, (pp. 355 –359).

Akyildiz, F., Lee, W. Y., Vuran, M. C., & Mohanty, S. (2006). NeXt generation dynamic spectrum access cognitive radio wireless networks: A survey. *Computer Networks. The International Journal of Computer and Telecommunications Networking*, *50*(13).

Arshad, K., Imran, M. A., & Moessner, K. (2010). Collaborative spectrum sensing optimization algorithms for cognitive radio networks. *International Journal of Digital Multimedia Broadcasting*, 2010.

Bixio, L., Cattoni, A. F., Regazzoni, C. S., & Varshney, P. K. (2008). *Embodied cognition-based distributed spectrum sensing for autonomic wireless systems. Autonomic Computing and Networking*. USA: Springer.

Bixio, L., Ciardelli, L., & Ottonello, M. Raffetto, Regazzoni, M., C., Alam, S. S., & Armani, C. (2011). *A transmit beamforming technique for MIMO cognitive radios*. Wireless Innovation Forum Conference on Communi- cations Technologies and Software Defined Radio, SDR'11.

Bixio, L., Oliveri, G., Ottonello, M., Raffetto, M., & Regazzoni, C. S. (2010). Cognitive radios with multiple antennas exploiting spatial opportunities. *IEEE Transactions on Signal Processing*, *58*(8), 4453–4459. doi:10.1109/TSP.2010.2048322

Bixio, L., Ottonello, M., Raffetto, M., & Regazzoni, C.S. (2011). Comparison among cognitive radio architectures for spectrum sensing. *EURASIP Journal on Wireless Communications and Networking*.

Cattoni, A. F., Minetti, I., Gandetto, M., Niu, R., Varshney, P. K., & Regazzoni, C. S. (2006). *A spectrum sensing algorithm based on distributed cognitive models*. SDR Forum Technical Conference, Orlando, Florida, USA.

Couturier, S., & Scheers, B. (2009). *The state of the art of dynamic spectrum access*. Military Communication and Information Systems Conference.

Dilmaghani, R. B., & Rao, R. R. (2006). *Hybrid wireless mesh network deployment: A communication testbed for disaster scenarios*. The 1st International Workshop on Wireless Network Testbeds, Experimental Evaluation & Characterization.

Erez, U., & ten Brink, S. (2004). A close to capacity dirty paper coding scheme. *IEEE Transactions on Information Theory*, *51*(10).

Farnham, T. Clemo, Haines, G., R., Seidel, E., Benamar, A., Billington, S., … Mangold, P. (2000). IST-TRUST: A perspective on the reconfiguration of future mobile terminals using software download. *IEEE International Symposium on Personal, Indoor and Mobile Radio Communications*, London, UK, (pp. 1054–1059).

Gandetto, M., & Regazzoni, C. S. (2007). Spectrum sensing: a distributed approach for cognitive terminals. *IEEE Journal on Selected Areas in Communications*, *25*(3). doi:10.1109/JSAC.2007.070405

Gardner, W. A. (1988). Signal interception: A unifying theoretical framework for feature detection. *IEEE Transactions on Communications*, *36*(8), 897–906. doi:10.1109/26.3769

Gardner, W. A. (1991). Exploitation of spectral redundancy in cyclostationary signals. *IEEE Signal Processing Magazine*, *8*(2), 14–36. doi:10.1109/79.81007

Hatfield, D., & Weiser, P. (2005). *Property rights in spectrum: Taking the next step*. The first IEEE Symposium on New Frontiers in Dynamic Spectrum Access Networks.

Hoang, A. T., & Liang, Y. C. (2008). Downlink channel assignment and power control for cognitive radio network. *IEEE Transactions on Wireless Communications*, *7*(8).

Hong, S. (2010). *Multi-resolution Bayesian compressive sensing for cognitive radio primary user detection*. IEEE Global Telecommunications Conference.

Hossain, E., & Bhargava, V. (2007). *Cognitive wireless communication networks*. Springer Publishers. doi:10.1007/978-0-387-68832-9

Hossain, E., Niyato, D., & Han, Z. (2009). *Dynamic spectrum access in cognitive radio network*. Cambridge University Press. doi:10.1017/CBO9780511609909

Hsien-Po, S., & Schaar, M. V. D. (2009). Distributed resource management in multihop cognitive radio networks for delay-sensitive transmission. *IEEE Transactions on Vehicular Technology*, *58*(2), 941–953. doi:10.1109/TVT.2008.925308

Leaves, P., Moessner, K., Tafazolli, R., Grandblaise, D., Bourse, D., Tonjes, R., & Breveglieri, M. (2004). Dynamic spectrum allocation in composite reconfigurable wireless networks. *IEEE Communications Magazine*, *42*, 72–81. doi:10.1109/MCOM.2004.1299346

Oner, M., & Jondral, F. (2007). On the extraction of the channel allocation information in spectrum pooling systems. *IEEE Journal on Selected Areas in Communications*, *25*(3), 558–565. doi:10.1109/JSAC.2007.070406

Pareek, U., Naeem, M., & Lee, D. C. (2010). *An efficient relay assignment scheme for multiuser cognitive radio networks with discrete power control*. IEEE 6th International Conference on Wireless and Mobile Computing, Networking and Communications (WiMob).

Pawełczak, P., Pollin, S., So, H. S. W., Motamedi, A., Bahai, A., Prasad, R. V., & Hekmat, R. (2008). *State of the art in opportunistic spectrum access medium access control design*. The 3rd International Conference on Cognitive Radio Oriented Wireless Networks and Communications.

Qin, H., & Su, J. (2009). Opportunistic scheduling with power control in cognitive radio networks. *International Symposium on Computer Network and Multimedia Technology*, (pp. 1-4). CNMT.

Saraydar, C. U., Mandayam, N. B., & Goodman, D. J. (2002). Efficient power control via pricing in wireless data networks. *IEEE Transactions on Communications*, *50*(2), 291–303. doi:10.1109/26.983324

Shi, Y., & Hou, Y. T. (2007). *Optimal power control for multi-hop software defined radio networks*. IEEE Infocom.

Shi, Y., & Hou, Y. T. (2008). A distributed optimization algorithm for multi-hop cognitive radio networks. *The 27th Conference on IEEE Computer Communications*, (pp. 1292-1300). INFOCOM.

Sun, C. H., Zhang, W., & Letaief, K. B. (2007). Cluster-based cooperative spectrum sensing in cognitive radio systems. *IEEE International Conference on Communications 2007*, (pp. 2511-2515).

Sutton, P., Nolan, K., & Doyle, L. (2008). Cyclostationary signatures in practical cognitive radio applications. *IEEE Journal on Selected Areas in Communications*, *26*(1), 13–24. doi:10.1109/JSAC.2008.080103

Sutton, P. D., Lotze, J., Nolan, K. E., & Doyle, L. E. (2007). Cyclostationary signature detection in multipath Rayleigh fading environments. *IEEE International Conference on Cognitive Radio Oriented Wireless Networks and Communications*.

Tan, L. T., & Kong, H. Y. (2010). *A novel and efficient mixed-signal compressed sensing for wide-band cognitive radio. International Forum on Strategic Technology*. IFOST.

Tkachenko, A., Cabric, D., & Brodersen, R. W. (2007). Cyclostationary feature detector experiments using reconfigurable BEE2. *IEEE International Symposium on New Frontiers in Dynamic Spectrum Access Networks*, (pp. 216–219).

Wang, F., & Wang, W. (2010). *Robust beamforming and power control for multiuser cognitive radio network*. IEEE Global Telecommunications Conference.

Wang, J., Ghosh, M., & Challapali, K. (2011). Emerging cognitive radio applications: A survey. *IEEE Communications Magazine*, 74–81. doi:10.1109/MCOM.2011.5723803

Wei, J., & Zhang, X. (2010). *Energy-efficient distributed spectrum sensing for wireless cognitive radio networks*. IEEE Conference on Computer Communications Workshops.

White, L. (2000). Propertyzing the electromagnetic spectrum: Why it is important, and how to begin. *Media Law and Policy*, *19*, 29–30.

Xu, L., Tonjes, R., Paila, T., Hansmann, W., Frank, M., & Albrecht, M. (2000). DRiVE-ing to the internet: Dynamic radio for IP services in vehicular environments. *The 25th Annual IEEE Conference on Local Computer Networks*, (pp. 281–289).

Yücek, T., & Arslan, H. (2009). A survey of spectrum-sensing algorithms for cognitive radio applications. *IEEE Communications Surveys & Tutorials*, *11*(1), 116–130. doi:10.1109/SURV.2009.090109

Zhao, Q., & Sadler, B. (2007). *A survey of dynamic spectrum access: Signal processing, networking, and regulatory policy. IEEE Signal Processing Magazine: Special Issue on Resource-Constrained Signal Processing*. Communications and Networking.

Zhao, Q., & Sadler, B. (2007). *A survey of dynamic spectrum access: Signal processing, networking, and regulatory policy. IEEE Signal Processing Magazine: Special Issue on Resource-Constrained Signal Processing*. Communications, and Networking.

Zhao, Q., & Swami, A. (2007). A survey of dynamic spectrum access: signal processing and networking perspectives. *IEEE International Conference on Acoustics, Speech, and Signal Processing (ICASSP): Special session on Signal Processing and Networking for Dynamic Spectrum Access*.

Chapter 2
Spectrum Sensing in Cognitive Radio: Aspects of Fading and Cooperation

Wei-Ho Chung
Academia Sinica, Taiwan

ABSTRACT

The cognitive radio has been widely investigated to support modern wireless applications. To exploit the spectrum vacancies in cognitive radios, the chapter considers the collaborative spectrum sensing by multiple sensor nodes in the likelihood ratio test (LRT) frameworks. In this chapter, the functions of sensors can be served through the cooperative regular nodes in the cognitive radio, or the specifically deployed sensor nodes for spectrum sensing. In the LRT, the sensors make individual decisions. These individual decisions are then transmitted to the fusion center to make the final decision, which provides better detection accuracy than the individual sensor decisions. The author provides the lowered-bounded probability of detection (LBPD) criterion as an alternative criterion to the conventional Neyman-Pearson (NP) criterion. In the LBPD criterion, the detector pursues the minimization of the probability of false alarm while maintaining the probability of detection above the pre-defined value. In cognitive radios, the LBPD criterion limits the probabilities of channel conflicts to the primary users. Under the NP and LBPD criteria, the chapter provides explicit algorithms to solve the LRT fusion rules, the probability of false alarm, and the probability of detection for the fusion center. The fusion rules generated by the algorithms are optimal under the specified criteria. In the spectrum sensing, the fading channels influence the detection accuracies. The chapter investigates the single-sensor detection and collaborative detections of multiple sensors under various fading channels and derives testing statistics of the LRT with known fading statistics.

DOI: 10.4018/978-1-4666-2005-6.ch002

2.1 INTRODUCTION

The cognitive radio can facilitate the increased utilization of the spectra. In conventional spectrum allocation schemes, the spectra are licensed to a certain group of users or systems, preventing other users from using the spectra. These conventional spectrum allocation schemes lead to low utilization of the spectra. To increase the utilization of the spectra, the concept of the cognitive radio is to allow the secondary users to use the spectra when the channel is not occupied by the primary (i.e., licensed) users.

In the cognitive radio system, the secondary users must seek spectrum vacancies, i.e., the spectrum not occupied, by using spectrum sensing techniques. The accuracies of the spectrum sensing techniques play a crucial role in the cognitive radio scheme. An accurate spectrum sensing facilitates the efficient utilization of the spectrum vacancy by the secondary users, and reduces the potential interferences and channel conflicts to the primary users. Several techniques (Cabric, 2004; Cabric 2006; Visotsky, 2005) have been used for spectrum sensing, such as energy detection (Kostylev, 2002; Digham, 2003), matched filter detection, and feature detection. These techniques are employed on the single-user detection scenario.

One approach to increase the sensing accuracies is to adopt cooperative spectrum sensing (Cabric, 2006; Visotsky, 2005), where the distributed secondary users form the sensor network and make independent detections (e.g., binary hypothesis tests), and then transmit their decisions to the fusion center (FC). The fusion center makes the final decision based on the decisions from the secondary users. The goal of the cooperative spectrum sensing is to make the decisions, whether to use the intended channel or not, for the secondary users based on the final decision of the fusion center. Thus, the fusion rule at the fusion center must be well designed (Mishra, 2006; Ghasemi, 2005; Lin, 2005; Chen, 2004), such that the probability of false alarm and the probability of detection of its final decision achieve the performance criteria. The cooperative sensing techniques in the cognitive radio belong to the category of the distributed detections (Viswanathan, 1997; Elias-Fuste, 1992; Chair, 1986; Barkat, 1989). We suggest the lowered-bounded probability of detection (LBPD) criterion to be more suitable than the Neyman-Pearson (NP) criterion for the cognitive radio problem. The algorithms are provided to explicitly compute the optimal decision fusion rules for the two criteria.

Since the fading channels are ubiquitous in wireless communications, we investigate the single-sensor detection and collaborative-detection under the influences of the fading channels. We formulate the pdfs of the fading channel gains into the fusion rules. Our approaches, called the likelihood ratio test with fading statistics (LRFS), incorporate the fading statistics into the likelihood ratio test framework. The receiver operating curves (ROCs) are evaluated. The ROCs show performance improvements when the fading statistics are incorporated in the detection frameworks.

2.2 COLLABORATIVE DECISION FUSION FOR SPECTRUM SENSING

The spectrum sensing scenario in the sensor networks is described as follows. A cognitive radio network with N secondary users is deployed. The N secondary users form a sensor network to detect the spectrum vacancies. Each secondary user conducts local detections using energy detection, matched filter detection, or feature detection. Each secondary user decides either H_0 (channel vacant) or H_1 (channel occupied). The probabilities of the local false alarm are denoted as $\{P_{FAi} \mid i = 1,2,\ldots,N\}$, where P_{FAi} represents the probability of false alarm of the *i-th* user. The probabilities of detection are denoted as $\{P_{Di} \mid i = 1,2,\ldots,N\}$, where P_{Di} represents the probability of detection of the *i-th* user. The local decisions of the users are denoted as $\{U_i \mid i = 1, 2, ..., N\}$. The U_i takes the value -1 when the local decision of the *i-th* user is the hypothesis H_0, meaning the channel is vacant. The U_i takes the value 1 when the local decision of the *i-th* user is the hypothesis H_1, meaning the channel is occupied by the primary users. The $\{P_{FAi} \mid i = 1, 2, ..., N\}$, $\{P_{Di} \mid i = 1, 2, ..., N\}$, and $\{U_i \mid i = 1, 2, ..., N\}$ are abbreviated as $\{P_{FAi}\}$, $\{P_{Di}\}$, and $\{U_i\}$ respectively. All possible realizations of $\{U_i\}$ constitute the space containing all possible *N*-bit sequences, denoted as $\{-1, 1\}^N$. The realization of $\{U_i \mid i = 1, 2, ..., N\}$ is denoted as $\{u_i \mid i = 1, 2, ...N\}$, abbreviated as $\{u_i\}$.

In this model, $\{P_{FAi}\}$ and $\{P_{Di}\}$ are fixed and known to the fusion center. When the cooperative spectrum sensing mechanism is activated, the FC obtains the realized local decisions, $\{u_i\}$. The FC makes the final decision, U_{fc}, which takes value -1 (i.e., H_0) or 1 (i.e., H_1), by the fusion rule $f_0(\{u_i\})$, defined as $f_0 : \{-1, 1\}^N \to \{-1, 1\}$. Since the randomized fusion rules include the deterministic fusion rule as the special case, we consider f_0 in the category of the randomized fusion rules. The proposed algorithm is to design the f_0 based on the given $\{P_{FAi}\}$ and $\{P_{Di}\}$, such that the false alarm probability of the U_{fc}, denoted as P_{FAfc}, and probability of detection of the U_{fc}, denoted as P_{Dfc}, fulfill the specified criteria. The criteria include Neyman-Pearson Criterion and Lower-Bounded Probability of Detection.

2.2.1 Detection Criterion of Neyman-Pearson Principle

The Neyman-Pearson criterion (NP criterion) is aimed to maximize the probability of detection under the condition that the probability of false alarm is below the predefined threshold, denoted as α. Applying the NP criterion to the cooperative spectrum sensing, the goal is to design a f_0 such that P_{Dfc} is maximized, under the condition that $P_{FAfc} \le \alpha$.

2.2.2 Detection Criterion of Lower-Bounded Probability of Detection

The possible interferences and channel conflicts to the primary users cause severe concerns. Specifically, when the channel is occupied by the primary users (i.e. H_1) but the secondary users decide H_0, the channel conflicts happen. When the channel is vacant (i.e. H_0) but the secondary users decide H_1, the secondary users miss the opportunity of using the spectrum vacancy. Therefore, from the viewpoint of limiting the probability of channel conflicts to protect the primary users, the desired criterion is to maintain the probability of detection above a pre-defined value, while minimizing the probability of false alarm (i.e. missed opportunities). Following this viewpoint, we suggest the lower-bounded probability of detection criterion (LBPD criterion). In the LBPD criterion, the goal is to minimize the probability of false alarm under the condition that the probability of detection is lower-bounded by the predefined threshold, β. This criterion guarantees the channel occupation by primary users being detected above the predefined probability. Therefore, the probability of channel conflicts is kept below a certain probability. Applying the LBPD criterion in the cooperative spectrum sensing, the goal is to design a f_0 that P_{FAfc} is minimized, under the condition that $P_{Dfc} \ge \beta$.

2.3 SPECTRUM SENSING USING DECISION FUSION WITHOUT FADING

In this section, we create an algorithm to compute the optimal fusion rule under the given $\{P_{FAi}\}$ and

Table 1. Pseudo-codes for computing the decision fusion rules. The notations are defined as $\varnothing$ meaning empty set, $\cup$ meaning set union, $:=$ meaning assigning the right-hand-side to the left-hand-side, and $\backslash$ meaning the set difference.

Algorithm for NP Criterion	Algorithm for LBPD Criterion
Input: α, $\{\gamma_j\}$, $\{P_{\gamma_j \mid H_0}\}$, $\{P_{\gamma_j \mid H_1}\}$, $\{\Lambda_j\}$.	**Input:** β, $\{\gamma_j\}$, $\{P_{\gamma_j \mid H_0}\}$, $\{P_{\gamma_j \mid H_1}\}$, $\{\Lambda_j\}$.
Output: S_0, S_1, ε, r.	**Output:** S_0, S_1, ε, r.
Initialization: $T_{FA} = 0$, $S_0 = \{\varnothing\}$, $S_1 = \{\varnothing\}$.	**Initialization:** $T_D = 0$, $S_0 = \{\varnothing\}$, $S_1 = \{\varnothing\}$.
1. k:= $\arg \max_j \{\Lambda_j\}$;	1. k:= $\arg \max_j \{\Lambda_j\}$;
2. **if** $(T_{FA} + P_{\gamma_k \mid H_0}) < \alpha$ **then**	2. **if** $(T_D + P_{\gamma_k \mid H_1}) < \beta$ **then**
$S_1 := S_1 \cup \gamma_k$;	$S_1 := S_1 \cup \gamma_k$;
Λ_k :=0;	Λ_k :=0;
$T_{FA} := T_{FA} + P_{\gamma_k \mid H_0}$;	$T_D := T_D + P_{\gamma_k \mid H_1}$;
Go to step 1;	**Go to** step 1;
end if	**end if**
3. $\varepsilon = \dfrac{\alpha - T_{FA}}{P_{\gamma_k \mid H_0}}$; $r = \gamma_k$; $S_0 := \{0,1\}^N \backslash \{S_1 \cup r\}$;	3. $\varepsilon = \dfrac{\beta - T_D}{P_{\gamma_k \mid H_1}}$; $r = \gamma_k$; $S_0 := \{0,1\}^N \backslash \{S_1 \cup r\}$;
4. **Return** S_0, S_1, ε, r.	4. **Return** S_0, S_1, ε, r.

$\{P_{Di}\}$ of the secondary users without fading influences. The f_0 is generated by separating the $\{-1,1\}^N$ into three mutually exclusive subsets, denoted as S_0, S_1, and r, where S_0 represents the set for H_0, S_1 represents the set for H_1, and r represents the sequence for the random decision. They must satisfy $S_0 \cup S_1 \cup r \equiv \{-1,1\}^N$. The probability, ε, associated with r, is used by f_0 to randomly generate the decision when the input is r. The f_0 is summarized as

$$f_0(\{u_i\}) = \begin{cases} -1, \text{ when } \{u_i\} \in S_0 \\ +1, \text{ when } \{u_i\} \in S_1 \\ \begin{cases} +1 \text{ with probability } \varepsilon \\ \text{-1 with probability 1-}\varepsilon \end{cases}, \text{ when } \{u_i\} \equiv r. \end{cases} \quad (1)$$

Each element in $\{-1,1\}^N$ represents a possible realization of the $\{U_i\}$. Thus, each element has its associated probabilities of occurrences under H_0 and H_1. For each element $\gamma_j \in \{-1,1\}^N$, we denote $P(\gamma_j \mid H_0)$ by $P_{\gamma_j \mid H_0}$ and $P(\gamma_j \mid H_1)$ by $P_{\gamma_j \mid H_1}$. The $P_{\gamma_j \mid H_0}$ and $P_{\gamma_j \mid H_1}$ can be computed by the given $\{P_{FAi}\}$ and $\{P_{Di}\}$. The likelihood ratio, associated with γ_j, is defined as $\Lambda_j = P_{\gamma_j \mid H_1} / P_{\gamma_j \mid H_0}$. The algorithms for computing the fusion rules under the NP criterion and the LBPD criterion are stated separately in Table 1.

The rationale of the algorithm for the NP criterion is to view the given α as the scarce resource, which must be consumed sparingly by choosing the γ_j with the largest Λ_j into S_1. Opposed to the NP criterion, the rationale of the algorithm for the

LBPD criterion is to view the given β as the threshold which must be achieved quickly by choosing the γ_j with the largest Λ_j into S_1. The randomization by the ε associated with r is to utilize the gap between the predefined threshold of the criteria and the achievable P_{FAfc} (NP criterion) or P_{Dfc} (LBPD criterion) by the deterministic fusion rules.

For the generated fusion rule, the probability of false alarm and the probability of detection at the fusion center are $P_{FAfc} = \alpha$ and $P_{Dfc} = \varepsilon \cdot P_{r|H_1} + \sum_{\{j|\gamma_j \in S_1\}} P_{\gamma_j|H_1}$ for the NP criterion, and $P_{FAfc} = \varepsilon \cdot P_{r|H_0} + \sum_{\{j|\gamma_j \in S_1\}} P_{\gamma_j|H_0}$ and $P_{Dfc} = \beta$ for the LBPD criterion. In other words, for the specific α, β, $\{P_{FAi}\}$, and $\{P_{Di}\}$, the algorithm generates a fusion rule and its associated pair of false alarm probability and detection probability, denoted as (P_{FAfc}, P_{Dfc}). The pair, (P_{FAfc}, P_{Dfc}), forms the receiver-operating-point (ROP) corresponding to that specific fusion rule.

To find all the feasible ROPs of the fusion center under the proposed algorithm, the algorithm can be routinely performed on values of α, incremented from 0 to 1. In other words, by gradually increasing the α from 0 to 1 and performing the algorithm repeatedly on each specific value of α, the feasible fusion rules and ROPs can be found. The set of the feasible ROPs forms the receiver operating curve (ROC). In the LBPD criterion, the ROC can be found by performing the same procedures with β incremented from 0 to 1. Our numerical results show that the ROCs generated by NP criterion are the same as the ROCs generated by the LBPD criterion, as demonstrated in Figure 1, 2, 3,4. Although the ROCs of the NP algorithm is the same as the ROCs of the LBPD algorithm, for arbitrary α and β, the two algorithms generate different fusion rules. The LBPD criterion fits the concept of the cognitive radio, whose first priority is to protect the primary users above a certain quality of service while allowing the secondary users to utilize the vacant channels.

2.4 SPECTRUM SENSING USING DETECTION AND DECISION FUSION WITH FADING STATISTICS

In this section, we first consider detection using a single sensor with fading. Besides the detection at the single sensor, collaborative decision fusions under the fading channels by multiple sensors are considered. Explicit expressions of the test statistics are derived.

2.4.1 Single Sensor Detection under Fading

We consider the scenario where the pdf of the channel gain is known at each secondary user. In this section, our goal is to incorporate the pdf of the fading channel gain in the likelihood ratio test of the local detector, resulting in the LRFS. We assume a slow-flat fading channel with the channel gain, denoted as κ, resulting in a multiplicative gain on the transmitted signal. The received signal at a single detector is given by

$$\underline{X} = \begin{cases} \underline{N}, H_0 \\ \kappa \underline{S} + \underline{N}, H_1 \end{cases} \tag{2}$$

where $\underline{X}$ is a $m \times 1$ vector of the received signal, $\underline{N}$ is $m \times 1$ white-Gaussian-noise vector with $m \times 1$ zero mean vector $\underline{0}$ and $m \times m$ covariance matrix $\sigma_n^2 I_m$ (σ_n^2 is the scalar noise variance and I_m is the $m \times m$ identity matrix), $\underline{S}$ is the $m \times 1$ transmitted signal known at the detector, and κ is the scalar of the fading amplitude with its pdf, $p(\kappa)$, known or estimated at the detector. In this chapter, $N(\underline{X}; \underline{a}, \underline{b})$ denotes the Gaussian pdf with

Figure 1. The receiver operating points for N=3

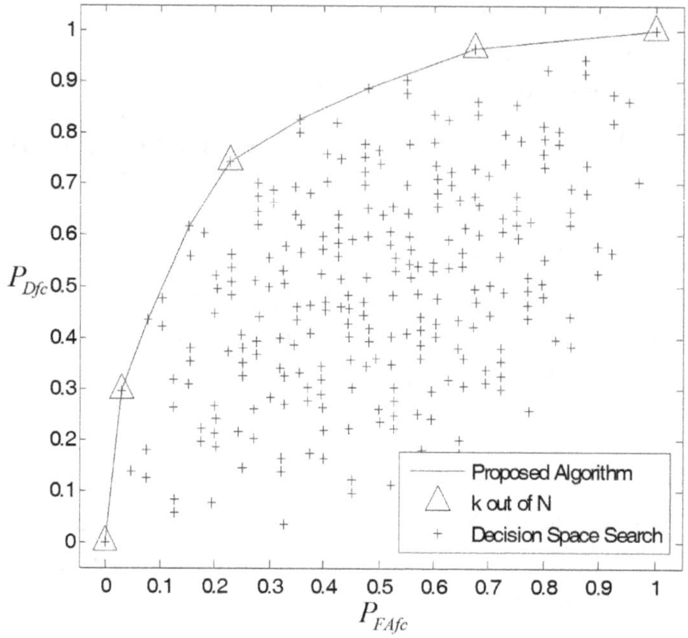

Figure 2. The receiver operating points for N=4

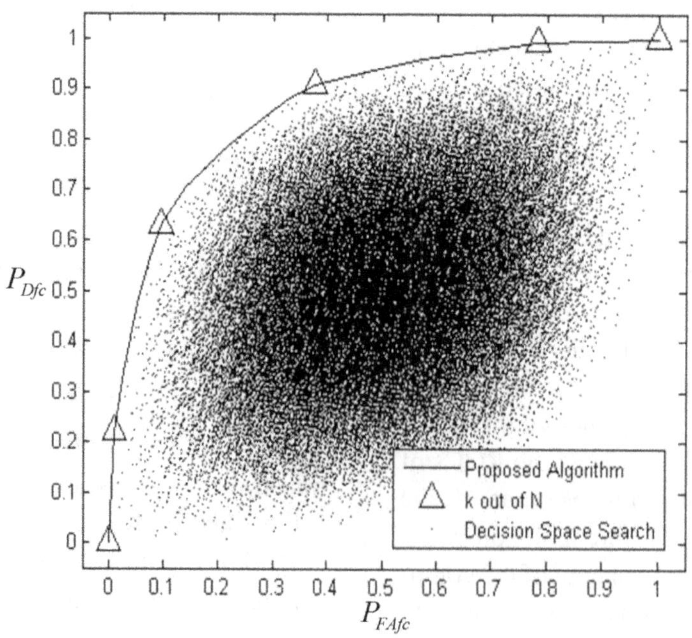

Figure 3. The receiver operating points for N=7

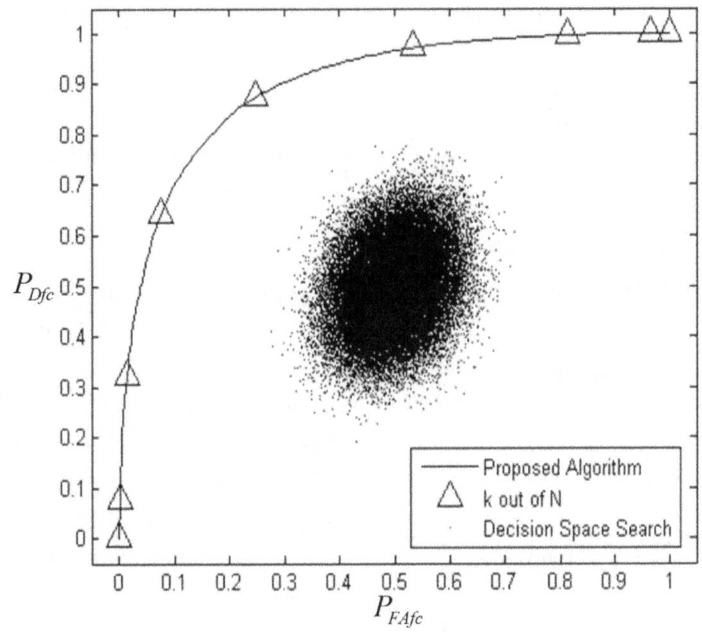

Figure 4. The receiver operating points for N=12

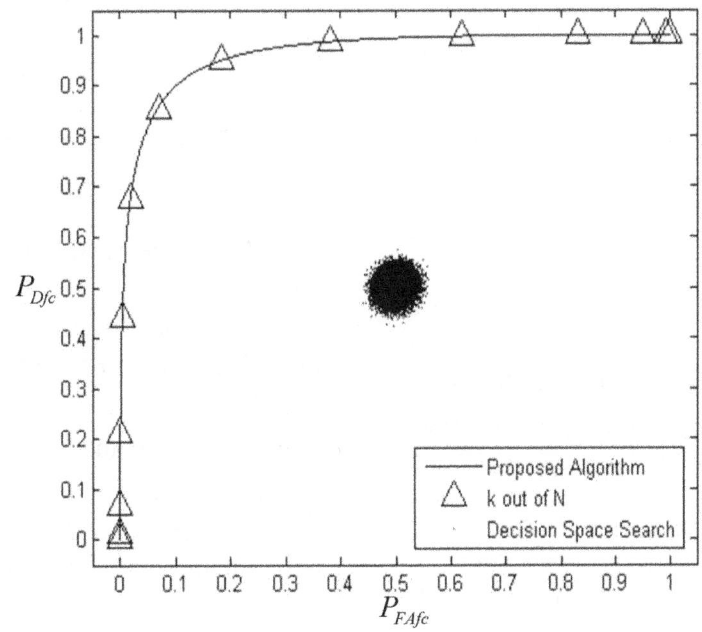

$m \times 1$ vector argument $\underline{X}$, $m \times 1$ mean vector $\underline{a}$, and the $m \times m$ covariance matrix $\underline{b}$. The likelihood function of $\underline{X}$ conditioned on H_0 is

$$p(\underline{X} \mid H_0) = N(\underline{X}; \underline{0}, \sigma_n^2 I_m) = \frac{1}{(2\pi\sigma_n^2)^{m/2}} e^{-\underline{X}^T \underline{X}/(2\sigma_n^2)} \tag{3}$$

where T is the transpose operator. The likelihood function of $\underline{X}$ conditioned on H_1 is

$$p(\underline{X} \mid H_1) = \int_0^\infty p(\underline{X} \mid \kappa, H_1) p(\kappa \mid H_1) d\kappa$$

$$= \int_0^\infty N(\underline{X}; \kappa\underline{S}, \sigma_n^2 I_m) p(\kappa) d\kappa$$

$$= \int_0^\infty \frac{1}{(2\pi\sigma_n^2)^{m/2}} e^{-(\underline{X}-\kappa\underline{S})^T(\underline{X}-\kappa\underline{S})/(2\sigma_n^2)} p(\kappa) d\kappa \tag{4}$$

In (4), we assume that κ is independent of H_1, i.e., $p(\kappa \mid H_1) = p(\kappa)$. By (3) and (4), the LRFS test statistic is expressed as

$$\Lambda_{LRFS}(\underline{X}) = \frac{p(\underline{X} \mid H_1)}{p(\underline{X} \mid H_0)} = \frac{\int_0^\infty \frac{1}{(2\pi\sigma_n^2)^{m/2}} e^{-(\underline{X}-\kappa\underline{S})^T(\underline{X}-\kappa\underline{S})/(2\sigma_n^2)} p(\kappa) d\kappa}{\frac{1}{(2\pi\sigma_n^2)^{m/2}} e^{-\underline{X}^T \underline{X}/(2\sigma_n^2)}} = \int_0^\infty e^{(2\kappa \underline{X}^T \underline{S} - \kappa^2 \underline{S}^T \underline{S})/(2\sigma_n^2)} p(\kappa) d\kappa \tag{5}$$

Denoting the test decision of the single detector as u, H_0 as -1, and H_1 as +1, the detection criterion with the threshold, λ_{LRFS}, is expressed as

$$\Lambda_{LRFS}(\underline{X}) \underset{u=-1}{\overset{u=+1}{\gtrless}} \lambda_{LRFS} \tag{6}$$

For a Rayleigh channel, the pdf of the channel gain is

$$p_{Ray}(\kappa; \sigma_R) = \kappa \exp\left(\frac{-\kappa^2}{2\sigma_R^2}\right) \Big/ \sigma_R^2 \tag{7}$$

Substituting (7) in (5) and using (Gradshtegn, 1980), we obtain the explicit expression of the LRFS test statistic under the Rayleigh channel as presented in Exhibit 1.

For a Rician channel, the pdf of the channel gain is

$$p_{Rician}(\kappa; \sigma_{Ri}, v) = \frac{\kappa}{\sigma_{Ri}^2} \exp\left(\frac{-(\kappa^2 + v^2)}{2\sigma_{Ri}^2}\right) I_0\left[\frac{\kappa v}{\sigma_{Ri}^2}\right] \tag{9}$$

Substituting (9) in (5), we obtain the expression of the LRFS test statistic under the Rician channel as

$$\Lambda_{LRFS}(\underline{X}) = \frac{1}{\sigma_{Ri}^2} \int_0^\infty \kappa e^{\frac{-v^2-\kappa^2}{2\sigma_{Ri}^2} + \frac{2\kappa \underline{X}^T \underline{S} - \kappa^2 \underline{S}^T \underline{S}}{\sigma_n^2}} I_0\left[\frac{\kappa v}{\sigma_{Ri}^2}\right] d\kappa \tag{10}$$

where $I_0[.]$ is the *Bessel* function of the first kind of order zero.

We numerically evaluate the ROCs of the Rayleigh and Rician LRFS statistics in (8) and (10). Since the pdfs of $\underline{X}$ under H_0 and H_1 can be derived to be Gaussian distributions from (2), together with (8) and (10), the pdfs of $\Lambda_{LRFS}(\underline{X})$ conditioned on H_0 or H_1 and the resulting ROCs can be obtained. Besides performance analyses,

Exhibit 1.

$$\Lambda_{LRFS}(\underline{X}) = \left[e^{\frac{\sigma_R^2(\underline{X}^T\underline{S})^2}{2(\sigma_R^2\sigma_n^2\underline{S}^T\underline{S}+\sigma_n^4)}} \sqrt{2\pi}\,\underline{X}^T\underline{S} + 2\sigma_n^2\sqrt{\frac{1}{\sigma_R^2}+\frac{\underline{S}^T\underline{S}}{\sigma_n^2}} + \sqrt{2\pi}\,\underline{X}^T\underline{S}\ Erf\left[\frac{\underline{X}^T\underline{S}}{\sigma_n^2\sqrt{\frac{2}{\sigma_R^2}+\frac{2\underline{S}^T\underline{S}}{\sigma_n^2}}}\right] e^{\frac{\sigma_R^2(\underline{X}^T\underline{S})^2}{2(\sigma_R^2\sigma_n^2\underline{S}^T\underline{S}+\sigma_n^4)}} \right] \Bigg/ 2\left(\sigma_R^2\underline{S}^T\underline{S}+\sigma_n^2\right)\sqrt{\frac{1}{\sigma_R^2}+\frac{\underline{S}^T\underline{S}}{\sigma_n^2}} \quad (8)$$

where *Erf[.]* is the *error* function.

the explicit expression in (8) also facilitates the implementation of the detection.

We denote $E[.]$ as the expectation of the random variable. In these examples, by the signal model of (2), the signal-to-noise-ratio (*SNR*) is expressed by $SNR = E[\kappa^2]\|s\|^2 / (m\sigma_n^2)$, where $\|\underline{S}\|$ is the l_2-norm of $\underline{S}$. Figure 5 shows the ROCs of *SNR* = 0, 5, and 10 dB, for Rayleigh and Rician channels. For evaluating the *SNR*s in these examples, we obtain $E[\kappa^2] = 2\sigma_R^2$ for a Rayleigh channel specified in (7), and $E[\kappa^2] = 2\sigma_{Ri}^2 + v^2$ for a Rician channel specified in (9).

For the performance comparison, we evaluate the conventional likelihood ratio test, which is

Figure 5. ROCs for LRFS and MF detectors for Rayleigh and Rician channels.

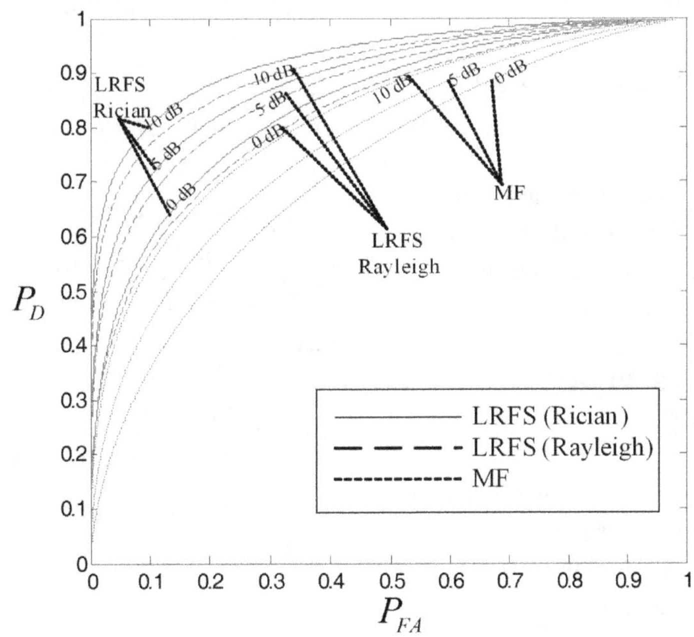

equivalent to the matched filter (MF) approach without incorporating the fading statistics. The testing statistic and the decision criterion of the MF approach is expressed by

$$\Lambda_{MF}(\underline{X}) = \underline{X}^T \underline{S} \underset{u=-1}{\overset{u=+1}{\gtrless}} \lambda_{MF}. \tag{11}$$

The likelihood function of the MF output conditioned on H_0 is

$$p(\Lambda_{MF}(\underline{X}) \mid H_0) = N(\Lambda; \underline{0}, \sigma_n^2 \|\underline{S}\|) \tag{12}$$

where Λ represents the argument for the pdf of $\Lambda_{MF}(\underline{X})$.

The MF output conditioned on H_1 is

$$(\Lambda_{MF}(\underline{X}) \mid H_1) = (\kappa \underline{S} + \underline{N})^T \underline{S} = \kappa \underline{S}^T \underline{S} + \underline{N}^T \underline{S} \tag{13}$$

In (13), under a Rayleigh channel, because $\kappa \underline{S}^T \underline{S}$ is independent from $\underline{N}^T \underline{S}$, we obtain

$$p(\Lambda_{MF}(\underline{X}) \mid H_1) = p_{Ray}(\Lambda; (\underline{S}^T \underline{S})\sigma_R) * N(\Lambda; \underline{0}, \sigma_n^2 \|\underline{S}\|) \tag{14}$$

where $*$ denotes the convolution operator and Λ represents the argument for the pdf. Therefore, the ROCs of the MF approach can be analytically evaluated from (12) and (14).

2.4.2 Multi-Sensor Decision Fusion under Fading

In this section, we consider the decision fusion scenario where the fusion center makes the final decision by fusing the decisions of the N detectors under fading channels. Following the same notations of the previous sections, the decisions of the N detectors are denoted as $\{u_i \mid i = 1, 2, ...N\}$, abbreviated as $\{u_i\}$. The u_i represents the decision of the *i-th* detector, with $u_i = +1$ for H_1, and $u_i = -1$ for H_0. The detectors operate at the probabilities of detection, $\{P_{Di} \mid i = 1, 2, ..., N\}$ (abbreviated as $\{P_{Di}\}$), and the probabilities of false alarm, $\{P_{FAi} \mid i = 1, 2, ..., N\}$ (abbreviated as $\{P_{FAi}\}$), known to the FC. The local decisions, $\{u_i\}$, are transmitted to the FC through the noisy and fading channels, with channel gains denoted as $\{\kappa_i, i = 1, ..., N\}$ (abbreviated as $\{\kappa_i\}$). We consider the fading statistics, i.e., the joint pdf of $\{\kappa_i\}$, are known at the FC. The FC receives the faded and noise-corrupted decisions, $\{y_i, i = 1, ..., N\}$, modeled (Lin, 2005; Chen, 2004) by $y_i = \kappa_i u_i + n_i$, where the n_i represents the noise with Gaussian pdf $N(0, \sigma_{n,i}^2)$. Denoting $\underline{Y} = \{y_1, y_2, ..., y_N\}$, the FC has to make the final decision, U_{fc}, based on the received $\underline{Y}$. The false alarm probability of the U_{fc} is denoted as P_{FAfc}, and probability of detection of the U_{fc} is denoted as P_{Dfc}. The LRFS statistic at the FC can be expressed as

$$\Lambda(\underline{Y}) = p(\underline{Y} \mid H_1) / p(\underline{Y} \mid H_0). \tag{15}$$

By the Bayes rule, we obtain the equation presented in Exhibit 2

From the independence of $\{u_i\}$ and $\{\kappa_i\}$, we have

$$\begin{aligned} &p(\{\kappa_i\}, \{u_i\} \mid H_1) \\ &= p(\{u_i\} \mid H_1) p(\{\kappa_i\} \mid H_1) \\ &= p(\{u_i\} \mid H_1) p(\{\kappa_i\}) \end{aligned} \tag{17}$$

Exhibit 2.

$$p(\underline{Y} \mid H_1) = \sum_{\forall\{u_i\}\in\{-1,+1\}^N} \int p(\underline{Y} \mid \{\kappa_i\},\{u_i\},H_1)p(\{\kappa_i\},\{u_i\} \mid H_1)d\kappa_1 d\kappa_2 ... d\kappa_N \quad (16)$$

Exhibit 3.

$$p(\underline{Y} \mid H_1) = \sum_{\forall\{u_i\}\in\{-1,+1\}^N} \int p(\underline{Y} \mid \{\kappa_i\},\{u_i\},H_1)p(\{u_i\} \mid H_1)p(\{\kappa_i\})d\kappa_1 d\kappa_2 ... d\kappa_N \quad (18)$$

Exhibit 4.

$$p(\underline{Y} \mid H_0) = \sum_{\forall\{u_i\}\in\{-1,+1\}^N} \int p(\underline{Y} \mid \{\kappa_i\},\{u_i\},H_0)p(\{u_i\} \mid H_0)p(\{\kappa_i\})d\kappa_1 d\kappa_2 ... d\kappa_N \quad (19)$$

Exhibit 5.

$$\Lambda(\underline{Y}) = \frac{\sum_{\forall\{u_i\}\in\{-1,+1\}^N} \int p(\underline{Y} \mid \{\kappa_i\},\{u_i\},H_1)p(\{u_i\} \mid H_1)p(\{\kappa_i\})d\kappa_1 d\kappa_2 ... d\kappa_N}{\sum_{\forall\{u_i\}\in\{-1,+1\}^N} \int p(\underline{Y} \mid \{\kappa_i\},\{u_i\},H_0)p(\{u_i\} \mid H_0)p(\{\kappa_i\})d\kappa_1 d\kappa_2 ... d\kappa_N} \begin{array}{c} u_{fc}=+1 \\ > \\ < \\ u_{fc}=-1 \end{array} \lambda_{fc} \quad (20)$$

Substituting (17) into (16), we obtain the equation presented in Exhibit 3

Following the same procedures, we obtain the equation presented in Exhibit 4

Substituting (18) and (19) into (15), we obtain the explicit expression of the statistic and the decision rule as presented in Exhibit 5

In (20), the $p(\underline{Y} \mid \{\kappa_i\},\{u_i\},H_1)$ and $p(\underline{Y} \mid \{\kappa_i\},\{u_i\},H_0)$ can be obtained by Gaussian distribution $N(\underline{Y};\underline{K},\sigma^2_{n,i}I_N)$, where I_N is the $N \times N$ identity matrix and

$$\underline{K} = \begin{bmatrix} \kappa_1 u_1 \\ \kappa_2 u_2 \\ \vdots \\ \kappa_N u_N \end{bmatrix} \quad (21)$$

The $p(\{u_i\} \mid H_1)$ and $p(\{u_i\} \mid H_0)$ of (20) can be obtained from $\{P_{Di}\}$ and $\{P_{FAi}\}$. The $p(\{\kappa_i\})$ in (20) is the joint distribution of channel gains.

2.5 NUMERICAL EXAMPLES

2.5.1 Numerical Examples under NP-Criterion and LBPD-Criterion

For the purpose of comparisons, the *k*-out-of-*N* rule, the decision space search rule, and the exhaustive search approach are introduced. In the *k*-out-of-*N* rule, the fusion decision is H_1 if the number of 1 in the $\{u_i\}$ is larger than *k*. The fusion decision is H_0 if the number of 1 in the $\{u_i\}$ is equal to or smaller than *k*. In the decision space search rule, all the elements in $\{0,1\}^N$ are assigned into S_0 or S_1 randomly. Each element has equal probability of being in S_0 or S_1. For the system with N users, there are 2^N elements in the domain of the fusion rule, $\{0,1\}^N$. The number of possible fusion rules is the number of elements in the power set of $\{0,1\}^N$. Therefore, it takes 2^{2^N} trials to exhaustively search all possible fusion rules for the *N*-user system. Because of the high complexity of 2^{2^N}, we only conduct exhaustive search for the examples of *N*=3 and 4 (Figure 1 and 2). For other cases, we randomly select 50000 fusion rules and plot their corresponding (P_{FAfc}, P_{Dfc}) pairs (Figure 3 and 4). In Figure 3 and 4, the 50000 (P_{FAfc}, P_{Dfc}) pairs from the random search are more concentrated with the increasing *N*. Compared with the ROPs of the exhaustive search, the ROCs of the proposed algorithm bound the ROPs of all decision rules by the exhaustive search, as shown in Figure 1 and 2. Therefore, the decision rule generated by the proposed algorithm is optimal among all possible decision rules.

2.5.2 Numerical Examples in Single-Sensor Detection and Multi-Sensor Decision Fusion under Fading

In Figure 5, the ROCs of the single detector under fading are evaluated and compared. From the ROCs at the same SNR value, we note the LRFS results outperform the conventional MF results. For example at *SNR* = 0 dB and P_{FA}= 0.05, Figure 5 shows the P_D= 0.24 for MF, P_D= 0.45 for the LRFS in the Rayleigh channel, and P_D= 0.47 for the LRFS in the Rician channel. These examples show the performance improvements, contributed by incorporating the fading statistics into the detection scheme.

We evaluate the ROCs of the fusion rule (20) through Monte Carlo simulations. In the simulations, the settings are $P_{D,i} = 0.7$ and $P_{FA,i} = 0.2$ for all *i*. In this example, the $\{\kappa_i\}$ are independent Rayleigh random variables with parameters $\{\sigma_{R,i}\}$. We evaluate the ROCs for 2-sensor and 3-sensor scenarios with the Rayleigh parameters of $\sigma_{R,i}$= 0.2, 0.5, and 0.8 for all *i*. The results are shown in Figure 6. At $\sigma_{R,i}$= 0.5 and P_{FAfc} = 0.05, Figure 6 shows the P_D =0.38 for the 2-sensor LRFS decision fusion, P_{Dfc} =0.46 for the 3-sensor LRFS decision fusion. The performance improvements by using more sensors can be evaluated by (20), as demonstrated in this example.

2.6 CONCLUSION

In cognitive radios, the cooperative spectrum sensing techniques increase the sensing accuracies compared to individual sensing. In this chapter, we suggest the LBPD criterion for the cognitive radio scenario. The LBPD criterion pursues the low probability of false alarm while maintaining probability of detection above a predefined level. Therefore, opposed to the conventional NP criterion, the LBPD criterion is aimed at protecting the primary users above a required probability of being detected while maximizing the secondary users' probabilities of utilizing spectrum holes.

Two algorithms are proposed for explicitly computing the fusion rules under the NP criterion and the LBPD criterion. The algorithms generate

Figure 6. ROCs for decision fusion with 2 and 3 detectors under Rayleigh channels.

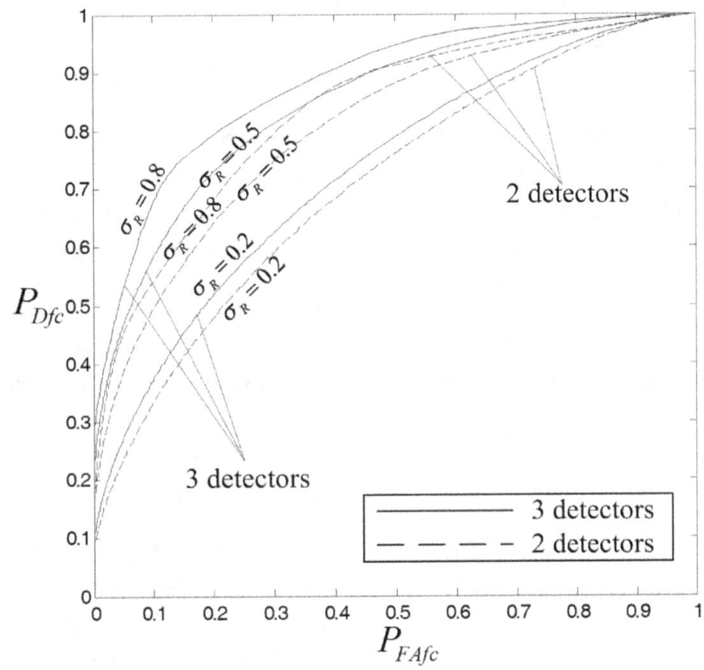

the fusion rules, the associated probability of false alarm, and the probability of detection for the final decision at the fusion center. The generated fusion rules are optimal under the specified criteria.

We investigated the spectrum sensing for fading channels and derive explicit expressions for LRFS. Since the fading effects are ubiquitous in wireless communications, the fading statistics provide the valuable information to improve the accuracies of detections. The ill effects due to fading can also be mitigated by using more sensors to improve detections.

REFERENCES

Barkat, M., & Varshney, P. K. (1989). Decentralized CFAR signal detection. *IEEE Transactions on Aerospace and Electronic Systems*, *25*(2), 141–149. doi:10.1109/7.18676

Cabric, D., Mishra, S. M., & Brodersen, R. W. (2004). Implementation issues in spectrum sensing for cognitive radios. *Conference Record of the Thirty-Eighth Asilomar Conference on Signals, Systems and Computers*, Vol. 1, (pp. 772-776).

Cabric, D., Tkachenko, A., & Brodersen, R. W. (2006). Experimental study of spectrum sensing based on energy detection and network cooperation. *Proceedings of the First International Workshop on Technology and Policy for Accessing Spectrum*, Article No. 12.

Chair, Z., & Varshney, P. K. (1986). Optimal data fusion in multiple sensor detection systems. *IEEE Transactions on Aerospace and Electronic Systems*, 98–101. doi:10.1109/TAES.1986.310699

Chen, B., Jiang, R., Kasetkasem, T., & Varshney, P. K. (2004). Channel aware decision fusion in wireless sensor networks. *IEEE Transactions on Signal Processing*, *52*(12), 3454–3458. doi:10.1109/TSP.2004.837404

Digham, F. F., Alouini, M.-S., & Simon, M. K. (2003). On the energy detection of unknown signals over fading channels. *IEEE International Conference on Communications*, Vol. 5, (pp. 3575-3579).

Elias-Fuste, A. R., Broquetas-Ibars, A., Antequera, J. P., & Yuste, J. C. M. (1992). CFAR data fusion center with inhomogeneous receivers. *IEEE Transactions on Aerospace and Electronic Systems*, *28*(1), 276–285. doi:10.1109/7.135453

Ghasemi, A., & Sousa, E. S. (2005). Collaborative spectrum sensing for opportunistic access in fadinenvironments. *2005 First IEEE International Symposium on New Frontiers in Dynamic Spectrum Access Networks*, (pp. 131-136).

Gradshtegn, I. S., & Ryzhik, I. M. (1980). *Table of integrals, series, and products*. New York, NY: Academic Press.

Kostylev, V. I. (2002). Energy detection of a signal with random amplitude. *IEEE International Conference on Communications*, Vol. 3, (pp. 1606-1610).

Lin, Y., Chen, B., & Varshney, P. K. (2005). Decision fusion rules in multi-hop wireless sensor networks. *IEEE Transactions on Aerospace and Electronic Systems*, *41*(2), 475–488. doi:10.1109/TAES.2005.1468742

Mishra, S. M., Sahai, A., & Brodersen, R. W. (2006). Cooperative sensing among cognitive-radios. *IEEE International Conference on Communications*, Vol. 4, (pp. 1658-1663).

Visotsky, E., Kuffner, S., & Peterson, R. (2005). On collaborative detection of TV transmissions in support of dynamic spectrum sharing. *2005 First IEEE International Symposium on New Frontiers in Dynamic Spectrum Access Networks*, (pp. 338-345).

Viswanathan, R., & Varshney, P. K. (1997). Distributed detection with multiple sensors I. Fundamentals. *Proceedings of the IEEE*, *85*(1), 54–63. doi:10.1109/5.554208

Chapter 3
Robust Performance of Spectrum Sensing in Cognitive Radio Networks

Shimin Gong
Nanyang Technological University, Singapore

Ping Wang
Nanyang Technological University, Singapore

Jianwei Huang
The Chinese University of Hong Kong, HKSAR, China

ABSTRACT

Harmonic coexistence of cognitive radio systems and licensed systems requires the secondary users to have the capability of sensing and keeping track of primary users' transmissions. While existing spectrum sensing methods usually assume known distributions of the primary signals, such an assumption is often not true in practice. As a result, applying existing sensing methods will often lead to unreliable detection performance in practical networks. In this chapter, the authors try to investigate the sensing performance under the distribution uncertainty of primary signals. They first investigate the performance bounds for single user detection with unknown distribution, and provide an analytical expression for the lower bound of detection probability. Moreover, they bring the distribution uncertainty into multi-user cooperative sensing. The authors formulate the optimal sensing design as a robust optimization problem, and propose an iterative algorithm to determine the optimal decision threshold for each user. Extensive simulations demonstrate the effectiveness of the proposed algorithm.

3.1 INTRODUCTION

Traditionally, wireless spectrum has been statically allocated to license holders for exclusive use, and the spectrum available for new applications has become increasingly scarce and pricey. On the other hand, field measurement report (Mchenry, 2003) has revealed that the actual utilization is often quite low, e.g., some radio bands for TV broadcasting and industrial, scientific and medical (ISM) radio bands are idle most of the time. Under this situation, Cognitive Radio (CR) (Haykin,

DOI: 10.4018/978-1-4666-2005-6.ch003

2005) is proposed as a promising technique to improve spectrum utilization. In a CR system, secondary users (SUs) are allowed to access the licensed spectrum bands when they are temporarily unoccupied by primary users (PUs). A key technology to achieve this is spectrum sensing, where SUs need to determine the presence of PUs based on the characteristics of the measured signals on the channels.

There have been extensive studies on the design and performance analysis of spectrum sensing mechanisms. Generally, spectrum sensing is formulated as a hypothesis testing problem, which takes measured signal samples (possibly embedded with primary signals) as inputs and makes a decision by comparing the samples with a predesigned decision threshold. The detection performance is usually measured by the receiver operating characteristic (ROC), which denotes the tradeoff between detection probability (i.e., probability of sensing a busy channel as busy) and false alarm probability (i.e., probability of sensing an idle channel as busy). Existing results usually assume that the received primary signal follows a Gaussian distribution, and express the detection probability in the Q -function (tail probability of the standard normal distribution) (Sousa, 2005) (Simon, 2007). Some other sensing schemes for multiuser cooperative sensing are analyzed through deterministic optimization problem (Sayed, 2008) (Taricco, 2011). For an overview study regarding the sensing techniques in cognitive radio networks, please refer to (Anant Sahai, 2004), (Sousa, 2005), (Li, 2008) and (Zhi Quan, 2008).

However, most of the existing studies assume that the statistical features (such as probability distribution function) of the primary signals received at SUs' receivers are precisely known beforehand, e.g., following Rayleigh, or Rician distributions. In fact, precise knowledge regarding the signal distribution is usually a very strong assumption, and does not necessarily lead to reliable decisions due to unpredictable factors under practical circumstance. On one hand, it is often hard to model a realistic distribution of the received signal in closed-form without significant simplifications. Even if a precise characterization is possible, a complex form usually makes the analysis computational demanding. On the other hand, a deterministic assumption may not always match the real distribution when spectrum environment fluctuates. For example, the received signals exhibit different distributions depending on whether there is line-of-sight between transmitter and receiver. Besides, uncertainty may also occur in the channel estimation. It further leads to unreliable distribution parameters and detection errors.

In this chapter, we will study robust spectrum sensing under distribution uncertainty of the primary signals. Specifically, we consider the case where the statistical features (i.e., mean and variance) of the primary signals are contaminated by estimation errors, and the actual probability density function (pdf) does not match what is often assumed in literature. To capture the distribution uncertainty, we define an uncertainty set that contains all possible signal distributions whose mean and variance are close to a nominal one, and then formulate the sensing design as a robust optimization problem. Solving the robust optimization problem in general is more difficult than solving its non-robust counterpart (also called nominal problem). Some researchers (Aharon Ben-Tal, 1998) (Aharon Ben-tal, 1998) (Teo, 2007) have studied the solvability of robust optimization and proposed some useful results that can be applied into practical problems.

Few prior results consider the issues of uncertainty and robustness in cognitive radio networks. Y. Zhang et.al. propose a probabilistic method that extracts constant information entropy of the white Gaussian noise, and use it against noise fluctuation (Y. Zhang, 2010). A nonparametric cyclic correlation estimator is proposed by by J. Lunden et al. to achieve robustness (J. Lunden, 2010). This correlation detector does not require explicit assumptions on the data or noise distributions, but requires prior information regarding the cyclic frequencies of the primary system. As far as we know, this is the first work that tackles the uncertainty issue for spectrum sensing from a viewpoint of robust optimization. Specifically, we formulate the robust design for both single user and multi-user cooperative spectrum sensing as a max-min robust optimization problem. For single user sensing, we obtain the lower and upper bounds for the detection probability. And for multi-user cooperative sensing, we propose an iterative algorithm that decomposes the problem into a series of semi-definite programs, which can be solved easily by conic optimization tools.

The rest of the chapter is organized as follows. Section 2 describes the detection structure, decision function and robust considerations. In Section 3, we consider the spectrum sensing of single user, and investigate the lower and upper bounds of detection probability. Further in section 4, we present the robust threshold design for multi-user cooperative sensing. Section 5 shows some numerical results demonstrating the effectiveness of our algorithm, and Section 6 concludes this chapter.

3.2 SYSTEM MODEL

We consider a CR network with one primary channel and N SUs. Let $\mathcal{N} = \{1,\dots,N\}$ denote the set of all SUs, which are equipped with local energy detectors, performing cooperative spectrum sensing (Li, 2008), (Sayed, 2008), (K. Ben Letaief, 2009). Each SU samples the primary channel and reports the sensing result to a fusion center, which makes a final decision and distributes the decision back to the cooperative users. We use a hard combination scheme (Sousa, 2005) at the fusion center and define a decision function as:

$$h(\lambda,\xi,k) = I\left(\sum_{i=1}^{N} I(\xi_i - \lambda_i \geq 0) \geq k\right),$$

where $\xi = \{\xi_i\}_{i=1}^{N}$ and $\lambda = \{\lambda_i\}_{i=1}^{N}$ denote the received signal and decision threshold at each user respectively. Indicator $I(A)$ equals 1 if A is true and 0 otherwise. Parameter k specifies the combining rule, i.e., $k=1$ means the OR-rule and $k = \lceil N/2 \rceil$ represents the majority-rule. When $h(\lambda,\xi,k)=1$, the fusion center reports the presence of the primary user (i.e., hypothesis H_1), otherwise reports an idle channel (i.e., hypothesis H_0). As the OR-rule provides better protection of PUs than the majority-rule and is commonly used in the fusion center, we consider $k=1$ in our framework and denote $h(\xi,\lambda,1)$ as $h(\xi,\lambda)$ for simplicity. Let $X = [X_1,\dots,X_N]$ denote the signal received by N SUs and $S \subset \mathcal{R}^N$ as the feasible set of X. Then joint distribution function of X under hypothesis H_1 (H_0, respectively) is given by $f_X^1(\xi)$ ($f_X^0(\xi)$, respectively). Finally, two performance metrics, i.e., the detection and false alarm probabilities can be represented as $Q_d = \int_{\xi\in S} h(\xi,\lambda) f_X^1(\xi) d\xi$ and $Q_f = \int_{\xi\in S} h(\xi,\lambda) f_X^0(\xi) d\xi$, respectively.

In practice, the signals transmitted by a PU are hard to know precisely in advance by SUs, as it is often time varying and experiences attenuation, shadowing, and multi-path fading before reaching the SUs' receivers. Thus the distribution f_X^1 of the test statistic under hypothesis H_1 would be very complicated to analyze. But we can obtain the primary signals' long-time averages in terms of the mean μ and variance Σ, which are regarded as the *nominal* statistics. Then we estimate the range of short-time fluctuations empirically according to a specific spectrum environment. It is reasonable to assume that the instant fluctuations of mean and variance should be within small ranges to their nominal values. Therefore, we define an uncertainty set for distribution f_X^1 using the nominal values of mean and variance in a similar way as (Erick Delage, 2010):

$$\mathcal{U} = \left\{ f_X^1 \middle| \begin{array}{c} \varepsilon\varepsilon^T \prec \gamma_1\Sigma, \\ \int_{\xi\in S}(\xi-\mu)(\xi-\mu)^T f_X^1(\xi)d\xi \prec \gamma_2\Sigma, \\ \int_{\xi\in S} f_X^1(\xi)d\xi = 1. \end{array} \right\}, \tag{1}$$

where $\varepsilon = \int_{\xi\in S}\xi f_X^1(\xi)d\xi - \mu$ denotes the deviation of instant mean from its nominal value μ. Parameters γ_1 and γ_2 regulate the uncertainty size and provide a way to evaluate the confidence in nominal μ and Σ, respectively. When there is no transmission from PUs, the background noise is also time varying (Y. Zhang, 2010), (J. Lunden, 2010). However, it only affects the false alarm probability but not the detection probability when decision threshold is fixed[1]. In this paper we focus on the choice of decision threshold and its impact on the detection probability, thus we will assume that noise follows a deterministic Gaussian distribution f_0, i.e., $f_X^0(\xi) = f_0(\xi)$. The distribution parameters of $f_0(\xi)$ can be obtained through field measurement or empirical data.

In terms of detection method, any hypothesis testing method based on known distribution function can be used to study the detection performance under distribution uncertainty. Based on the Neyman-Pearson criterion which is widely studied in literature (Sousa, 2005), (Taricco, 2011), our robust design is to find the decision threshold λ that maximizes the detection probability Q_d, subject to the constraint that the false alarm Q_f is no larger than α, and the distribution function f_X^1 comes from the uncertainty set $\mathcal{U}$. The robust optimization problem can be expressed as follows.

$$\begin{aligned} \max_{\boldsymbol{\lambda}} \min_{f_{\mathbf{X}}^1(\boldsymbol{\xi})} & \int_{\boldsymbol{\xi}\in S} h(\boldsymbol{\xi},\boldsymbol{\lambda}) f_{\mathbf{X}}^1(\boldsymbol{\xi})d\boldsymbol{\xi} \\ s.t.\quad & f_{\mathbf{X}}^1(\boldsymbol{\xi}) \in \mathcal{U} \\ & \int_{\boldsymbol{\xi}\in S} h(\boldsymbol{\lambda},\boldsymbol{\xi}) f_0(\boldsymbol{\xi})d\boldsymbol{\xi} \le \alpha. \end{aligned} \tag{2}$$

Note that solving such a max-min problem will provide the optimal worst-case performance, i.e., the optimal decision threshold vector that maximizes the detection probability when the worst-case distribution[2] of X is considered.

3.3 DETECTION BOUNDS OF SINGLE USER

We first consider the robust scenario for single user detection. Problem (2) is now in scalar form. In a Neyman-Pearson detector, the decision threshold of single user is uniquely determined by the false alarm probability α. Then the optimization over decision threshold does not exist anymore. But we can investigate the detection bounds for an uncertainty set defined in (1). Note that the knowledge of the bounds of detection probability would be beneficial for real-world decision making, and can be used as an important reference to make appropriate detection decision. With a fixed decision threshold, to obtain the lower bound of detection probability is to solve the following minimization problem:

$$\begin{aligned} \min_{f_X^1} & \int_{\xi\in S} h(\xi,\lambda) f_X^1(\xi)d\xi \\ s.t.\quad & \left(\int_{\xi\in S}\xi f_X(\xi)d\xi - \mu\right)^2 \le \gamma_1\sigma^2 \\ & \int_{\xi\in S}(\xi-\mu)^2 f_X(\xi)d\xi \le \gamma_2\sigma^2 \\ & \int_{\xi\in S} f_\xi d\xi = 1, \end{aligned} \tag{3}$$

where μ and σ^2 are nominal mean and variance of the received primary signal strength by a single user. γ_1, γ_2 are defined as in (1). Without loss of generality, we assume $\gamma_1 < 1 < \gamma_2$. To simplify the problem (3), we can show the following result:

Theorem 1: *Lower bound problem (3) can be equivalently solved in the following form:*

$$\begin{aligned} \max_{\nu,r,\omega} \quad & (\mu^2 - \gamma_2\sigma^2)\nu + 2\mu\omega - r - 2\sqrt{\gamma_1}\sigma\,|\,\omega\,| \\ s.t. \quad & (\mu\nu + \omega)^2 - \nu(r+1) \le 0, \\ & (\lambda^2 - 2\mu\lambda)\nu - 2\lambda\omega + r \ge 0, \\ & \mu + \frac{\omega}{\nu} \ge \lambda, \quad \nu \ge 0. \end{aligned} \tag{4}$$

The proof follows the Lagrangain duality method; details can be found in (S Gong P. W., 2012). Similarly, the upper bound of detection probability can be obtained by solving problem (5) as follows,

$$\begin{aligned} \min_{\nu,r,\omega} \quad & (\gamma_2\sigma^2 - \mu^2)\nu + r - 2(\mu + \sqrt{\gamma_1}\sigma)\omega \\ s.t. \quad & (\mu\nu + \omega)^2 - \nu r \le 0, \\ & (\lambda^2 - 2\mu\lambda)\nu - 2\lambda\omega + r - 1 \ge 0, \\ & \mu + \frac{\omega}{\nu} \le \lambda, \quad \nu \ge 0. \end{aligned} \tag{5}$$

The proof is similar to that of Theorem 1 and thus omitted here. An optimal distribution to lower bound problem (4) or upper bound problem (5) is called an extremal distribution. We will discuss some properties regarding the extremal distribution for lower bound. Similar properties are also applied to the upper bound.

Corollary 1: Define set $S_e = \{\xi \in S | I(\xi > \lambda) = -\nu^*\xi^2 + 2(\omega^* + \mu\nu^*)\xi - r^*\}$, where ν^*, ω^*, and r^* are optimal to problem (4) for a fixed decision threshold λ, thus an *extremal distribution* $f_e(\bullet)$ that achieves the lower bound has the following properties: $f_e(\xi) \ge 0$, $\forall \xi \in S_e$, and $f_e(\xi') = 0$, $\forall \xi' \notin S_e$.

The result in Corollary 2 is fairly understandable by looking back into the Lagrangian dual process. The formal proof can be derived by referring to (Isii, 1962), (Lieven Vandenberghe, 2007). Actually, we find out that $S_e = \{\lambda, \mu + \omega^* / \nu^*\}$ in our problem, thus a discrete distribution can be constructed, i.e., $\Pr ob(\xi_1 = \lambda) = 1 - P_L$, and $\Pr ob(x_2 = m + w^* / n^*) = P_L$, where $P_L = (\mu^2 - \gamma_2\sigma^2)\nu^* + 2\mu\omega^* - r^* - 2\sqrt{\gamma_1}\sigma\,|\,\omega^*\,|$ denotes the optimal detection probability in problem (4). With this discrete distribution, the lower bound can be achieved.

Theorem 2: *The extremal distribution achieving the lower bound is with mean* $\mu_e = \mu - \sqrt{\gamma_1}\sigma$ *and variance* $\sigma_e^2 = (\gamma_2 - \gamma_1)\sigma^2$. *Moreover, the lower bound can be expressed in an analytical form, i.e.,*

$$P_L(\lambda) = \left[1 + \sigma_e^2 / (\mu_e - \lambda)^2\right]^{-1},$$

for a given decision threshold $\lambda \le \mu_e$.

Extremal distribution assigns nonnegative measures on the points $\{\lambda, \mu + \omega^* / \nu^*\}$, which satisfies the equation

$$I(\xi > \lambda) = -\nu^*\xi^2 + 2(\omega^* + \mu\nu^*)\xi - r^*,$$

thus the inequality constraints in (4) turn to be active at the optimum. Details can be found in (S Gong P. W., 2012). Theorem 2 provides an easy way to measure the lower bound, instead of solving the optimization problem in (4). Figure 1 gives an illustration for the lower and upper bounds, when we set $\gamma_1 = 0.01$, $\gamma_2 = 1.2$, $\mu = 3$ and $\sigma^2 = 0.5$. We also plot the detection probabilities when the received primary signal is assumed to be Gaussian distribution with different parameters (the mean and variance are indicated in the legend of Figure 1). We first set the mean to be $\mu - \sqrt{\gamma_1}\sigma$ (or $\mu + \sqrt{\gamma_1}\sigma$) and variance to be $\gamma_2\sigma^2$ which

Figure 1. Lower and upper bounds of the detection probability

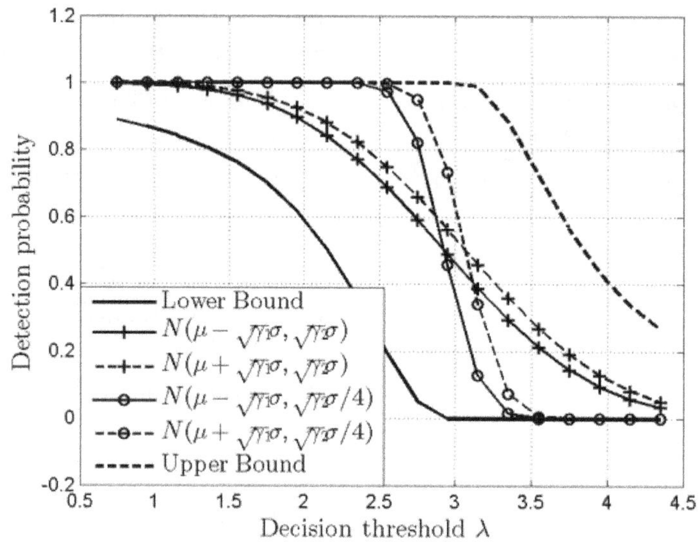

correspond to the boundary points of uncertainty set $\mathcal{U}$, and then we keep the same mean but choose a smaller variance $\gamma_2\sigma^2 / 16$. We can see that the detection probabilities under different normal distributions are all well bounded by the lower and upper bounds.

3.4 ROBUST DESIGN FOR COOPERATIVE SENSING

In the multi-user cooperative sensing, our goal in (2) is to design the threshold vector (which contains the decision threshold for each cooperative user) that maximizes the worst-case detection. This constrained optimization is hard to analyze directly, since it involves the manipulation of function integrations (i.e., constraints on mean and variance). To ease the analysis, we propose an equivalent transformation that reduces the function integrations, and turns the max-min problem into a maximization problem: (see Theorem 3).

Theorem 3 can also be proved using the Lagrangian duality theory. We omit the details due to space limit. Here $\otimes$ denotes the Frobenius inner product[3]. Variables Z, Q and r are penalties to the constrained optimization problem if corresponding moment constraints in the original max-min problem are violated. The main difficulty of solving (6)-(10) is that constraint (10) defines a non-convex set. This motivates us to solve (6)-(10) using an iterative algorithm. In each iteration, we will separate (10) from the solvable structure (6)-(9), i.e., we solve sub-problem (6)-(9) with a *fixed* decision threshold vector λ. Then we will iteratively improve the objective by adjusting decision threshold in the feasible set $\mathcal{D} = \{\lambda | \int_{\xi \in S} h(\lambda, \xi) f_0(\xi) d\xi \le \alpha\}$ defined by (10). The algorithm continues until a stable point is reached. Although we cannot prove that such a stable point is always the global optimal solution, simulation results show that such result leads to a much robust performance comparing with the algorithms where the uncertainty is not explicitly taken into consideration.

Theorem 3. The proposed max-min problem is equivalent to (6)-(10) as follows:

$$\max_{\lambda,Z,Q,r} (\mu\mu^T - \gamma_2\Sigma)\otimes Q - \Sigma \otimes P + 2\mu^T p - \gamma_1 s - r \quad (6)$$

$$s.t. \quad 1 + r - (p + Q\mathbf{u})^T Q^{-1}(p + Q\mu) \geq 0 \quad (7)$$

$$\xi^T Q\xi - 2\xi^T(p + Q\mu) + r \geq 0, \forall \xi \prec \lambda \quad (8)$$

$$Q \succeq 0, Z = \begin{bmatrix} P & p \\ p^T & s \end{bmatrix} \succeq 0 \quad (9)$$

$$\int_{\xi \in S} h(\lambda, \xi) f_0(\xi) d\xi \leq \alpha \quad (10)$$

where Z and Q are symmetric positive semi-definite matrices.

3.4.1 Initial Decision Threshold (IDT)

In order to solve problem (6)-(9), we need to first set an IDT in the feasible set (10). Choosing a conservative (i.e., too large) IDT will start the iterative algorithm in an interior point of the set (10), leading to the false alarm probability strictly below α and thus suboptimal detection probability. Hence we prove the following proposition that will assist our choice for a good initial decision threshold λ.

Proposition 1: *An optimal decision threshold λ^* to the problem (6)-(10) satisfies (10) with equality, that is*

$$\int_{\xi \prec \lambda^*} f_0(\xi) d\xi = 1 - \alpha. \quad (11)$$

Note that (11) is a necessary but not a sufficient condition for the optimality of Problem (6)–(10).

Detailed proof can be found in (S Gong P. W., 2012). It implies that an IDT can be chosen from the set $\mathcal{D}_c \triangleq \{\lambda | \int_{\xi \prec \lambda} f_0(\xi) d\xi = 1 - \alpha\}$. Moreover, we can guarantee the feasibility of false alarm probability constraint by updating the decision threshold λ within the same set $\mathcal{D}_c$ in every iteration. In this way we separate constraint (10) from the problem (6)-(9). Note that set $\mathcal{D}_c$ is not empty, since we can always find a decision threshold that balances equation (11).

For convenience, we choose an equal IDT[4] that assigns the same value for all SUs' decision thresholds, i.e., $\lambda_1 = \cdots = \lambda_N = \lambda_0$. We further assume that the noise at each cooperative SU is i.i.d., then (11) turns into

$$\int_{\xi \leq \lambda_0} f_0(\xi) d\xi = (1 - \alpha)^{1/N}.$$

Since we have assumed the noise pdf $f_0(\xi)$ to be a known Gaussian distribution with cdf $F_0(\lambda_0) = \int_{\xi \leq \lambda_0} f_0(\xi) d\xi$,

then the equal IDT will set $\lambda_0 = F_0^{-1}\left((1-\alpha)^{1/N}\right)$ to all SUs.

3.4.2 α-Preserving Adjustment

The values of IDT will be adjusted iteratively in the algorithm, such that objective (7) is improved and the new threshold is still within set $\mathcal{D}_c$. In other words, such adjustments always satisfy the α-Preserving condition.

Definition: *An adjustment to the decision threshold is called* α-Preserving adjustment *if the new decision threshold maintains the same false alarm probability, and the equality condition (11) is an* α-Preserving condition.

When some SU i changes its decision threshold λ_i by Δ_i, we need to change one or several other users' decision thresholds correspondingly so that the new decision threshold is α-Preserving. Specifically, we propose two strategies:

- 1-to-1 strategy: Choose another user $j \neq i$, and change its threshold by Δ_j;
- 1-to-m strategy: Choose a user subset $\mathcal{I} = \{i_1, \ldots, i_m\}$ with $\mathcal{I} \subset \mathcal{N}$, $|\mathcal{I}| = m$ and $i \notin \mathcal{I}$, then for each SU i_n in $\mathcal{I}$, change its thresholds by Δ_{i_n}.

Note that Δ_j can be both positive and negative, indicating an increment or deduction to λ_j. Corresponding to these two strategies, strict and relaxed conditions are provided to achieve an α-Preserving adjustment as follows:

Proposition 2: *In the 1-to-1 strategy, two users change their decision threshold by Δ_i and Δ_j, respectively, if*

$$\Delta_j = F_0^{-1}\left(F_0(\lambda_j)\frac{F_0(\lambda_i)}{F_0(\lambda_i + \Delta_i)}\right) - \lambda_j, \qquad (12)$$

then it is an α-Preserving adjustment.

Proposition 3: *In the 1-to-m strategy, user i_0 changes its decision threshold λ_0 by Δ_0, if the adjustment $\mathbf{\Delta}_{\mathcal{I}} = \{\Delta_{i_1}, \ldots, \Delta_{i_m}\}$ for users in set $\mathcal{I}$ satisfies the following linear equation*

$$\mathbf{a}_{\mathcal{I}}^T \mathbf{\Delta}_{\mathcal{I}} = -\Delta_0 \qquad (13)$$

where $a_{\mathcal{I}} = \frac{F_0(\lambda_0)}{f_0(\lambda_0)}\left[\frac{f_0(\lambda_{i_1})}{F_0(\lambda_{i_1})}, \frac{f_0(\lambda_{i_2})}{F_0(\lambda_{i_2})}, \ldots, \frac{f_0(\lambda_{i_m})}{F_0(\lambda_{i_m})}\right]^T$, *then $\Delta_{\mathcal{I}}$ is an approximate α-Preserving adjustment.*

Proof: In order to maintain the same false alarm probability with the 1-to-1 strategy, the adjustments for decision thresholds of any two different users i and j should satisfy $F_0(\lambda_i)F_0(\lambda_j) = F_0(\lambda_i + \Delta_i)F_0(\lambda_j + \Delta_j)$, where Δ_i and Δ_j are changes made to λ_i and λ_j respectively. Since $F_0(\bullet)$ denotes the known cdf function, we get the result (12) directly. In 1-to-m strategy, there are $1+m$ decision thresholds changed simultaneously. Similarly, we have (adding log function for simplicity):

$$\sum_{i=0}^{m} \log F_0(\lambda_i) =$$
$$\log F_0(\lambda_0 + \Delta_0) + \sum_{i=1}^{m} \log F_0(\lambda_i + \Delta_i)$$

By the Taylor approximation, i.e.,

$$F_0(\lambda_i + \Delta_i) \approx F_0(\lambda_i) + \Delta_i F'(\lambda_i),$$

we have

$$\log\left(\frac{F_0(\lambda_0)}{F_0(\lambda_0) + \Delta_0 F_0'(\lambda_0)}\right) \approx \sum_{i=1}^{m} \log\left(\frac{F_0(\lambda_i) + \Delta_i F_0'(\lambda_i)}{F_0(\lambda_i)}\right).$$

If Δ_0 is small enough, apply a linear approximation (i.e., $\log(1+x) \approx x$, if $x \ll 1$), we get

$$\sum_{i=1}^{m} \Delta_i \frac{f_0(\lambda_i)}{F_0(\lambda_i)} = -\Delta_0 \frac{f_0(\lambda_0)}{F_0(\lambda_0)},$$

where $f_0(\bullet) = F_0'(\bullet)$. Then (13) can be easily obtained. For the 1-to-1 strategy, which is a special case of the 1-to-*m* strategy, we can get the exact adjustment Δ_j for any user $j \neq i$. However, for the 1-to-m strategy, we obtain an approximate adjustment through a linear equation. We observe that the solution to (13) is not unique, but we prefer the solution with the least total change[5] (i.e., $\|\Delta_{\mathcal{I}}\|$ is the smallest). Therefore, we propose the following least norm problem,

$$\begin{aligned} \min_{\Delta_{\mathcal{I}}} \quad & \|\Delta_{\mathcal{I}}\| \\ s.t. \quad & a_{\mathcal{I}}^T \Delta_{\mathcal{I}} = -\Delta_0. \end{aligned} \tag{14}$$

The optimal solution to (14) is given by

$$\Delta_{\mathcal{I}} = -\Delta_0 (a_{\mathcal{I}}^T a_{\mathcal{I}})^{-1} a_{\mathcal{I}}. \tag{15}$$

3.4.3 Solving Sub-Problem (6)-(9) with Fixed Thresholds

Sub-problem (6)-(9) can be solved in polynomial time if and only if a separation problem (Erick Delage, 2010) can be solved in polynomial time. In the separate problem, we will check whether (8) is satisfied for all $\xi \prec \lambda$ when (Q,p,r) is fixed, if not we need to provide a violated case. However, the separation problem is proved to be NP-hard in (Katta G. Murty, 1987). The main difficulty is that we need to consider a free variable ξ when judging the non-negativity of the quadratic polynomial in LHS of (8). The proposed α-Preserving condition provides a way to eliminate ξ and transform the separation problem (8) into the following form:

Proposition 4: *α-Preserving condition ensures that (8) is equivalent to*

$$\lambda^T Q \lambda - 2\lambda^T (p + Q\mu) + r \geq 0$$

if λ is optimal to problem (6)-(10).

The proof is given in Appendix. Proposition 4 removes ξ from the sub-problem, and thus reduces (8) to a linear constraint with respect to (Q,p,r). Furthermore, we reformulate sub-problem (6)-(9) into an equivalent semi-definite program, which can be efficiently solved by the interior-point method (Y. E. Nesterov, 1994). (see Theorem 2.)

Proof: *Let* $z = (\mu\mu^T - \gamma_2\Sigma) \otimes Q - \Sigma \otimes P + 2\mu^T p - \gamma_1 s - r$ *and* $q \geq 2\sqrt{\gamma_1 p^T \Sigma p} = 2\sqrt{\gamma_1}\, \| Lp \|_2$, *where L is a triangular matrix denoting the Cholesky decomposition of Σ, i.e., $\Sigma = L^T L$. Since* $Z = \begin{bmatrix} P & p \\ p^T & s \end{bmatrix} \succeq 0$, *i.e.,* $P \succeq \frac{1}{s} pp^T$ *we have*

Theorem 2. Sub-problem (6)-(9) has an equivalent form in (16)-(20) as follows:

$$\begin{aligned} \max_{\lambda,Q,p,q,r,z} \quad & z && (16)\\ s.t. \quad & \boldsymbol{\lambda}^T Q \boldsymbol{\lambda} - 2\boldsymbol{\lambda}^T (p + Q\boldsymbol{\mu}) + r \geq 0 && (17)\\ & (\boldsymbol{\mu}\boldsymbol{\mu}^T - \gamma_2\Sigma) \otimes Q + 2\mu^T p \geq z + r + q && (18)\\ & q \geq 2\sqrt{\gamma_1}\, \| Lp \|_2 && (19)\\ & \begin{bmatrix} r+1 & (p+Q\mu)^T \\ (p+Q\mu) & Q \end{bmatrix} \succeq 0. && (20) \end{aligned}$$

Algorithm 1. Decision Threshold Adjustment Algorithm

Initialization:

Set stopping criterion $\Gamma = 200$ and $\epsilon = 10^{-4}$

Set initial $\lambda_i(t) = F_0^{-1}\left((1-\alpha)^{1/N}\right)$ for all SUs

While $\Delta_{i^+} \geq \epsilon$ and $t \leq \Gamma$ **do**

Solve sub-problem (16)-(20) with $\lambda(t)$ fixed and return the optimizer $y = [vec(Q), p^T, q, r, z]$

Set adjustment vector $G(t) = Q \bullet \lambda(t) - Q \bullet \mu - p$

Update $\Delta_{i^+} = \max_i G_i(t) - \min_j G_j(t)$

Set $i^+ = \arg\min_i G_i(t)$ and calculate adjustment vector $\Delta_{\mathcal{I}}$

Update decision threshold:

$$\lambda_{i^+}(t+1) = \lambda_{i^+}(t) + \Delta_{i^+},$$

$$\lambda_{i_n}(t+1) = \lambda_{i_n}(t) + \Delta_{i_n}, \ \forall n \leq m.$$

Set t=t+1

End while

$$\begin{aligned} z &\leq (\mu\mu^T - \gamma_2\Sigma) \otimes Q - \frac{1}{s} p^T \Sigma p + 2\mu^T p - \gamma_1 s - r \\ &\leq (\mu\mu^T - \gamma_2\Sigma) \otimes Q + 2\mu^T p - r - 2\sqrt{\gamma_1} \,\| Lp \|_2 \\ &\leq (\mu\mu^T - \gamma_2\Sigma) \otimes Q + 2\mu^T p - r - q. \end{aligned}$$

Therefore, the objective in (6) can be written as (16) with a linear constraint (18) and a quadratic constraint (19). Constraint (7) is written in its matrix form (20), and (8) is equivalent to (17) according to Proposition 4. When λ is fixed, (16)-(20) is a conic optimization problem in the standard dual form, and can be easily solved by a conic optimization package such as SeDuMi (Sturm, 1999). We denote its solution by $y = [vec(Q), p^T, q, r, z]^T$, where $vec(Q)$ transforms the variable matrix Q into a row vector.

3.4.4 Decision Threshold Update

Given the solution $y = [vec(Q), p^T, q, r, z]^T$ from a semi-definite program, we show how the decision thresholds λ can be updated iteratively such that the objective (16) is improved.

Proposition 5: *Given the parameters Q, p, q, and r in (16)-(20), the objective z will be improved if $\lambda(t+1) = \lambda(t) + \tau G(t)$, where $G(t) = Q \bullet \lambda(t) - Q \bullet \mu - p$ and $\tau > 0$ is the step-size of the adjustment.*

Proof: *From (17) and (18), we have*

$$\begin{aligned} z \leq \lambda^T Q \lambda - 2\lambda^T(p + Q\mu) + (\mu\mu^T - \gamma_2\Sigma) \otimes Q \\ + 2\mu^T p - q. \end{aligned}$$

The last term does not depend on λ and can be viewed as a constant with respect to λ. In order to find another λ that improves the objective z, we can update λ in its gradient direction, i.e., $\lambda(t+1) = \lambda(t) + \tau G(t)$. Positive τ is a step size, and $G(t)$ is the gradient and given by $Q \bullet \lambda(t) - Q \bullet \mu - p$. However, from the proof of Proposition 4 we know that $\lambda \prec \mu + Q^{-1}p$, which implies a non-positive gradient $G(t) \prec 0$. If the decision threshold $\boldsymbol{\lambda}$ is adjusted as in Proposition 5, then it will decrease in every iteration. This is intuitive since lowering the decision thresholds can increase detection probability in hypothesis

Figure 2. Comparison with existing method

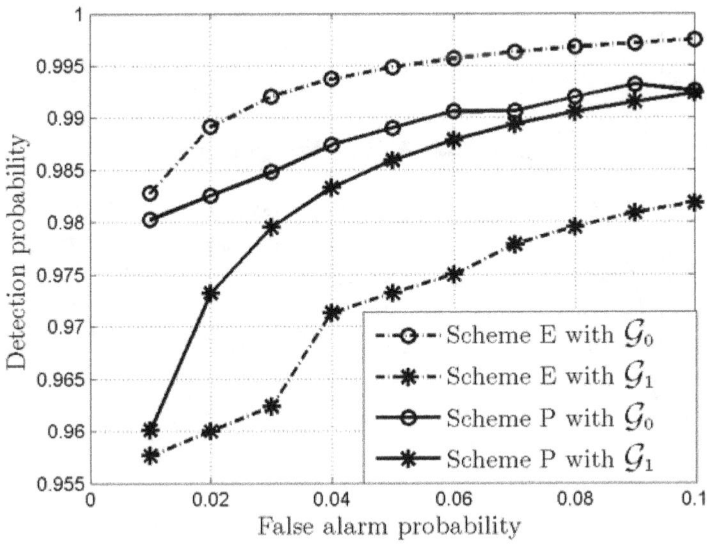

testing, but the false alarm probability cannot be maintained at the prescribed level. This motivates us to increase the decision threshold of the user i^+ *who corresponds to the minimum term in* $G(t)$. *We choose this term because we prefer minimum adjustment to the decision threshold, and this term impacts the objective (16) mostly if the same adjustment is applied to all SUs. Meanwhile we apply an* α*-Preserving adjustment to the user in subset* $\mathcal{I}$, *such that the false alarm probability is maintained at the same level. The decision threshold adjustment algorithm is summarized in Algorithm 1, where* Δ_{i^+} *denotes the difference between the maximum and minimum terms in* $G(t)$, *and is used to control the convergence. The algorithm terminates when* Δ_{i^+} *is small enough or the number of iterations exceeds a predefined threshold* Γ.

3.5 NUMERICAL AND SIMULATION RESULTS

In the simulation, we consider $N = 3$ SUs and they receive primary signals with different nominal mean values (denoted by the vector μ). We set $\mu = [3.7, 4.0, 4.3]$ and $a=0.1$. The signals received at the SUs are independent with each other with variances $\Sigma = 2I_N$, where I_N is an identity matrix with size N. The distribution uncertainty parameters in set (1) are $\gamma_1 = 0.02$ and $\gamma_2 = 1.2$.

First, we compare the performance of our proposed robust sensing algorithm with an existing method. In the existing scheme (denoted as Scheme E in Figure 2) SUs will optimize the sensing thresholds assuming that the primary received signals follow Gaussian distribution $\mathcal{G}_0$ with mean $\mu = [3.7, 4.0, 4.3]$. However, the actually distribution may follow either $\mathcal{G}_0$ or $\mathcal{G}_1$, where $\mathcal{G}_1$ is another Gaussian distribution with the mean $\mu = [3.0, 3.5, 4.0]$. For our proposed scheme (i.e., Scheme P in Figure 2), we consider such distribution uncertainty.

As shown in Figure 2, Scheme E works very well when the actual distribution of primary signals is $\mathcal{G}_0$ (and thus matches with the assumption). However, when the actual distribution is $\mathcal{G}_1$ (and

Figure 3. Detection probabilities within different strategies

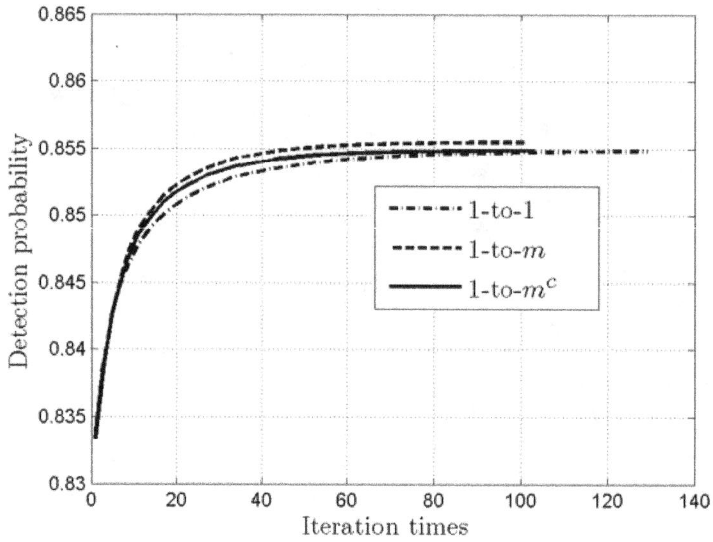

thus there is a mismatch), Scheme E's performance degrades significantly. While for Scheme P, though the detection performance is not as good as that of Scheme E in the matched case, it is more robust and has a much better performance when the distribution is mismatched. This shows that our proposed algorithm achieves a good tradeoff between robustness and performance. In fact, the optimization under our Scheme P is performed with respect to the uncertainty between $\mathcal{G}_0$ and $\mathcal{G}_1$, instead on a particular distribution. In the following, we further investigate various properties of Scheme P.

3.5.1 Comparison between 1-to-1 and 1-to-m Strategies

From Proposition 2, we obtain the exact adjustment Δ_j for the 1-to-1 strategy, or an approximate adjustment $\Delta_{\mathcal{I}}$ for the 1-to-m strategy as in Proposition 3. Since we have assumed $N = 3$ cooperative users, thus we simply set $m = 2$ in the 1-to-m strategy. Figure 3 plots the detection probability for different strategies applied in our algorithm. The detection probability converges to almost the same value for both the 1-to-1 and 1-to-m strategies as shown in Figure 3.

Figure 4 also shows that the false alarm probability is strictly maintained at the prescribed level with the 1-to-1 strategy, while it is slightly higher than the prescribed value with the 1-to-m strategy due to the errors in approximation. However, the increment is actually acceptable since we only observe 1% increase over the prescribed level. Besides, solving the linear approximation (1.13) greatly reduces the computation complexity in finding the exact change vector $\Delta_{\mathcal{I}}$. Moreover, 1-to-m strategy adapts more than one user's decision threshold simultaneously, it takes less iterations to converge to the stable level, i.e., the number of iterations for the 1-to-m strategy is about 100 times compared to 130 times for the 1-to-1 strategy in the current parameter setting as shown in Figure 3 and Figure 4. In the case when false alarm probability is strictly imposed, i.e., the threshold α should not be exceeded; we propose a corrected 1-to-m strategy (denoted by 1-to-m^c) in order to reduce the false alarm probability. We found that the observed increase of

Figure 4. False alarm probabilities within different strategies

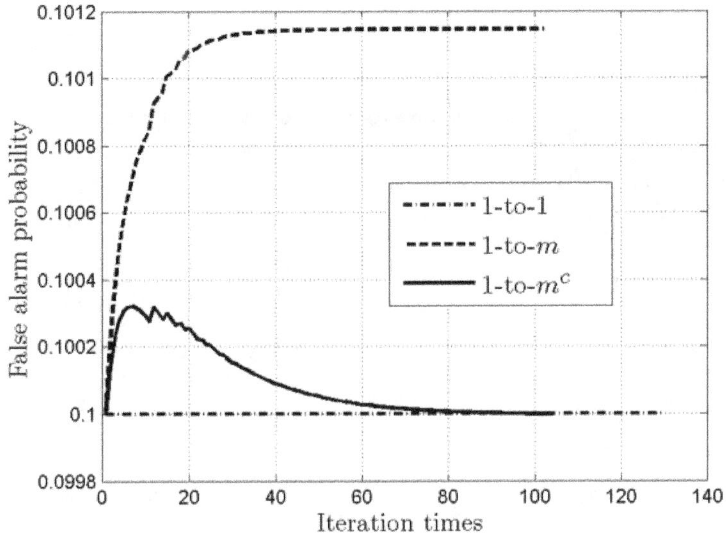

the false alarm probability is due to a smaller decision threshold chosen in 1-to- m strategy as compared with 1-to-1 strategy. Therefore we obtain the final increment vector $\Delta_{\mathcal{I}}^{c}$ by adding a positive quantity to $\Delta_{\mathcal{I}}$. Figure 4 shows that the corrected 1-to- m strategy can reduce the false alarm probability without sacrificing much detection performance.

3.5.2 Different Initial Decision Thresholds (IDTs)

We have assumed that each SU chooses the same value for initial decision threshold. Here, we show that such an assumption is not crucial to the performance and convergence of the algorithm. For comparison, we examine three cases, i.e., $\lambda_0^e = [1.82, 1.82, 1.82]$, $\lambda_0^1 = [2.12, 1.82, 1.63]$, and $\lambda_0^2 = [2.42, 1.82, 1.55]$. Note that for all parameter choices, false alarm probability is kept at 0.1 as required.

Figure 5 plots the iteration results for the detection probabilities as IDT is set to λ_0^e, λ_0^1 and λ_0^2, respectively. Though the initial detection probabilities are not the same, the gaps quickly diminish as the iteration increases. To further check the consistence of the final decision thresholds given different IDTs, we illustrate the dynamics of each user's decision threshold in Figure 6, where λ_j^i denotes the decision threshold for user j when the IDT is set to λ_0^i. For illustration purpose, we plot the cases for λ_0^e and λ_0^1 with the 1-to- m^c strategy. It is observed that for a particular user, the eventual decision threshold does not depend on the initial choice.

3.5.3 Impact of Different Mean Values μ

Now we have demonstrated that the 1-to-1 and 1-to- m strategies provide similar detection performance, and the choice of IDT is irrelevant with the final solution. In the following, we focus on the 1-to- m strategy, and investigate its performance with various μ. Looking at the result of Figure 6 obtained when $\mu = [3.7, 4.0, 4.3]$, we notice that the convergent decision threshold for each user follows an ordering, i.e., $\lambda_3 > \lambda_2 > \lambda_1$, which is consistent with the order of elements

Figure 5. Detection probabilities converge with different IDTs

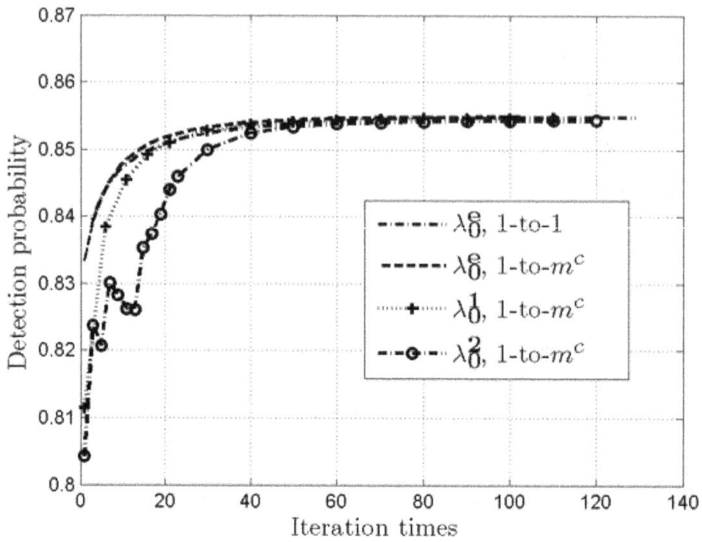

in μ, i.e., $\mu_3 > \mu_2 > \mu_1$. This result is consistent with our expectation that SU would have a higher decision threshold if the received primary signal is stronger.

Moreover, we consider another two cases, i.e., $\mu_2 = [3.7, 4.0, 4.0]$, $\mu_3 = [3.7, 3.7, 4.0]$. IDT is set to $\lambda_0^1 = [2.22, 1.82, 1.60]$ for both cases. Note that μ_2 and μ_3 are specially designed such that they are partially ordered, i.e., $\mu_1 \pm \mu_2 \pm \mu_3$. The convergent detection probabilities exhibit a consistent order as shown in Figure 7. Though uncertainty exists in practical detection system, a

Figure 6. Thresholds converge with different IDTs

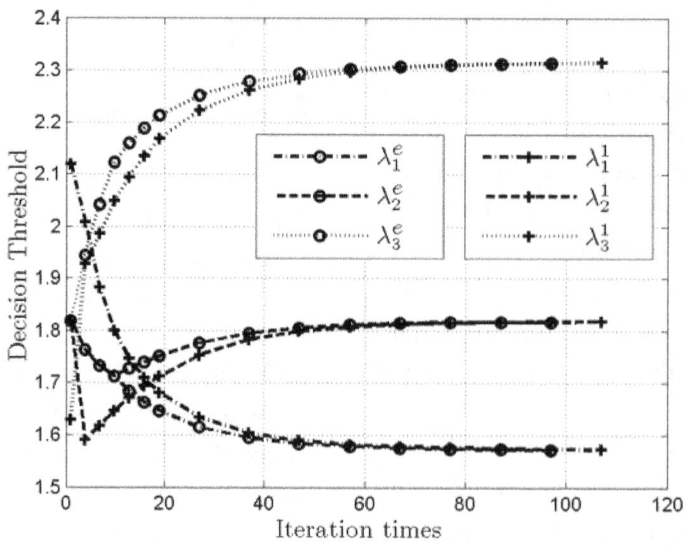

Figure 7. Detection probabilities with different μ

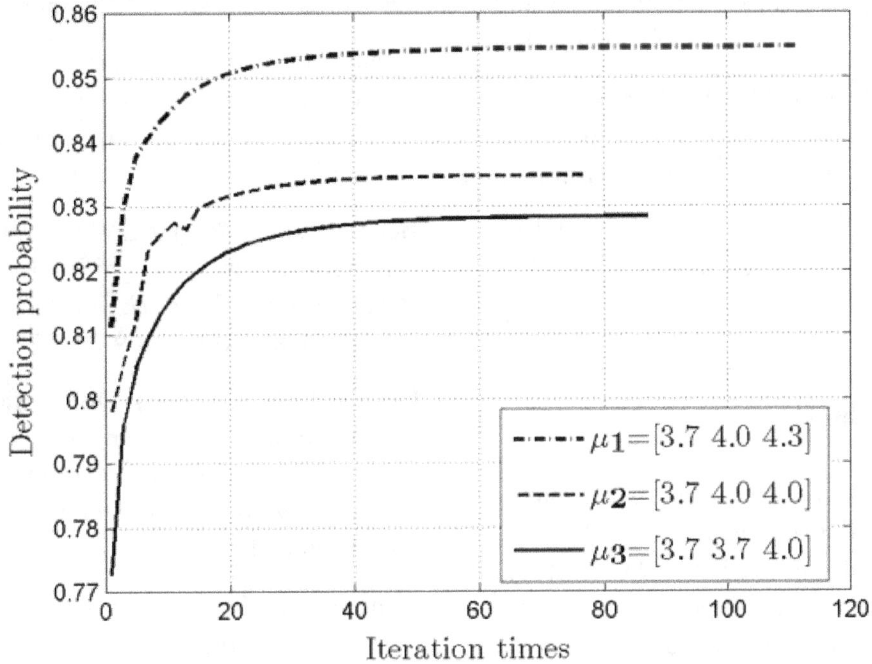

Figure 8. Convergent thresholds have the same ordering with μ

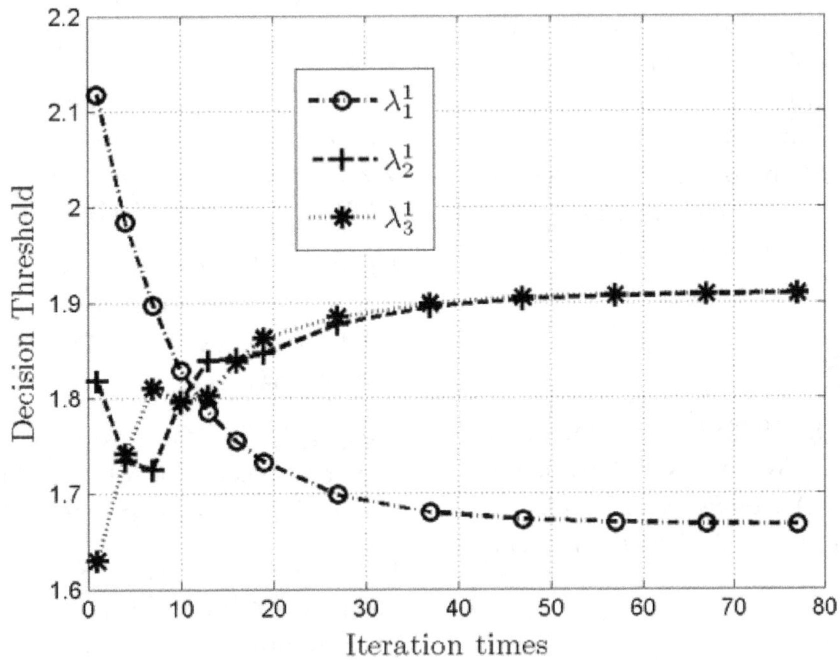

better channel condition (thus stronger received primary signal) always leads to a better detection performance. Figure 8 also plots the threshold dynamics for $\mu_2 = [3.7\,4.0\,4.0]$. Note that $\mu_1 < \mu_2 = \mu_3$, consequently λ_2 and λ_3 converge to the same level, which is higher than λ_1 as shown in Figure 8. A similar result is also shown for the case $\mu_3 = [3.7\,3.7\,4.0]$ where λ_1 and λ_2 converge to the same level, and is lower than λ_3.

3.6 CONCLUSION

In this paper, we consider the robust performance for spectrum sensing in CR systems. Max-min robust problem is formulated to find the optimal worst-case performance, given an uncertainty set of the distribution function of the received primary signal strength. For single user case, the max-min problem is degenerated to find the lower bound of detection probability. We also provide an analytical expression for the lower bound. For multi-user case, the max-min robust problem is very hard to optimally solve. We find a way to separate the false alarm probability constraint, and propose a threshold adjustment procedure that maintains the prescribed false alarm probability. Moreover, we determine the robust threshold for each cooperative SU by our proposed iterative algorithm, and its performance is demonstrated through simulations from various aspects.

REFERENCES

Aharon Ben-tal. L. E. (1998). Robust semidefinite programming. In H. Wolkowics, R. Saigal, & L. Vandenberghe (Eds.), *Handbook on semidefinite programming*, (pp. 139-162). Kluwer Academis Publishers.

Aharon Ben-Tal, A. N. (1998). Robust convex optimization. *Mathematics of Operations Research*, *23*(4), 769–805. doi:10.1287/moor.23.4.769

Anant Sahai, N. H. (2004). *Some fundamental limits on cognitive radio*. Forty-second Allerton Conference on Communication, Control, and Computing.

Ben Letaief, K., & Zhang, W. (2009). Cooperative communications for cognitive radio networks. *Proceedings of the IEEE*, *97*(5), 878–893. doi:10.1109/JPROC.2009.2015716

Delage, E. Y. Y. (2010). Distributionally robust optimization under moment uncertainty with application to data-driven problems. *Operations Research*, *58*(3), 595–612. doi:10.1287/opre.1090.0741

Gong, S. P. W. (2012). *Spectrum sensing under distribution uncertainty in cognitive radio networks*. IEEE ICC 2012 - Cognitive Radio and Networks Symposium (ICC'12 CRN), Ottawa, Canada.

Gong, S. P. W. (2012). *Robust threshold design for cooperative sensing in cognitive radio networks*. The 31st Annual IEEE International Conference on Computer Communications (INFOCOM'2012 -- Mini-Conference), Orlando, Florida, USA.

Haykin, S. (2005). Cognitive radio: Brain-empowered wireless communications. *IEEE Journal on Selected Areas in Communications*, *23*(2), 201–220. doi:10.1109/JSAC.2004.839380

Isii, K. (1962). On sharpness of tchebycheff-type inequalities. *Annals of the Institute of Statistical Mathematics*, *14*(1), 185–197. doi:10.1007/BF02868641

Iyengar, E. E. (2004). Ambiguous chance constrained problems and robust optimization. *Mathematical Programming*, *107*(1), 37–61.

Katta, G., & Murty, S. N. (1987). Some NP-complete problems in quadratic and nonlinear programming. *Mathematical Programming*, *39*(2), 117–129. doi:10.1007/BF02592948

Knuth, R. M. (1996). On the lambert w function. *Advances in Computational Mathematics*, *5*, 329–359. doi:10.1007/BF02124750

Li, J. M. (2008). Soft combination and detection for cooperative spectrum sensing in cognitive radio networks. *IEEE Transactions on Wireless Communications*, 4502–4507.

Lieven Vandenberghe, S. B. (2007). Generalized Chebyshev bounds via semidefinite programming. *SIAM Review*, *49*(1), 52–64. doi:10.1137/S0036144504440543

Lunden, J., Kassam, S. A., & Koivunen, V. (2010). Robust nonparametric cyclic correlation-based spectrum sensing for cognitive radio. *IEEE Transactions on Signal Processing*, *58*(1), 38–52. doi:10.1109/TSP.2009.2029790

Mchenry, M. (2003). *Spectrum white space measurements*. New America Foundation Broadband Forum.

Nesterov, Y., & Nemirovskii, A. (1994). *Interior-point polynomial algorithms in convex programming*. SIAM Studies in Applied and Numerical Mathematics. doi:10.1137/1.9781611970791

Quan, Z., Shuguang, C., & Sayed, A. H. (2008). Optimal linear cooperation for spectrum sensing in cognitive radio networks. *IEEE Journal of Selected Topics in Signal Processing*, *2*(1), 28–40. doi:10.1109/JSTSP.2007.914882

Simon, F. F.-S. (2007). On the energy detection of unknown signals over fading channels. *IEEE Transactions on Communications*, *5*, 21–24.

Sousa, A. G. (2005). Collaborative spectrum sensing for opportunistic access in fading environments. *IEEE International Symposium on New Frontiers in Dynamic Spectrum Access Networks*, (pp. 131-136).

Sturm, J. F. (1999). Using SeDuMi 1.02, a {MATLAB} toolbox for optimization over symmetric cones. *Optimization Methods and Software*, *11*(1), 625–653. doi:10.1080/10556789908805766

Taricco, G. (2011). Optimization of linear cooperative spectrum sensing for cognitive radio networks. *IEEE Journal of Selected Topics in Signal Processing*, *5*(1), 77–86. doi:10.1109/JSTSP.2010.2055537

Teo, K. M. (2007). *Nonconvex robust optimization*. Massachusetts Institute of Technology.

Zhang, Y., Zhang, Q., & Wu, S. (2010). Entropy-based robust spectrum sensing in cognitive radio. *IET Communications*, *4*(4), 428–436. doi:10.1049/iet-com.2009.0389

ENDNOTES

1. Given the decision threshold, $Q_f = \int_{\xi \in S} h(\xi,\lambda) f_X^0(\xi) d\xi$ only relates to noise distribution $f_{\mathbf{X}}^0$, while $Q_d = \int_{\xi \in S} h(\xi,\lambda) f_X^1(\xi) d\xi$ is independent of noise distribution.
2. In the formulation of problem (2)-(4), the worst distribution, denoted by f_λ^w, is an extreme-case distribution that results in minimum detection probability for a given decision threshold λ, i.e., $f_\lambda^w = \arg\min_{f_X^1(\xi)} \int_{\xi \in S} h(\xi,\lambda) f_X^1(\xi) d\xi$.}
3. The Frobenius inner product is the component-wise inner product of two matrices, e.g., $A \otimes B = trace(AB^T)$, A and B are two matrices.
4. The IDT can be set in different ways. However, an equal IDT is easy to implement and does not affect the algorithm convergence, which will be demonstrated in the simulation results.
5. A large change to λ_i can always be counteracted by another large change to λ_j, $j \neq i$. The least total increment is the minimum change required to maintain the same level of false alarm probability.

Chapter 4
Throughput Modeling, Analysis, and Optimization in Cognitive Wireless Networks

Alireza Babaei
Auburn University, USA

Prathima Agrawal
Auburn University, USA

Bijan Jabbari
George Mason University, USA

ABSTRACT

In cognitive wireless networks, secondary nodes are required to sense the spectrum and to be cognizant of their interference on the primary receiving nodes. In this chapter, the authors show that the throughputs in the primary and the secondary networks have an entangled relationship and need to be considered jointly for optimization. Assuming that the primary and secondary networks use ALOHA as their medium access control protocol, the chapter shows that the throughputs in primary and secondary networks are influenced by the choices of transmission attempt probabilities in both of these networks. Authors show that the interference constraint limits the feasible space of transmission probabilities over which these networks can optimize their throughputs. They propose a progressive transmission probability optimizations algorithm and show through simulation that the algorithm converges to the optimum values of throughput in few time slots. Future research directions and the application of the constrained game theory to the throughput optimization problem are also discussed.

4.1 INTRODUCTION

In cognitive wireless networks, a primary (licensed) network coexists and shares the spectrum with a secondary (unlicensed) network. The exclusive right to access the spectrum belongs to the primary network and the secondary network's access is contingent on limiting their aggregate interference to the primary receiving nodes such that it does not hinder the primary receivers in decoding their data. In this scenario, a secondary device has node-level as well as network-level obligations to

DOI: 10.4018/978-1-4666-2005-6.ch004

avoid interference on the primary receiving nodes. At the node-level, it is only allowed to transmit if it does not detect data reception from its nearby primary neighbors. This is typically satisfied by the secondary nodes performing spectrum sensing. The secondary nodes must be alert and vacate the channel they are using, upon sensing the primary signals on that channel (Cabric, Mishra & Brodersen, 2004). Secondary nodes have to make sure that the aggregate secondary interference power satisfies an interference constraint on active primary receiving nodes. The interference constraint specifies the maximum interference tolerable at the primary receivers. The node-level spectrum sensing, however, may be inadequate to achieve this goal as the aggregate interference coming from a large secondary network may still be significant. Therefore, network-wide interference avoidance strategies must also be considered.

The node-level spectrum sensing and the network-level interference awareness can be achieved by employing cognitive radios in the secondary network. Cognitive radios differ from conventional radios by employing advanced features such as spectrum sensing and adaptability (Haykin, 2005).

Given that the primary and secondary wireless networks share the space and the spectrum, throughput in both of these networks is limited by not only the intra-network interference, i.e., the interference among the nodes of the same network, but also by the inter-network interference, i.e., the aggregate interference originated from transmitting nodes in one network on the receiving nodes of the other network (Babaei & Jabbari, 2010a). On the other hand, due to the factors like randomness in the locations of the primary and the secondary nodes, the type of MAC layer protocols and the scheduling algorithms used in these networks which determine the simultaneous transmitters, as well as the fading and the shadowing effects, the intra-network and inter-network and their cumulated effect are random in nature. Therefore, statistical characterization of interference is an important prerequisite for modeling and optimization of throughput in the primary and secondary networks.

The throughput maximization problem in a homogeneous random wireless ad hoc networks (i.e., when the wireless nodes all belong to the same network) with nodes distributed according to a Poisson point process and using ALOHA as the medium access control scheme is considered in (Venkataraman & Haenggi, 2004). Assuming that the transmission attempt probability is p and with a half-duplex assumption (i.e. the nodes can either transmit or receive data at a given time), the probabilistic throughput[1] (τ) can be found as

$$\tau = p(1-p)\Pr\left\{\mathrm{SINR} > \theta\right\},$$

where SINR is the ratio of the desired signal power to the interference and noise power, θ is the threshold required for successful decoding of data and $p(1-p)$ is probability that the sources node is in the transmission mode and the destination node is in the reception mode (due to the half-duplex assumption)[2]. With a Gaussian probability distribution assumption for interference, the transmission attempt probability parameter used in the ALOHA protocol is shown to influence on the mean and the variance of interference and through that on the probabilistic throughput. The optimum value of transmission attempt probability to maximize the throughput is found in (Venkataraman & Haenggi, 2004) using numerical evaluation of throughput.

In cognitive radio networks, the primary and secondary throughputs have an intricate relationship. In such networks, the secondary nodes are bound to avoid interference on the primary receiving nodes. For this purpose, they perform spectrum sensing and abstain from transmission if they detect active primary receiving nodes in their vicinity. The effective transmission probability in the MAC layer is therefore influenced by the possible abstention due to the presence of active primary receivers. On the other hand, the number of actively receiving primary nodes implicitly relates to probability of successful reception and the probabilistic throughput in the primary network. The higher the primary throughput is, therefore, the higher is the probability that secondary nodes sense the spectrum to be busy and avoid transmission. This can be done in the secondary network by reducing their transmission power level or the MAC layer transmission attempt probability which impact on the throughput in secondary network. Primary and secondary networks' throughputs are therefore inter-related and attempts toward throughput improvement in these networks need to consider their joint optimization (see (Babaei & Jabbari, 2010a), for example).

In this chapter, assuming that the primary and the secondary networks are random wireless networks and use a random medium access control technique like ALOHA, we attempt to jointly optimize their throughput. We first discuss some preliminary results from the Theory of point processes and use them for modeling the intra and inter-network interference. We then use theses results for modeling the primary and secondary networks' throughputs. Considering the effect of interference constraint as well as the primary network's throughput on the feasible space for the secondary network's transmission attempt probability, we find the constraint space for primary and secondary throughput optimization. We show that Game Theory with constraints on multi-strategies is the right discipline to solve the throughput optimization problem (Pensavalle and Pieri, 2000). We propose a distributed algorithm to jointly optimize the throughput in the primary and secondary networks.

4.2 STATISTICAL MODELING OF INTERFERENCE

Due to the aggregate interference constraint from the secondary network on the primary nodes, knowledge of statistical properties of secondary network's interference is crucial for satisfying this constraint and successful coexistence of primary and secondary networks. Moreover, statistical models for primary and secondary networks' intra-network and inter-network interference are necessary for modeling the parameters of interest, like throughput, in these networks.

In a wireless network, where nodes are distributed randomly and transmit information according to the medium access regulations, statistics of interference depend on factors like the power levels chosen by the interferers (i.e., the nodes simultaneously transmitting with the desired transmitter), the density and the locations of the interferers, along with random propagation effects, like shadowing and fading. Due to randomness in the locations of nodes, tools from stochastic geometry and particularly the Theory of point processes have proven useful for modeling of random networks (Sousa & Silvister, 1990; Ilow & Hatzinakos, 1998; Haenggi, Andrews, Baccelli, Dousse & Franceschetti, 2009). Using these tools, statistical models for inter-nodal distances, which affect on the path loss and the received signal strength, can be obtained. In addition, MAC protocols parameters like transmission attempt probability, traffic load and density of nodes affect the interference statistics as well. In this section, we derive statistical models for aggregate interference originated from primary and secondary random wireless networks.

4.2.1 Random Networks and Stochastic Geometry

Most of the wireless networks provide the freedom of mobility for their users. Due to mobility, nodes change their locations randomly and at random times. The snapshot of the nodes distribution can therefore be assumed a random point process (see the formal definition below). In addition, in the essentially non-mobile wireless networks, like wireless sensor networks, nodes may be deployed randomly in an operational region. Therefore, a realistic modeling of wireless networks must take into account the randomness in the spatial distribution of nodes. In this section, we review some of the mathematical tools necessary to study the statistical properties of such random networks.

- **Point Processes:** Point processes are random collection of nodes that fall in some space. The points can represent the time or location of an event. Well-known one-dimensional examples of point processes are the renewal processes which are defined in the time domain. The events in such processes are called the arrival times (Cox, 1970). Spatial point processes are defined in $\Re^d$, $d \geq 2$ and a particular realization of such a process is often referred to as a Point Pattern.
  Spatial point processes have found applications in disciplines like seismology, ecology, and forestry (Diggle, 2003; Ludwig & Reynolds, 1998). Point processes have been considered recently for modeling the spatial distribution of nodes in random wireless networks (Haenggi, Andrews, Baccelli, Dousse & Franceschetti, 2009).
- **Poisson Point Process (PPP):** The homogeneous Poisson point process (PPP) is a well-adopted model for distribution of nodes in random wireless networks. The model is valid when the nodes in a mobile network move in an uncorrelated fashion. In fixed wireless networks, like wireless sensor networks, the model is also valid when the nodes are deployed randomly in the operation region (for example when sensors are dropped in large numbers from an airplane) which reduces the probability of clustering of nodes.
  We give a definition of the Poisson point process in the Euclidean space: $\Re^d$, $d \leq 3$ with density $\lambda > 0$ [3]. Let us define a counting variable $N(A)$ for every $A \subset \Re^d$ as the number of points falling in A. The Poisson point process defined over $S \subset \Re^d$ must have following properties:
  - **Property 1:** For any bounded closed set $A \subseteq S$, $N(A)$ is a Poisson random variable with mean $\lambda\mu(A)$ where $\mu(A)$ is the length, area or volume of A depending on the dimension of the space.
  - **Property 2:** If $A_1, \ldots, A_m$ are disjoint subsets of S, then $N(A_1), \ldots, N(A_m)$ are independent.

  The Poisson point process also has the property that it remains a Poisson point process under probabilistic thinning. In other words, if the points of a PPP with density λ are retained independently and with some fixed probability p, the retained process will remain a PPP but with a smaller density of λp. In Figure 1, a sample 2-dimensional PPP with density $\lambda = 1$ in a disc with radius 5 units is shown. Alongside, a thinned version of the same sample process with $p = 0.2$ is shown.
- **Inter-nodal distance distribution:** When modeling random wireless networks using a point process, the statistical characterization of the distances between the points is important in signal and interference modeling. This is due to the fact that the attenuations in signal and interference power levels relate to the distances between the receiver and the transmitters (Schwartz, 2005). Haenngi (2005) shows that the distribution of the distances between a point and its nth nearest neighbor (denoted by R_n) in a homogeneous Poisson process defined in $\Re^m$ and with density λ is generalized gamma. The probability distribution function (PDF) of R_n is found to be

Figure 1. Sample PPP with $\lambda = 1$ in a disc with radius 5 and its thinned version with $p = 0.2$.

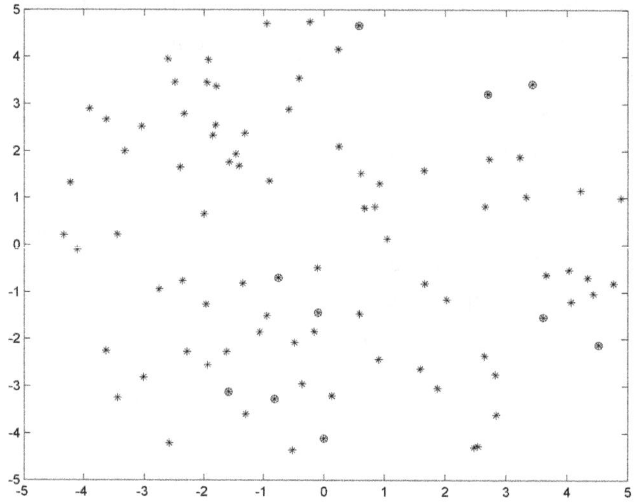

$$f_{R_n}(r) = e^{-\lambda c_m r^m} \frac{m(\lambda c_m r^m)^n}{r\Gamma(n)},$$

where

$$c_m = \begin{cases} \dfrac{\pi^{m/2}}{(m/2)!} & \text{for even } m \\ \dfrac{\pi^{\frac{m-1}{2}} 2^m (\frac{m-1}{2})!}{(m/2)!} & \text{for odd } m \end{cases}$$

which makes $c_m r^m$ the volume of m-dimensional ball of radius r and $\Gamma(.)$ denotes the gamma function. Using this PDF, the moments can also be found as (Haenggi, 2005; Babaei & Jabbari, 2010b)

$$E\{R_n^\alpha\} = \frac{\Gamma(n + \frac{\alpha}{m})}{(\lambda c_m)^{\alpha/m}\Gamma(n)},$$

where α can be any real number such that $n + \alpha / m > 0$.

- *Random Sums*: Let $f(x)$ be a real valued function. The sum

$$F = \sum_{x \in \Pi} f(x), \tag{1}$$

is a random variable. The mean of F for any point process defined in the region R is known to reduce to an integral over the region R as

$$E\{F\} = E\left\{\sum_{x \in \Pi} f(x)\right\} = \int_R f(x)\lambda(x)dx.$$

The higher moments are not generally known for an arbitrary point process. For the case of Poisson point processes, however, characteristic function of the random sum is known in closed form. Higher order statistics of random sums can be obtained from the characteristic function. This result is due to Campbell (1909). Random sums have many applications in physics, signal processing and communication problems.
- *Campbell's Theorem*: Campbell's Theorem gives the characteristic function of a ran-

dom sum F defined in (1) over a Poisson point process. The characteristic function of F is given in (Kingman, 1993) as

$$\psi_F(\omega) = E\left\{e^{j\omega F}\right\} = \exp\left(\int_R \left(\exp\left(j\omega f(x)\right) - 1\right)\lambda(x)dx\right).$$

Consequently, the n th cumulant of F can be found as

$$\kappa_n = \frac{1}{j^n}\left[\frac{\partial^n \ln \psi_F(\omega)}{\partial \omega^n}\right]_{\omega=0} = \int_R f^n(x)\lambda(x)dx. \tag{2}$$

The cumulants and moments can be determined uniquely from one another (Papoulis & Pillai, 2002).

4.2.2 Interference Modeling

Let us assume that the spatial distribution of interferers in a wireless network can be modeled as a Poisson point process Φ in the 2-D Euclidean space. In this section, we aim to obtain statistics of the aggregate interference coming from this Poisson field of interferers (Gulati, Evans, Andrews & Tinsley, 2010). Let us consider an arbitrary location in the Euclidean space and focus on the aggregate received interference power at this location. Since we are considering an infinite-area point process, we can focus on the origin without loss of the generality. This location may or may not correspond to one of the points in Φ . It is know from Slivnyak's Theorem (Stoyan, 1996) that by conditioning on a point in the PPP, the statistical properties of the PPP remains the same[4] and therefore whether the point that the interference is measured belongs to Φ or not does not change the distribution of interference.

Assume that the near field cut-off radius is d_0 and no interferer within radius d_0 around destination can transmit. With a transmission power of P and a power law path loss model, the received interference power from this interferer is

$$I_i = PR_i^{-\alpha}x_i$$

where α is the path loss exponent. The aggregate interference power will be

$$I = \sum_{i:R_i>d_0} I_i = \sum_{i:R_i>d_0} PR_i^{-\alpha}x_i^2. \tag{3}$$

We can see that I is a random sum over Φ and this suggests the use of the Campbell's Theorem (See Section 2.1) to find its statistics. Let us define $h_i = x_i^2$ the fading power gain. The square of a Rayleigh random variable is exponentially distributed and therefore h_i is an exponential random variable with mean 1 and $E\left\{h_i^n\right\} = n!$ (Papoulis & Pillai, 2002). From (2) and (3), we have

$$\Psi_I(\omega) = \exp\left(2\pi\int_h \int_{d_0}^{\infty}\left[\exp\left(j\omega Phr^{-\alpha}\right) - 1\right]\lambda f_h(h)rdrdh\right)$$

where we have used integration in the polar coordinate. Note that the effect of fading is considered as thinning of Φ and as a result, the effective density is $\lambda f_h(h)$ (Ghasemi & Sousa, 2008). We also have

$$\kappa_n(I) = \frac{1}{j^n}\left[\frac{\partial^n \ln \psi_I(\omega)}{\partial \omega^n}\right]_{\omega=0} = 2\pi\lambda P^n E\left\{h^n\right\}\frac{d_0^{2-n\alpha}}{n\alpha - 2}$$

Effect of MAC layer on Interference Statistics: The MAC protocol used in a network determines the simultaneous transmitters and thereby the set of the interferers. The MAC protocol therefore influences on the statistics of interferers by thinning the point process corresponding to the wireless network. We assume that the MAC protocol used is ALOHA with transmission attempt probability of p. Using this MAC protocol, the field of interferers will be a probabilistically thinned version of the original point process with density λp (Haenggi & Ganti, 2008). With an ALOHA MAC protocol, the interference cumulants will be

$$\kappa_n(I) = 2\pi\lambda p P^n E\left\{h^n\right\}\frac{d_0^{2-n\alpha}}{n\alpha - 2},$$

and specifically the mean and variance of interference will be

$$E\left\{I\right\} = 2\pi\lambda p P\frac{d_0^{2-\alpha}}{\alpha - 2},$$

$$\text{var}\{I\} = 2\pi\lambda p P\frac{d_0^{2-2\alpha}}{\alpha - 1}.$$

Other MAC protocols like CSMA also result in the thinning of wireless nodes but the resulting point process may or may not be a Poisson process. Haenggi (2011) shows that when a CSMA network is modeled as hard core process of type II the point process can be closely approximated as a Poisson point process outside the hard core radius while for the hard core process of type I this approximation is not correct (See (Matérn 1986) for definitions of hard core processes.).

4.3 THROUGHPUT MODELING

In the previous section, we found that different parameters including the power levels, transmission attempt probabilities and the density of nodes in the primary and the secondary networks influence on the statistics of intra and inter-network interference. In this section, we use these obtained statistics and show that throughputs in the primary and the secondary networks are non-trivial functions of the transmission attempt probabilities in both of these networks. The obtained throughput models will be used in the next section for throughput optimization.

Network Model

The primary and secondary networks are considered to be independently distributed in a region according to the Poisson point processes with densities λ_p and λ_s, respectively. We denote the distances between a primary node and its nth nearest secondary and primary neighbors as $R_{PS,n}$ and $R_{P,n}$ respectively. Similarly, $R_{SP,n}$ and $R_{S,n}$ denote the distances between a secondary node and its nth nearest primary and secondary neighbors. It is known from Slivnyak's Theorem (Stoyan, 1996), that the distribution of distances from secondary nodes to a given secondary node or an arbitrary point (which can correspond to a primary node) is the same and consequently, we have $R_{PS,n} \overset{d}{=} R_{S,n}$, where $\overset{d}{=}$ denotes equality in distribution. Similarly, we have $R_{SP,n} \overset{d}{=} R_{P,n}$.

Channel Model

We consider a power law path loss model as $r^{-\alpha}$, where r is the distance and α is the path loss exponent (Schwartz, 2005). We also consider a Rayleigh fading component (x) with $E\left\{x^2\right\} = 1$, which is assumed to be independent and identically distributed (I.I.D.) across the interferers. The power therefore decays as $x^2 r^{-\alpha}$. We use the notations: $x_{SP,n}$, $x_{P,n}$, $x_{PS,n}$ and $x_{S,n}$ to denote the fading component along the corresponding inter-nodal links.

Spectrum Sensing

We assume that each primary receiving node sends a beacon signal on a control channel indicating that it is actively engaged in receiving primary data (Hulbert, 2005). Secondary nodes receive the aggregate of beacon signals transmitted by all of the primary receivers on this channel. A secondary node that wishes to transmit data senses this channel first and may transmit only if the SNR of the aggregate of beacons is below a threshold. The other way would be that a secondary node detects individual beacons and abstains from transmission if at least one primary beacon is detected. This brings about the contention problem in the control channel. Due to the exponential decaying of power of beacons, the aggregate power of beacons will be significant only if a secondary node receives beacon signals from its nearby primary nodes. These primary nodes are in fact mostly affected by the secondary node's transmission. The aggregate beacons detection, as opposed to individual beacon detection, obviates the contention problem that arises in the control channel and simplifies the protocol.

Access Model

We consider the slotted ALOHA to be the access method in both systems. The transmission attempt probability in the primary network is assumed to be p_1. In the event that a secondary node does not detect the beacon signal (assume that the probability of this event is ρ), it attempts to transmit with probability p_2'. The effective transmission probability of the secondary nodes is therefore $p_2 = p_2'\rho$.

Interference and Throughput

Assuming P_P and P_S are the transmission power levels of the primary and secondary nodes respectively, we define following variables as the interference power inflicted by each system on its own nodes or on the nodes of the other system:

$$I_P = \sum_n P_P \phi_{P,n} x_{P,n}^2 R_{P,n}^{-\alpha} \tag{4}$$

$$I_{P\to S} = \sum_n P_P \phi_{P,n} x_{SP,n}^2 R_{SP,n}^{-\alpha} \tag{5}$$

$$I_S = \sum_n P_S \phi_{S,n} x_{S,n}^2 R_{S,n}^{-\alpha} \tag{6}$$

$$I_{S\to P} = \sum_n P_S \phi_{S,n} x_{PS,n}^2 R_{PS,n}^{-\alpha} \tag{7}$$

where I_P and I_S are the intra-network interference in the primary and secondary networks respectively and $I_{P\to S}$ and $I_{S\to P}$ are the variables for the inter-network interference between primary and secondary networks. Also, $\phi_{P,n}$ and $\phi_{S,n}$ are Bernoulli random variables with parameters p_1 and p_2 respectively which account for transmission attempt probabilities in the primary and secondary networks respectively. We can see from the above definitions that $I_P \overset{d}{=} I_{P\to S}$. This is because $x_{P,n} \overset{d}{=} x_{SP,n}$ and $R_{P,n} \overset{d}{=} R_{SP,n}$. Similarly, we have $I_S \overset{d}{=} I_{S\to P}$. Assuming a noise power of σ^2, the Signal power to Interference and Noise power Ratio (SINR) at a destination which is R meters away from its primary or secondary transmitter can be written as:

$$\text{SINR}^P = \frac{P_P x_P^2 R^{-\alpha}}{I_P + I_{S\to P} + \sigma^2}, \tag{8}$$

$$\text{SINR}^S = \frac{P_S x_S^2 R^{-\alpha}}{I_S + I_{P\to S} + \sigma^2}. \tag{9}$$

With a half-duplex assumption (i.e., a node can either transmit or receive data), the probabilistic throughput (i.e., the probability that the transmitter is in the transmission mode, the intended receiver is in the reception mode and the link SINR is above a threshold) in the primary and secondary networks will be

$$\tau_P = p_1(1-p_1)\Pr\left\{\text{SINR}^P > \theta\right\} \tag{10}$$

$$\tau_S = p_2(1-p_2)\Pr\left\{\mathrm{SINR}^S > \theta\right\} \quad (11)$$

respectively, where θ is the SINR threshold for successful detection which is assumed to be the same in both systems. The transmitter is assumed to have no channel information and cannot therefore estimate the SINR values at the receivers.

Statistical Models: τ_P, given in (10), is the probability that a primary node receives successfully and can be interpreted as the probability that a primary node is actively receiving data. Primary receiving nodes therefore form a thinned Poisson process with density $\lambda' = \lambda_P \tau_P$. We denote the distance between a secondary node and its ith nearest primary receiving node as R_i, the fading level on this link as x_i, the power of the beacon signal as P_b and the power of aggregate beacon signals as Y. With these definitions, we have

$$Y = P_b \sum_i x_i^2 R_i^{-\alpha}. \quad (12)$$

Note that the intra- and inter-network interference given in equations (4)-(7) and the aggregate beacons power in (12) are random sums over Poisson point process and have the same form. The exact distribution of interference in a network with a Poisson field of interferers is not known and approximation methods have been proposed in the literature (Gulati, Evans, Andrews & Tinsley, 2010; Sousa & Silvister, 1990; Ilow & Hatzinakos, 1998). The cumulants can however be found as we discussed in Section 2.2. We assume that the power of interference or beacon signals add up independently and therefore Central Limit Theorem can be used to model them as Gaussian random variables. Mean and variance therefore suffices to characterize these random variables.

In Section 2.2, we show that for a random Poisson network with density λ, transmission attempt probability of p, transmit power level of P and Rayleigh fading channel with $E\left\{x^2\right\} = 1$, the mean and variance of interference is found as

$$m = 2\pi\lambda p P \frac{d_0^{2-\alpha}}{\alpha - 2}, \quad (13)$$

$$v^2 = 2\pi\lambda p P \frac{d_0^{2-2\alpha}}{\alpha - 1}. \quad (14)$$

The distribution of $I_{S\to P}$ (or I_S) and $I_{P\to S}$ (or I_P) can be obtained by substituting the corresponding parameters of the secondary and primary networks in (13) and (14).

Calculation of ρ

We have defined ρ as the probability that a secondary node does not sense the aggregate beacons on the control channel. This happens when the power of aggregate beacons is smaller than a threshold. If at least one primary receiving node exists within the radius d_0 of a secondary node (which happens with probability $1 - e^{-\lambda'\pi d_0^2}$), we assume that the secondary node will detect the beacon and therefore $\rho = 0$. Otherwise, it abstains from transmission if the SNR of aggregate beacons is above a threshold (γ_t). We therefore have

$$\rho = 1 - e^{-\lambda'\pi d_0^2} \Pr\left\{\frac{Y}{N} < \gamma_t\right\},$$

where Y is the power of aggregate beacons, N is the noise power and γ_t is the SNR threshold. The parameters of the distribution of Y (i.e., (m_y, v_y^2)) can be found by substituting the corresponding parameters in (13) and (14), i.e., P_b and λ' for P and λ and setting $p = 1$ (Note that we have excluded the possibility of contention in the control channel by considering the aggregate of beacons). Therefore, we can calculate ρ as

$$\rho = 1 - e^{-\lambda'\pi d_0^2}\left(1 - Q\left(\frac{\gamma_t N - m_y}{v_y}\right)\right). \qquad (15)$$

where $Q(.)$ is the standard Gaussian Q function (Papoulis & Pillai, 2002).

Throughput Calculation

We denote the mean and variance of $I_{S\to P}$ (or I_S) as (m_S, v_S^2) and those of $I_{P\to S}$ (or I_P) as (m_P, v_P^2). Due to the independence assumption of networks, $I_P + I_{S\to P}$ in (8) (or $I_S + I_{P\to S}$ in (9)) will be Gaussian random variable with parameters $(m_P + m_S, v_P^2 + v_S^2)$. From (13) and (14), mean and variance depend on both p_1 and p_2.

Let us denote $\text{SINR} = \dfrac{Px^2R^{-\alpha}}{I + \sigma^2}$, where x is a Rayleigh random variable with $E\{x^2\} = 1$ and I is a Gaussian random variable with parameters (m, v^2). We have

$$\Pr\{\text{SINR} > \theta\} = \Pr\left\{\frac{x^2}{I + \sigma^2} > \frac{\theta R^\alpha}{P}\right\}.$$

Denoting $Z = X^2$,

$$\Pr\{\text{SINR} > \theta\} = E\left\{F_{c,Z}\left(\frac{\theta(I + \sigma^2)R^\alpha}{P}\right)\right\} = E\left\{\exp\left(\frac{-\theta(I + \sigma^2)R^\alpha}{P}\right)\right\}, \qquad (16)$$

where $F_c(.)$ denotes the Complementary Cumulative Distribution Function (CCDF). Above result is found by noting that Z is an exponential random variable and $F_{c,Z}(z) = e^{-z}$. We can write (16) as

$$\Pr\{\text{SINR} > \theta\} = \exp\left(-\frac{\theta\sigma^2R^\alpha}{P}\right)E\left\{\exp\left(\frac{-\theta IR^\alpha}{P}\right)\right\}.$$

With a Gaussian assumption for interference (with mean m and variance v^2) and considering only the positive values for I (interference power cannot be negative), we have

$$E\left\{\exp\left(\frac{-\theta IR^\alpha}{P}\right)\right\} = \int_0^\infty \exp\left(\frac{-\theta yR^\alpha}{P}\right)\frac{1}{\sqrt{2\pi v}}\exp\left(\frac{-(y-m)}{2v^2}\right)dy = \exp\left(\frac{-\theta m}{P}\right)\exp\left(\frac{\theta^2v^2}{2P^2}\right)Q\left(\frac{\theta v}{P} - \frac{m}{v}\right)$$

and

$$\Pr\{\text{SINR} > \theta\} = \exp\left(\frac{-\theta(m + \sigma^2)}{P}\right)\exp\left(\frac{\theta^2v^2}{2P^2}\right)Q\left(\frac{\theta v}{P} - \frac{m}{v}\right).$$

Using this result, $\Pr\{\text{SINR}^{\text{P}} > \theta\}$ and $\Pr\{\text{SINR}^{\text{S}} > \theta\}$ can be found by substituting the corresponding parameters. Consequently, τ_P and τ_S can be found from (10) and (11).

4.4 THROUGHPUT OPTIMIZATION

From the previous section, τ_P and τ_S depend on both p_1 and p_2 and we can explicitly write them as $\tau_P(p_1, p_2)$ and $\tau_S(p_1, p_2)$. The choice of p_1 by primary node and p_2' (and thereby p_2 through $p_2 = p_2'\rho$) by the secondary network affects on the throughput of both systems.

Interference Constraint

The probability that secondary interference on primary receiving nodes, i.e. primary nodes which are actively receiving data, is larger than a threshold should be small. The interference constraint at a primary node can be written as:

$$\Pr\left\{I_{S\to P} > \eta,\ \text{primary node is actively receiving}\right\} < \varepsilon \quad (17)$$

where (η, ε) are system-defined values. The two events in (17) are not independent. (17) can also be written as:

$$p_1(1-p_1)\Pr\left\{I_{S\to P} > \eta, \mathrm{SINR}^P > \theta\right\} < \varepsilon \quad (18)$$

Denoting the constraint space obtained through (18), i.e. the (p_1, p_2) for which (18) is satisfied, as C, we have $C \in [0,1]^2$. The primary and secondary networks choose p_1 and p_2 respectively such that $(p_1, p_2) \in C$ and each try to optimize its own throughput. The primary and secondary nodes can compete for transmission as long as the chosen (p_1, p_2) belongs to C. This is a game with constraints on multi-strategies (Pervasalle & Pierri, 2000).

Throughput Optimization of Coexisting Networks in Unlicensed Bands

In the previous section, we formulated the problem as a game with constraints on multi-strategies. Obtaining the constraint space through (18) is an involved task which may not lead into a practical result.

In this section, we discuss a sub-problem which is amenable to a practical solution. We consider a single primary network which is working in its optimal transmission probability and think of the secondary network as a network superimposed on the primary network. The secondary network, gradually, i.e., in a slot-by-slot fashion, increases its attempt probability while making sure that the interference constraint is not violated. Since the optimization is done slot by slot, the secondary interference in each time slot is assumed to be independent from the successful primary reception in the previous time slot and (18) can be written as:

$$\Pr\left\{I_{S\to P} > \eta\right\}\Pr\left\{\text{primary node is actively receiving}\right\} < \varepsilon$$

or,

$$\Pr\left\{I_{S\to P} > \eta\right\}\tau_p < \varepsilon.$$

Defining $g(p_2) \triangleq \Pr\left\{I_{S\to P} > \eta\right\}$, we have

$$g(p_2) = Q\left(\frac{\eta - m_S}{v_S}\right),$$

which is dependent on p_2 through m_S and v_S.

The optimization algorithm consists of few steps. At each step, optimization is performed either in the primary or in the secondary network. At the end of each step, some information needs to be communicated to the other network. We propose that at least for the time duration that the algorithm is being executed (the algorithm converges quite fast as we will see in Section 5), primary and secondary networks establish a common control channel to communicate this information. This is beneficial to both networks as both will reach their optimal throughput and deviation from optimum transmission probabilities will degrade throughput in both networks [5]. When the secondary network is laid over the primary network, it uses its knowledge of p_1 and τ_p from the primary network and finds the optimum value p_2 which does not violate the interference constraint and also optimizes its throughput. This causes a secondary interference which was not present beforehand for the primary nodes. In the next time

Algorithm 1. Progressive transmission probability optimization algorithm

Set $p_2 = 0$

$p_1^* = \arg\max_{0 \le p_1 \le 1} \tau_p(p_1, 0)$, find $\tau_p^* = \tau_p(p_1^*, 0)$

repeat

TIME SLOT i : Secondary Optimization

from constraint $\tau_p^* g(p_2) < \varepsilon$, find the range of p_2 : $0 \le p_2 \le p_{,\max}$

$p_2^* = \arg\max_{0 \le p_2 \le p_{2,\max}} \tau_s(p_1^*, p_2)$

BROADCAST p_2^*

TIME SLOT $i+1$: Primary Optimization

from constraint $\tau_p^* g(p_2) < \varepsilon$, find the range of τ_p : $0 \le \tau_p \le \tau_{p,\max}$

$p_1^* = \arg\max_{0 \le \tau_p \le \tau_{p,\max}} \tau_p(p_1, p_2^*)$, find $\tau_p^* = \tau_p(p_1^*, p_2^*)$

BROADCAST p_1^* and τ_p^*

until sufficient convergence of p_1^* and p_2^*

$\lambda' = \tau_p^* \lambda_p$, calculate ρ from (15), $p_2' = p_2^* / \rho$

slot, primary network uses its knowledge of p_2 to optimize its throughput by finding the optimal value of p_1. The new choice of p_1 and the resulting τ_p will affect the secondary throughput and the secondary network tries to re-optimize its throughput in the next time slot. Secondary and primary networks sequentially adjust their transmission probabilities one after the other until convergence takes place (See Algorithm 1).

4.5 ANALYTICAL AND SIMULATION RESULTS

We consider primary and secondary ad hoc wireless networks independently distributed according to Poisson point processes with parameters $\lambda_p = 1$ and $\lambda_s = 1$ respectively. The transmission power levels in secondary and primary networks are assumed to be $P_S = 100\mu W$ and $P_P = 200\mu W$, respectively. The noise power is considered to be $\sigma^2 = 5fW$ and the SINR threshold for signal detection is assumed to be $\theta = 7$ dB. The system-defined values for (η, ε) in the interference constraint (given in equation (17)) are assumed to be $(5 \times 10^{-4}, 0.001)$. The near-field cut off radius, d_0, is 1m and R is set to be 0.5 m.

In Figure 2, for the path loss exponent $\alpha = 4$, we use the progressive transmission probability optimization algorithm, as described in Section 4, to find the optimum transmission attempt probabilities. The results show that the primary transmission probability gradually decreases (starting from $p_1 = 0.366$) and the secondary transmission probability gradually increases (starting from $p_2 = 0$) to reach their stable values of $p_1 = 0.3470$ and $p_1 = 0.366$ after 8 time slots. The values of $p_2' = p_2 / \rho$ are also shown in the same graph.

Figure 2. Transmission probabilities vs. time slot progression

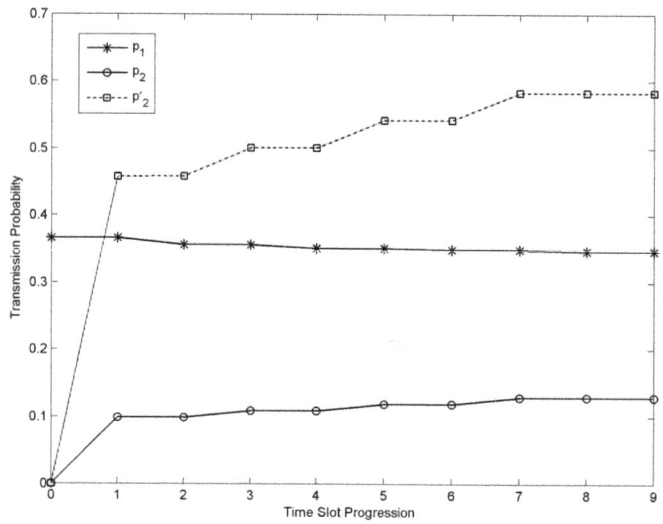

Figure 3. Throughput versus time slot progression

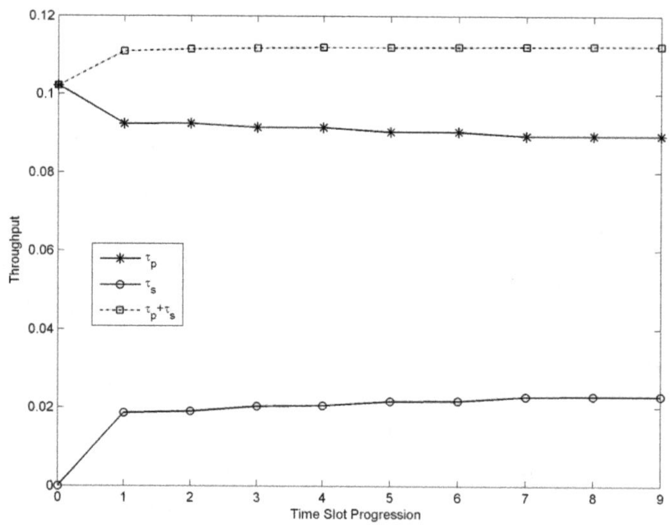

In Figure 3, we have shown the corresponding values of throughput in each time slot. The primary throughput gradually decreases (starting from $\tau_p = 0.1022$) and the secondary throughput gradually increases (starting from $\tau_S = 0$) until they reach their stable values of $\tau_p = 0.0894$ and $\tau_S = 0.0229$. It indicates that at the optimum transmission probabilities, the primary network is deviated slightly from its optimum throughput, while the secondary network gains a considerable amount of throughput. The sum of primary and secondary optimum throughput is also shown to be larger than the optimum primary throughput when the secondary network is not present.

In Figure 4, optimum throughput is plotted versus η in the interference constraint when it varies from 10^{-4} to 10^{-3}. As η increases, i.e., as

Figure 4. Optimum throughput vs. η *(interference threshold)*

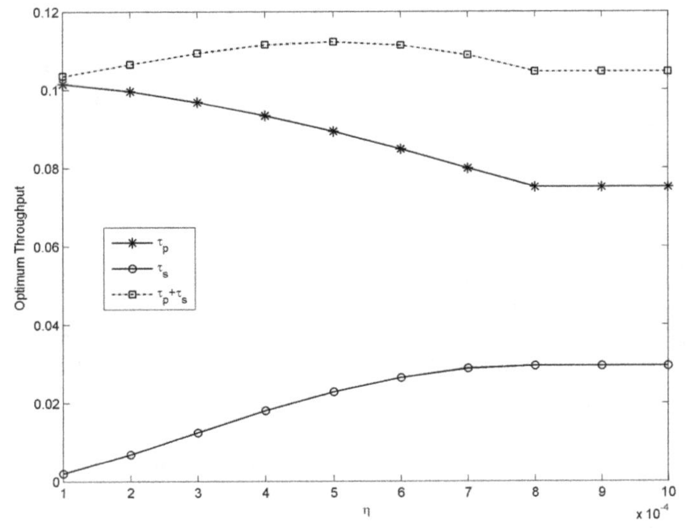

Figure 5. Optimum throughput vs. ε

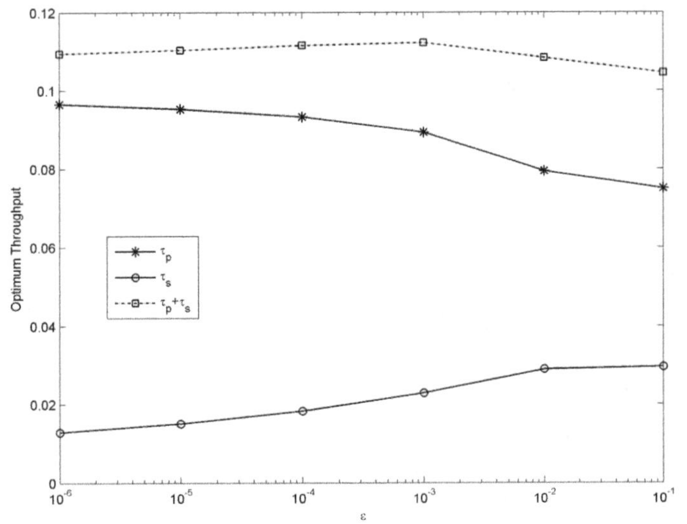

the constraint become looser, primary network looses more throughput and the secondary network gains more throughput. The sum throughput on the other hand has less variation. The same observation can be seen in Figure 5, where optimum throughput is plotted versus ε in the interference constraint.

In Figure 6, the optimum transmission probabilities are shown versus the path loss exponent (α). In the same graph, the values of $p_2' = p_2 / \rho$ are shown. The corresponding values of throughput (using the converged values of p_1 and p_2) are shown in Figure 6. The results indicate that the optimum transmission probabilities and the values of throughput increase with path loss exponent.

Figure 6. Optimum transmission probabilities vs path loss exponent

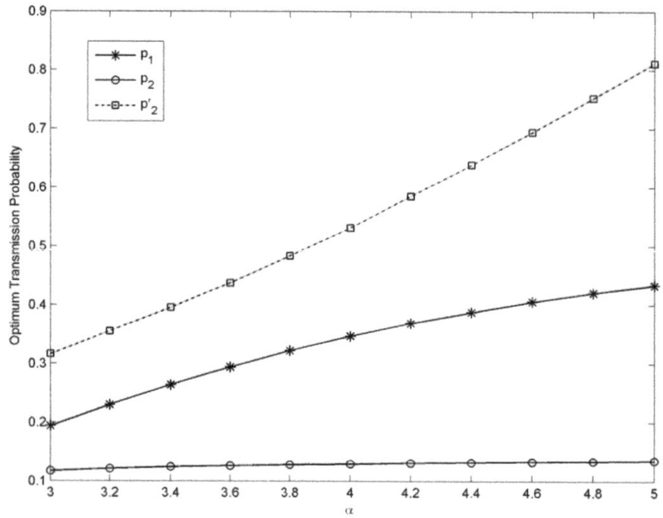

Figure 7. Optimum throughput vs path loss exponent

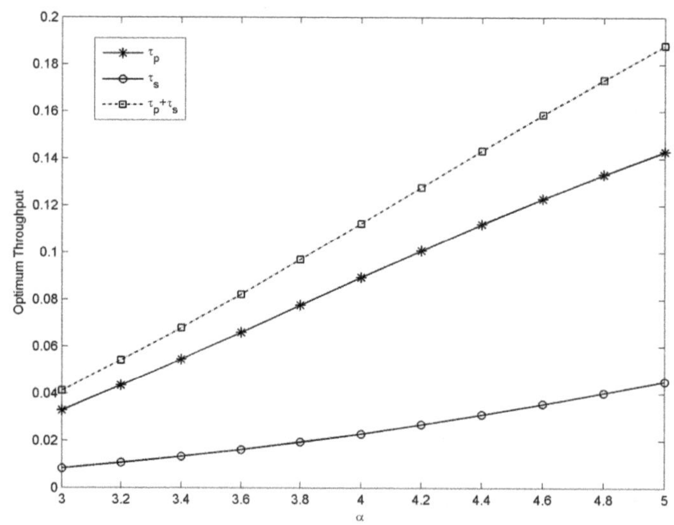

4.6 FUTURE RESEARCH DIRECTIONS

As we show in Section 4, the interference constraint limits the feasible space of transmission attempt probabilities for both the primary and the secondary networks. We can formulate the throughput optimization problem as a game where the primary and the secondary networks are the players of the game, the actions of each player is its choice of transmission probability and the throughput in each network is its utility (Osborne & Rubinstein, 1994). Due to the limitation in the action space (i.e. the transmission attempt probabilities), this is a game with constraints on multi-strategies (Pensavalle & Pieri, 2000). Constrained Game Theory has recently been considered to solve the resource allocation problems in wireless communications (See (Scutari, Barbarossa & Palomar, 2006)). Pensavalle & Pieri (2000) give the sufficient conditions on the existence of an n -player game with constraints on the multi-strategies. We believe that, due to the interference constraint, the constrained Game Theory is the right discipline to solve the resource allocation and optimization problems associated with cognitive wireless networks. Moreover, the game formulation can be extended to include both power levels and transmission probabilities as the action spaces of the primary and the secondary networks.

4.7 CONCLUSION

The throughputs in the coexisting primary and secondary ad hoc networks have a close interrelation. Throughputs in these networks are dependent on the choices of transmission probabilities in both networks. Meanwhile, the secondary network needs to be alert in its choice of transmission probability so that it does not violate the interference constraint on the primary nodes. We consider a simple Gaussian model for interference and incorporate a sensing mechanism in the secondary network to avoid interference with the primary receiving nodes. We assume that the secondary network is superimposed into a primary network which is already operating at its optimum parameters and propose a progressive transmission probability optimization algorithm which progressively optimizes the primary and secondary transmission probabilities. Simulation results indicate that the proposed algorithm converges very fast (after few time slots) and at the optimum transmission probabilities, the primary network is deviated slightly from its optimum throughput, while the secondary network gains a considerable amount of throughput. In addition, the sum throughput is found to increase by using these optimum transmission probabilities.

REFERENCES

Babaei, A., & Jabbari, B. (2010a). Throughput optimization in cognitive random wireless ad hoc networks. *Proceedings of IEEE Global Communications Conference*, (pp. 1-5).

Babaei, A., & Jabbari, B. (2010b). Distance distribution of bivariate Poisson network nodes. *IEEE Communications Letters*, *14*(9), 848–850. doi:10.1109/LCOMM.2010.080610.100668

Baddeley, A. (2007). Spatial point processes and their applications. *Lecture Notes in Mathematics*, *1892*, 1–75. doi:10.1007/978-3-540-38175-4_1

Cabric, D., Mishra, S. M., & Brodersen, R. W. (2004). Implementation issues in spectrum sensing for cognitive radios. *Proceedings of Asilomar Conference on Signals, Systems, and Computers*, (pp. 772–776).

Campbell, N. (1909). The study of discontinuous phenomena, mathematica. *Proceedings of the Cambridge Philosophical Society*, *15*, 117–136.

Cox, D. (1970). *Renewal theory*. Methuen.

Diggle, P. J. (2003). *Statistical analysis of spatial point patterns* (2nd ed.). Arnold.

Fremlin, D. H. (2003). *Measure theory*. Torres Fremlin.

Ghasemi, A., & Sousa, E. S. (2008, February). Interference aggregation in spectrum-sensing cognitive wireless networks. *IEEE Journal on Selected Areas in Communications*, *2*(1), 41–56.

Gulati, K., Evans, B. L., Andrews, J. G., & Tinsley, K. R. (2010, December). Statistics of co-channel interference in a field of Poisson and Poisson-Poisson clustered interferers. *IEEE Transactions on Signal Processing*, *58*(12). doi:10.1109/TSP.2010.2072922

Haenggi, M. (2005, October). On distances in uniformly random networks. *IEEE Transactions on Information Theory*, *51*(10), 3584–3586. doi:10.1109/TIT.2005.855610

Haenggi, M. (2011). Mean interference in hard-core wireless networks. *IEEE Communications Letters*, *15*(8). doi:10.1109/LCOMM.2011.061611.110960

Haenggi, M., Andrews, J. G., Baccelli, F., Dousse, O., & Franceschetti, M. (2009). Stochastic geometry and random graphs for the analysis and design of wireless networks. *IEEE Journal on Selected Areas in Communications*, *27*(7), 1029–1046. doi:10.1109/JSAC.2009.090902

Haenggi, M., & Ganti, R. K. (2008). Interference in large wireless networks, foundations and trends in networking. *NOW Publishers*, *3*(2), 127–248.

Haykin, S. (2005). Cognitive radio: Brain-empowered wireless communications. *IEEE Journal on Selected Areas in Communications*, *32*(2), 201–220. doi:10.1109/JSAC.2004.839380

Hulbert, A. P. (2005). Spectrum sharing through beacons. *Proceedings of IEEE International Symposium in Personal, Indoor and Mobile Radio Communications (PIMRC)*, Berlin, Germany, (pp. 989–993).

Ilow, J., & Hatzinakos, D. (1998). Analytical alpha-stable noise modeling in a poisson field of interferers or scatterers. *IEEE Transactions on Signal Processing*, *46*(6), 1601–1611. doi:10.1109/78.678475

Kingman, J. (1993). *Poisson processes*. Oxford University Press.

Ludwig, J. A., & Reynolds, J. F. (1998). *Statistical ecology: A primer on methods and computing*. New York: John Wiley & Sons.

Matérn, B. (1986). *Spatial variation* (2nd ed.). Springer Lecture Notes in Statistics.

Osborne, M. J., & Rubinstein, A. (1994). *A course in game theory*. MIT Press.

Papoulis, A., & Pillai, S. U. (2002). *Probability, random variables and stochastic processes* (4th ed.). McGraw Hill.

Pensavalle, C., & Pierri, G. (2000). On existence of Nash equilibria of games with constraints on multistrategies. *Journal of Optimization Theory and Applications*, *107*(3), 601–613. doi:10.1023/A:1026403417009

Schwartz, M. (2005). *Mobile wireless communications*. Cambridge University Press.

Scutari, G., Barbarossa, S., & Palomar, D. (2006). Potential games: A framework for vector power control problems with coupled constraints. *Proceedings of IEEE International Conference on Acoustics, Speech, and Signal Processing (ICASSP)*.

Sousa, E. S., & Silvester, J. A. (1990). Optimum transmission ranges in a direct-sequence spread-spectrum multihop packet radio network. *IEEE Journal on Selected Areas in Communications*, *8*(5), 762–771. doi:10.1109/49.56383

Stoyan, D., Kendall, W., & Mecke, J. (1996). *Stochastic geometry and its applications*. John Wiley.

Venkataraman, J., & Haenggi, M. (2004). Optimizing the throughput in random wireless Ad hoc networks. *Proceedings of 42st Annual Allerton Conference on Communication, Control, and Computing*.

ENDNOTES

1. The probabilistic throughput is the probability that the transmission from the transmitter to the receiver is successful.

2 With the full-duplex assumption, the receiver can transmit and receive at the same time and $\tau = p\,\Pr\left\{\mathrm{SINR} > \theta\right\}$.

3 The formal definition of the Poisson point process in a general space requires some background from measure Theory (Fermlin, 2003).

4 More technically, the distribution of the original process is equal to its reduced Palm distribution (Baddeley, 2007).

5 This is in fact the equilibrium of the constrained game described in Section 4.

Chapter 5
Distributed Multicell Precoding for Network MIMO

Winston W. L. Ho
Institute for Infocomm Research, Singapore

Tony Q. S. Quek
Singapore University of Technology and Design, Singapore

Robert W. Heath Jr.
The University of Texas at Austin, USA

ABSTRACT

Interference is a performance limiting factor in dense cellular networks with aggressive frequency reuse. Cooperation among base stations (BSs) is a promising approach for improving data rates by eliminating or mitigating interference. For the downlink, the highest spectral efficiency gains are achieved through precoding with full coordination, which requires complete channel state information (CSI) and data sharing among BSs at the cost of significant utilization of the backhaul. In this chapter, the authors introduce distributed precoding techniques for the multicell MIMO downlink. Each BS designs its own precoder without requiring data or downlink CSI of links from other BSs. The authors also study the effect of imperfect CSI on these precoders and introduce a robust precoder in the presence of CSI uncertainty. Simulations show that these methods enjoy a rate increase with SNR similar to multicell joint dirty paper coding in the high SNR regime due to effective interference mitigation.

5.1 INTRODUCTION

In next generation cellular networks, the increased density of base stations (BSs) and the aggressive reuse of frequencies are seen as the solution to the pressing problem of scarce frequency spectrum. As a result, cellular systems are going to be interference limited. Multicell or network MIMO promises to reduce or eliminate the interference problem through cooperation or coordination between BSs (Viswanathan, Venkatesan, & Huang, 2003; Foschini, Karakayali, & Valenzuela, 2006; Andrews, Choi, & Heath Jr., 2007; Boudreau et al., 2009; H. Zhang & Dai, 2004; Tölli, Pennanen, & Komulainen, 2009; Cai, Quek, & Tan, 2011;

DOI: 10.4018/978-1-4666-2005-6.ch005

Cai, Quek, Tan, & Low, n.d.). Multicell MIMO refers to a technique where BSs cooperate and use multiple antenna transmission to increase the performance of a cellular system. Multiple antennas may exist on the BSs, on the users, or both. As a result, the spatial dimension can be exploited to boost the users' rates or reduce the transmission power required, even with the same time and frequency resources.

Joint encoding with DPC among cooperating BSs has been shown to achieve the capacity or maximum sum rate of the multicell MIMO downlink. Joint encoding is hard to achieve in practical systems due to the requirement of precise time and phase synchronization of the transmitted signals from involved BSs (Andrews et al., 2007). Besides DPC, zero-forcing beamforming (Somekh, Simeone, Bar-Ness, Haimovich, & Shamai (Shitz), 2009) and game theoretic optimization for scheduling across frequency and time (Gesbert, Kiani, Gjendemsjø & Øien, 2007; Oteri & Paulraj, 2008) have also been proposed. Linear block diagonalization methods null out inter-cell interference using linear zero-forcing (ZF) techniques to create a block diagonal effective channel from the BSs to the users (Shim, Kwak, Heath Jr., & Andrews, 2008; Kaviani & Krzymien, 2008). Chae, Kim, and Heath Jr. (2009) proposed linear algorithms for full and clustered broadcast channel scenarios that approach the sum capacity of multicell DPC. Bhagavatula and Heath Jr. (2011) proposed a feedback-bit partitioning algorithm to ensure a manageable load on the finite-capacity backhaul. Ng, Evans, Hanly, & Aktas (2008) interpretsthe multicell downlink as a factor graph with local message passing between neighboring BSs. They proposed a downlink beamforming algorithm based on belief propagation. H. Huang et al. (2009) analyzed the effect of the coordination cluster size through cellular system simulations. Marsch and Fettweis (2008)proposed a framework for optimizing the downlink where multicell joint transmission is possible but constrained by a limited backhaul infrastructure between sites. Jing et al. (2008)considered single-class and double-class networks and analyzed schemes involving cell-breathing, cophasing, superposition coding, as well as other hybrid strategies.Related work on the uplink can be found in the papers by Lee, Je, Shin, & Lee (2009); Simeone, Somekh, Poor, & Shamai (Shitz) (2009).

Most techniques assume a high level of cooperation among BSs in which (processed) data for an intended user needs to be available at multiple BSs. In such a scenario, multiple cooperating BSs act as a large virtual antenna array to beamform to a particular user. Clearly, system performance is improved at the cost of significant system overhead related to channel state information (CSI) feedback to the central processor and data exchange among the BSs. The concept of maximizing the signal-to-leakage-plus-noise ratio (SLNR) has been used to reduce the complexity of the transmitter processing (Sadek, Tarighat, & Sayed, 2007; H. Zhang, Mehta, Molisch, Zhang, & Dai, 2008; Lee, Je, Sohn, Shin, & Lee, 2008). Sadek et al. (2007) applied it in the single-cell multiuser MIMO downlink. H. Zhang et al. (2008) extended it to the multicell scenario using joint transmission. Lee et al. (2008) considered a single user per cell and applied the SLNR concept for decentralized precoding. SLNR-based precoding can be seen as a balance between the altruistic and egoistic beamforming strategies in the game theory of interference channels (Larsson, Jorswieck, Lindblom, & Mochaourab, 2009; Z. K. M. Ho, Kaynia, & Gesbert, 2010; Z. K. M. Ho & Gesbert, 2010; Björnson & Ottersten, 2009; Zakhour, Ho, & Gesbert, 2009; Zakhour & Gesbert, 2009; Björnson, Zakhour, Gesbert, & Ottersten, 2009).

In this chapter, we introduce two decentralized precoding methods for network MIMO downlink channels, where multi-antenna BSs serve multiple users equipped with one or more antennas. Each BS performs coordinated processing in a distributed fashion with only the CSI of links originating from itself and only the data intended for users in its own coverage region (intra-cell users). If time division duplex (TDD) is used, the CSI of the channels from the BS to the intra-cell and out-of-cell users can be obtained implicitly during the uplink phase using channel reciprocity without the need for explicit feedback. Neighboring BSs may need to coordinate their served users to perform channel sounding at different times for the training to be orthogonal. With frequency division duplex (FDD), the user may need to feedback the downlink CSI from the neighboring BS via backhaul to the said BS.

The proposed precoders tackle both intra-cell and out-of-cell interference. In particular, this chapter describes a nonlinear precoder called the leakage-DPC and a linear precoder called the multicell leakage suppression. The leakage-DPC obtains a higher sum rate than the multicell leakage suppression, while the later has lower complexity because it is a linear technique. The leakage-DPC method performs a channel precoding based on maximizing a cell-based SLNR followed by DPC for intra-cell users. To enhance the robustness of the leakage-DPC method with respect to CSI uncertainty, a robust variation of the leakage-DPC strategy is also obtained. The multicell leakage suppression method is a linear precoding method based on the SLNR.

Unlike prior work, which either had backhaul CSI sharing as a prerequisite or could only support a single user per cell, both proposed precoders handle multiple intra-cell and out-of-cell users in a decentralized manner. In other words, a central processor among the BSs is not required. We examine the performance of these precoders with respect to the amount of out-of-cell interference, SNR, numbers of antennas and users as well as CSI uncertainty. With sufficient number of transmit antennas, these methods enjoy a rate growth with SNR similar to multicell joint DPC in the high SNR regime, due to the interference mitigation. At high out-of-cell channel gains, increasing the out-of-cell channel gain has diminishing negative effects on the sum rate performances because these precoders are able to mitigate the out-of-cell interference effectively.

This chapter is organized as follows. Section 2 describes the multicell downlink system model. Sections 3 and 4 introduce the nonlinear and linear precoders respectively. Numerical simulations are given in Section 5. Finally, the discussion and conclusion are presented in Sections 6 and 7 respectively. The following notation is used. Bold lowercase letters $\mathbf{v}$ are used to denote column vectors and bold uppercase letters $\mathbf{M}$ are used to denote matrices. $\min(a,b)$ is the minimum of two real numbers a and b. $(\cdot)^T$ and $(\cdot)^H$ denote the matrix transpose and conjugate transpose operations respectively. $E[\cdot]$ stands for statistical expectation. $C^{P\times Q}$ denotes the space of complex $P\times Q$ matrices. The distribution of a circularly symmetric complex Gaussian (CSCG) vector with mean vector $\mathbf{m}$ and covariance matrix $\mathbf{R}$ is denoted by $\mathcal{CN}(\mathbf{m},\mathbf{R})$, and $\sim$ means "distributed as". $\mathbf{R}_{\mathbf{x}} = E[\mathbf{x}\mathbf{x}^H]$ is the covariance matrix of a vector $\mathbf{x}$. $\|\cdot\|_2$ denotes the vector Euclidean norm, while $\mathbf{I}_N$ denotes the $N\times N$ identity matrix. $\mathrm{Tr}(\mathbf{A})$ stands for the trace of a matrix $\mathbf{A}$. $\det(\mathbf{A})$ denotes the determinant of $\mathbf{A}$. $[\mathbf{A}]_{i,j}$ is the scalar entry of $\mathbf{A}$ in the i-th row and j-th column. $\mathrm{vec}(\mathbf{A})$ is a column vector composed of the entries of $\mathbf{A}$ taken column-wise. $\mathrm{diag}(\mathbf{A})$ represents the diagonal matrix with the same diagonal as the matrix $\mathbf{A}$. $\mathrm{blkdiag}(\mathbf{A}_1,\mathbf{A}_2,\ldots,\mathbf{A}_U)$ denotes a block diagonal matrix whose block diagonal elements are $\mathbf{A}_u, u=1,\ldots,U$.

5.2 SYSTEM MODEL

Consider a downlink cellular network with M cells of N_{T} transmit antennas each, serving U users with N_{R} receive antennas each. For notational simplicity, we assume a synchronous multicell system. To facilitate the rate analysis later, the received signal vector of all users in the system is written in a block system format as

$$\mathbf{y}_{\text{sys}} = \mathbf{H}_{\text{sys}}\mathbf{x}_{\text{sys}} + \mathbf{z}_{\text{sys}} \tag{1}$$

where $\mathbf{y}_{\text{sys}} = \text{vec}([\mathbf{y}_1, \mathbf{y}_2, \ldots, \mathbf{y}_M]) \in C^{MUN_\text{R}\times 1}$ is the system receive signal vector such that $\mathbf{y}_m \in C^{UN_\text{R}\times 1}$ is for the m-th cell. $\mathbf{x}_{\text{sys}} = \text{vec}([\mathbf{x}_1, \mathbf{x}_2, \ldots, \mathbf{x}_M]) \in C^{MN_\text{T}\times 1}$ is the system transmit signal vector such that $\mathbf{x}_m \in C^{N_\text{T}\times 1}$ is the transmit signal vector from the m-th BS with average power constraint $P_m = E[\text{Tr}(\mathbf{x}_m\mathbf{x}_m^H)]$. $\mathbf{z}_{\text{sys}} \sim \mathcal{CN}(\mathbf{0}, N_0\mathbf{I}_{MUN_\text{R}})$ is an additive CSCG noise vector. The instantaneous downlink channel matrix $\mathbf{H}_{\text{sys}} \in C^{MUN_\text{R}\times MN_\text{T}}$ is given by

$$\mathbf{H}_{\text{sys}} = \begin{bmatrix} \mathbf{H}_1 & \mathbf{H}_{2\to 1} & \cdots & \mathbf{H}_{M\to 1} \\ \mathbf{H}_{1\to 2} & \mathbf{H}_2 & \cdots & \mathbf{H}_{M\to 2} \\ \vdots & \vdots & \ddots & \vdots \\ \mathbf{H}_{1\to M} & \mathbf{H}_{2\to M} & \cdots & \mathbf{H}_M \end{bmatrix} \tag{2}$$

where $\mathbf{H}_m = \mathbf{H}_{m\to m} \in C^{UN_\text{R}\times N_\text{T}}$ denotes the channel matrix from the m-th BS to all its served users and $\mathbf{H}_{m\to n}$ denotes the channel matrix from the m-th BS to all the U users in the n-th cell. For the precoders described in the following sections, the m-th BS would only require at most the CSI of $\mathbf{H}_{m\to n}, 1 \le n \le M$. By denoting $\mathbf{y}_m = \text{vec}([\mathbf{y}_{m,1}, \mathbf{y}_{m,2}, \ldots, \mathbf{y}_{m,U}]) \in C^{UN_\text{R}\times 1}$, the received signal of the u-th user at the m-th cell is given by

$$\mathbf{y}_{m,u} = \mathbf{H}_{m,u}\mathbf{x}_m + \sum_{n\neq m}^{M}\mathbf{H}_{n\to m,u}\mathbf{x}_n + \mathbf{z}_{m,u} \tag{3}$$

where $\mathbf{H}_{m,u} \in C^{N_\text{R}\times N_\text{T}}$ is the channel matrix from the m-th BS to the u-th intra-cell user such that $\mathbf{H}_m = [\mathbf{H}_{m,1}^T, \mathbf{H}_{m,2}^T, \ldots, \mathbf{H}_{m,U}^T]^T \in C^{UN_\text{R}\times N_\text{T}}$ in (2). $\mathbf{H}_{n\to m,u} \in C^{N_\text{R}\times N_\text{T}}$ is the channel matrix from the n-th BS to the u-th user in the m-th cell such that $\mathbf{H}_{n\to m} = [\mathbf{H}_{n\to m,1}^T, \mathbf{H}_{n\to m,2}^T, \ldots, \mathbf{H}_{n\to m,U}^T]^T$ in(2). $\mathbf{z}_{m,u} \sim \mathcal{CN}(\mathbf{0}, N_0\mathbf{I}_{N_\text{R}})$ is the additive noise. To accommodate the transmission of multiple data streams, we allow each user to receive d data streams from its own serving BS where $d \le N_\text{R}$.

For the m-th BS, if an out-of-cell channel $\mathbf{H}_{m\to n,u}$ for some cell $n \neq m$ and some user u has small path gains compared to $\mathbf{H}_{m\to m}$, as in the small co-channel interference scenario, this channel need not be known because the interference effect is negligible. Calibration of the isolation of the out-of-cell users from the m-th BS may be required (Marsch & Fettweis, 2008). The benefit is that in TDD systems, the BS would only need to perform uplink channel estimations for users with significant path gains. Suppose there are $\bar{U}$ such out-of-cell users. Let $\bar{\mathbf{H}}_m \in C^{\bar{U}N_\text{R}\times N_\text{T}}$ represent the channel towards these $\bar{U}$ users. Also, define $\tilde{\mathbf{H}}_{m,u} = [\mathbf{H}_{m,1}^T, \ldots, \mathbf{H}_{m,u-1}^T, \mathbf{H}_{m,u+1}^T, \ldots, \mathbf{H}_{m,U}^T, \bar{\mathbf{H}}_m^T]^T \in C^{(U-1+\bar{U})N_\text{R}\times N_\text{T}}$ as the channel from this BS to its $U-1$ intra-cell users other than the u-th user as well as those $\bar{U}$ out-of-cell users. In the following sections, we design precoding matrices for each BS by taking into account both intra-cell and out-of-cell interference.

The following channel model shall be used to incorporate CSI uncertainty (Yoo & Goldsmith, 2006)

$$\mathbf{H} = \hat{\mathbf{H}} + \mathbf{E} \tag{4}$$

where $\mathbf{E}$ is the probabilistic additive error component with independent and identically distributed (i.i.d.) elements $[\mathbf{E}]_{i,j} \sim \mathcal{CN}(0, \sigma_e^2)$. σ_e^2 is a parameter that captures the quality of the channel estimation and is assumed to be known at the transmitter. $\hat{\mathbf{H}}$ is the estimate of the channel available at the transmitter with elements distributed as $\mathcal{CN}(0, 1-\sigma_e^2)$ and $\mathbf{H}$ is the actual channel with elements distributed as $\mathcal{CN}(0,1)$.

5.3 LEAKAGE-DPC STRATEGY

In this section, we introduce a nonlinear technique called leakage-DPC. This approach ensures no interference between the signals to different users by applying a block diagonal DPC processing. For this to work, a projection operation is applied beforehand as though the base station is transmitting on multiple eigenmodes to the users collectively, not individually as in the approach ofSadek et al. (2007). The cell index subscript m is sometimes dropped for notational simplicity because the BS processing is decentralized. The m-th BS requires at most the CSI of $\mathbf{H}_{m \rightarrow n}, 1 \leq n \leq M$.

5.3.1 Algorithm Description

The objective is to find an orthogonal precoder that maximizes the intra-cell signal strength and minimizes the out-of-cell leakage. It should also have sufficient rank to support multiple intra-cell users. The idea is to group the intra-cell users as one virtual desired user and to group the out-of-cell users as one virtual undesired user. The first step is to find an orthogonal precoder that maximizes the virtual desired user's signal relative to the virtual undesired user's signal and noise. This is the projection step. Given this channel projection, the next step is to eliminate the interference between the intra-cell users via a combination of dirty paper coding and block diagonal processing.

Consider the hypothetical case in which the BS performs beamforming via a semi-unitary matrix $\mathbf{V} \in C^{N_T \times N_V}$ where $\mathbf{V}^H \mathbf{V} = \mathbf{I}_{N_V}$ (which does not change the transmit power). $\mathbf{V}$ and N_V are to be determined. The stacked desired signal vector received by the intra-cell users is given by $\mathbf{y}^{\text{sig}} \in C^{UN_R \times 1}$ where

$$\mathbf{y}^{\text{sig}} = \sqrt{P / N_V}\,\mathbf{H}_m \mathbf{V}\mathbf{s} \tag{5}$$

and $\mathbf{s}$ is the (uncoded) data symbol vector. Assume the stacked noise vector received by these users is $\mathbf{z} \sim \mathcal{CN}(\mathbf{0}, N_0 \mathbf{I}_{UN_R})$. The leakage to the out-of-cell users is given by $\mathbf{y}^{\text{leak}} \in C^{\bar{U}N_R \times 1}$ where

$$\mathbf{y}^{\text{leak}} = \sqrt{P / N_V}\,\bar{\mathbf{H}}_m \mathbf{V}\mathbf{s}. \tag{6}$$

The covariance matrices of the desired signal, leakage, and noise are given respectively by

$$\mathbf{R}_{\mathbf{y}^{\text{sig}}} = \left(P / N_V\right) \mathbf{H}_m \mathbf{V}\mathbf{V}^H \mathbf{H}_m^H$$

$$\mathbf{R}_{\mathbf{y}^{\text{leak}}} = \left(P / N_V\right) \bar{\mathbf{H}}_m \mathbf{V}\mathbf{V}^H \bar{\mathbf{H}}_m^{\ H}$$

$$\mathbf{R}_{\mathbf{z}} = N_0 \mathbf{I}_{UN_R}.$$

The cell SLNR ζ_C is defined as

$$\zeta_C = \frac{\text{Tr}\left(\mathbf{R}_{\mathbf{y}^{\text{sig}}}\right)}{\text{Tr}\left(\mathbf{R}_{\mathbf{y}^{\text{leak}}}\right) + \text{Tr}\left(\mathbf{R}_{\mathbf{z}}\right)} = \frac{\text{Tr}\left(P\mathbf{H}_m \mathbf{V}\mathbf{V}^H \mathbf{H}_m^H\right)}{\text{Tr}\left(P\bar{\mathbf{H}}_m \mathbf{V}\mathbf{V}^H \bar{\mathbf{H}}_m^{\ H}\right) + UN_R N_0 N_V}$$
$$= \frac{\sum_{i=1}^{N_V} \mathbf{v}_i^H \left(P\mathbf{H}_m^H \mathbf{H}_m\right) \mathbf{v}_i}{\sum_{i=1}^{N_V} \mathbf{v}_i^H \left(P\bar{\mathbf{H}}_m^{\ H} \bar{\mathbf{H}}_m + UN_R N_0 \mathbf{I}_{N_T}\right) \mathbf{v}_i}$$
$$= \frac{\sum_{i=1}^{N_V} \mathbf{v}_i^H \mathbf{G}_A \mathbf{v}_i}{\sum_{i=1}^{N_V} \mathbf{v}_i^H \mathbf{G}_B \mathbf{v}_i} \tag{7}$$

where $\mathbf{G}_\mathrm{A} = \rho \mathbf{H}_m^H \mathbf{H}_m, \mathbf{G}_\mathrm{B} = \rho \bar{\mathbf{H}}_m{}^H \bar{\mathbf{H}}_m + U N_\mathrm{R} \mathbf{I}_{N_\mathrm{T}}$,and $\rho = P / N_0$. For a given N_V, finding the optimal $\mathbf{V}$ to maximize ζ_C is analytically challenging. Therefore, we maximize a lower bound ζ_L of the cell SLNR given by H. Zhang et al. (2008)

$$\zeta_\mathrm{L} = \min_{i=1,\ldots,N_\mathrm{V}} \frac{\mathbf{v}_i^H \mathbf{G}_\mathrm{A} \mathbf{v}_i}{\mathbf{v}_i^H \mathbf{G}_\mathrm{B} \mathbf{v}_i}. \tag{8}$$

Maximizing the lower bound in (8) maximizes the smallest generalized Rayleigh quotient $\frac{\mathbf{v}_i^H \mathbf{G}_\mathrm{A} \mathbf{v}_i}{\mathbf{v}_i^H \mathbf{G}_\mathrm{B} \mathbf{v}_i}$ among all the BS's transmission eigenmodes. Consequently,

$$\zeta_{\mathrm{L,max}} = \max_{\mathbf{V}^H\mathbf{V}=\mathbf{I}_{N_\mathrm{V}}} \min_{i=1,\ldots,N_\mathrm{V}} \frac{\mathbf{v}_i^H \mathbf{G}_\mathrm{A} \mathbf{v}_i}{\mathbf{v}_i^H \mathbf{G}_\mathrm{B} \mathbf{v}_i} = \max_{\mathcal{V}:\dim(\mathcal{V})=N_\mathrm{V}} \min_{\mathbf{v}\in\mathcal{V}} \frac{\mathbf{v}^H \mathbf{G}_\mathrm{A} \mathbf{v}}{\mathbf{v}^H \mathbf{G}_\mathrm{B} \mathbf{v}}. \tag{9}$$

According to the generalized Courant-Fischer max-min theorem (Horn & Johnson, 1990, Chap. 4), $\zeta_{\mathrm{L,max}}$ is equal to the N_V-th largest eigenvalue of $\mathbf{G}_\mathrm{S}$ where $\mathbf{G}_\mathrm{S} = \mathbf{G}_\mathrm{B}^{-1}\mathbf{G}_\mathrm{A}$. $\mathbf{V}$ is then given by the N_V orthonormal basis vectors of the space spanned by the N_V dominant eigenvectors of $\mathbf{G}_\mathrm{S}$ as follows fromH. Zhang et al. (2008) assuming $\mathrm{rank}(\mathbf{G}_\mathrm{S}) \geq N_\mathrm{V}$. This solution maximizes ζ_L. (By the generalized Rayleigh-Ritz theorem (Horn & Johnson, 1990, Chap. 4), for the case where $\mathbf{G}_\mathrm{B} = \mathbf{I}$, the proposed $\mathbf{V}$ maximizes ζ_C.) After obtaining $\mathbf{V}$, apply an initial precoding via a projection matrix $\mathbf{V}\mathbf{V}^H$ to the intra-cell channel $\mathbf{H}_m$ to get $\mathbf{H}_{m,\perp} = \mathbf{H}_m \mathbf{V}\mathbf{V}^H$. This precoding step increases the desired signal power directed to the intra-cell users and reduces the interference power towards the out-of-cell users. The reason is due to the objective of maximizing the cell SLNR in (7). The overall effect is an altruistic cooperation among BSs that serves to increase the overall system rate.

The second step of the leakage-DPC method involves block diagonal DPC processing on $\mathbf{H}_{m,\perp}$ to transmit to the intra-cell users. By multiplying $\mathbf{H}_{m,\perp}$ by $\mathbf{W}_m$ and $\mathbf{Q}_m$ where $\mathbf{W}_m$ and $\mathbf{Q}_m$ have orthonormal columns, it is possible to create a lower triangular equivalent channel $\mathbf{L}_m$ in which users employing multiple data streams can have equal channel gain per stream. The block diagonal geometric mean decomposition (BD-GMD) introduced by Lin, Ho, & Liang (2008) provides the required matrices $\mathbf{W}_m$, $\mathbf{L}_m$, and $\mathbf{Q}_m$ such that

$$\mathbf{W}_m^H \mathbf{H}_{m,\perp} \mathbf{Q}_m = \mathbf{L}_m. \tag{10}$$

$\mathbf{W}_m$ is the receive equalization matrix, $\mathbf{Q}_m$ is the transmit pre-equalization matrix, and $\mathbf{L}_m$ is the resultant lower triangular channel matrix for the m-th cell. The BD-GMD is a nonlinear ZF technique for the single-cell multiuser MIMO broadcast channel. Its MMSE variant, the block diagonal uniform channel decomposition, achieves the single-cell broadcast channel capacity. The BD-GMD has a lower complexity of $\mathcal{O}(UN_T^3)$ (Liu, Jiang, & He, 2010). The advantage of the BD-GMD processing is that all the users' data streams are decoupled, making it amenable to efficient power control techniques to satisfy user quality of service (QoS). For example, to ensure QoS for the users' real-time traffic, power allocation can be performed to minimize the BS transmit power subject to users' rate requirements (W. W. L. Ho & Liang, 2008). In this way, the receive power of the users are adjusted based on their channel gains to compensate for different transmission distances from the BS. When users have a single antenna each, the BD-GMD is equivalent to ZF-DPC. As there is some residual inter-cell interference after the leakage-based precoding, a performance margin for co-channel

Figure 1. Block diagram of the leakage-DPC MIMO downlink transmission that implements THPfor a cell with U users

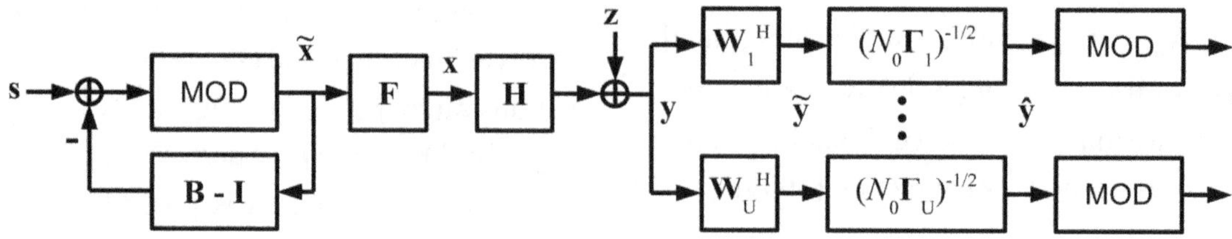

interference effects should be considered just as in conventional cellular network planning.

From (10), $\mathbf{W}_m^H \mathbf{H}_m \mathbf{Q}_m = \mathbf{L}_m$ follows directly from $\mathbf{H}_{m,\perp} \mathbf{Q}_m = \mathbf{H}_m \mathbf{Q}_m$ (W. W. L. Ho, Quek, & Sun, 2010). The transmit pre-equalization matrix is $\mathbf{F}_m = \mathbf{Q}_m \mathbf{\Omega}_m$ where $\mathbf{\Omega}_m$ is the diagonal power allocation matrix for the intra-cell users. The block diagonal receive beamforming matrix is $\mathbf{W}_m = \text{blkdiag}(\mathbf{W}_{m,1}, \mathbf{W}_{m,2}, \ldots, \mathbf{W}_{m,U})$. The block diagram for the transceiver that uses the leakage-DPC is shown in Figure 1. Tomlinson-Harashima precoding (THP) is used as a low-complexity suboptimal implementation of DPC, where $\mathbf{B}_m = \mathbf{\Omega}_m^{-1} diag(\mathbf{L}_m)^{-1} \mathbf{L}_m \mathbf{\Omega}_m$ is the monic lower triangular matrix for THP interference pre-subtraction. $\mathbf{\Gamma}_m = \frac{1}{N_0} diag(\mathbf{L}_m) \mathbf{\Omega}_m$ is the diagonal matrix of SNRs for each data stream in the m-th cell.

A short description of the THP interference pre-cancelation process is provided. Let the width of the $\bar{M}$-QAM constellation normalized to unit average energy be given by $\overline{\Delta}_{\bar{M}\text{-QAM}}$. The vector $\tilde{\mathbf{x}}$ can be evaluated from $\tilde{x}_1$ to $\tilde{x}_{Kd}$ using

$\tilde{x}_1 = s_1$

for $i = 2:1:Kd$

$$\tilde{x}_i = \text{mod}\left[s_i - \sum\nolimits_{j=1}^{i-1} [\mathbf{B}]_{i,j} \tilde{x}_j\right]$$

end

where "mod" represents the modulo operation over $\overline{\Delta}_{\bar{M}\text{-QAM}}$. Due to the THP processing, there is a slight increase in the average transmit power by a factor of $\bar{M} / (\bar{M} - 1)$ for $\bar{M}$-QAM symbols, called the *precoding loss*.

To perform the BD processing, express the matrices in (10) as follows:

$$\mathbf{W} = \begin{bmatrix} \mathbf{W}_1 & \mathbf{0} \\ \mathbf{0} & \breve{\mathbf{W}}_2 \end{bmatrix}, \ \mathbf{H} = \begin{bmatrix} \mathbf{H}_1 \\ \breve{\mathbf{H}}_2 \end{bmatrix}, \ \mathbf{Q} = \begin{bmatrix} \mathbf{Q}_1 & \breve{\mathbf{Q}}_2 \end{bmatrix}, \text{ and } \mathbf{L} = \begin{bmatrix} \mathbf{L}_1 & \mathbf{0} \\ \check{\mathbf{z}}_1 & \breve{\mathbf{L}}_2 \end{bmatrix}. \tag{11}$$

$\breve{\mathbf{W}}_k$, $\breve{\mathbf{H}}_k$, $\breve{\mathbf{Q}}_k$, and $\breve{\mathbf{L}}_k$ refer respectively to the receive beamforming, channel, transmit beamforming, and equivalent channel matrices of users k to K combined. For the first user, $\mathbf{W}_1^H \mathbf{H}_1 \mathbf{Q}_1 = \mathbf{L}_1$. SVD or GMD can be applied to the matrix $\mathbf{H}_1$ (Jiang, Li, & Hager, 2005). If subchannel selection is employed, the matrices $\mathbf{W}_1$ and $\mathbf{Q}_1$ would not be square. However, they always have orthonormal columns.

Next, to make sure that $\mathbf{W}_1^H \mathbf{H}_1 \breve{\mathbf{Q}}_2 = \mathbf{0}$, the projection matrix $\mathbf{I} - \mathbf{Q}_1 \mathbf{Q}_1^H$ is used on $\breve{\mathbf{H}}_2$. The equivalent channel is, therefore,

$$\breve{\mathbf{W}}_2^H \breve{\mathbf{H}}_2 (\mathbf{I} - \mathbf{Q}_1 \mathbf{Q}_1^H) \breve{\mathbf{Q}}_2 = \breve{\mathbf{L}}_2 .$$

Since this equation has the same form as (10), the algorithm to calculate the matrices can proceed recursively. This method is referred to as BD processing.

Algorithm 1. Leakage-DPC

1	Calculate $\mathbf{G}_{\mathrm{S}} = \left(\rho \bar{\mathbf{H}}_m{}^H \bar{\mathbf{H}}_m + U N_{\mathrm{R}} \mathbf{I}_{N_{\mathrm{T}}}\right)^{-1} \rho \mathbf{H}_m^H \mathbf{H}_m$.
2	Form the matrix $\mathbf{V} \in C^{N_{\mathrm{T}} \times N_{\mathrm{V}}}$ in which the columns are the orthonormal basis vectors of the space spanned by the first N_{V} dominant eigenvectors of $\mathbf{G}_{\mathrm{S}}$.
3	Pre-multiply the channel: $\mathbf{H}_{m,\perp} = \mathbf{H}_m \mathbf{V}\mathbf{V}^H$.
4	Apply the block diagonal processing on $\mathbf{H}_{m,\perp}$ to get $\mathbf{W}_m^H \mathbf{H}_{m,\perp} \mathbf{Q}_m = \mathbf{L}_m$ as in (10).
5	Perform DPC on the equivalent channel $\mathbf{L}_m$ with pre-equalization matrix $\mathbf{F}_m = \mathbf{Q}_m \boldsymbol{\Omega}_m$ and receive beamforming matrices $\mathbf{W}_{m,u}^H$ as in Figure 1.

Ξ_1 is given by $\Xi_1 = \breve{\mathbf{W}}_2^H \breve{\mathbf{H}}_2 \mathbf{Q}_1$. As an example, if there are 3 intra-cell users, the equivalent channel, $\mathbf{L}$, is given by

$$\mathbf{L} = \begin{bmatrix} \mathbf{L}_1 & \mathbf{0} & \\ \Xi_1 & \mathbf{L}_2 & \mathbf{0} \\ & \Xi_2 & \mathbf{L}_3 \end{bmatrix} \tag{12}$$

where $\Xi_2 = \mathbf{W}_3^H \mathbf{H}_3 \mathbf{Q}_2$.

The question of what value of N_{V} to use will now be addressed. It is clear that $N_{\mathrm{V}} \geq Ud$ in order to support the U users. Also, $N_{\mathrm{V}} \leq \min(N_{\mathrm{T}}, UN_{\mathrm{R}})$ due to the dimensions of the system. As N_{T} increases, it is better to use a larger value of N_{V}. For example, N_{V} may be chosen to be $N_{\mathrm{T}} - \bar{U}N_{\mathrm{R}}$ in order to have sufficient degrees of freedom (DoF) for suppressing the interference to the $\bar{U}N_{\mathrm{R}}$ antennas of the out-of-cell users. A heuristic value of N_{V} is thus given by

$$N_{\mathrm{Q}} = \min\left(\max\left(Ud, N_{\mathrm{T}} - \bar{U}N_{\mathrm{R}}\right), UN_{\mathrm{R}}, N_{\mathrm{T}}\right). \tag{13}$$

This value of $N_{\mathrm{V}} = N_{\mathrm{Q}}$ has been shown to work well in the simulations. The leakage-DPC method works best when the BS has sufficient transmit antennas, i.e. $\varphi_{\mathrm{a}} = N_{\mathrm{T}} - Ud \geq \varphi_{\mathrm{r}}$, where φ_{r} is the rank of $\bar{\mathbf{H}}_m$. Algorithm 1 summarizes the leakage-DPC strategy.

5.3.2 User Rates with Perfect CSI

Assuming $N_{\mathrm{T}} \geq U$ for data transmission to the U intra-cell users, denote $\tilde{\mathbf{y}}_m = \mathbf{W}_m^H \mathbf{y}_m$ as the (stacked) signal vector received by the users inside the m-th cell after equalization. The length of $\tilde{\mathbf{y}}_m$ is given by the total number of data streams of the intra-cell users, whereby each user can have multiple data streams. We can express $\tilde{\mathbf{y}}_m$ as a sum of the signal, interference, and noise terms

$$\tilde{\mathbf{y}}_m = \tilde{\mathbf{y}}_m^{\mathrm{sig}} + \tilde{\mathbf{y}}_m^{\mathrm{int}} + \tilde{\mathbf{z}}_m \tag{14}$$

where

$$\tilde{\mathbf{y}}_m^{\mathrm{sig}} = \mathbf{W}_m^H \mathbf{H}_m \mathbf{F}_m \tilde{\mathbf{x}}_m$$

$$\tilde{\mathbf{y}}_m^{\text{int}} = \sum_{n \neq m}^{M} \tilde{\mathbf{y}}_{n \to m}, \quad \tilde{\mathbf{y}}_{n \to m} = \mathbf{W}_m^H \mathbf{H}_{n \to m} \mathbf{F}_n \tilde{\mathbf{x}}_n$$

$$\tilde{\mathbf{z}}_m = \mathbf{W}_m^H \mathbf{z}_m.$$

$\tilde{\mathbf{x}}_m$ is the Tomlinson-Harashima precoded data symbol vector for the m-th BS. Due to independence of transmit signals from different BSs, the covariance matrix of the interference after equalization is

$$\mathbf{R}_{\tilde{\mathbf{y}}_m^{\text{int}}} = \sum_{n \neq m}^{M} \mathbf{R}_{\tilde{\mathbf{y}}_{n \to m}}. \tag{15}$$

After applying the leakage-DPC, the rate for each data stream is given by

$$r_i = \log_2 \left(1 + \frac{N_0 \left[\mathbf{\Gamma}_m \right]_{i,i}}{\left[\mathbf{R}_{\tilde{\mathbf{y}}_m^{\text{int}}} \right]_{i,i} + N_0} \right) \tag{16}$$

assuming that the transmitted signals and noise are Gaussian. The rate for each user can be obtained by summing up over the data streams of the user.

5.3.3 User Rates with Imperfect CSI

In this subsection, we analyze the effect of CSI uncertainty on the sum rate for the leakage-DPC method introduced earlier. The transmit and receive beamforming matrices $\mathbf{Q}_m$ and $\mathbf{W}_m$ are derived from $\hat{\mathbf{H}}_m$ because the BS processing is based on the CSI estimate. A signal model for THP with channel uncertainty and without out-of-cell interference can be found in the work of M. Huang, Zhou, & Wang (2008). Consider the case where there is out-of-cell interference as in a cellular system. The received signal in the m-th cell can be expressed as the sum of desired signal, error, interference, and noise components as follows:

$$\tilde{\mathbf{y}}_m = \tilde{\mathbf{y}}_m^{\text{sig}} + \tilde{\mathbf{y}}_m^{\text{err}} + \tilde{\mathbf{y}}_m^{\text{int}} + \tilde{\mathbf{z}}_m \tag{17}$$

where

$$\tilde{\mathbf{y}}_m^{\text{sig}} = \mathbf{W}_m^H \hat{\mathbf{H}}_m \mathbf{F}_m \tilde{\mathbf{x}}_m$$

$$\tilde{\mathbf{y}}_m^{\text{err}} = \mathbf{W}_m^H \mathbf{E}_m \mathbf{F}_m \tilde{\mathbf{x}}_m$$

$$\tilde{\mathbf{y}}_m^{\text{int}} = \sum_{n \neq m}^{M} \tilde{\mathbf{y}}_{n \to m}, \quad \tilde{\mathbf{y}}_{n \to m} = \mathbf{W}_m^H \mathbf{H}_{n \to m} \mathbf{F}_n \tilde{\mathbf{x}}_n$$

$$\tilde{\mathbf{z}}_m = \mathbf{W}_m^H \mathbf{z}_m.$$

In (17), $\mathbf{H}_{n \to m} \in C^{UN_R \times N_T}$ is the 'actual' channel matrix from the n-th BS to the users in the m-th cell. (This is the actual as opposed to estimated channel matrix.) $\mathbf{E}_m$ is the submatrix of the CSI uncertainty matrix $\mathbf{E}$ in (4) corresponding to the m-th cell, each element having variance σ_e^2. Due to the independence and zero mean of transmit signals from different BSs, $\mathbf{E}_m$, and $\tilde{\mathbf{z}}_m$,

$$\mathbf{R}_{\tilde{\mathbf{y}}_m} = \mathbf{R}_{\tilde{\mathbf{y}}_m^{\text{sig}}} + \mathbf{R}_{\tilde{\mathbf{y}}_m^{\text{err}}} + \mathbf{R}_{\tilde{\mathbf{y}}_m^{\text{int}}} + \mathbf{R}_{\tilde{\mathbf{z}}_m}. \tag{18}$$

The covariance matrix of the error component, $\tilde{\mathbf{y}}_m^{\text{err}}$, is given by

$$\begin{aligned} \mathbf{R}_{\tilde{\mathbf{y}}_m^{err}} &= E_E E_{\tilde{x}} \left[\mathbf{W}_m^H \mathbf{E}_m \mathbf{Q}_m \mathbf{\Omega}_m \tilde{\mathbf{x}}_m \tilde{\mathbf{x}}_m^H \mathbf{\Omega}_m \mathbf{Q}_m^H \mathbf{E}_m^H \mathbf{W}_m \right] \\ &= E_E \left[\mathbf{W}_m^H \mathbf{E}_m \mathbf{Q}_m \mathbf{\Omega}_m \mathbf{R}_{\tilde{\mathbf{x}}_m} \mathbf{\Omega}_m \mathbf{Q}_m^H \mathbf{E}_m^H \mathbf{W}_m \right] \\ &= \mathbf{W}_m^H \sigma_e^2 \mathrm{Tr} \left(\mathbf{Q}_m \mathbf{\Omega}_m \mathbf{R}_{\tilde{\mathbf{x}}_m} \mathbf{\Omega}_m \mathbf{Q}_m^H \right) \mathbf{I}_{UN_R} \mathbf{W}_m \\ &= \sigma_e^2 P_m \mathbf{I}_{Ud}. \end{aligned} \tag{19}$$

A lower bound on the achievable rate for each data stream is given by

$$r_{L,i} = \log_2\left(1 + \frac{N_0\left[\boldsymbol{\Gamma}_m\right]_{i,i}}{\left[\mathbf{R}_{\tilde{\mathbf{y}}_m^{\mathrm{err}}}\right]_{i,i} + \left[\mathbf{R}_{\tilde{\mathbf{y}}_m^{\mathrm{int}}}\right]_{i,i} + \left[\mathbf{R}_{\tilde{\mathbf{z}}_m}\right]_{i,i}}\right) = \log_2\left(1 + \frac{N_0\left[\boldsymbol{\Gamma}_m\right]_{i,i}}{\sigma_e^2 P_m + \left[\mathbf{R}_{\tilde{\mathbf{y}}_m^{\mathrm{int}}}\right]_{i,i} + N_0}\right) \quad (20)$$

assuming that the transmitted signals and noise are Gaussian. (Assuming Gaussian input, the worst case noise is Gaussian (Hassibi & Hochwald, 2003).) The rate per stream is calculated as in (20) due to the decoupling of the data streams by the block diagonal processing in (10). The lower bound rate for each user can be obtained by summing up over the data streams of the user.

5.3.4 Robust Version

The leakage-DPC is sensitive to channel uncertainty, thus a robust version that incorporates the variance of the uncertainty can be obtained. Assume σ_e^2 is known at the transmitter. The effective additive noise due to the receiver noise and the CSI error is $\bar{N}_0 = N_0 + \sigma_e^2 P$. Accounting for the CSI error, the following value of ρ is used.

$$\rho = P / \bar{N}_0. \quad (21)$$

Algorithm 1 is applied using the estimated values of $\mathbf{H}_m$, $\bar{\mathbf{H}}_m$, and the ρ given above. The lower bound rate can be calculated as in Section 3.3.

5.4 MULTICELL LEAKAGE SUPPRESSION

If low complexity is essential, it may be practical to use linear transmission techniques. In this section, a linear precoding method called multicell leakage suppression is introduced. Unlikeprior work, this approach allows for an arbitrary number of data streams per user and mitigates out-of-cell interference to boost the cell-edge users' rates.The work of Sadek et al. (2007) focuses on single-cell processing where each desired user's signal strength is maximized relative to all other users' signal and noise. This section introduces a more general precoder for the multicell setting. The prior work can be seen like a special case. A slight difference is that the columns of the beamformer in the paper of Sadek et al. (2007) are not orthogonal whereas they are orthogonal for the beamformer proposed here. The precoder in the paper of Lee et al. (2008) can also be seen as a special case where there is only 1 user per cell.

5.4.1 Algorithm Description

First consider a single data stream per user. Let the transmit power to each intra-cell user be P_m / U. The signal component transmitted by the m-th BS intended for its u-th user is given by $\mathbf{q}_{m,u}\sqrt{P_m / U}s_{m,u}$, where $\mathbf{q}_{m,u}$ is a unit norm beamforming vector. Considering this transmission alone, the desired signal received by this user is

$$\mathbf{y}_{m,u}^{\mathrm{sig}} = \mathbf{H}_{m,u}\mathbf{q}_{m,u}\sqrt{P_m / U}s_{m,u}. \quad (22)$$

The *leakage* directed away from this user is given by

$$\mathbf{y}_{m,u}^{\mathrm{leak}} = \bar{\mathbf{H}}_{m,u}\mathbf{q}_{m,u}\sqrt{P_m / U}s_{m,u}. \quad (23)$$

Algorithm 2. Multicell leakage suppression

1	**for** each intra-cell user i.e. $u = 1$ to U **do**
2	Form the matrix $\mathbf{G}_{m,u} = \left(\bar{\mathbf{H}}_{m,u}^H \bar{\mathbf{H}}_{m,u} + (UN_R N_0 / P_m)\mathbf{I}_{N_T}\right)^{-1} \mathbf{H}_{m,u}^H \mathbf{H}_{m,u}$.
3	Obtain $\mathbf{q}_{m,u} = \mathbf{q}_{m,u}^o$, the unit norm eigenvector corresponding to the maximum eigenvalue of $\mathbf{G}_{m,u}$.
4	**end for**
5	BS transmits $\mathbf{x} = \sum_{u=1}^{U} \mathbf{q}_{m,u} \sqrt{P_m / U} s_{m,u}$.

Recall from Section 2 that $\bar{\mathbf{H}}_{m,u}$ is the matrix concatenation of the intra-cell and out-of-cell channel gains. The signal-to-leakage-plus-noise ratio (SLNR) $\zeta_{m,u}$ is defined as

$$\zeta_{m,u} = \frac{\left\|\mathbf{y}_{m,u}^{\text{sig}}\right\|_2^2}{\left\|\mathbf{y}_{m,u}^{\text{leak}}\right\|_2^2 + N_R N_0} = \frac{\mathbf{q}_{m,u}^H \mathbf{H}_{m,u}^H \mathbf{H}_{m,u} \mathbf{q}_{m,u}}{\mathbf{q}_{m,u}^H \left(\bar{\mathbf{H}}_{m,u}^H \bar{\mathbf{H}}_{m,u} + (UN_R N_0 / P_m)\mathbf{I}_{N_T}\right)\mathbf{q}_{m,u}}. \quad (24)$$

The generalized Rayleigh quotient $\zeta_{m,u}$ is maximized when $\mathbf{q}_{m,u}$ is the generalized eigenvector corresponding to the maximum generalized eigenvalue of the matrix pencil

$$\left(\mathbf{H}_{m,u}^H \mathbf{H}_{m,u}, \bar{\mathbf{H}}_{m,u}^H \bar{\mathbf{H}}_{m,u} + (UN_R N_0 / P_m)\mathbf{I}_{N_T}\right). \quad (25)$$

Since the second matrix argument is invertible, the solution is given by $\mathbf{q}_{m,u} = \mathbf{q}_{m,u}^o$, where $\mathbf{q}_{m,u}^o$ is the unit norm eigenvector corresponding to the maximum eigenvalue of the matrix

$$\left(\bar{\mathbf{H}}_{m,u}^H \bar{\mathbf{H}}_{m,u} + (UN_R N_0 / P_m)\mathbf{I}_{N_T}\right)^{-1} \mathbf{H}_{m,u}^H \mathbf{H}_{m,u}. \quad (26)$$

The resulting beamforming vectors to each user reduce the total interference leakage to other intra-cell users as well as out-of-cell users. The extension to the case of multiple data streams per user can be obtained by applying a similar approach as described in Section 3.1. Algorithm 2 summarizes the multicell leakage suppression strategy.

5.4.2 User Rates with Perfect CSI

We now calculate the achievable rate for each user when applying the multicell leakage suppression. The signal received by the u-th user in the m-th cell is given by

$$\mathbf{y}_{m,u} = \mathbf{y}_{m,u}^{\text{sig}} + \mathbf{y}_{m,u}^{\text{int}} + \mathbf{z}_{m,u} \quad (27)$$

where $\mathbf{y}_{m,u}^{\text{sig}}$ is the desired signal (22) and $\mathbf{z}_{m,u} \sim \mathcal{CN}(\mathbf{0}, N_0 \mathbf{I}_{N_R})$ is the noise. The received interference is

$$\mathbf{y}_{m,u}^{\text{int}} = \mathbf{H}_{m,u} \sum_{k \neq u}^{U} \mathbf{q}_{m,k} \sqrt{P_m / U} s_{m,k} + \sum_{n \neq m}^{M} \mathbf{H}_{n \to m,u} \sum_{k=1}^{U} \mathbf{q}_{n,k} \sqrt{P_n / U} s_{n,k}. \quad (28)$$

The first summand is the intra-cell interference while the second summand is the out-of-cell interference. Using the notation $\mathbf{R}_{\mathbf{x}} = E[\mathbf{x}\mathbf{x}^H]$ as the covariance matrix of a given vector $\mathbf{x}$, the achievable rate for the u-th user in the m-th cell is given by

$$r_{m,u} = \log_2 \det\left(\mathbf{R}_{\mathbf{y}_{m,u}}\left(\mathbf{R}_{\mathbf{y}^{\text{int}}_{m,u}} + \mathbf{R}_{\mathbf{z}_{m,u}}\right)^{-1}\right) = \log_2 \det\left(\mathbf{R}_{\mathbf{y}_{m,u}}\left(\mathbf{R}_{\mathbf{y}_{m,u}} - \mathbf{R}_{\mathbf{y}^{\text{sig}}_{m,u}}\right)^{-1}\right). \quad (29)$$

5.4.3 User Rates with Imperfect CSI

We now derive the rate for the multicell leakage suppression method in the presence of CSI errors. Following the same channel uncertainty model in (4), the received signal of the u-th user in the m-th cell is given by

$$\mathbf{y}_{m,u} = \mathbf{y}^{\text{sig}}_{m,u} + \mathbf{y}^{\text{err}}_{m,u} + \mathbf{y}^{\text{int}}_{m,u} + \mathbf{z}_{m,u} \quad (30)$$

where

$$\mathbf{y}^{\text{sig}}_{m,u} = \hat{\mathbf{H}}_{m,u}\mathbf{q}_{m,u}\sqrt{P_m / U}s_{m,u}$$

$$\mathbf{y}^{\text{err}}_{m,u} = \mathbf{E}_{m,u}\mathbf{q}_{m,u}\sqrt{P_m / U}s_{m,u}$$

$$\mathbf{y}^{\text{int}}_{m,u} = \mathbf{H}_{m,u}\sum_{k\neq u}^{U}\mathbf{q}_{m,k}\sqrt{P_m / U}s_{m,k} + \sum_{n\neq m}^{M}\mathbf{H}_{n\to m,u}\sum_{k=1}^{U}\mathbf{q}_{n,k}\sqrt{P_n / U}s_{n,k}$$

and $\mathbf{z}_{m,u} \sim \mathcal{CN}(\mathbf{0}, N_0\mathbf{I}_{N_{\text{R}}})$ is the noise. The desired signal is $\mathbf{y}^{\text{sig}}_{m,u}$, the error component is $\mathbf{y}^{\text{err}}_{m,u}$, and the received interference is $\mathbf{y}^{\text{int}}_{m,u}$. $\mathbf{H}_{n\to m,u}$ is the actual channel from the n-th BS to the u-th user in the m-th cell. Due to the independence and zero mean of the transmit signals from different BSs, $\mathbf{E}_m$, and $\tilde{\mathbf{z}}_m$,

$$\mathbf{R}_{\mathbf{y}_{m,u}} = \mathbf{R}_{\mathbf{y}^{\text{sig}}_{m,u}} + \mathbf{R}_{\mathbf{y}^{\text{err}}_{m,u}} + \mathbf{R}_{\mathbf{y}^{\text{int}}_{m,u}} + \mathbf{R}_{\mathbf{z}_{m,u}}. \quad (31)$$

The covariance matrix of the error component $\mathbf{y}^{\text{err}}_{m,u}$ is given by

$$\begin{aligned}\mathbf{R}_{\mathbf{y}^{\text{err}}_{m,u}} &= E_{\mathbf{E}}E_{s_{m,u}}\left[(P_m / U)|s_{m,u}|^2\,\mathbf{E}_{m,u}\mathbf{q}_{m,u}\mathbf{q}^H_{m,u}\mathbf{E}^H_{m,u}\right] \\ &= E_{\mathbf{E}}\left[(P_m / U)\mathbf{E}_{m,u}\mathbf{q}_{m,u}\mathbf{q}^H_{m,u}\mathbf{E}^H_{m,u}\right] \\ &= \left(\sigma_{\text{e}}^2 P_m / U\right)\operatorname{Tr}\left(\mathbf{q}_{m,u}\mathbf{q}^H_{m,u}\right)\mathbf{I}_{N_{\text{R}}} \\ &= \left(\sigma_{\text{e}}^2 P_m / U\right)\mathbf{I}_{N_{\text{R}}}.\end{aligned} \quad (32)$$

A lower bound on the achievable rate for the u-th user in the m-th cell is obtained as

$$r_{\text{L},m,u} = \log_2 \det\left(\mathbf{R}_{\mathbf{y}_{m,u}}\left(\mathbf{R}_{\mathbf{y}^{\text{err}}_{m,u}} + \mathbf{R}_{\mathbf{y}^{\text{int}}_{m,u}} + \mathbf{R}_{\mathbf{z}_{m,u}}\right)^{-1}\right) \quad (33)$$

assuming that the transmitted signals and noise are Gaussian. (Assuming Gaussian input, the worst case noise is Gaussian (Hassibi & Hochwald, 2003).)

5.5 MULTICELL LEAKAGE SUPPRESSION

Consider a circular variant of the linear Wyner model (Wyner, 1994) depicted in Figure 2, in which $M = 4$ cells are arranged in a circle. The circular setup is homogenous and provides symmetry between the BSs. For a large number of cells, the circular setup and the more conventional linear setup are equivalent (Somekh et al., 2009; Jing et al., 2008). By using a 4-cell simulation and adjusting the parameters of interest, we

Figure 2. Circular Wyner model used in the simulations with M=4 cells and U=2 users per cell

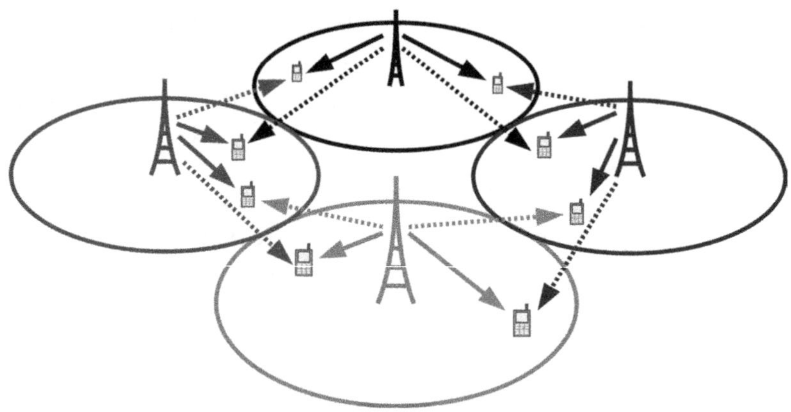

are able to concisely demonstrate the effect that physical phenomena such as relative channel gains have on these proposed precoders.

For each BS, the channel towards the intra-cell users is represented by the solid arrows and each element of the channel matrix $\mathbf{H}_m$ is i.i.d. as $\mathcal{CN}(0,1)$. The channel towards the out-of-cell users is represented by the dashed arrows. Each element of the channel matrix $\bar{\mathbf{H}}_m$ to these out-of-cell users is i.i.d. $\mathcal{CN}(0,\alpha^2)$ where α is the co-channel interference factor (Jing et al., 2008; Shamai (Shitz) & Zaidel, 2001). Each BS has N_T transmit antennas and each cell has U intra-cell users and $\bar{U} = U$ out-of-cell users. All users have N_R receive antennas each. The input-output relationship for the system is given in (1).

On the one hand, channel inversion power control enforces user fairness but may not be desirable from the service provider's point of view as it reduces the system sum rate. On the other hand, maximizing the sum rate via water-filling by loading even higher powers for the users with good channels may adversely affect the data rates for users with poor channel conditions. Therefore, in the simulations, equal power loading is utilized for all the downlink methods. For the purpose of comparison, ignore the THP precoding loss $\bar{M} / (\bar{M} - 1)$ for $\bar{M}$-QAM constellations. This will result in $\mathbf{R}_{\tilde{\mathbf{x}}} = \mathbf{I}_{Ud}$. In practice, $\mathbf{R}_{\tilde{\mathbf{x}}} \approx \mathbf{I}_{Ud}$ for large $\bar{M}$.

The simulation graphs include the curve for 'no cooperation' which is the non-cooperative BD-GMD scenario with single-cell processing. The curve for 'orthogonal transmission' refers to single-cell processing with a frequency reuse factor of $f_\mathrm{reuse} = 1/2$. For the 4-cell topology, 2 orthogonal channels are sufficient to create the idealized scenario of zero inter-cell interference. In frequency reuse, the rate becomes scaled by f_reuse. The graphs also show the curves for 'projected channel DPC'(W. W. L. Ho et al., 2010), which is a multiuser extension of the projected channel SVD (R. Zhang & Liang, 2008). The BS performs ZF nulling towards the out-of-cell users and applies block diagonal DPC to transmit to the intra-cell users (W. W. L. Ho et al., 2010). This precoding strategy is also decentralized. Unless otherwise indicated, the default system settings for the simulations are as follows. $N_\mathrm{T} = 8$, SNR $P / N_0 = 15\,\mathrm{dB}$, $\alpha = 0.5$, and $U = 4$. 1000 Monte Carlo runs are used. Table 1 compares the complexities of the schemes tested.

The BD-GMD is an efficient ZF-DPC technique with low complexity (Liu et al., 2010). However, multicell joint BD-GMD processing requires CSI of all the inter-cell links which may make it difficult for implementation due to the

Table 1. Complexities of the various schemes

	Processing per cell excluding BD-GMD			BD-GMD processing
	Matrix multiplications	**Matrix inversions**	**Other**	
No cooperation	0	0	-	single-cell
Orthogonal transmission	0	0	-	single-cell
Projected channel DPC	2	0	1 SVD	single-cell
Multicell leakage suppression	$3U$	U	U eigenvector evaluations	none
(Robust) Leakage-DPC	5	1	1 eigenvector matrix evaluation, 1 SVD	single-cell
Multicell joint DPC	0	0	-	multicell

backhaul requirement. Except for the multicell scheme, all other approaches are distributed so the CSI matrix sizes are reasonable, at most $(U+\bar{U})N_{\mathrm{R}}$ or N_{T} per dimension. Therefore all proposed techniques can be considered of low complexity.

In Figure 3, the sum rate per cell is plotted against the number of BS antennas N_{T}. When $N_{\mathrm{T}}=4$, the projected channel DPC and the leakage-DPC coincide with the case of no cooperation. This is because there are insufficient transmit antennas to do any projection of the channel. As N_{T} increases, all three methods as well as the no cooperation case improve. The rate improvement of the three methods over the no cooperation case also increases with N_{T} due to the increase in the DoF. Although orthogonal transmission has the highest sum rate for $N_{\mathrm{T}}=4$, the proposed methods obtain significantly higher rates for $N_{\mathrm{T}}\geq 6$.

The effect of SNR on the sum rate is shown in Figure 4. As long as there are sufficient transmit antennas, i.e. $N_{\mathrm{T}}\geq Ud+\bar{U}N_{\mathrm{R}}$, the sum rates per cell for the projected channel DPC and leakage-DPC are shown to increase linearly with SNR for high SNR. For the multicell leakage suppression, $N_{\mathrm{T}}\geq (U-1+\bar{U})N_{\mathrm{R}}+d$ is required for sum rate to increase linearly at high SNR. Otherwise, the sum rates will plateau due to insufficient DoF. The rate with no cooperation plateaus at about 12 bps/Hz. At low SNR, the system is noise-limited so the projected channel DPC performs poorly due to unnecessary channel projections. The multicell leakage suppression has a high sum rate as it takes noise variance into account when designing the beamforming vectors. The leakage-DPC approaches this curve at low SNR due to the precoding based on the SLNR. At high SNR, the system is interference-limited so the multicell leakage suppression has a relatively lower sum rate due to the linear processing. The leakage-DPC approaches the curve for the projected channel DPC at high SNR because the best strategy is to eliminate all interference and the leakage-DPC method tends to eliminate all the interference in that regime. Therefore, we see that the leakage-DPC method obtains the benefits of both the multicell leakage suppression and the projected channel DPC. There is an SNR gap of 7dB between multicell joint DPC and leakage-DPC at high SNR because multicell joint DPC makes use of out-of-cell channel gains by sharing data and CSI among BSs.

Next, in Figure 5, the performance of the three techniques are compared when the interference

Figure 3. Sum rate per cell vs number of BS transmit antennas N_T for $U=4, N_R=d=1$, SNR=15dB, and $\alpha = 0.5$

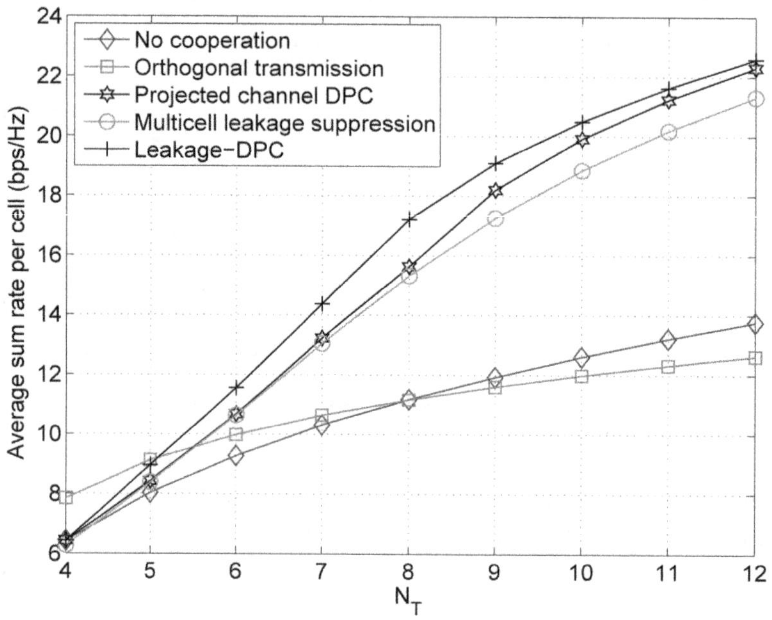

factor α is varied. As expected, increasing α does not degrade the performance of orthogonal transmission. The rate for projected channel DPC is also invariant to α because of the channel projection that completely nulls the interference to the out-of-cell users. All the other methods suffer rate loss as α increases. If the number of transmit antennas are not enough, the projected channel DPC will also suffer considerable rate loss. The multicell leakage suppression suffers more rate loss for high α as the system is interference-limited. The leakage-DPC strategy consistently performs better than all the other methods. When $\alpha = 0$, the leakage-DPC method is as good as the case with no cooperation. The reason is that the leakage term in the cell SLNR expression (7) is zero. Consequently, the channel after projection $\mathbf{H}_{\perp} = \mathbf{H}\mathbf{V}\mathbf{V}^H$ is identical to $\mathbf{H}$. As the interference factor α increases, the degradation due to interference becomes small for the decentralized precoders because the number of antennas at the BS is sufficient. Eventually, marginal increases in α have negligible effects on the decentralized strategies. The rate for multicell joint DPC is observed to increase without bounds as the interference factor increases because multicell joint DPC always benefits from increasing channel gains across cells. These gains may be unrealistic because joint processing demands exchange of data and CSI among BSs.

The effect of channel uncertainty on the sum rates of the proposed methods are examined in Figure 6. To model the CSI uncertainty (Yoo & Goldsmith, 2006), the following is used to generate the channels:

$$[\hat{\mathbf{H}}]_{i,j} = \sqrt{[\boldsymbol{\Psi}]_{i,j}(1-\sigma_e^2)}[\mathbf{H}_w]_{i,j}$$

$$[\mathbf{E}]_{i,j} = \sqrt{[\boldsymbol{\Psi}]_{i,j}\sigma_e^2}[\mathbf{E}_w]_{i,j}$$

$$\mathbf{H} = \hat{\mathbf{H}} + \mathbf{E} \qquad (34)$$

Figure 4. Sum rate as a function of SNR P/N_0 for N_T=8, $\alpha = 0.5$, U=2, and N_R=d=2. A similar trend can be seen for U=4, and N_R=d=1.

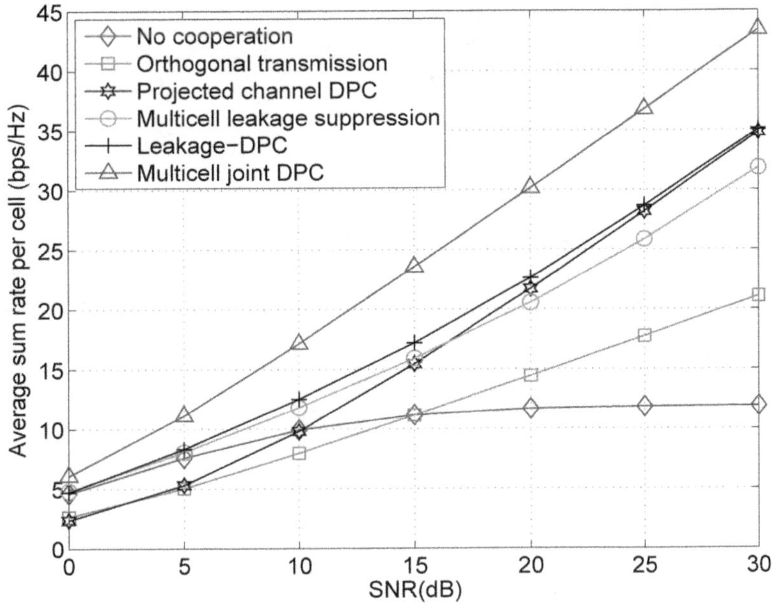

where $[\mathbf{H}_w]_{i,j} \sim \mathcal{CN}(0,1)$ and $[\mathbf{E}_w]_{i,j} \sim \mathcal{CN}(0,1)$. The variance of each channel element $[\mathbf{H}]_{i,j} \sim \mathcal{CN}(0,[\mathbf{\Psi}]_{i,j})$ is set to 1 for intra-cell links. For the curve of the robust leakage-DPC, the CSI error variance σ_e^2 is assumed to be available at the transmitter. It can be seen that the projected channel DPC is more sensitive to CSI error than the multicell leakage suppression due to the sensitivity of eigenspace projections. As the precoding of the leakage-DPC is based on the cell SLNR metric, it obtains a higher rate than these two methods. Also, it performs better than all other strategies for $\sigma_e^2 \le -7$ dB. The robust leakage-DPC is observed to recover some of the rate loss of the leakage-DPC method as σ_e^2 increases. The robust leakage-DPC provides 1 bps/Hz improvement over the leakage-DPC at high estimation errors $\sigma_e^2 = -5$ dB. It would be interesting to look at ways to find precoders that are more robust to CSI uncertainty.

Figure 7 plots the sum rate per cell versus the number of users per cell U. The proposed methods have better performance than the case of 'orthogonal transmission' when $1 \le U \le 4$ due to the ability of the BS to perform interference mitigation. For the proposed methods, there is a significant peak at $U = 4$. This is the point when the the number of BS antennas is just sufficient to suppress the interference to the outside-cell users' antennas and also to support the inside-cell users as $N_T = Ud + \bar{U}N_R = 8$. It is observed that these methods are not suitable for spatial multiplexing of more than 4 users as the BS does not have enough number of antennas to suppress the outside-cell interference. Therefore, care must be taken to ensure that the dimensionality constraint is satisfied. Conventional scheduling in time can be performed to enable the BS to support more users in its cell at the same sum rate when using the leakage-DPC strategy. In other words, the leakage-DPC could in practice outperform the 'orthogonal transmission' even when $U = 8$.

Figure 5. Effect of the interference factor α on the sum rate for N_T=8, U=4, N_R=d=1, and SNR=15dB

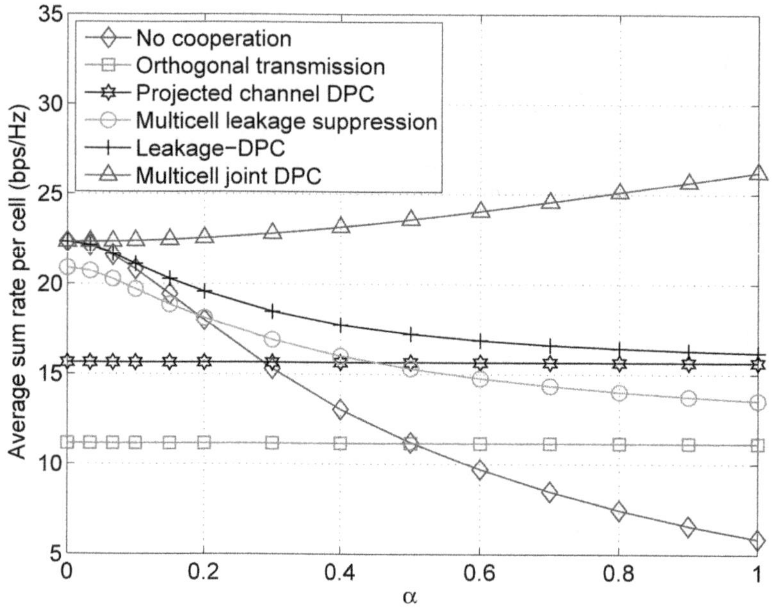

Figure 6. Effect of channel uncertainty on the sum rates of various methods for N_T=8, U=4, N_R=d=1, $\alpha = 0.5$, and SNR=15dB

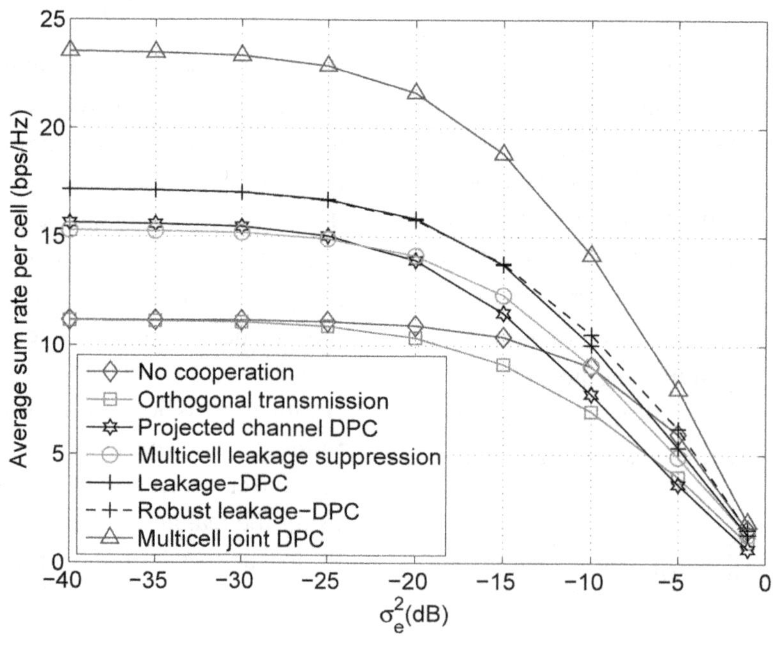

Figure 7. Performance when number of users per cell U ranges from 1 to 8, where N_T=8, N_R=d=1, SNR=15dB, and $\alpha = 0.5$

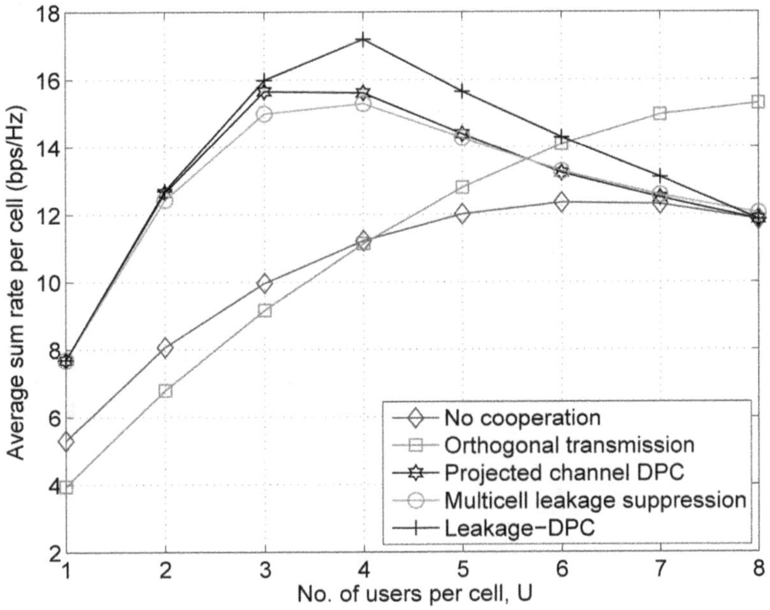

The circular Wyner model had been used to illustrate the effects of a few parameters. We now examine a more comprehensive $M = 19$ cell simulation using the "Urban Macro NLOS" model (3GPP, TR 36.814, n.d.). The cells are arranged hexagonally with inter-site distance of 500m. A log-distance path loss model with Rayleigh fading is used, with the path loss exponent $\eta = 3.9$. The path loss at the reference distance of 10m is 58.7dB. The minimum separation of the users from the BS is also 10m. The noise power density is taken to be -174dBm/Hz. Shadow fading is not considered. The transmit power of each BS is -24dBm/Hz. We also simulate a fully loaded hard fractional frequency reuse (FFR) system with 16 subcarriers, 7 of which are used for the inner band and 3 are used for the outer band in each cell(Lei, Zhang, Zhang, & Yang, 2007). The radius for the inner band is set as 150m. In all the schemes compared, there are $U = 2$ users per subcarrier per cell. Figure 8 shows the cumulative distribution function (CDF) of the user rates for a system with $N_T = 8$ and $N_R = d = 1$. The plot is obtained by evaluating user rates over 25000 independent channel sets. For the 50-th percentile rate, the leakage-DPC performs the best, followed by multicell leakage suppression, projected channel DPC (W. W. L. Ho et al., 2010) with $\varphi = 3$ dominant eigenmodes used for projection, FFR, 'no cooperation', 'frequency reuse 1/3', in this order. The 10-th percentile rate for FFR is 0.1bps/Hz higher than the leakage-DPC, indicating greater isolation of the cell-edge users from interference. The 90-th percentile rate for leakage-DPC is 3bps/Hz higher than FFR due to universal frequency reuse of the proposed method. The leakage-DPC has a higher rate than the projected channel DPC because the latter nulls the interference even though the out-of-cell channel may be very weak, but the leakage-DPC balances the projection according to the out-of-cell channel gains. The difference in rates between the 'no cooperation' and projected channel DPC decreases as the CDF increases from 10% to 90% because the rates are dominated by

Figure 8. Cumulative distribution function (CDF) of user rates for N_T=8, U=2, N_R=d=1

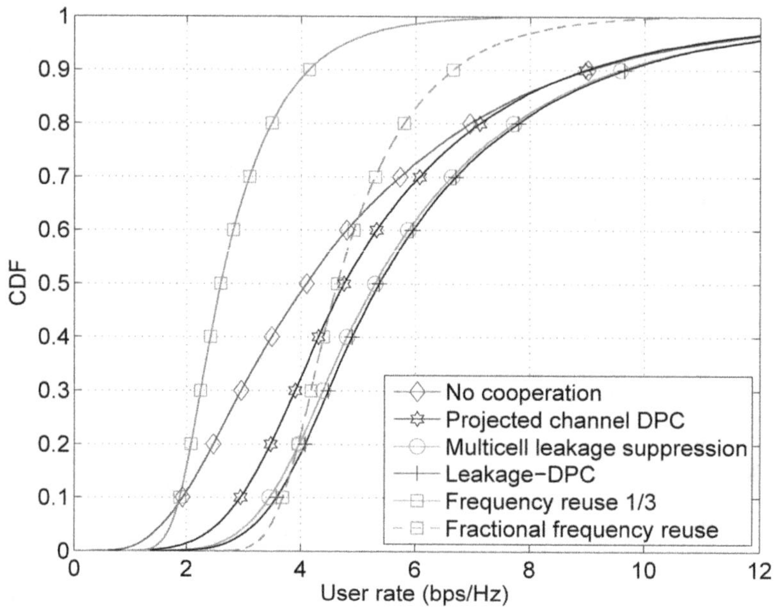

the cell-center users which observe minimal benefit of the interference avoidance technique.

5.6 DISCUSSION

Multicell or network MIMO is also known as coordinated multi-point transmission/reception (CoMP) (3GPP, TR 36.814, n.d.), which has been identified as one of the potential technologies for satisfying the quality of service (QoS) and rate requirements of the 3rd Generation Partnership Project (3GPP) Long Term Evolution (LTE) Release 10 / "LTE-Advanced" or future wireless standards (3GPP, TR 36.913, n.d.). This chapter introduces two techniques for network MIMO communication. The first is theleakage-DPC which is a nonlinear method that mitigates out-of-cell interference by applying a precoding based on the cell SLNR and block diagonal DPC for the intra-cell users. The second isthe multicell leakage suppression which is a linear method that performs beamforming to each user to maximize its SLNR. The leakage-DPC strategy attains higher sum rates than the multicell leakage suppression over varying SNR, interference conditions, and channel imperfectness. The higher spectral efficiency is because of the use of DPC. On the other hand, the multicell leakage suppression has a lower complexity than the leakage-DPC due to the use of linear transmit and receive processing.

The number of the BS antennas required for the proposed methods is expected to be larger than for conventional cellular systems because additional BS antennas are required for transmission to intra-cell users and interference mitigation for out-of-cell users. We see this requirement as reasonable due to the relaxed hardware and processing power constraints at the BSs. $\bar{U}$ out-of-cell users impose a requirement of $\bar{U}N_R$ more transmit antennas at the BS for interference suppression. However, this requirement can be reduced by considering the aggregate of the channel matrix and the out-of-cell users' receive beamforming. Thus, only $\bar{U}d$ additional transmit antennas would be required for interference mitigation. This method is beyond the scope of this chapter, as we assume that the receive matrices of out-of-cell users are unknown to the BS. It would be good to examine the practicality of such an implementation, with respect to feedback, channel estimation, and oscillation issues.

In this chapter, the techniques for uplink channel estimation or channel state feedback was not considered. Multi-user detection or limited feedback (Bhagavatula & Heath Jr., 2011) to reduce uplink overhead are also interesting topics for future work. While the precoders proposed in this chapter are suitable for same/single frequency networks, they are also applicable to cellular systems employing frequency reuse, especially those with smaller cell sizes which indicate high co-channel interference. In this case, each BS only needs to consider the users in the co-channel cells.

Conventional frequency reuse requires careful frequency planning to ensure reliable service. In FFR, further planning is also required to demarcate the handover regions. The FFR scheme is transparent to the standards and can be considered as an implementation issue (Giuliano, Monti, & Loreti, 2008). It involves the division of the spectrum to provide full frequency reuse for cell-center users and isolation of the cell-edge users from interference. According to the localized topology of each cell, the division of the spectrum, the power allocations, and the physical boundary between the inner and outer bands for hard/soft FFR need to be evaluated.In contrast, themethods described in this chapter are able to work with single frequency networks, thereby alleviating the need for frequency planning. While a well designed FFR scheme can result in greater isolation of cell-edge users from interference, the proposed precoders may be a simpler alternative for higher overall user rates.The 3GPP LTE-Advanced has specified to support as many as eight antennas at the BS, where these methods have already shown significant capacity gain. Furthermore, with even higher number of antennas, more gains can be achieved. In cases of insufficient transmit antennas, user scheduling can always be implemented together with these techniques.

The leakage-DPC and multicell leakage suppression methods need only the covariance matrix for the interfering channel and not that channel itself. For example, the leakage-DPC requires $\bar{\mathbf{H}}_m^H \bar{\mathbf{H}}_m$ instead of $\bar{\mathbf{H}}_m$, and the multicell leakage suppression requires $\bar{\mathbf{H}}_{m,u}^H \bar{\mathbf{H}}_{m,u}$ instead of $\bar{\mathbf{H}}_{m,u}$. Therefore these can be approximated at the BS assuming equal transmit powers for the users' uplink transmission when TDD is used, without the need for orthogonal channel sounding. The same applies to the 'projected channel DPC' (W. W. L. Ho et al., 2010). For FDD, to avoid the exchange of CSI over the backhaul, the BS may estimate the uplink channel covariance matrix and derive the downlink channel covariance matrix using a technique called frequency calibration (Liang & Chin, 2001; W. W. L. Ho & Liang, 2006). It would be interesting to explore this in future work.

5.7 CONCLUSION

To improve cellular system throughput, there is a trend towards a more aggressive reuse of frequencies and higher cell densities. Due to increased inter-cell interference, network MIMO or BS cooperation becomes an appealing solution to handle interference and raise spectral efficiencies. Most existing techniques involve coordination with heavy exchange of CSI or data over backhaul links. This chapter has presented a nonlinear and a linear distributed precoder for the multicell multiuser MIMO downlink. These precoders are designed to tackle both out-of-cell and intra-cell interference in a decentralized manner. Each BS constructs its own precoder without requiring downlink CSI from other BSs or data of out-of-cell users. This chapter studied the general scenario where each cell contains a multi-antenna BS that serves multiple users equipped with one or more antennas.

Simulations show that with sufficient transmit antennas, the precoders presented enjoy a linear rate growth with SNR similar to multicell joint DPC in the high SNR regime due to the interference mitigation. These methods also show significant improvements over orthogonal transmission for various interference scenarios provided there are sufficient transmit antennas at each BS relative to the number of users per cell. With high channel gains between cells, an increase in the co-channel interference factor α has diminishing negative effect on the sum rate performances of these precoders as they are able to mitigate the out-of-cell interference effectively. Furthermore, this chapter analyzed the rate of the precoders in the presence of CSI error. The robust version of the leakage-DPC was observed to retrieve some of the rate loss due to the CSI uncertainty.

REFERENCES

Bhagavatula, R., & Heath, R. W. Jr. (2011). Adaptive limited feedback for sum-rate maximizing beamforming in cooperative multicell systems. *IEEE Transactions on Signal Processing*, *59*(2), 800–811. doi:10.1109/TSP.2010.2090346

Björnson, E., & Ottersten, B. (2009). On the principles of multicell precoding with centralized and distributed Cooperation. *Proceedings of IEEE Wireless Communication Signal Processing (WCSP)*, (pp. 1-5).

Björnson, E., Zakhour, R., Gesbert, D., & Ottersten, B. (2009). Distributed multicell and multiantenna precoding: characterization and performance evaluation. *Proceedings of the IEEE Global Communication Conference (GLOBECOM)*, (pp. 1-6).

Boudreau, G., Panicker, J., Guo, N., Chang, R., Wang, N., & Vrzic, S. (2009). Interference coordination and cancellation for 4G networks. *IEEE Communications Magazine*, *47*(4), 74–81. doi:10.1109/MCOM.2009.4907410

Cai, D. W. H., Quek, T. Q. S., & Tan, C. W. (2011). Coordinated max-min SIR optimization in multicell downlink -duality and algorithm. *Proceedings of the IEEE International Conference on Communication (ICC)*, (pp. 1-6).

Cai, D. W. H., Quek, T. Q. S., Tan, C. W., & Low, S. (2012). Coordinated max-min SINR optimization in multicell downlink - Duality and algorithms. *IEEE Transactions on Signal Processing*, 60.

Chae, C.-B., Kim, S.-H., & Heath, R. W. Jr. (2009). Network coordinated beamforming for cell-boundary users: linear and nonlinear approaches. *IEEE Journal of Selected Topics in Signal Processing*, *3*(6), 1094–1105. doi:10.1109/JSTSP.2009.2035857

Foschini, G. J., Karakayali, K., & Valenzuela, R. A. (2006). Coordinating multiple antenna cellular networks to achieve enormous spectral efficiency. *IEE Proceedings. Communications*, *153*(4), 548–555. doi:10.1049/ip-com:20050423

Gesbert, D., Kiani, S. G., Gjendemsjø, A., & Øien, G. E. (2007). Adaptation, coordination, and distributed resource allocation in interference-limited wireless networks. *Proceedings of the IEEE*, *95*(12), 2393–2409. doi:10.1109/JPROC.2007.907125

Giuliano, R., Monti, C., & Loreti, P. (2008). WiMAX fractional frequency reuse for rural environments. *IEEE Wireless Communication Magazine*, *15*(3), 60–65. doi:10.1109/MWC.2008.4547524

3GPP, TR 36.913. (n.d.). *Requirements for further advancements for evolved universal terrestrial radio access (E-UTRA)*. Andrews, J. G., Choi, W., & Heath Jr., R. W. (2007). Overcoming interference in spatial multiplexing MIMO cellular networks. *IEEE Wireless Communications Magazine, 14*(6), 95-104.

3GPP, TR 36.814. (n.d.). *Further advancements for E-UTRA physical layer aspects*.

Hassibi, B., & Hochwald, B. M. (2003). How much training is needed in multiple-antenna wireless links? *IEEE Transactions on Information Theory, 49*(4), 951–963. doi:10.1109/TIT.2003.809594

Ho, W. W. L., & Liang, Y.-C. (2006). Precoder design for MIMO systems with transmit antenna correlations. *Proceedings of the IEEE Vehicular Technology Conference (VTC)*, (pp. 1382-1386).

Ho, W. W. L., & Liang, Y.-C. (2008). User ordering and subchannel selection for power minimization in MIMO broadcast channels using BD-GMD. *Proceedings of the IEEE Vehicular Technology Conference (VTC)*, (pp. 1-5).

Ho, W. W. L., Quek, T. Q. S., & Sun, S. (2010). Decentralized base station processing for multiuser MIMO downlink CoMP. *Proceedings of the IEEE Vehicular Technology Conference (VTC)*, (pp. 1-5).

Ho, Z. K. M., & Gesbert, D. (2010). Balancing egoism and altruism on interference channel: The MIMO case. *Proceedings of the IEEE International Conference on Communication (ICC)*, (pp. 1-5).

Ho, Z. K. M., Kaynia, M., & Gesbert, D. (2010). Distributed power control and beamforming on MIMO interference channels. *Proceedings of IEEE European Wireless.*

Horn, R. A., & Johnson, C. R. (1990). *Matrix analysis*. Cambridge University Press.

Huang, H., Trivellato, M., Hottinen, A., Shafi, M., Smith, P. J., & Valenzuela, R. (2009). Increasing downlink cellular throughput with limited network MIMO coordination. *IEEE Transactions on Wireless Communications, 8*(6), 2983–2989. doi:10.1109/TWC.2009.080179

Huang, M., Zhou, S., & Wang, J. (2008). Analysis of Tomlinson-Harashima precoding in multiuser MIMO systems with imperfect channel state information. *IEEE Transactions on Vehicular Technology, 57*(5), 2856–2867. doi:10.1109/TVT.2007.914051

Jiang, Y., Li, J., & Hager, W. W. (200). Joint transceiver design for MIMO communications using geometric mean decomposition. *IEEE Transactions on Signal Processing, 53*(10), 3791–3803. doi:10.1109/TSP.2005.855398

Jing, S., Tse, D. N. C., Soriaga, J. B., Hou, J., Smee, J. E., & Padovani, R. (2008). Multicell downlink capacity with coordinated processing. *EURASIP Journal on Wireless Communications and Networking, 2008*, 1–19. doi:10.1155/2008/586878

Kaviani, S., & Krzymien, W. A. (2008). Sum rate maximization of MIMO broadcast channels with coordination of base stations. *Proceedings of IEEE Wireless Communication and Networking Conference (WCNC)*, (pp. 1079-1084).

Larsson, E. G., Jorswieck, E. A., Lindblom, J., & Mochaourab, R. (2009). Game theory and the flat-fading Gaussian interference channel. *IEEE Signal Processing Magazine, 26*(5), 18–27. doi:10.1109/MSP.2009.933370

Lee, B. O., Je, H. W., Shin, O.-S., & Lee, K. B. (2009). A novel uplink MIMO transmission scheme in a multicell environment. *IEEE Transactions on Wireless Communications, 8*(10), 4981–4987. doi:10.1109/TWC.2009.081690

Lee, B. O., Je, H. W., Sohn, I., Shin, O.-S., & Lee, K. B. (2008). Interference-aware decentralized precoding for multicell MIMO TDD systems. *Proceedings of the IEEE Global Communication Conference (GLOBECOM)*, (pp. 1-5).

Lei, H., Zhang, L., Zhang, X., & Yang, D. (2007). A novel multi-cell OFDMA system structure using fractional frequency reuse. *Proceedings of the IEEE International Symposium on Personal, Indoor and Mobile Radio Communication (PIMRC)*, (pp. 1-5).

Liang, Y.-C., & Chin, F. P. S. (2001). Downlink channel covariance matrix (DCCM) estimation and its applications in wireless DS-CDMA systems. *IEEE Journal on Selected Areas in Communications, 19*(2), 222–232. doi:10.1109/49.914500

Lin, S., Ho, W. W. L., & Liang, Y.-C. (2008). Block diagonal geometric mean decomposition (BD-GMD) for MIMO broadcast channels. *IEEE Transactions on Wireless Communications, 7*(7), 2778–2789. doi:10.1109/TWC.2008.070171

Liu, F., Jiang, L., & He, C. (2010). Advanced joint transceiver design for block-diagonal geometric-mean-decomposition-based multiuser MIMO systems. *IEEE Transactions on Vehicular Technology, 59*(2), 692–703. doi:10.1109/TVT.2009.2032555

Marsch, P., & Fettweis, G. (2008). On multicell cooperative transmission in backhaul-constrained cellular systems. *Annals in Telecommunication, 63*(5), 253–269. doi:10.1007/s12243-008-0028-3

Ng, B. L., Evans, J. S., Hanly, S. V., & Aktas, D. (2008). Distributed downlink beamforming with cooperative base stations. *IEEE Transactions on Information Theory, 54*(12), 5491–5499. doi:10.1109/TIT.2008.2006426

Oteri, O., & Paulraj, A. (2008). Multicell optimization for diversity and interference mitigation. *IEEE Transactions on Signal Processing, 56*(5), 2050–2061. doi:10.1109/TSP.2007.909338

Sadek, M., Tarighat, A., & Sayed, A. H. (2007). A leakage-based precoding scheme for downlink multi-user MIMO channels. *IEEE Transactions on Wireless Communications, 6*(5), 1711–1721. doi:10.1109/TWC.2007.360373

Shamai (Shitz). S., & Zaidel, B. M. (2001). Enhancing the cellular downlink capacity via co-processing at the transmitting end. *Proceedings of IEEE Vehicular Technology Conference (VTC)*, (pp. 1745-1749).

Shim, S., Kwak, J. S., Heath, R. W. Jr, & Andrews, J. G. (2008). Block diagonalization for multi-user MIMO with other-cell interference. *IEEE Transactions on Wireless Communications, 7*(7), 2671–2681. doi:10.1109/TWC.2008.070093

Simeone, O., Somekh, O., & Poor, H. V., & Shamai (Shitz), S. (2009). Local base station cooperation via finite-capacity links for the uplink of linear cellular networks. *IEEE Transactions on Information Theory, 55*(1), 190–204. doi:10.1109/TIT.2008.2008151

Somekh, O., Simeone, O., Bar-Ness, Y., & Haimovich, A. M., & Shamai (Shitz), S. (2009). Cooperative multicell zero-forcing beamforming in cellular downlink channels. *IEEE Transactions on Information Theory, 55*(7), 3206–3219. doi:10.1109/TIT.2009.2021371

Tölli, A., Pennanen, H., & Komulainen, P. (2009). SINR balancing with coordinated multi-cell transmission. *Proceedings of IEEE Wireless Communication and Networking Conference (WCNC)*, (pp. 1-6).

Viswanathan, H., Venkatesan, S., & Huang, H. (2003). Downlink capacity evaluation of cellular networks with known-interference cancellation. *IEEE Journal on Selected Areas in Communications, 21*(5), 802–811. doi:10.1109/JSAC.2003.810346

Wyner, A. D. (1994). Shannon-theoretic approach to a Gaussian cellular multiple-access channel. *IEEE Transactions on Information Theory, 40*(6), 1713–1727. doi:10.1109/18.340450

Yoo, T., & Goldsmith, A. (2006). Capacity and power allocation for fading MIMO channels with channel estimation error. *IEEE Transactions on Information Theory, 52*(5), 2203–2214. doi:10.1109/TIT.2006.872984

Zakhour, R., & Gesbert, D. (2009). Coordination on the MISO interference channel using the virtual SINR framework. *Proceedings of the ITG Workshop on Smart Antennas (WSA)*, (pp. 75-81).

Zakhour, R., Ho, Z. K. M., & Gesbert, D. (2009). Distributed beamforming coordination in multicell MIMO channels. *Proceedings of the IEEE Vehicular Technology Conference (VTC)*, (pp. 1-5).

Zhang, H., & Dai, H. (2004). Cochannel interference mitigation and cooperative processing in downlink multicell multiuser MIMO networks. *EURASIP Journal on Wireless Communications and Networking, 2*, 222–235.

Zhang, H., Mehta, N. B., Molisch, A. F., Zhang, J., & Dai, H. (2008). Asynchronous interference mitigation in cooperative base station systems. *IEEE Transactions on Wireless Communications, 7*(1), 155–165. doi:10.1109/TWC.2008.060193

Zhang, R., & Liang, Y.-C. (2008). Exploiting multi-antennas for opportunistic spectrum sharing in cognitive radio networks. *IEEE J. Sel. Topics Signal Process., 2*(1), 88–102. doi:10.1109/JSTSP.2007.914894

Chapter 6
Application of Fuzzy Logic Power Control for Cognitive Radio Networks

Abbas Mohammed
Blekinge Institute of Technology, Sweden

Elisabeth Rakus-Andersson
Blekinge Institute of Technology, Sweden

Wlodek Kulesza
Blekinge Institute of Technology, Sweden

ABSTRACT

In this chapter, the authors present the application of fuzzy logic for power control in spectrum sharing cognitive radio network. The investigated network consists of a pair of primary users (PUs) and a pair of cognitive radio users (CRs). In order to compute an optimal power scale for the CR link, a fuzzy-based optimal power control strategy is investigated based on the Mamdani fuzzy control model using two input variables: the PU's signal-to-noise ratio (SNR) and PU's interference channel gain (channel gain from the cognitive radio users to primary users). Applying the fuzzy-based optimal power control strategy, the desired quality of service (QoS) is assured on the PU link, and the efficiency of spectrum usage is also maximized on the CR link in the spectrum sharing network. In addition, the bit error rate (BER) is also improved as compared with the spectrum sharing network without power control.

6.1 INTRODUCTION INTO THE NATURE OF FUZZY SET THEORY

We are accustomed to traditional tools of reasoning being strict and precise. In conventional binary logic a statement can be true or false, and there is no place for even a little uncertainty in this judgment. By studying the nature of sets, we can note that an element either belongs to a set or does not. We call these kinds of sets *crisp sets*. In practice we often experience those real situations that are represented by crisp sets, as impossible to describe accurately. If we assign a truth-value of 1 to the element that is included in the set, and a

DOI: 10.4018/978-1-4666-2005-6.ch006

truth-value comparable to 0 to such an element that lies outside the set, then we will create the range of two-valued logic. This sort of logic assumes that precise symbols must be employed, and it is therefore not applicable to the real existence but only to an imagined existence.

If we consider the characteristic features of real world systems, we will conclude that real situations are very often uncertain or vague in a number of ways. If the information demanded by a system is lacking, the future state of such a system may not be known completely. This type of uncertainty has been handled by probability theories and statistics, and it is called stochastic uncertainty. The vagueness, concerning the description of the semantic meaning of the events, phenomena, or statements themselves, is called fuzziness (Zadeh, 1965; Zimmermann, 2002).

Let us thus yield the description of the concept of a fuzzy set, whose creator is Professor Lotfi Zadeh from Berkeley University. The crisp set A, established in the universe $X = \{x\}$, is characterized by a membership function $\mu_A : X \rightarrow \{0,1\}$. It means that $x \in A$ is related to A with the strongest truth degree equaling 1. Otherwise, element x does not reveal any connection with A when its membership is determined as 0. In fuzzy set theory this property is generalized. Therefore, in fuzzy set A, it is not necessary that either $x \in A$ or $x \notin A$ (Buckley & Eslami, 2002; Lowen, 1996; Pedrycz & Gomide, 1998; Rakus-Andersson, 2007; Zadeh, 1965; Zimmermann, 2002).

The generalization is assumed as follows. According to the above statement for any crisp set A it is possible to define a membership function μ_A: $X \rightarrow \{0, 1\}$. In fuzzy set theory, the above membership function is generalized to a form that assigns every $x \in X$ a value from the unit interval [0, 1] instead of being assigned from the two-element set {0, 1}. The set A, defined on the basis of a generalized membership function, is called a fuzzy set. The membership function μ_A of fuzzy set $A \subset X$ is a function defined as μ_A: $X \rightarrow [0, 1]$. Every element $x \in X$ has thus *a membership degree* $y = \mu_A(x) \in [0, 1]$, which, via its value, provides information about strength of the relation between the element and the definition of the set. The fuzzy set A is finally completely determined by the set of pairs as $A = \left\{(x, y) = (x, \mu_A(x)), x \in X\right\}$. The important part of fuzzy set A is *a support* denoted by supp(A) and defined as a non-fuzzy set $\text{supp}(A) = \left\{x \in X : \mu_A(x) > 0\right\}$. As imprecision of information is inserted into fuzzy sets then supports of sets closely defined create non-empty intersections.

6.2 BACKGROUND AND PROBLEM STATEMENT

Efficient use of the available licensed radio spectrum is becoming increasingly difficult as the demand and usage of the radio spectrum increases. This usage of the spectrum is not uniform within the licensed band but concentrated in certain frequencies of the spectrum while other parts of the spectrum are inefficiently utilized.

Among the efforts taken, by regulators worldwide, in order to achieve better usage of spectrum is the promotion of secondary markets. In a secondary usage context, the spectrum owned by the license owner, also called primary user (PU), can be shared by a non-licensee referred to as a secondary or cognitive radio (CR) user. The term Cognitive Radio was first introduced by Mitola in 1999 (Mitola & Maguire, 1999).

In cognitive radio environments the primary users are allocated licensed frequency bands while secondary cognitive users can be dynamically allocated the empty frequencies within the licensed frequency band, according to their requested quality of service (QoS) specifications. In this chapter we are interested in the spectrum sharing mode of cognitive radio systems (Kang et. al., 2009). In spectrum sharing, the PU and CR links utilize spectrum simultaneously. The CR link utilizes spectrum with an appropriate power scale under a given interference constraint. Thus, the application of optimal power control strategies is an important research issue in spectrum sharing in order to protect the PU transmission (Kang et. al., 2009) and the application of fuzzy-base power control is the main focus of this chapter.

The authors in (Vu et. al., 2009a) proposed the concept of primary exclusive region and devised a relationship between the outage probability and primary exclusive region. The authors in (Cho & Andrew, 2009) investigated the concept of interference radius and formulated a relationship between the interference of the PU link and relative distance; according to this concept the CR can transmit with an opportunistic power scale depending on its relative distance to the PU link. The authors in (Vu & Tarokh, 2009b) introduced the power scaling law for single-hop cognitive radio networks. In (Kang et. al., 2009) and (Chen et. al., 2008), the authors proposed an opportunistic power control strategy for spectrum sharing CR networks where the CR can transmit with its peak power when the PU link is considered idle or robust. Otherwise, when the PU link is considered as sensitive to interference, the CR can transmit with a fraction of its peak power. Using this opportunistic power control strategy, the efficiency of spectrum usage is optimized and the target transmission rate is assured on the PU link. The authors in (Zhou et. al., 2008) proposed a probability-based power control strategy for dynamic spectrum access where the CR link only utilizes the spectrum when it is not utilized by the PU link. In essence this strategy optimizes the sensing and intersensing duration of the CR link in order to maximize the opportunistic spectrum usage. In (Le & Liang, 2007), the authors proposed a fuzzy-based power control scheme in which a CR is is able to opportunistically adjust its transmit power in response to the changes of the interference level to PU, the distance to PU and its received power difference at the base station (BS) while satisfying the requirement of sufficiently low interference to PU. The authors in (Le & Ly, 2008) proposed a fuzzy-based opportunistic spectrum access strategy employing Mamdani fuzzy control model where the CR link considers three input variables (spectrum utilization efficiency, degree of mobility and relative distance) and selects an optimal spectrum band to utilize.

In this chapter we present an optimal power control strategy in cognitive radio networks by considering the transition between transmission states. The transition can be viewed as the partial state in mathematical modeling and the authors propose to use fuzzy logic as an effective technique in this situation. Accordingly, we apply Mamdani fuzzy control model to differentiate between the transmission states in order to optimize the transmit power of CR link in the spectrum sharing network. With the fuzzy-based power control strategy, the desired QoS on the PU link is guaranteed and the efficiency of spectrum usage is also maximized on the CR link in the spectrum sharing network.

The organization of the remainder of this chapter is as follows. In section 6.3, we introduce the spectrum sharing system model. In section 6.4, the fuzzy-based power control strategy and system model are presented and analyzed. In section 6.5, the fuzzy optimal power control strategy is validated and the simulation results are compared to that without power control. Finally, the conclusions and directions for future work are drawn in section 6.6.

6.3 SYSTEM MODEL

In this chapter we consider a spectrum sharing cognitive radio network consisting of a pair of primary users (PUs) and a pair of cognitive radio users (CRs) as shown in Figure 1. The pair of PUs establishes a wireless link as the PU link. The pair of CRs establishes a wireless link as the CR link. The PU link and CR link utilize spectrum simultaneously with different priorities. The PU

Figure 1. The spectrum sharing cognitive radio network

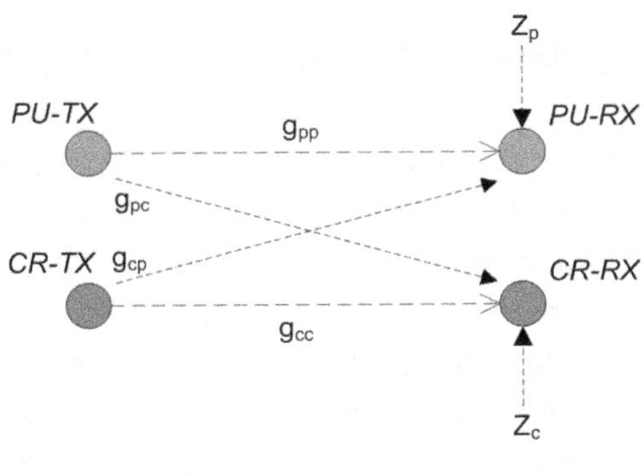

link has a higher priority to utilize spectrum with respect to the CR link. When the PU link utilizes spectrum, a desired quality of service (QoS) is specified and the CR utilizes spectrum with an optimal power scale under this constraint and assuring the desired QoS on the PU link.

In Figure 1, the PU link comprises a primary transmitter (PU-TX) and a primary receiver (PU-RX). The CR link also comprises a cognitive transmitter (CR-TX) and a cognitive receiver (CR-RX). Inside the spectrum sharing network, all channel state information (CSI) on each side is assumed to be available to the other. The additive noises at PU-RX and CR-RX are assumed to be independent circularly symmetric complex Gaussian random variables Z_p and Z_c, respectively, with a zero mean and variance of σ^2 All the instantaneous channel gains (g_{pp}, g_{pc}, g_{cc} and g_{cp}) are assumed to be independent random variables. The PU link has the priority to utilize the spectrum with respect to the CR link. When the PU link utilizes the spectrum, a desired QoS should be assured on the PU link. This imply the CR link should allocate an appropriate transmit power to guarantee the desired QoS on the PU link when both links utilize the spectrum simultaneously.

The signal-to-noise ratio (SNR) and signal-to-interference-plus-noise ratio (SINR) are used as performance measures to assess the QoS of the system (Vu & Tarokh, 2009b). Before spectrum sharing, the PU's SNR without the CR link is defined as

$$\alpha = \frac{P_p g_{pp}}{\sigma^2} \tag{1}$$

where P_p denotes the PU's transmit powers. During spectrum sharing, the PU's SINR with the CR link is given by

$$\beta = \frac{P_p g_{pp}}{P_c g_{cp} + \sigma^2} \tag{2}$$

where P_c are the CR's transmit powers. Assume that the CR has a peak power scale P_{cmax} and the power allocation is controlled by the peak power scale ratio K, then equation (2) can be written as

$$\beta(K) = \frac{P_p g_{pp}}{K P_{c\max} g_{cp} + \sigma^2}. \tag{3}$$

6.4 THE FUZZY POWER CONTROL POLICY

6.4.1 Power Control Principles for Spectrum Sharing

Assume the threshold for the desired QoS is β_t, so to guarantee this value the PU's SINR $\beta(K)$ should be greater than the threshold β_t during spectrum sharing. Thus, the CR should allocate a peak power scale ratio K that ensure the PU's SINR $\beta(K)$ is greater than threshold β_t (Kang et. al., 2009; Chen et. al., 2008). From equation (3) the value of K is influenced by the PU's SNR, α, and PU's interference channel gain, g_{cp}. Thus, the power control principles for spectrum sharing networks in the propagation environment can be structured from two different perspectives: the PU's SNR and PU's interference channel gain which represents the channel gain from the cognitive radio users to primary users.

The Perspective of PU's SNR

- When the PU's SNR is well below the threshold then the desired QoS is not assured. The PU link is already in an outage before spectrum sharing and the spectrum utilization of the CR link will not cause harmful influence on the PU link. Thus, the CR user can transmit with its peak power.
- When the PU's SNR is below but close to the threshold, then the desired QoS is not assured but is likely in a transition state, nearing to be assured. In this case, the PU link is sensitive to interference and spectrum utilization of the CR link will cause interference to the PU link. Thus, the CR user can transmit with a fraction of its peak power.
- When the PU's SNR is above but close to the threshold, then the desired QoS is just assured. The PU link is sensitive to interference and spectrum utilization of the CR link will downgrade the desired QoS. Thus, the CR user can transmit with a fraction of its peak power.
- When the PU's SNR is far above the threshold, then the desired QoS is highly assured. The PU link is robust to interference and spectrum utilization of the CR link will not degrade the desired QoS. Accordingly, the CR user can transmit with its peak power.

The Perspective of PU's Interference Channel Gain

When the PU's interference channel gain g_{cp} is high, then the received interference intensity caused by the CR link is high and the CR user should transmit with low power scale in this case. On the other hand, when the PU's interference channel gain g_{cp} is low, the received interference intensity caused by the CR link is low and thus the CR user can transmit with high power scale.

6.4.2 Mamdani Fuzzy Control Model

The tools of fuzzy set theory are particularly applicable to form data systems involving human experience. We thus determine IF...THEN logical rules, called knowledge-based rules, which allow computing with words instead of performing operations on numerical data; e.g., if we cannot mathematically formulate the dependency between the set of independent variables and the dependent variable then we will utilize the connections among verbally expressed levels of variables. The levels, in turn, are replaced by fuzzy sets. The communication between an expert, who wishes to solve his technical problem and a mathematician, is conducted by stating the verbal data at the former stage. At the latter steps the mathematician uses fuzzy techniques to find the solutions.

The fuzzy-based optimal power control strategy can be illustrated by the Mamdani fuzzy control model shown in Figure 2. The PU's SNR channel gain ratio α/β_t and PU's interference channel gain ratio g_{cp}/g_{cp_max} are chosen as the two input variables (fuzzy antecedents). The value of K is chosen as the output variable (fuzzy consequence). Each of these variables is translated into its fuzzy representation with membership functions.

6.4.3 Fuzzification

Fuzzy sets help us to state soft borders for such descriptions of levels as, e.g., "low", "medium" or "high", which cover parts of a reference set $[x_{min}, x_{max}]$ typical of the common variable "height". We are not able to set strict boundary values between levels mentioned above. We, instead, make a softer partition of the reference set, in which elements of $[x_{min}, x_{max}]$ can belong to more sets than one with different membership degrees. This effect of the sets' overlapping characterizes fuzziness. The

Figure 2. The Mamdani fuzzy control model

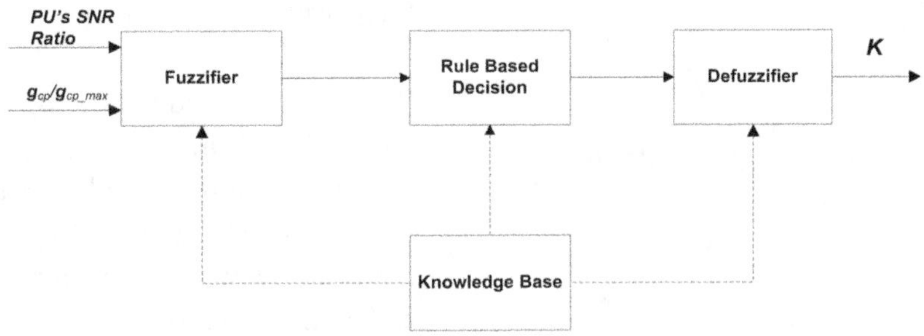

values of membership degrees in fuzzy sets are computed on the basis of the formulas of membership functions assigned to the sets.

The simplest shapes of membership functions consist of segments of straight lines. The membership function of the leftmost term "low" should start with a part $\mu_{"low"}(x) = 1$ since the first elements of [x_{min}, x_{max}] totally correspond to definition "low". After exceeding the first threshold of full certainty in "low" the rest of its elements should belong to "low" with descending values of membership degrees between 0 and 1 until we find the final threshold *x* possessing membership 0. The equation of the second part of "low" membership function is derived as an equation of the line passing through two characteristic points (first threshold, 1) and (final threshold, 0). The values of thresholds can be determined either experimentally or intuitionally. These are also generated by computer programs. We note that the second segment of the line $\mu_{"low"}(x) < 1$ introduces the uncertain relation between the value of *x*, *x* ∈ (first threshold, final threshold) and the description "low".

The membership function of the rightmost level "high" constitutes the inverted pattern of "low". The expression "medium" stands for the middle level; therefore its membership function can contain three segments. The first and the third linear parts reflect uncertainty, thus they have to satisfy the conditions $\mu_{"medium"}(x) < 1$. The part in the middle can be extended to a bit of line $\mu_{"medium"}(x) = 1$ or can contain only one *x* value assisted by membership 1. The first left branch of "medium" membership design is ascending to meet the middle part, whereas the third part starts at the end of the second part to go down to membership 0.

By remembering the general conditions referred to leftmost, in the middle and rightmost levels we can extend a list of expressions due to actual needs. We can also replace the designs of linear membership functions by more sophisticated patterns.

For the concrete problem of power control in cognitive radio nwtworks, in the fuzzification process the two input variables are used as the antecedents and the output variable is used as the consequence in the Mamidani fuzzy control. We formulate the two input variables and the output variable into different membeship functions as will be explained below. In this case, antecedent 1 is the PU's SNR channel gain ratio and antecedent 2 is the PU's interference channle gain ratio and the consequence is the CR's peak power scale ratio.

Antecedent 1: The PU's SNR Channel Gain Ratio

For the sake of simplicity, assume a preset value of PU's SNR, *α*, is given as a threshold for the desired QoS before spectrum sharing, then the ratio of PU's SNR, *α*, to β_t can be used to judge the transmission state of the PU link which is defined as

Figure 3. The membership functions for PU's SNR channel gain ratio

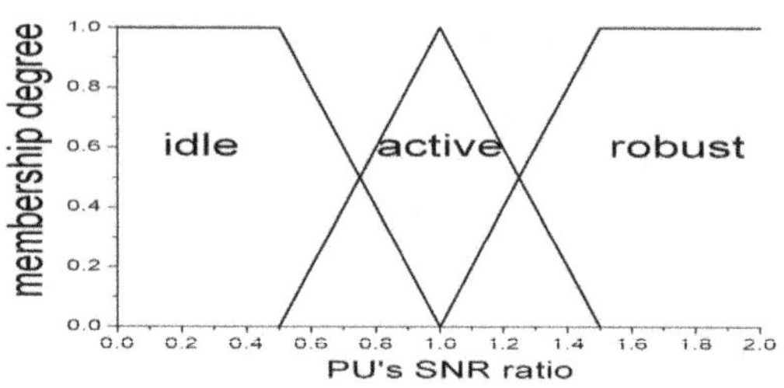

$$x = \frac{\alpha}{\beta_t} \tag{4}$$

From (4), we differentiate the PU's transmission states into various levels signified by the linguistic variable x as (X_1 = *idle*, X_2 = *active* and X_3 = *robust*) states defined by the limited membership functions in equations (5), (6), and (7), respectively.

$$\mu_{idle}(x) = \begin{cases} 1 & x < 0.5 \\ -2x + 2 & 0.5 \le x < 1 \\ 0 & x \ge 1 \end{cases} \tag{5}$$

$$\mu_{active}(x) = \begin{cases} 0 & x < 0.5 \\ 2x - 1 & 0.5 \le x < 1 \\ -2x + 3 & 1 \le x < 1.5 \\ 0 & x \ge 1.5 \end{cases} \tag{6}$$

$$\mu_{robust}(x) = \begin{cases} 0 & x < 1 \\ 2x - 2 & 1 \le x < 1.5 \\ 1 & x \ge 1.5 \end{cases} \tag{7}$$

These membership functions are plotted in Figure 3.

Antecedent 2: The PU's Interference Channel Gain Ratio

Similar to antecedent 1, suppose a predetermined value of PU's interference channel gain g_{cp_max} is specified as a threshold. The ratio of PU's interference channel gain g_{cp} to g_{cp_max} is used to judge the scale of PU's interference channel gain and so can be expressed as

$$y = \frac{g_{cp}}{g_{cp\_\max}} \tag{8}$$

Based on equation (8), we differentiate the PU's interference channel gain g_{cp} into various levels signified by the linguistic variable y as (Y_1 = *low*, Y_2 = *medium* and Y_3 = *high*) states defined by the limited membership functions in equations (9), (10), and (11), respectively.

$$\mu_{low}(y) = \begin{cases} -2y + 1 & 0 \le y < 0.5 \\ 0 & y \ge 0.5 \end{cases} \tag{9}$$

$$\mu_{medium}(y) = \begin{cases} 2y & 0 \le y < 0.5 \\ -2y + 2 & 0.5 \le y < 1 \end{cases} \tag{10}$$

$$\mu_{high}(y) = \begin{cases} 0 & y < 0.5 \\ 2y - 1 & 0.5 \le y < 1 \end{cases} \tag{11}$$

These membership functions are shown in Figure 4.

Consequence: CR's Peak Power Scale Ratio

The CR's peak power scale ratio is defined as

$$K = \frac{P_c}{P_{c\max}} \tag{12}$$

Figure 4. The membership functions for PU's interference channel gain ratio

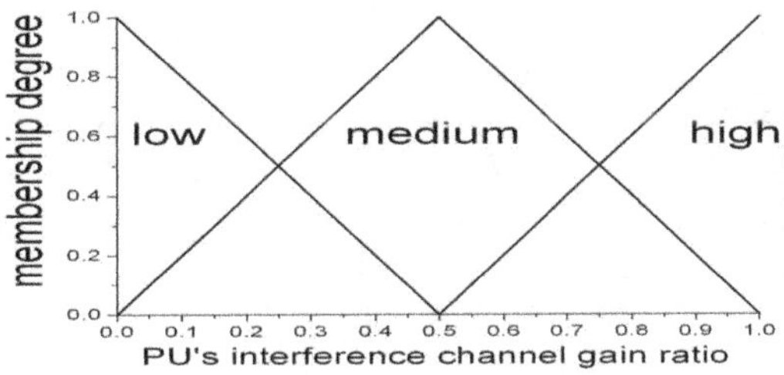

The CR's peak power scale ratio K is differentiated into three intensity levels as (K_1 = *low,* K_2 = *medium* and K_3 = *high*) states defined by the limited membership functions in equations (13), (14), and (15), respectively.

$$\mu_{low}(K) = \begin{cases} 1 & K < 0.2 \\ -5K + 2 & 0.2 \le K < 0.4 \\ 0 & K \ge 0.4 \end{cases} \quad (13)$$

$$\mu_{medium}(K) = \begin{cases} 0 & \text{K} < 0.2 \\ \text{5K-1} & 0.2 \le \text{K} < 0.4 \\ 1 & 0.4 \le \text{K} < 0.6 \\ \text{-5K} + 4 & 0.6 \le \text{K} < 0.8 \\ 0 & \text{K} \ge 0.8 \end{cases} \quad (14)$$

$$\mu_{high}(K) = \begin{cases} 0 & K < 0.6 \\ 5K - 3 & 0.6 \le K < 0.8 \\ 1 & K \ge 0.8 \end{cases} \quad (15)$$

These membership functions are illustrated in Figure 5.

6.4.4 Fuzzy Control Rules

Following the proposed power control principles for the spectrum sharing cognitive radio network, the fuzzy control rules are established as shown in Table 1.

Let us now state the mathematical basis of the Mamdani controller by listing the steps of the algorithm (Mamdani & Assilian, 1973; Rakus-Andersson, Zerrervall & Forrssell, 2011; Zerrervall, Rakus-Andersson & Forrssell, 2011; Zerrervall, 2011; Rakus-Andersson, 2012).

Suppose that we would like to make the K prognosis for an arbitrary pair $\left(x, y\right)$ – with other words we want to evaluate $K = f\left(x, y\right)$ when assuming that the *f*-formula is not developed.

Furthermore, x often belongs to more than one fuzzy sets X_i, i = 1,2,3, with different membership degrees equaling $\mu_{X_i}\left(x\right)$. Element y associated to x can be a member of some fuzzy sets Y_j, j = 1,2.3, in which it takes membership degrees $\mu_{Y_j}\left(y\right)$.

By means of IF-THEN statements grounded on the basis of Table 1, we can determine the contents of rules by attaching the pair of input variable levels to a level of the output variable according to

$$\text{Rule } R_{(x,y):r} : \text{If } x \text{ is } X_{i:r} \text{ and } y \text{ is } Y_{j:r}, \text{ then } K \text{ is } K_{k:r}, \quad (16)$$

where r is the rule number. The expressions $X_{i:r}$, $Y_{j:r}$ and $K_{k:r}$ denote the fuzzy sets X_i, Y_j and K_k, i, j, k = 1,2,3, assisting rule number r. $R_{(x,y):r}$ is determined for actual x and y.

To evaluate the influence of the input variables on the output consequences we need an estimate

Figure 5. The membership fucntions for CR's peak power scale ratio

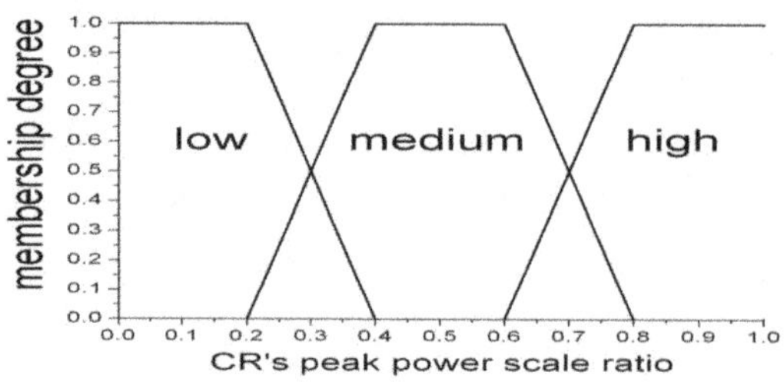

$\alpha_{(x,y):r}$ computed by performing the minimum operation

$$\alpha_{(x,y):r} = \min\left(\mu_{X_i:r}\left(x\right), \mu_{Y_j:r}\left(y\right)\right) \qquad (17)$$

for each $X_{i:r}$ and $Y_{j:r}$ concerning the choice of (x,y).

We use $\alpha_{(x,y):r}$ and the minimum operator to determine consequences of all rules $R_{(x,y):r}$ for the output. Fuzzy sets $R_{(x,y):r}^{conseq}$, stated in the output space K, will have the membership functions

$$\mu_{R_{(x,y):r}^{conseq}}\left(z\right) = \min\left(\alpha_{(x,y):r}, \mu_{K_k:r}\left(K\right)\right) \qquad (18)$$

In the last step of the processing part we aggregate the consequence sets $R_{(x,y):r}^{conseq}$ in one common set $conseq_{(x,y)}$ allocated in K over a continuous interval $[K_0, K_n]$. To derive the membership function of $conseq_{(x,y)}$ we prove the action of the maximum operator in the form of

$$\mu_{conseq_{(x,y)}}\left(z\right) = \max_r\left(\mu_{R_{(x,y):r}^{conseq}}\left(K\right)\right) \qquad (19)$$

In order to assign a crisp value K to the selected pair (x, y) we defuzzify the consequence fuzzy set in K.

As a defuzzification rule we select the centre of gravity method (COG). This model of computing is clearly interpretable as seen in Ex-

Table 1. The fuzzy control rules for the spectrum sharing cognitive radio network

Rule	α/β_t	g_{cp}/g_{cp_max}	K
1	idle	low	high
2	idle	medium	high
3	idle	high	high
4	active	low	high
5	active	medium	medium
6	active	high	low
7	robust	low	high
8	robust	medium	high
9	robust	high	high

Exhibit 1.

$$K = f\left(x, y\right) = \frac{\int_{K_0}^{K_n} K \cdot \mu_{conseq_{(x,y)}}\left(K\right) dK}{\int_{K_0}^{K_n} \mu_{conseq_{(x,y)}}\left(K\right) dK} = \frac{\int_{K_0}^{K_1} K \cdot \mu_{conseq_{(x,y)}}\left(K\right) dK + \cdots + \int_{K_{n-1}}^{K_n} K \cdot \mu_{conseq_{(x,y)}}\left(K\right) dK}{\int_{K_0}^{K_1} \mu_{conseq_{(x,y)}}\left(K\right) dK + \cdots + \int_{K_{n-1}}^{K_n} \mu_{conseq_{(x,y)}}\left(K\right) dK} \quad (20)$$

hibit 1. We expand COG as with the inner borders $K_1,\ldots,K_{n-1}$ being either K-coordinates of intersection points between adjacent branches of the $conseq_{(x,y)}$ membership function or characteristic support values of fuzzy sets included in $conseq_{(x,y)}$.

From Table 1, we can state the following:

- The PU's SNR α/β_t level is categorized into three transmission states: idle, active and robust;
- The PU's interference channel gain ratio g_{cp}/g_{cp_max} is classified into three scales: low, medium and high;
- The CR's peak power ratio K is divided into three power scales: low, medium and high.

When the PU's transmission state is recognized as idle (Rule 1-3) or robust (Rule 7-9), the CR user can transmit with its peak power. When the PU's transmission state is recognized as active, the CR user should transmit with a fraction of its peak power depending on the scale of the PU's interference channel gain ratio g_{cp}/g_{cp_max}. The relationship is outlined as follows:

- When the PU's SNR ratio α/β_t is active and the PU's interference channel gain ratio g_{cp}/g_{cp_max} is low, the CR user can transmit with high power (Rule 4).
- When the PU's SNR ratio α/β_t is active and the PU's interference channel gain ratio g_{cp}/g_{cp_max} is medium, the CR user can transmit with medium power (Rule 5).
- When the PU's SNR ratio α/β_t is active and the PU's interference channel gain ratio g_{cp}/g_{cp_max} is high, the CR user can transmit with low power (Rule 6).

6.4.5 Defuzzification

As stated earlier, COG method is used in the defuzzification process in order to compute a crisp value of CR's peak power scale ratio K, which is associated with the output consequence fuzzy set found for the crisp pair (x, y) as shown in Figure 6, where x denotes the PU's SNR ratio and y is the PU's interference channel gain ratio.

From Figure 6 we can note the following:

- When the PU's SNR α/β_t level is close to 0 (*idle state*), the value of K is always high no matter what the PU's interference channel gain ratio g_{cp}/g_{cp_max} is— close to 1 (*high*) or close to 0 (*low*).
- When the PU's SNR α/β_t level is close to 1 (*active state*), the value of K is turning to high as the PU's interference channel gain ratio g_{cp}/g_{cp_max} is turning to 0 (*PU's interference channel gain ratio is low*).
- When the PU's SNR α/β_t level is close to 2 (*robust state*), the value of K is always

Figure 6. The CR's peak power scale ratio as a function of PU's SNR and PU's interference channel gain ratio

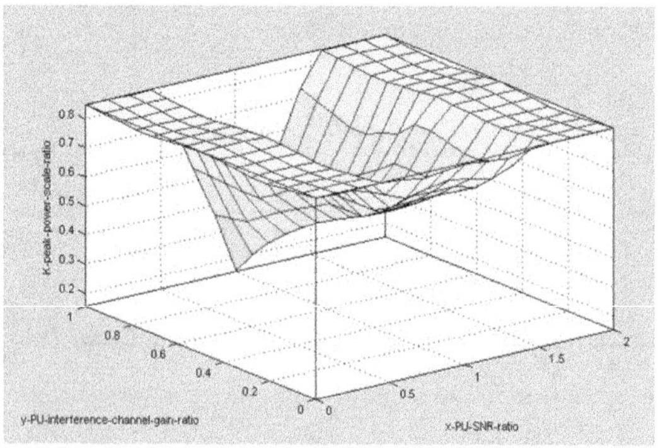

high no matter what the PU's interference channel gain ratio g_{cp}/g_{cp_max} is— close to 1 (*high*) or close to 0 (*low*).

6.5 PERFORMANCE EVALUATION

The investigated fuzzy-based optimal power control strategy presented in the previous sections is validated by simulations and the results are compared with spectrum sharing networks without power control strategy. The performance measure used is the bit error rate *(BER)* as a function of energy per bit to noise power spectral density *(E_b/N_0)* using 16-Quadrature Amplitude Modulation (16-QAM) format. The fuzzy-based optimal power control strategy is based on two input variables, the PU's SNR ratio α/β_t and PU's interference channel gain ratio g_{cp}/g_{cp_max}. In the verification model, we fix the variable of PU's interference channel gain ratio g_{cp}/g_{cp_max} at three distinct scales: low, medium and high, respectively, and let PU's SNR α/β_t level as the main variable.

The BER as a function of E_b/N_0 is plotted in Figures 7-9 for three different values of PU's interference channel gain ratio: low (g_{cp}/g_{cp_max} = 0.2), medium (g_{cp}/g_{cp_max} = 0.5) and high (g_{cp}/g_{cp_max} = 0.8), respectively. It can be clearly seen from these figures that the spectrum sharing network with the fuzzy-based optimal power control strategy has a lower BER compared to that without power control. It is also noticeable that the BER decreases as the E_b/N_0 increases. In addition, when the PU's interference channel gain is higher, the fuzzy-based optimal power control strategy has better performance since the BER is lower and simultaneously the QoS is assured on the PU link. The low computational complexity is another advantage of the fuzzy-based optimal power control strategy.

Figure 7. BER as a function of E_b/N_0 when PU's interference channel gain ratio is low

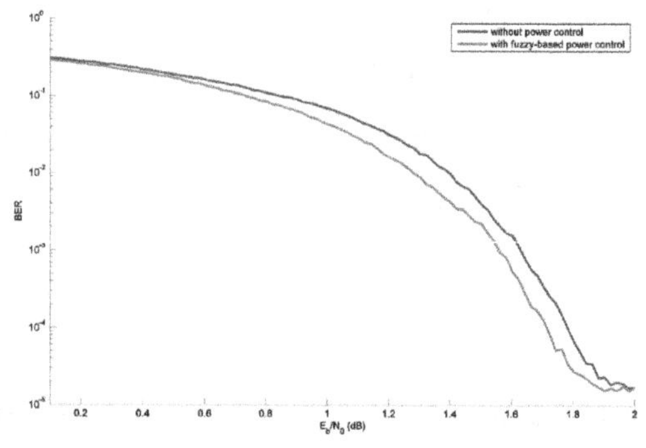

6.6 CONCLUSION

In this chapter we presented a fuzzy-based optimal power control strategy applied to spectrum sharing cognitive radio networks consisting of a pair of PU links and a pair of CR link and highlighted its advantages compared to that without power control. The CR can allocate an optimal power scale with a lower BER on the PU link as a function of PU's SNR and the interference channel gain, and simultaneously guarantee the QoS on the PU link. In addition, the fuzzy-based optimal power control has a low computational complexity. For future work, this or fuzzy-based power

Figure 8. BER as a function of E_b/N_0 when PU's interference channel gain ratio is medium

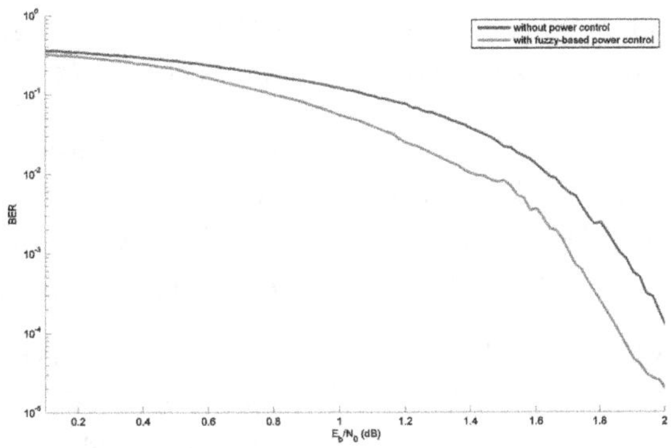

Figure 9. BER as a function of E_b/N_0 when PU's interference channel gain ratio is high

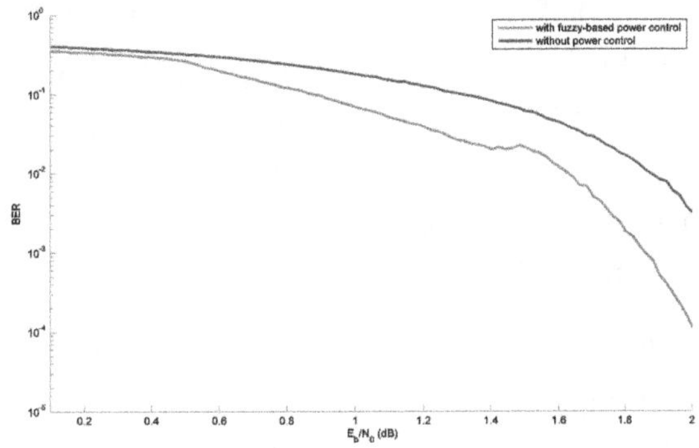

control strategy or a modified version of it can be extended to other fuzzy controllers, propagation environments and to mobile cooperative cognitive radio networks with multiple PU and CR links.

REFERENCES

Buckley, J., & Eslami, E. (2002). *An introduction to fuzzy logic and fuzzy sets*. Berlin, Germany: Springer-Verlag.

Chen, Y., Yu, G., Zhang, Z., Chen, H. H., & Qiu, P. (2008). On cognitive radio networks with opportunistic power control strategies in fading channels. *IEEE Transactions on Wireless Communications*, *7*(7), 2752–2761. doi:10.1109/TWC.2008.070145

Cho, H., & Andrews, J. G. (2009). Upper bound on the capacity of cognitive radio without cooperation. *IEEE Transactions on Wireless Communications*, *8*(9), 4380–4385. doi:10.1109/TWC.2009.071106

Kang, X., Zhang, R., Liang, Y. C., & Garg, H. K. (2009). Optimal power allocation for cognitive radio under primary user's outage loss constraint. *IEEE International Conference on Communications, ICC'09,* (pp. 1-5).

Le, H.-S. T., & Liang, Q. (2007). An efficient power control scheme for cognitive radios. *8th IEEE Wireless Communications and Networking Conference, WCNC 2007,* (pp. 2559-2563).

Le, H.-S. T., & Ly, H. D. (2008). Opportunistic spectrum access using fuzzy logic for cognitive radio networks. *2nd International Conference on Communications and Electronics, ICCE 2008*, (pp. 240-245).

Lowen, R. (1996). *Fuzzy set theory: Basic concepts, techniques and bibliography*. Dordrecht, The Netherlands: Kluwer Academic Publishers.

Mamdani, E. H., & Assilian, S. (1973). An experiment in linguistic synthesis with a fuzzy logic controller. *International Journal of Man-Machine Studies*, *7*, 1–13. doi:10.1016/S0020-7373(75)80002-2

Mitola, J., & Maguire, G. (1999). Cognitive radio: Making software radios more personal. *IEEE Personal Communications*, *6*(4), 13–18. doi:10.1109/98.788210

Pedrycz, W., & Gomide, F. (1998). *An introduction to fuzzy sets: Analysis and design*. MIT Press.

Rakus-Andersson, E. (2007). *Fuzzy and rough techniques. Medical diagnosis and medication*. Berlin, Germany: Springer-verlag.

Rakus-Andersson, E. (2011 in press). The Mamdani controller with modeled families of constraints in evaluation of cancer patient survival length. In Ramanna, S., Jain, L. C., & Howlett, R. J. (Eds.), *Emerging paradigms in machine learning*.

Rakus-Andersson, E., Zettervall, H., & Forssell, H. (2011). Fuzzy controllers in evaluation of survival length in cancer patients. In *Recent Advances in Fuzzy Sets, Intuitionistic Fuzzy Sets, Generalized Nets and Related Topics,* Vol. 2: Applications, Polish Academy of Sciences, System Research Institute, Warsaw, (pp. 203-222).

Vu, M., Devroye, N., & Tarokh, V. (2009a). On the primary exclusive region of cognitive networks. *IEEE Transactions on Wireless Communications*, *8*(7), 3380–3385. doi:10.1109/TWC.2009.080454

Vu, M., & Tarokh, V. (2009b). Scaling law of single-hop cognitive networks. *IEEE Transactions on Wireless Communications*, *8*(8), 4098–4097. doi:10.1109/TWC.2009.080453

Zadeh, L. A. (1965). Fuzzy sets. *Information and Control*, *8*, 338–353. doi:10.1016/S0019-9958(65)90241-X

Zettervall, H. (2011). *Fuzzy and rough theory in the treatment of elderly gastric cancer patients*. Licentiate Dissertation, Karlskrona, Sweden.

Zettervall, H., Rakus-Andersson, E., & Forssell, H. (2011). The Mamdani controller in prediction of the survival length in elderly gastric patients. *Proceedings of Bioinformatics 2011*, Rome, (pp. 283–286).

Zhou, X., Ma, J., Li, Y., Kwon, Y. H., Soong, A. C. K., & Zhao, G. (2008). *Probability based transmit power control for dynamic spectrum access*. IEEE 3rd Symposium on New Frontiers in Dynamic Spectrum Access Networks.

Zimmermann, H. J. (2002). *Fuzzy set theory and its applications* (4th ed.). Boston, MA: Kluwer Academic Publishers.

Chapter 7
Game Theory for Cognitive Radio

Hany Kamal
Rakedet, Egypt

Abbas Mohammed
Blekinge Institute of Technology, Sweden

ABSTRACT

Demands on high data rate applications are increasing and consequently demands on spectral resources are increasing as well. Although electromagnetic spectrum is recently said to be in a scarcity situation, several studies have shown that this scarcity is mainly due to the legacy command-and-control regulation rather than due to physical scarcity of spectrum. For this reason, researchers have started investigating techniques to better manage the usage of spectrum. Among these techniques there exist the methods allowing the primary/secondary usage of spectrum, or secondary market. Secondary market techniques mainly manage sensing, accessing, and aborting the spectrum usage by the secondary users. Techniques developed for secondary market context are also referred to as algorithms for Cognitive Radio (CR) networks. Regulators worldwide took measures to promote the deployment of primary/secondary context. In this chapter, the authors give an illustrative discussion on CR and on the application of game theory to overcome the spectrum scarcity problem. Game theory is a field of applied mathematics that describes and analyzes scenarios with interactive decisions. In recent years, there has been considerable interest in adopting game theoretic approaches to model many communications and networking problems such as radio resource management and routing. Nowadays, game theory is also used to model interactive situations for CR terminals.

7.1 INTRODUCTION

We start the chapter by giving a brief about the efforts carried out by regulators to introduce the primary/secondary context. We give an overview on game theory and we discuss the motivations for game theory usage in CR networks modeling. We illustrate the concept of "Nash equilibrium" and we discuss both the convergence and the efficiency of the equilibrium. We also present for the reader examples from the literature on game theory application in cognitive radio networks.

DOI: 10.4018/978-1-4666-2005-6.ch007

Several studies, initiated recently by the US regulator Federal Communications Commission (FCC), have shown that the frequency spectrum is underutilized and inefficiently exploited: some bands are highly crowded, at some day hours or in dense urban areas, while others remain poorly used. Regulators worldwide are beginning to recognize that the traditional way of managing the electromagnetic spectrum, called Fixed Spectrum Access (FSA), in which the licensing method of assigning fixed portions of spectrum, for very long periods, is inefficient (Liu, 2011).

Among the efforts taken, by regulators worldwide, in order to achieve better usage of spectrum is the introduction (promotion) of secondary markets. In a secondary usage context, the spectrum owned by the license owner (also called primary user) can be shared by a non-licensee referred to as a secondary user. As an example of these efforts, FCC in USA proposes two models of spectrum leasing to promote the secondary markets. In the proposed models the constraints on the *license owner* and on the *duration of the license* are relaxed[1]. The two models are named De-facto transfer and spectrum manager. In the former model the lessee is responsible of reporting back to FCC the rules compliance with the original license terms. In the latter model, the primary owner is responsible of reporting to FCC.

Regulators in Europe have also taken measures to overcome the spectrum scarcity problem. In France, a secondary market has been recently allowed in the band 3.4–3.6 GHz, where two licenses have been awarded to deploy WiMAX based Wireless Local Loop (WLL) services. It is possible for the license owner to resell the whole license or a part of it. Reselling a part of the license may concern the geographical area, the frequency band, or the license duration. It is also possible for the license owner to keep the license while authorizing another entity to use the frequencies. In this case the primary user is responsible to fulfill the license obligations (Coupechux, Godlewski & Kumar, 2007). The British regulator, Ofcom, proposes a similar framework named spectrum trading. The main difference from the French framework is that the spectrum trading can be applied to several bands, such as cellular, Private Mobile Radio (PMR), or WLL, as given in (OFCOM, 2002).

Besides the promotion for secondary markets, we are currently experiencing rapid evolutions of Software Defined Radio (SDR) techniques. Such techniques allow reconfigurable wireless transceivers to change their transmission/reception parameters, such as the operating frequency that can be modified over a very wide band, according to the network or users' demands. The efforts taken by regulators in order to make better usage of spectrum, in particular the promotion for secondary market, together with the rapid evolution of the SDR techniques, have led to the development of opportunistic Cognitive Radio (CR) systems.

The term Cognitive Radio (CR) was first introduced by Mitola in 1999, and it was defined as: "*The point in which wireless personal digital assistants (PDAs) and the related networks are sufficiently computationally intelligent about radio resources and related computer-to-computer communications to detect user communications needs as a function of use context and to provide radio resources and wireless services most appropriate to those needs.*" (Mitola & Maguire, 1999). CR generally refers to a radio device that has the ability to sense its Radio Frequency (RF) environment and modify its spectrum usage based on what it detects (Sherman et. al., 2008).

7.1.1 Game Theory

Game theory is a decision-making mathematical tool that studies and analyzes interactions among decision makers, whether in conflict-of-interests, or in cooperation situations. It is a tool that provides outcome expectations in complex interactive situations. Game theory was first introduced by John Von Neumann and Morgenstern in 1944 upon publishing "*The Theory of Games and Economic*

Behaviour," however modern game theory has started with the papers of John Nash and with the definition of *equilibrium* in 1950. The theory was extensively and exclusively applied to economics before being applied to other fields such as politics, sociology and also biology where competition between species is modeled.

Game theory has demonstrated its ability to model situations in which participants interact and affect each others' outcomes. In contrast to the optimization perspective; the optimization process is about optimizing an objective function, as seen by one of the participants, independently without considering the reactions of the other users. Nevertheless, interactions among the participants have their influence on the outcome of each of them, and after all other participants made their actions (Srivastava et. al., 2005). Mainly due to this property, game theory was found to be appealing when it comes to networks and telecommunications. It has been applied to resource management, for example to admission control, power control, rate adaptation, congestion control. It has also been applied to routing and load balancing problems.

The fundamental unit in game theory is the definition of the interactive situation, referred to as a *game*. A game is defined by three sets of parameters, i.e. set of finite decision makers or *players*, set of actions or *strategies* available for each player and the final set is the expected outcome, said *payoff*, obtained by each player upon taking a certain decision (action). The payoff obtained by a player does not only depend on the action performed by the player herself, but also depends on the actions performed by the other players (usually considered as opponents) in the game. The payoff is defined using an objective function, or *utility function,* that represents the performance metrics, such as throughput, delay, or capacity, by modeling and taking into account the environment of the game. The decision maker (or the player) is not necessary an individual. The player could be a group of individuals who are having a common interest and are making decision together. As an example, we can imagine a file-sharing situation between a pair of nodes who are competing for resources in a large network. In this case the player can be considered to be the pair of nodes, not an individual node.

Games are divided into two main categories, i.e. cooperative and non-cooperative games. *Cooperative* games require a sort of coordination among the participants. In telecommunications filed, this coordination is translated into signaling protocols. Such protocols add complexity when it comes to the implementation. *Non-Cooperative* games make the most used category of game theory. Non-cooperative games model the conflict of interests among the players who are considered to be rational and selfish in the sense that each of them is mainly interested in maximizing her payoff.

7.1.2 Motivation for Cognitive Radio Modeling using Game Theory

CR networks have characteristics that made them particular and made their studies challenging;

- The distributed and dynamic nature of CR nodes
- The nodes' behavior; cooperative, non-cooperative, selfish or rational.

We briefly discuss below these characteristics. In a centralized network approach a central entity manages the network by setting the appropriate parameters of operation to all nodes of the network, with the target of optimizing the network performance, such as capacity, throughput, or delay. In a centralized architecture, the optimization process usually has a global point of view for the social benefit, where the interaction among the different nodes is hardly modeled. On the other hand, in a distributed approach the central entity does not exist and the nodes take their own decision independently. A global optimization is hard to obtain, where each node acts selfishly trying to maximize its own benefits.

CR networks are characterized by their distributed architecture and by the dynamicity of the CR nodes that are capable of taking decisions as well as adapting the operating parameters, such as power level or frequency of transmission, according to the surrounding environment (e.g., perceived interference level, capacity requirement at a certain instant, etc.). This situation involves high level of interaction among the nodes. Talking for example about the dynamic spectrum sharing issue; with the introduction of secondary markets, primary/secondary access context is usually referred to as CR networks. In these networks the secondary users whether perform an opportunistic access to the spectrum or negotiate the access to the spectrum. For both cases game theory provides appropriate analysis tool to tackle the dynamic spectrum sharing issue whether from a negotiation or from an opportunistic approach.

Game theory suits to the most the modeling and studying of the interactions (cooperation or conflict) among cognitive radio terminals, in particular the *non-cooperative* games category, where efficient distributed algorithms can be driven using local information. One of the advantages of game theory is that it can be used to provide incentives for players to cooperate and behave in a constructive manner.

In most of the studies dealing with CR networks, that made use of the game theoretical framework, the CR networks are modeled as games where radio terminals are considered to be the players, while the strategies of the players has varied to take several forms; for example in references (Etkin, Parekh & Tse, 2007) and (Giupponi & Ibars, 2009) the power level of the radio terminals is considered to be the players' strategies. In other studies, the choice of the spectrum band (channel) accessed by each terminal (Nie & Comaniciu, 2005), is considered to be the players' strategies. Though independently of the choice of the strategies' the utility function should represent the metric of interest to the players while taking into consideration the parameters affecting the payoff (interest) of the nodes, e.g., frequency, power, etc.

7.2 NASH EQUILIBRIUM: CONVERGENCE AND EFFICIENCY

7.2.1 Introduction

Formally a game G can be defined as, $G = (P, S, U)$, where P is the set of players, S is the strategy (action) set for each player, and U is the payoff (utility) obtained by each player given the strategy S. Rationally, each player i chooses a strategy $s_i \in S$ in order to maximize her utility u_i. Let $u_i(s_i, s_{-i})$ be the utility of player i given her strategy s_i, and the strategy s_{-i} of the opponent players. The strategy profile $s*$ is a strict Nash Equilibrium (NE) strategy if, for each player i,

$$u_i(s_i^*, s_{-i}^*) > u_i(s_i, s_{-i}^*), \forall s_i \in S_i, s_i \neq s_{-i}^* \quad (1)$$

In other words, the equilibrium point is the steady-state where none of the selfish players has an incentive to unilaterally deviate from her strategy. In case each player plays his *best-response* strategy against the opponent player. The best-response function $br_i(s_i)$ of player i to the opponents' strategies s_{-i} is denoted by,

$$br_i(s_i) = \max u_i(s_i, s_{-i}), \forall s_i \in S_i \quad (2)$$

In other words, it is the strategy s_i played by player i that maximizes her own payoff u_i knowing that the opponents play the strategy s_{-i}. We present below the most well-known game, the Prisoners Dilemma (PD).

The PD game tells the story of two arrested persons for a crime. Police put them in separated rooms for investigation. Each of them has two options, whether to "confess" the crime or to

Table 1. Outcomes of the PD game

		Prisoner 1 (P1)	
		Confess (C)	Remain Silent (RS)
Prisoner 2 (P2)	Confess (C)	P1: 3 and P2: 3	P1: 5 and P2: 1
	Remain Silent (RS)	P1: 1 and P2: 5	P1: 2 and P2: 2

Table 2. Normal form of the battle of the sexes (BoS) game

		Boy	
		Theatre (T)	Football game (F)
Girl	Theatre (T)	(10,8)	(0,0)
	Football game (F)	(0,0)	(8,10)

"remain silent." We use in this section the abbreviations (C) for "confess" and (RS) for "remain silent." Police authorities offer them a deal; if one prisoner confesses, while the other remains silent, the cooperating (with the authorities) person will be awarded by serving only one year of prison, while the other person will be sentenced five years of prison. If they both confess, each one will be sentenced 3 years in jail (instead of five). Nevertheless, the prisoners know that if they both remain silent, the police will not be able to prove charges against them, and hence each of them will be sentenced two years in jail. What should be the choices of the prisoners? Obviously, the PD game has four outputs. Table 1 summarizes the outputs (payoffs) of the game, represented by the number of years the prisoners will be served in jail according to their strategies. The way of representing a game in the form of a matrix, such as the case of Table 1, is called *normal form.* The *extensive form* is a second method of representing games, however, in a more graphical appearance.

According to the payoffs in Table 1, the strategy (C, C), where both the prisoners confess the crime to the authorities, is the unique stable strategy. That is because while adopting (C, C) strategy none of the players has the intensive to unilaterally change her strategy in order to obtain better payoff. As an example, considering the strategy (RS, RS), each player has the possibility to obtain better payoff if she changes her strategy from "remain silent" to "confess," while the other player remains silent. This change will allow the prisoner to serve one year in jail instead of two years. It is to be noted that the strategy (C, C) is Nash equilibrium (NE) for the PD game. As we can see from Table 1, the NE strategy is not the best choice for the prisoners. If both prisoners choose "remain silent" they are better off, on the other hand the (RS, RS) strategy is not stable. We highlight here that the strategy (RS, RS) is a Pareto Optimal (PO) for the PD game. We illustrate later in this chapter the definition of Pareto optimality.

Another well-known game is the Battle of the Sexes (BoS). In this game a couple decide to meet, but they are not sure about where to spend the evening. The girl prefers to go to the theatre while the boy prefers to go to the football game. Nevertheless, they both prefer to spend the evening together rather than spending it separately. Table 2 gives the payoff matrix of the BoS game.

We see from Table 2 that two strategies (T, F) and (F, T) are not interesting for the couple. We can also see that the other two options (T, T) and (F, F) are NE. It is to be noted that the BoS game is categorized among a type of games said to be *coordination games*.

The strategies (T, T), and (F, F) of the BoS game, and the strategy (C, C) of the PD game are called *pure* NE strategies. Pure NE is a deterministic NE where the players are choosing to play their strategies with probability 1. On the contrary, in *mixed* strategies, players choose to play a certain strategy (let's say strategy C in the PD game) with a certain probability (say probability q), while playing the other strategies (RS in the PD game) with different probability $(1 - q)$. The mixed strategies are characterized by a set of probabilities assigned to the pure strategies. All coordination games have mixed strategies NE.

Before using game theory as a tool to model and analyze CR networks (or model any other interactive situation), the reader needs to pay attention to the following inquiries:

- Does the game have a NE?
- Is the NE unique?
- Is the NE efficient?
- How long the players need to converge to NE?

Nash equilibrium concept is usually associated with convergence as well as efficiency notations. In this section we illustrate both notations, while giving examples from the literature.

7.2.2 Convergence

Few types of games are known to have Nash equilibrium such as supermodular games and potential games. A game is said to be a supermodular game if its utility function satisfies the following condition:

$$\frac{\partial^2 u_i(s)}{\partial^2 s_i \partial^2 s_j} \geq 0 \forall j \neq i \in P \tag{3}$$

Supermodular games with players deploying best-response algorithm do converge to pure NE (Topkis, 1998). This framework has been used to analyze power control issue in wireless networks (Saraydar, Mandayam & Goodman, 2002). The authors in (Saraydar, Mandayam & Goodman, 2002) have modelled the power control problem through the formulation of a non-cooperative game where the wireless terminals are the players. Each player (wireless terminal) aims at maximizing its utility. The utility function takes into account both the power level of the terminal and the perceived Signal to Interference Ratio (SIR) level. This way the payoff obtained by each terminal does not only depend on its chosen power level but also depends on the power level chosen by the other players. The players adopt *best-response* strategies such that the power level chosen by each player forms a best response to the power levels chosen by the opponent players.

A game is said to be an Exact Potential Game (EPG) if there exists a function F such that

$$u_i(s_i, s_{-i}) - u_i(s_i', s_{-i}) = F(s_i, s_{-i}) - F(s_i', s_{-i}), \forall s_i, s_i' \in S_i. \tag{4}$$

Giupponi and Ibars have formulated the channel and power allocation problem for secondary users into Bayesian potential game (Giupponi & Ibars, 2009). The adopted strategy of the secondary users, according to (Giupponi & Ibars, 2009), relies in the selection of both the frequency channel and the transmission power level in order to maximize their throughput, while not causing interference for the primary users. As the EPGs are shown to converge to pure Nash when best-response algorithm is deployed, the authors have demonstrated that the formulated Bayesian potential game converges to pure Bayesian equilibrium.

An important issue for the implementation of game-theoretical algorithms in CR terminals is the convergence to NE. The choice of the utility function has its major impact on the convergence issue. While the utility function should give an appropriate physical representation of the players' payoff, it also needs to have the mathematical properties that ensure convergence to NE.

In case the game has more than one NE point (NE is not unique), none of the equilibrium points dominate the others, and if the players are deploying best-response method, the game will not converge to a stable point, making the players' actions oscillate. As an example, we refer to the reference (Kamal, Coupechoux & Godlewski, 2009) where a non-zero sum game is formulated to tackle the Dynamic Spectrum Access (DSA) problem in cellular networks. In (Kamal, Coupechoux & Godlewski, 2009), cellular operators are considered to share a common pool of spectral resources in a specific region. The common pool is divided into spectrum blocks of equal size, and each cellular operator operates one radio network, e.g. one 2G network or one 3G network. The operators do not own the spectrum but rather share the common pool. The operators access the pool in a direct and simultaneous manner, according to the load variations of their radio networks. In the game formulation the operators are the players, and the strategy of each player is the number of spectrum blocks it accesses from the pool. It is shown in (Kamal, Coupechoux & Godlewski, 2009) that for specific arrival rate values (loads) the formulated game has more than one pure NE point where none of them dominates the others. Simulations show that in 62% of the cases the operators converge to NE (they oscillate in 38% of the cases), depending on the initial strategy of both players. The authors in (Kamal, Coupechoux & Godlewski, 2009) consider that the players while playing best-response strategies, and at each play instant, they consider to maximize their reward for the next play, only for the next play, without considering future rewards. The reader needs to note that in order to take into consideration the future rewards, *repeated game* model is to be used.

A *repeated game* (or a multi-stage repeated game) is a game in which the players have the possibility of playing several rounds (or trials). It is the repetition of a game. Each of the trails can be represented by the normal form or a matrix as given in Section 2.1. Repeated-game analysis is not the focus of this chapter; the reader interested in more details can consult the references (Richard & Anantharam, 2002) and (Hongyan, Yang & Dexian, 2010).

Figure 1 gives the simultaneous plays of the operators as given in (Kamal, Coupechoux & Godlewski, 2009) for a game with two NE points. The figure shows two different cases where the players have different initial strategies for each case. In one of the cases (on the left side of Figure 1) the operators converge to one of the NE points, illustrated by the number of spectrum blocks accessed by each player. Though in the other case, they do not converge but they oscillate between a mixes of the two NE strategies.

Smoothed best-response technique with mixed strategies can be used to ensure convergence to NE. A smoothed best response will avoid the sudden alternate between pure strategies and ease the convergence to NE. Mixed strategies are characterized by a set of probabilities assigned to the pure strategies.

7.2.3 Efficiency

The existence of Nash equilibrium (or the steady state of a game) does not guarantee its efficiency. We have seen this issue clearly in the PD game. The NE points are not always efficient compared to Pareto optimal points (Ji & Liu, 2007). Traditionally, Pareto concept is used to evaluate the efficiency of Nash equilibrium usually using the Price of Anarchy (PoA) notation. The PoA is the ratio of payoff obtained at the NE point compared to the payoff obtained at the PO point. A strategy is Pareto-optimal when no Pareto improvements can be made. The strategy *s*'' is Pareto-superior to the strategy profile *s* if, for at least one player *i*,

$$u_i(s_i'', s_{-i}'') > u_i(s_i, s_{-i}), \forall s_i \in S_i, \quad (5)$$

Figure 1. Number of simultaneous plays performed by the players

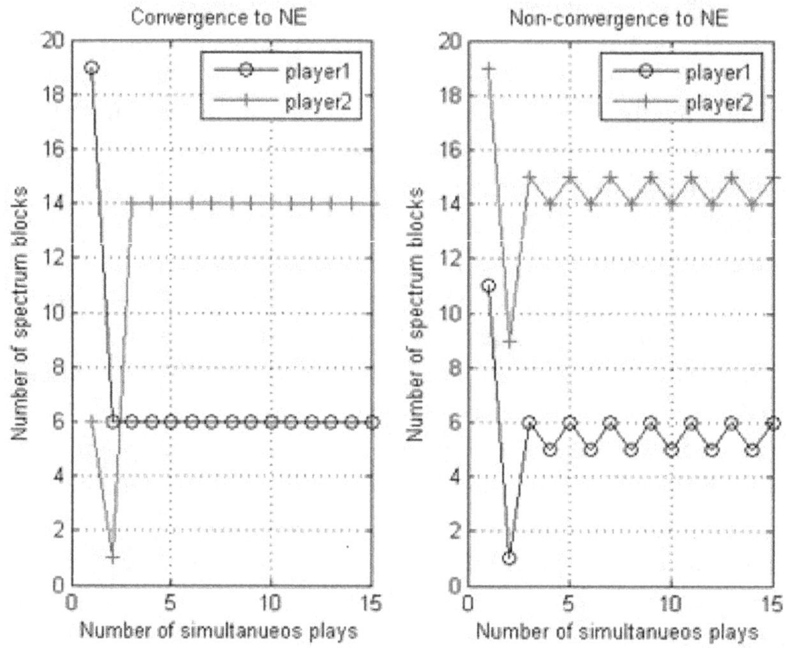

without making another player worse off (Felegyhazi & Hubaux, 2007).

It is likely to have more than one PO point for the same game. In this case, all Pareto optimal points form the Pareto frontier. A Pareto frontier is the set of points that are Pareto efficient. Figure 2 illustrates the idea of Pareto frontier in a game where NE is not efficient. We can see that all points on the Pareto frontier curve have higher payoffs (for both players) than the payoff obtained at NE.

Although points A and B are Pareto efficient, however we can notice that for point A player 1 is better off than player 2. The opposite is correct for point B. In this case a selection criterion (or fairness criterion) is needed in order to choose a unique PO point. As an example, the authors in (Kamal, Coupechoux & Godlewski, 2009) have chosen to maximize the sum of operators' utilities (social welfare maximization) in case the game solution gives more than one PO point. In reference (Kamal, Coupechoux & Godlewski, 2009), a game theoretical framework, based on the Pareto optimality concept, is proposed for the DSA problem among cellular operators.

7.3 EXAMPLES OF GAME THEORY APPLICATION IN COGNITIVE RADIO

Game theory is being used to formulate several problems related to cognitive radio networks. In this section we present to the reader few examples from the literature, trying from our side to provide clear illustration on how game theory can be used as a tool to tackle research challenges in CR networks. We start by illustrating the game formulated by the authors in (Etkin, Parekh & Tse, 2007). The authors made use of game theory to analyze the power allocation problem of peer- to-peer systems, such as Bluetooth or 802.11 networks, in unlicensed bands. The player is considered to be a system formed by a single transmitter-receiver pair. The objective is to determine the power allocations of each system in order to maximize a global utility function, while satisfying the power

Figure 2. Pareto frontier in a two-player game

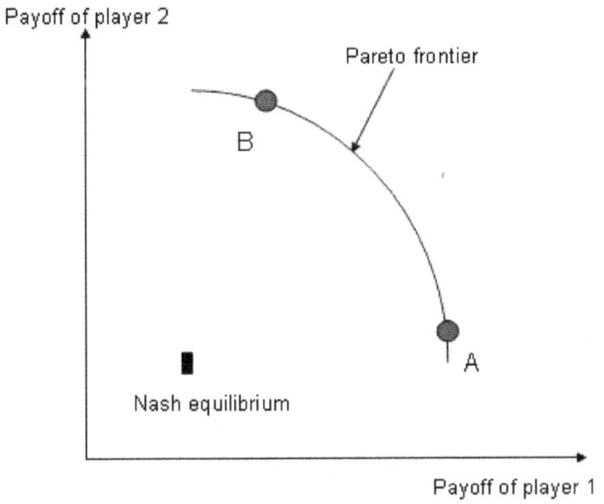

constraints. Two forms are considered for the global utility function of the system, both related to the achieved bit rate. The first one represents the sum of the achieved players' bit rates, while the second utility function represents the sum of the *log* of the achieved rates. Applying the *log*(.) function to the bit rates promotes the system in disadvantage, as a sort of fairness criterion, according to (Kelly, Maulloo & Tan, 1998). In a first part of the study, the authors have implicitly assumed that the players cooperate to maximize the global utility function by choosing appropriate power allocations. A system model of two players has been studied in (Etkin, Parekh & Tse, 2007), see Figure 3. The antenna size in Figure 3 represents the transmission power while the distance between the antennas is directly related to the propagation conditions (the channel gains).

Etkin, Parekh and Tse have represented the bit rate R_i obtained by system *i*, in a spectrum sharing situation with system *j*, as a function of: (1) the used bandwidth *W*, (2) the noise variance N_0, (3) the transmission power level of both systems p_i and p_j, and (4) the cross channel gains $c_{i,j}$ and $c_{j,i}$. The authors studied a game scenario using the following set of parameters: $N_0 = 1$, $W = 1$, $c_{1,2} = 0.5$, $c_{2,1} = 10$ and the power constraints of both systems P_1 and P_2 have been set to *10* and *1* respectively. The authors presented a plot for the obtained Pareto frontier curve for this game (Figure 2 in Etkin, Parekh & Tse, 2007); this curve is similar to the one we plotted in Figure 2 of Section 2.3 where the payoff in this case is the bit rate of the two players. On the curve of the figure presented in (Etkin, Parekh & Tse, 2007), the authors make clear the two Pareto points corresponding to the different considered utility functions, i.e. simple summation of rates, and summation of the rates' logarithms (proportional fair).

In coherence to what we have mentioned in Section 2.3, in case the non-cooperative game results in several achievable rates, hence a selection (fairness) criterion is needed. The authors in (Etkin, Parekh & Tse, 2007) mention that "*the operating point is chosen by the protocol to achieve efficiency and fairness. We can think of this protocol as a widely known standard that the systems can choose to follow or as a set of spectrum sharing rules imposed by the regulation authority.*"

The reader needs to keep in mind that in order to perform a cooperative frame, all the information is supposed to be known to all cooperative players. Since that in a spectrum sharing scenario transceivers are competing to gain access to the shared spectrum, assuming selfish behavior is more realistic. For this reason, Etkin, Parekh & Tse have also analyzed non-cooperative behavior of the players, i.e. selfish and rational, and their effect on the achievable rates. For more details about the non-cooperative model and the related issues of efficiency, fairness and compatibility, the reader is invited to consult the mentioned reference.

The second reference example we give in this section is presented by Nie and Comaniciu, where they have used game theory to analyze node-conflicts in CR networks. The transmitting-receiving pairs of nodes are considered to be the players of the game (Nie & Comaniciu, 2005). The objective of the players is to maximize their

Figure 3. Two CR systems sharing the same band in an asymmetric situation, i.e. asymmetric Tx power and asymmetric cross channel gains according to (Etkin, Parekh & Tse, 2007)

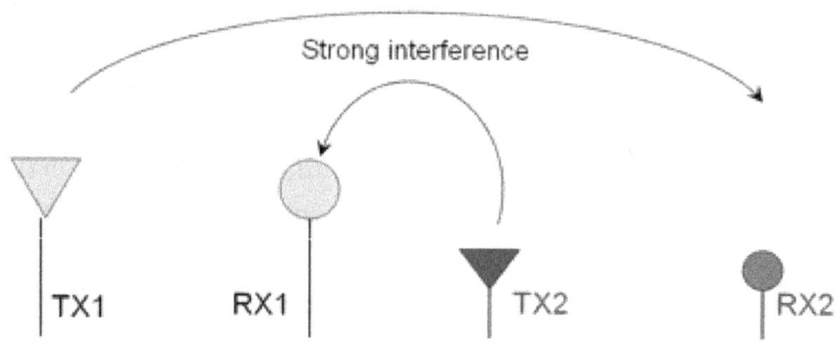

utility (payoff) by determining the appropriate operating channel (frequency) in a spectrum sharing context. In such a system where the number of node-pairs is larger than the number of available frequencies, a frequency reuse situation is inevitable. Figure 4 gives a simple example of a spectrum sharing system as adopted and illustrated by Nie and Comaniciu.

According to the example given in Figure 4, we can note a system that contains a number of $N = 9$ players (node-pairs) sharing a number of $K = 4$ frequencies. Each used frequency is represented by a different line-color in Figure 4. The strategy of the players is represented by the choice of the transmitting frequency. For a given modulation format and channel coding, a target Signal-to-Interference Ratio (SIR) can be determined for a certain application or Bit Error Rate (BER) requirement. The players' utility is directly related to the channel quality. The utility functions considered in (Nie & Comaniciu, 2005) take into consideration the level of interference generated from the frequency reuse performed by the nodes. The authors have proposed two utility functions in order to model different levels of cooperation. The first utility function models the "selfish" users while evaluating the chosen frequency channel based on the received interference level. The second utility function models the "cooperative" users while evaluating the interference seen by the users upon using a particular channel at the same time as considering the interference this particular channel causes on the neighboring nodes. The authors used the second utility function to formulate an exact potential game, which fundamentally converges to pure NE solutions when the players adopt best-response strategies (as we have mentioned in Section 2.2).

In (Nie & Comaniciu, 2005), the authors have simulated a system model with $N = 30$ players, randomly distributed over a $200m$ x $200m$ area, and sharing $K = 4$ frequency channels. The authors have provided plots for the selected pure NE frequency assignments by the nodes 1, 8, 16 and 24 (Figure 4 in the reference). The plots represents the strategies, or the chosen frequencies by the nodes, versus the number of plays (trials) performed by the nodes. According to the result's plots, the players hop among different frequency channels as they play, for a number of play-rounds, until they converge, i.e. their chosen frequency channels remain fixed and do not change versus the number of trails. The plots show that the players converge to pure NE solutions and that the convergence speed is very fast that equals to around 30 trials (or play-rounds).

For more examples on game theory applications, the reader can consult the reference (Wang & Wassell, 2002). A distributed channel allocation algorithm has been proposed by Wang and Wassell in (Wang & Wassell, 2002) for Broadband Fixed Wireless Access (BFWA) networks. The

Figure 4. A simple illustrative example of the studied system of nodes according to (Nie & Comaniciu, 2005)

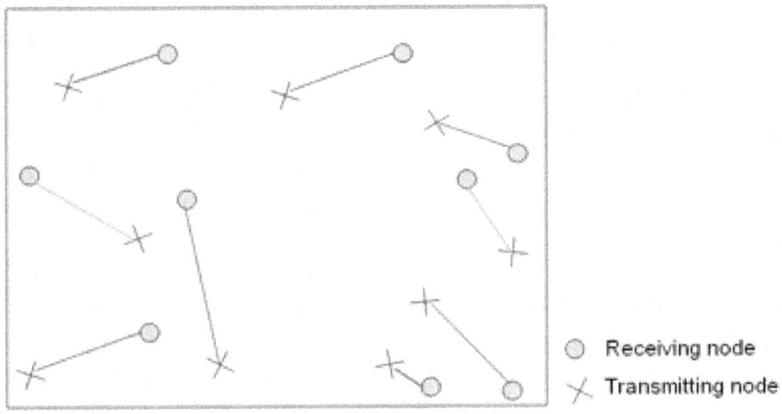

proposed method is supposed to replace the regular frequency planning method. The algorithm is based on a mixed strategy game.

7.4 CONCLUSION

In this chapter, we have presented the different efforts carried out by regulators worldwide to overcome the spectrum scarcity situation: i.e. mainly promoting the introduction of primary/secondary spectrum usage. We gave an overview on game theory including the definition of a "game" and the presentation of two well known games, i.e. the Prisoner's Dilemma (PD) and the Battle of the Sexes (BoS) games. We discussed the motivations for game theory application in cognitive radio: The distributed and dynamic nature of CR nodes along with the nodes' behavior. We gave the definition of Nash Equilibrium (NE) concept and we discussed the associated notations: the convergence and the efficiency. Concerning the convergence, we presented the definition of games' types who are known to have Nash equilibrium; i.e. supermodular and Exact Potential Games (EPG). Regarding the efficiency we presented the definition of Pareto optimality. We finally gave examples from the literature on game theory application for cognitive radio networks. The examples are treating with spectrum sharing situations for CR networks. In one of the examples, the player's strategy is about choosing the appropriate power level. In the second presented example, the strategy of the players lies behind the choice of the frequency channel. The presented examples in this chapter give the reader an illustrative view on how game theory can be used to tackle research challenges in cognitive radio.

REFERENCES

Coupechoux, M., et al. (2007). *Generic architecture of resource management and related protocols: spectrum occupancy: State of the art.* URC (Urban Planning for Radio Communication) Project, (Deliverable 1.3.1).

Etkin, R., Parekh, A., & Tse, D. (2007). Spectrum sharing for unlicensed bands. *IEEE Journal on Selected Areas in Communications*, *25*(3), 517–528. doi:10.1109/JSAC.2007.070402

Felegyhazi, M., & Hubaux, J. (2007). *Game theory in wireless networks: A tutorial.* (EPFL Technical report: LCA-REPORT-2006-002).

Giupponi, L., & Ibars, C. (2009). Bayesian potential games to model cooperation for cognitive radios with incomplete information. *Proceedings of 2009 IEEE ICC*, June 14-18, (pp. 1-6).

Hongyan, L., Yang, L., & Dexian, Z. (2010). Dynamic spectrum access for cognitive radio systems with repeated games. *Proceedings of IEEE Wireless Communications, Networking and Information Security (WCNIS)*, (pp. 59-62).

Ji, Z., & Liu, K. (2007). Dynamic spectrum sharing: A game theoretical overview. *IEEE Communications Magazine, 45*(5), 88–94. doi:10.1109/MCOM.2007.358854

Kamal, H., Coupechoux, M., & Godlewski, P. (2009). Inter-operator spectrum sharing for cellular networks using game theory. *Proceedings of IEEE PIMRC*, September 13-16, (pp. 425-429).

Kelly, F. P., Maulloo, A., & Tan, D. (1998). Rate control in communication networks: Shadow prices, proportional fairness and stability. *The Journal of the Operational Research Society, 49*(3).

Liu, K. (2011, April). Cognitive radio and game theory. *IEEE Spectrum*.

Mitola, J., & Maguire, G. (1999). Cognitive radio: making software radios more personal. *IEEE Personal Communications, 6*(4), 13–18. doi:10.1109/98.788210

Nie, N., & Comaniciu, C. (2005). Adaptive channel allocation spectrum etiquette for cognitive radio networks. *Proceedings of IEEE DySPAN*, (pp. 269-278).

OFCOM. (2002). *Trading guidance notes: A practical user guide for spectrum trading.* (OFW224, OFCOM, Technical Report).

Richard, J. L., & Anantharam, V. (2002). Optimal routing control: Repeated game approach. *IEEE Transactions on Automatic Control, 47*(3), 437–450. doi:10.1109/9.989076

Saraydar, C., Mandayam, N. B., & Goodman, D. J. (2002). Efficient power control via pricing in wireless data networks. *IEEE Transactions on Communications, 50*(2), 291–303. doi:10.1109/26.983324

Sherman, M., Mody, A. N., Martinez, R., Rodriguez, C., & Reddy, R. (2008). IEEE standards supporting cognitive radio and networks, dynamic spectrum access and coexistence. *IEEE Communications Magazine, 46*(7), 72–79. doi:10.1109/MCOM.2008.4557045

Srivastava, V., Neel, J., Mackenzie, A. B., Menon, R., Dasilva, L. A., & Hicks, J. E. (2005). Using game theory to analyze wireless ad hoc networks. *IEEE Communications Magazine, 7*(4), 46–56.

Topkis, D. (1998). *Supermodularity and complementarity*. Princeton University Press.

Wong, S., & Wassell, I. (2002). Application of game theory for distributed dynamic channel allocation. *Proceedings of IEEE VTC*, (pp. 404-408).

ENDNOTE

1 It is to be noted that in the licensing method actually deployed by the regulators and during the license period, the primary user (license owner) has no-right to resell the spectrum nor to change the type of service originally obtained the license for.

Chapter 8
Distributed Coalition Formation and Resource Allocation in Cognitive LTE Femto-Cells

Omid Namvar Gharehshiran
University of British Columbia, Canada

Alireza Attar
University of British Columbia, Canada

Vikram Krishnamurthy
University of British Columbia, Canada

ABSTRACT

This chapter investigates resource allocation in a Universal Mobile Telecommunication System (UMTS) Long Term Evolution (LTE) network. Users form coalitions, each exploiting resources in a particular femto-cell, while occupying optimal resources in the sense that total network throughput is maximized. Users in each cell collaborate to increase network throughput and simultaneously attempt to increase their own payoffs subject to a fairness criterion. Payoffs to the users are defined as the monetary equivalent of the individual users' achievable throughput in the specified coalition structure. A distributed game-theoretic resource allocation mechanism is studied whereby users autonomously decide which sub-channel in which coalition to join. It is proved that if each user operates according to the proposed algorithm, the sum throughput of all links converges almost surely to its maximum feasible value.

8.1 INTRODUCTION

8.1.1 Game Theory

Next generation wireless communication systems are evolving towards more autonomous and distributed architectures with shorter communication links so as to provide truly mobile broadband connectivity services. The key motivation to adopt such a distributed networking approach is the possibility to substantially reduce transmission power while enhancing system capacity, e.g., through localizing (spatial offloading of) the traffic demand, exploiting cooperative diversity schemes and utilizing cognitive mechanisms.

DOI: 10.4018/978-1-4666-2005-6.ch008

Several emerging standards manifesting such architectural paradigm shifts are currently in final preparation phases to achieve the goal of shaping the so-called 4G communications. One such candidate technology is the universal mobile telecommunication system (UMTS) long term evolution (LTE) and LTE-Advanced (LTE-A), developed in the framework of 3rd Generation Partnership Project (3GPP).

Amongst important features of the UMTS LTE standard, we emphasize the introduction of *X2* interface to facilitate direct interaction of base-stations (BSs) (3rd Generation Partnership Project, 2009). This novel architecture evolution alleviates the need for radio network controller (RNC) to centrally coordinate BSs as in previous generations of cellular technologies. 3GPP standards further support the deployment of low-power BSs, commonly known as *femto-cell BS (fBS)* (Knisely, Yoshizawa, & Favichi, 2009). Deployment of fBS sets the scene for inclusion of inexpensive reduced-scale access points with simple plug-and-play characteristic, primarily geared towards indoor applications. These developments yield a mixed wireless architecture comprised of randomly located hot-spot femto-cells within traditional macro-cells. Thus, macro-BSs and fBSs require intelligent resource allocation and sophisticated self-organizing capability in an interference-limited capacity regime.

Besides, it is now well understood that there are very few spectral ranges (particularly in the prime spectrum bands such as 300-3000 MHz) available to be dedicated to the next generation technologies. The scarcity of available frequency bands has excessively increased the cost of licensed spectrum, which mandates a considerable amount of investment from service providers. As a result, service providers are shifting towards *universal frequency reuse (UFR)* in network planning in order to maximally utilize their licensed spectrum. The choice of UFR, however, calls for adaptive scheduling methods to mitigate co-channel interference (CCI). In the LTE/LTE-A context, due to the provisioned architecture, coordination of resources among different BSs in a multi-cell scenario (as well as fBSs in an isolated cell) needs to be realized in a distributed manner. A promising tool to realize such distributed interference management solutions is *Cognitive Radio (CR)*. It is thus not surprising that several studies have investigated the potentials of cognitive fBSs in recent years (Lien, Tseng, Chen, & Su, 2010; Attar, Krishnamurthy, & Namvar Gharehshiran, 2011). In this chapter, it is assumed that fBSs are equipped with CR capability so as to avoid interfering with neighbouring cells.

As shown in Figure 1, users in both the macro-cell and femto-cells utilize resources belonging to the same service provider[1]. It is thus natural to assume that the service provider puts in place mechanisms to guide such entities to operate collaboratively so as to maximally exploit the licensed spectrum. At times, such egalitarian policy opposes individual user's desire of maximizing their own achievable throughput, hence, mandating a level of cooperation amongst users.

The main theme of this chapter is to motivate adopting collaborative strategies in LTE/LTE-A networking scenarios by illustrating superior long-term performance, as opposed to short-term greedy schemes such as opportunistic scheduling. We further embark on developing a novel distributed coalition formation and resource allocation scheme such that the *network throughput* is maximized subject to the local access constraints. Two questions that arise are: (i) what are the optimal coalition structures and resource allocations (in each coalition) to maximally utilize resources designated to the service provider? (ii) How can users, in a distributive fashion, form collaborative coalitions and be allocated optimal resources in the cells to ensure such a goal?

The above questions can be addressed within the framework of coalition formation in a cooperative game. The abstract formulation we consider is a non-superadditive cooperative game. The term non-superadditive means that the *grand* coali-

Figure 1. A typical isolated-cell LTE network, comprised of: (i) one macro BS, referred to as eNode-B (eNB) in 3GPP terminology, (ii) several fBSs, referred to as Home Node-B (HNB), and (iii) end users, referred to as User Equipment (UE)

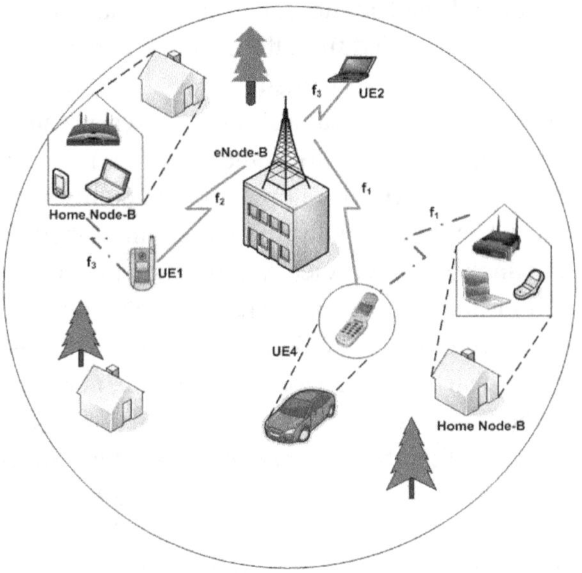

tion (the coalition comprising all users utilizing resources in the macro-cell) is not optimal. Users (as the players of the game) form collaborative coalitions in each femto-cell to increase cell throughput and are rewarded with payoffs proportional to their achievable throughput in the allocated sub-channel.

8.1.2 Why Cooperative Game Theory?

Cooperative game theory provides an expressive and flexible framework for modeling collaboration in multi-agent systems. This is appropriate for resource allocation in the CR-enabled LTE networking scenario where users encounter limited and restricted resources. In such settings, cooperation can lead to superior performance gains and fair allocation of resources. This is in contrast to greedy strategies where users attempt to achieve the highest feasible throughput by occupying sub-channels with the highest channel gains. The distributed and dynamic nature of scheduling mechanisms required in the LTE context also necessitates the use of more sophisticated analytical tools such as game theory to model interaction amongst users. In this chapter, we embark on analyzing the globally-optimum resource allocation using a special class of cooperative games, namely coalition formation games (Kahan & Rapoport, 1984). Canonical cooperative games seek to reach a globally optimum grand coalition to which all players join. However, the *local* constraints, pertaining to the resource allocation problem in a "closed access" scenario, prohibit existence of such optimal grand coalitions in the LTE network. The coalition formation game formulation helps to identify optimal localized coalitions to which only a subset of the users subscribe such that the *overall* network throughput is maximized. In general, finding the optimal coalition structure and resource allocations is an NP-hard problem. This is captured by the fact that one needs to search among all possible coalition structures for which the computational cost is proportional to the Kth *Bell number* (Harris, Hirst, & Mossinghoff, 2008) in a network with K users. This motivates the use of randomized adaptive search schemes in devising coalition formation algorithms.

8.1.3 Main Focus of Chapter

This chapter focuses on:

1. **Formulation of the Resource Allocation Problem as a Coalition Formation Game:** The resource allocation problem is formulated as a maximization problem on the total LTE network utility, defined as the sum of cell throughputs. Given that the revenue associated to the achievable throughput in each cell is divided amongst its subscribers, a fairness criterion is defined on the individual users' utilities. The problem is then interpreted as a coalition formation game. The modified *core*

is proposed as the equilibrium notion for this game, which will be shown to represent the solution to the resource allocation problem.

2. **Distributed Coalition Formation Algorithm:** A distributed coalition formation algorithm is devised, whereby users myopically maximize their expected utilities for the next period by choosing the optimal (f) BS and sub-channel. It is proved that if each user follows this algorithm, the entire LTE network eventually reaches the maximum feasible throughput, which corresponds to the core of the defined coalition formation game.
3. **Randomized Exploration Scheme:** To reduce the computational and memory overhead (as explained in Section 6.1) involved in searching for blocked users (users aware of the chance for further utility improvement in future), a randomized exploration method is presented. Convergence to the core is established taking into account the fact that, employing the randomized exploration method, the full set of blocked users may not be available at each iteration. The proposed exploration scheme provides a trade-off between the computational cost and the convergence rate of the algorithm.
4. **Numerical Examples:** Finally, numerical examples are provided to illustrate behavior and performance of the proposed algorithm. It will be shown – via extensive simulations – that it outperforms different opportunistic scheduling schemes (see Section 7 for the details of these schemes) in terms of network throughput, or equivalently the average throughput per user.

8.1.4 Related Work

Developing solutions to facilitate coexistence of femto-cells in a distributed manner has recently gained significant attention (Calin, Claussen, & Uzunalioglu, 2010). In (Chandrasekhar & Andrews, 2009), the authors investigate distributed scheduling mechanisms with minimum QoS guarantee and verify dependence of the performance results on the QoS requirement, hot-spot density and the co-channel interference from both macro-cell and femto-cells. In this chapter, the problem is tackled from the different perspective of distributed mechanisms geared at network throughput maximization. Using a Bayesian game-theoretic approach, the authors in (He, Debbah, & Altman, 2010) develop a distributed resource allocation strategy with a unique equilibrium point and analyze the efficiency characteristics of this equilibrium. However, extension of such solutions from a centralized medium access channel setting to the distributed network architecture, pertinent to the framework of this chapter, is not trivial.

In the literature, there exist only a few works (Konishi & Ray, 2003; Agastya, 1999; Apt & Witzel, 2009) that investigate the coalition formation as a dynamic process. In (Agastya, 1999), a dynamic social learning model is studied where each player observes a random sample of demand vectors and adjust his demand based on a best-reply rule. This work only suits supperadditive games. In addition, it completely abstracts from the coalition formation and focuses on allocations. In (Apt & Witzel, 2009), a generic approach is proposed for coalition formation through simple *merge and split* operations. These operations take place when they result in an improvement with respect to some given comparison relation. This approach, unlike (Agastya, 1999), can be utilized in both supperadditive and non-superadditive games. However, (Apt & Witzel, 2009) departs from the work presented here in the sense that it is focused on the coalition structure generation process and ignores bargaining (payoff allocation) process. The algorithm presented in this chapter is based on (Arnold & Schwalbe, 2002) and focuses on both the allocations and coalition formation for both supperadditive and non-superadditive games. Results of this chapter differ from (Arnold & Schwalbe, 2002) in that convergence to the core of the game is established assuming that only partial information about the set of blocked players is available at each iteration.

8.2 COOPERATIVE GAME THEORY

Game theory provides a formal analytical framework to investigate complex interactions among rational players. Throughout the past decades, game theoretic methods have been a research focal point in a large class of disciplines such as engineering, economics, political science and philosophy (Myerson, Game Theory: Analysis of Conflict, 1997). Recently, there has been a significant increase in employing game theory both as an analysis and a synthesis tool in wireless networks. This is mainly due to: (i) the emergence of large-scale, distributed and heterogeneous wireless networks; (ii) dramatic improvement in computation power, which makes it possible for network entities to make independent and rational strategic decisions; and (iii) the need for low complexity distributed algorithms that can efficiently represent competitive or cooperative scenarios between network entities (Saad W., Han, Debbah, Hjørungnes, & Basar, 2009).

There exist two main branches in game theory:

1. **Non-Cooperative Game Theory:** This branch provides analytical tools to study the interactions among *competing* players. Each player chooses its strategy independently such that its own payoff improves. In non-cooperative games, players cannot make binding commitments. There exists several equilibrium notions for non-cooperative games, among the leading ones are *Nash equilibrium* and *correlated equilibrium* (Başar & Jan Olsder, 1999).
2. **Cooperative Game Theory:** This branch investigates behavior of rational players when they *collaborate*. Cooperative games are mainly focused on the formation of co-operating groups of players, referred to as *coalitions* (Myerson, Game Theory: Analysis of Conflict, 1997). Hence, the game is a competition between coalitions of players, rather than between individual players.

As this chapter focuses on coalition formation games as a solution to the resource allocation problem in LTE networks, in what follows, we briefly introduce this class of cooperative games.

8.2.1 Coalition Formation Games

A *coalition formation game* is uniquely defined by the pair $(\mathcal{N}, v)$, where $\mathcal{N} = \{1, 2, \cdots, \mathrm{N}\}$ denotes the set of players (e.g. network entities) who seek to form groups in order to collaborate with each other. Any nonempty subset $\mathcal{C} \subseteq \mathcal{N}$ is called a *coalition*. Coalitions comprising a single player ($|\mathcal{C}| = 1$, where $|\mathcal{C}|$ denotes the cardinality of the set $\mathcal{C}$), are called *singleton coalitions*. The coalition comprising all player $\mathcal{N}$ is also called the *grand coalition*. A coalition represents an agreement between coalition members to act as a single entity (e.g. pursue the same goal, perform the same task). The set of all coalitions in a game is called *coalition structure* and is denoted by $\mathcal{S}$, i.e. $\bigcup \mathcal{C}_k = \mathcal{S}$. . In addition, v denotes the coalition payoff which quantifies the worth of a coalition in a game. Coalition formation games can be categorized into three different classes based on the definition of coalition payoff as follows:

8.2.1.1 Games in Characteristic Form

This class is the most common form of coalition formation games in which the payoff to each coalition $\mathcal{C}$ is exclusively determined by the players in that coalition. That is, the coalition payoff is independent from all players outside that coalition $\mathcal{N} \setminus \mathcal{C}$ and how they are structured. Games in characteristic form are classified as:

1. **Games with Transferable Utility (TU):** This class of cooperative games in characteristic form was first introduced by Von Neuman and Morgenstern (Von Neumann & Morgenstern, 1944). The payoff function in these games is defined by a mapping $v : 2^{\mathcal{N}} \rightarrow \mathbb{R}$, where $2^{\mathcal{N}}$ denotes the *power set* of players. This payoff function is referred to as *characteristic function* and associates with every coalition $\mathcal{C} \subseteq \mathcal{N}$ the maximal total payoff for that coalition. The TU property implies that the total payoff $v(\mathcal{C})$ gained by coalition $\mathcal{C}$ can be distributed in any manner between the coalition members $i \in \mathcal{C}$. However, this payoff is commonly distributed using an appropriate *fairness* rule. Each player's share of the total payoff is denoted by x_i which forms a vector $\mathbf{x} = (x_1, \cdots, x_N) \in \mathbb{R}^N$. This vector is referred to as *payoff allocation* or simply *allocation* vector. In this chapter, this class of coalition formation games is used to formulate the resource allocation problem in LTE networks.
2. **Games with Non-transferable Utility (NTU):** This class of coalition formation games was first introduced by Aumann and Peleg (Aumann & Peleg, 1960). In an NTU game, there exist rigid constraints on the distribution of the coalition payoffs and the characteristic function represents a mapping to a vector space, i.e. $v(\mathcal{C}) \subset \mathbb{R}^{|\mathcal{S}|}$.

Furthermore, the share of the total payoff for each player i is a function of the joint actions of other players in that coalition $\mathcal{C} \setminus \{i\}$. A TU game can thus be considered as a special case of the NTU game (Myerson, Game Theory: Analysis of Conflict, 1997).

Coalition formation games in characteristic form (both TU and NTU) constitute the most important type of cooperative games and their application in wireless networks is progressively increasing. As some instances, the interested reader is referred to (Mathur, Sankar, & Mandayam, 2008; Saad W., Han, Debbah, Hjørungnes, & Basar, Coalitional games for distributed collaborative spectrum sensing in cognitive radio networks, 2009).

8.2.1.2 Games in Partition Form

This class of coalition formation games was first introduced by Thrall and Lucas (Thrall & Lucas, 1963). In this class, unlike the games in characteristic form, the payoff of each coalition $\mathcal{C}$ depends on how players outside the coalition $\mathcal{N} \setminus \mathcal{C}$ are structured.

The following Definition introduces the notion of *collections* and *partitions* on the set of players.

Definition 1: *A collection in* $\mathcal{N}$ *is any family of mutually disjoint coalitions* $\mathbb{C} := \{\mathcal{C}_1, \cdots, \mathcal{C}_M\}$, *i.e.,* $\mathcal{C}_j \cap \mathcal{C}_k = \varnothing$ *for all* $j \neq k$. M *is called the size of the collection. If additionally* $\bigcup_{k=1}^{M} \mathcal{C}_k = \mathcal{N}$ *the collection* $\mathbb{C}$ *is called a partition of* $\mathcal{N}$.

Using the above definition, a coalition structure $\mathcal{S}$ is a partition of the grand coalition $\mathcal{N}$. In these games, the payoff to each coalition $\mathcal{C}_k \in \mathcal{S}$ is defined as $v(\mathcal{C}_k, \mathcal{S})$ which imposes the dependence on the structure of other players in the game. Games in partition form are intrinsically difficult to solve.

8.2.1.3 Games in Graph Form

This class of coalition formation games was first introduced by Myerson (Myerson, Graphs and cooperation in games, 1977; Jackson & Wolinsky, 1996). In some settings (Saad W., Han, Debbah,

Hjørungnes, & Basar, A game-based self-organizing uplink tree for VoIP services in IEEE 802.16j networks, 2009), players are interconnected and communicate through pair-wise links in a graph. Games in characteristic and partition form, however, cannot model the interconnectivity in such scenarios. Games in graph form consider this interconnection by defining the coalition payoffs based on the connectivity structure of the graphs. In these games, given a graph $\mathbf{G}_{\mathcal{C}}$ with the vertices defined as the members of coalition $\mathcal{C}$, the coalition payoff is defined by $v(\mathbf{G}_{\mathcal{C}})$.. As a result, in this class, a particular coalition $\mathcal{C}$ can be assigned different payoffs for different connectivity structures, i.e., $v(\mathbf{G}_{\mathcal{C}}^1) \neq v(\mathbf{G}_{\mathcal{C}}^2)$. These graphs can be both directed and undirected and the payoff of each coalition can also depend on the connectivity structure of other coalitions $\mathbf{G}_{\mathcal{N}\setminus\mathcal{C}}$.

8.2.2 Solution Concept

In coalition formation games, the aim is to find a coalition structure $\mathcal{S}^{sc}$ and an allocation vector $\mathbf{x}^{sc}$ such that no group of player has the incentive to leave their current coalition. This coalition structure is referred to as a *stable coalition structure* and, together with the corresponding allocation vector, is considered as the *solution* to the game. Simply put, a solution concept for a game in characteristic form is a *stable coalition structure* under which the highest payoff is achieved and payoffs are distributed in a *fair* fashion among players.

In the literature, there exist formal solutions for coalition formation games in characteristic form with supperadditive characteristic function. Below, we define *superadditivity* as a property of the characteristic function in TU games as presented in Exhibit 1.

Definition 2: *In a TU game, the characteristic function v is referred to as supperadditive if (see Exhibit 1)*

Exhibit 1.

$$v(\mathcal{C}_1 \cup \mathcal{C}_2) \geq v(\mathcal{C}_1) + v(\mathcal{C}_2), \quad \forall \mathcal{C}_1, \mathcal{C}_2 \subseteq \mathcal{N}, \mathcal{C}_1 \cap \mathcal{C}_2 = \varnothing. \tag{1}$$

In other words, a TU game is superadditive if cooperation is always beneficial. It is thus guaranteed that if two disjoint coalitions merge and form a single coalition, they will at least receive the same payoff as by acting separately. In superadditive games, the grand coalition $\mathcal{N}$ forms the stable coalition structure.

The most celebrated solution concept for the coalition formation games with supperadditive characteristic functions is the *core* (Owen, 1995). Other prominent solutions include *Shapley value*, *kernel*, and *Nucleolus*. These solution concepts are only defined for superadditive TU games (extensions for NTU superadditive games are only formalized for the core and Shapley value (Myerson, Game Theory: Analysis of Conflict, 1997)). The literature dealing with non-supperadditive games has modified these solution concepts for particular game settings(Arnold & Schwalbe, 2002; Apt & Witzel, 2009; Sandholm, Larson, Andersson, Shehory, & Tohmé, 1999; Ray, 2007). However, a general definition covering all non-superadditive games is yet to be worked out. Therefore, unlike the coalitional games with superadditive characteristic functions for which formal solutions exist, the solution for a non-superadditive game is not straightforward and mainly depends on the game under study.

In order to define the core for a non-superadditive TU game, we need the following definitions from the cooperative game theory.

Definition 3: *An allocation vector* $\mathbf{x}$ *is called individually rational if* $x_i \geq v(\{i\})$ *for all* $i \in \mathcal{N}$.

In other words, an allocation vector is individually rational if no player can do better by acting alone.

Definition 4: *In a superadditive TU game, an allocation* $\mathbf{x}$ *is called feasible if*

$$\sum_{i \in \mathcal{N}} x_i \leq v(\mathcal{N}), \tag{2}$$

and is called group rational or efficient if equality holds.

Simply put, an allocation is feasible if sum of the payoffs allocated to the players in the grand coalition does not exceed the total payoff achievable by the grand coalition. Suppose an allocation is proposed. If a group of players can form a coalition where players are guaranteed to achieve higher payoffs, the new coalition will *block* the proposed allocation.

Definition 5: *In a TU game, an allocation* $\mathbf{x}$ *is said to be blocked by a coalition* $\mathcal{C}$ *if*

$$\sum_{i \in \mathcal{C}} x_i < v(\mathcal{C}). \tag{3}$$

This property is independent of the superadditive characteristic of the game and acts as the fairness rule on the payoffs allocated to the players. We now proceed to define the core of a supperadditive TU game.

Definition 6: *If the TU game is superadditive, an allocation* $\mathbf{x}^c$ *is in the core if it is both efficient, i.e.,*

$$\sum_{i \in \mathcal{N}} x_i^c = v(\mathcal{N}), \tag{4}$$

and non-blocking, i.e.,

$$\sum_{i \in \mathcal{S}'} x_i^c \geq v(\mathcal{C}') \qquad \forall \mathcal{C}' \subseteq \mathcal{N}. \tag{5}$$

The core of a superadditive game, by definition, constitutes the grand coalition $\mathcal{N}$ and an allocation which guarantees no player (or group of players) is willing to leave the grand coalition. Hence, reaching the core directly implies stability of the grand coalition.

In many scenarios, superadditivity of the characteristic function is a quite restrictive requirement. An example includes settings in which a charge is incurred for the exchange of information or bargaining process amongst players. To deal with these scenarios, the feasibility concept is redefined for non-superadditive games as follows (Arnold & Schwalbe, 2002):

Definition 7: *Let* $\mathcal{P}$ *denote the set of all possible coalition structures. In non-superadditive TU games, an allocation* $\mathbf{x}$ *is called feasible if*

$$\sum_{i \in \mathcal{N}} x_i \leq \max_{\mathcal{S} \in \mathcal{P}} \sum_{\mathcal{C} \in \mathcal{S}} v(\mathcal{C}), \tag{6}$$

and is called efficient if the equality holds.

In other words, an allocation is called feasible if the total payoff allocated to the players does not exceed the highest possible outcome. Note that if v is supperadditive,

$$\max_{\mathcal{S} \in \mathcal{P}} \sum_{\mathcal{C} \in \mathcal{S}} v(\mathcal{C}) = v(\mathcal{N}), \tag{7}$$

and (6) reduces to (2). In many practical applications, players cannot be allocated continuous-valued payoffs. Let Δ be the smallest accounting unit. Then, the players' allocations are restricted to integral multiples of Δ in the interval $[v(\{i\}), \max_{\mathcal{S} \in \mathcal{P}} \sum_{\mathcal{C} \in \mathcal{S}} v(\mathcal{C})]$. This new set is denoted by $\mathcal{X}_i$. Considering this restriction, together with the modified definition of feasibility, the core can be redefined for non-superadditive games as follows:

Definition 8: *For any non-superadditive TU game, an allocation vector* $\mathbf{x}^c$ *is in the core if*

$$\sum_{i\in\mathcal{N}} x_i^c \le \max_{S\in\mathcal{P}} \sum_{C\in S} v(C) < \sum_{i\in\mathcal{N}} x_i^c + \Delta, x_i^c \in \mathcal{X}_i \tag{8}$$

$$\sum_{i\in C'} x_i^c \ge v(C') \qquad \forall C' \subseteq \mathcal{N}. \tag{9}$$

It is worth emphasizing that in non-supperadditive games, unlike superadditive games where the core is reached in the grand coalition, the core constitutes several disjoint coalitions of players. In the rest of this chapter, resource allocation amongst users in an LTE network is formulated as a non-superadditive coalition formation game and the modified core is employed as the solution concept.

8.3 LTE NETWORK MODEL AND CELL UTILITY

8.3.1 Notation and Terminology

Let

$$\mathcal{B} = \{BS_0, BS_1, BS_2, \ldots, BS_M\}$$

and

$\mathcal{K} = \{1, 2, \ldots, K\}$ denote the set of BSs and active users in an isolated LTE cell, respectively. We designate BS_0 to represent the macro-cell BS and $BS_i, i \ne 0$, to represent the ith femto-cell BS. The goal is to find the optimal resource allocation to the users, where resources are N OFDM sub-channels, denoted by the set $\mathcal{F} = \{f_1, f_2, \ldots, f_N\}$. Coalitions are defined as the set of users who are associated with a particular BSs and are indexed as $\mathcal{C}_{BS_i} \subseteq \mathcal{K}$, where $BS_i \in \mathcal{B}$. Subsets of users comprising one user $\{k\}$, which are not associated with any BS, are called *singleton coalitions*. The *coalition structure* comprising all coalitions in the LTE cell (both singleton and non-singleton) is also denoted by $\mathcal{S}$. By definition, each coalition structure forms a partition on $\mathcal{K}$. Finally, the set of all possible coalition structures (the set of all possible partitions on $\mathcal{K}$) is denoted by $\mathcal{P}$, the cardinality of which is given by the Kth Bell number (Harris, Hirst, & Mossinghoff, 2008).

Each coalition $\mathcal{C}_{BS_i}$ is identified by a tuple $(\mathbf{x}_{BS_i}, \mathbf{y}_{BS_i})$ where $\mathbf{x}_{BS_i} \in \{0,1\}^K$ is defined as

$$x_{BS_i}(k) = \begin{cases} 1, & \text{if } k \in \mathcal{C}_{BS_i}, \\ 0, & \text{if } k \notin \mathcal{C}_{BS_i}. \end{cases} \tag{10}$$

The vector $\mathbf{y}_{BS_i} \in \left(\mathbb{Z}^+\right)^N$, where $\mathbb{Z}^+$ denotes the set of non-negative integers, also indicates the sub-channel to which each user is connected in $\mathcal{C}_{BS_i}$.

$$y_{BS_i}(j) = \begin{cases} k, & \text{if } f_j \text{ in } BS_i \text{ is occupied by user } k \\ 0, & \text{if } f_j \text{ in } BS_i \text{ is unoccupied} \end{cases} \tag{11}$$

Finally, the indicator function $\Im(\cdot)$ returns the sub-channel index to which each user is connected:

$$\Im(k) = \begin{cases} f_j, & \text{if } \exists i \in \{0, 1, \ldots, M\} : y_{BS_i}(j) = k, \\ 0, & \text{otherwise.} \end{cases} \tag{12}$$

8.3.2 Closed Access Assumption

It is assumed that all users can be admitted to connect to the macro-cell BS, i.e., BS_0, whereas only a specific subset $\mathcal{A}_{BS_i}$, $i \ne 0$, are allowed access to a given femto-cell BS_i. Note that

$\mathcal{A}_{BS_i} \cap \mathcal{A}_{BS_j} = \varnothing$ for $i, j \in \{1, \cdots, M\}$ and $i \neq j$. Each user is thus allowed access to a maximum of two BSs: macro-cell BS and possibly an fBS. This restriction is referred to as *"closed access"* in the literature (Chandrasekhar & Andrews, 2009) and is due to the fact that indoor users may not be willing to share the capacity of their femto-cell BS with outdoor or other neighboring indoor users. Note that due to the consideration of outdoor users in this chapter, it would generally be the case that $\bigcup_{i=1}^{M} \mathcal{A}_{BS_i} \neq \mathcal{K}$.

8.3.3 Channel Propagation and Access Model

A resource block in the UMTS LTE is a time-frequency block comprising 12 sub-carriers (with 15 KHz sub-carrier spacing) and one time slot of 0.5 ms. Each 0.5 ms time slot, depending on cyclic prefix length, can cover 6 or 7 OFDM symbols. Furthermore, a sub-frame is defined as two 0.5 ms time slots and 10 equally-sized sub-frames constitute a radio frame. The scheduling process is performed on a sub-frame basis (3rd Generation Partnership Project, 2008).

The set of N sub-channels $\mathcal{F}$, is shared amongst all licensed BSs, including femto-cell BSs. This choice reflects the fact that femto-cell BSs, belonging to the same service provider, utilize the same licensed band as the macro-cell BS. It is, however, assumed that all femto-cell BSs have CR capability (Attar, Krishnamurthy, & Namvar Gharehshiran, 2011) so as to sense the utilized channels by the macro-cell BS and neighbouring femto-cell BSs and avoid interfering with such occupied channels. Due to very limited range of coverage for femto-cell BSs, it is expected that frequency channel reuse amongst femto-cell BSs can be realized over relatively small radius. However, to develop self-organizing capability in the proposed resource allocation strategy, no static network planning as regards the available resources in each cell is assumed. Note that the dynamics of RF environment is considerably more complex than traditional cellular systems as femto-cell BSs can appear or disappear in the network at any time due to the plug-and-play characteristic of their operation. For tractability of analysis, it is assumed that femto-cell BSs can reliably detect available resources in their locale via their CR capability[2], and correspondingly initialize their available resource set prior to the coalition formation process. Throughout, the set of available resources to the ith femto-cell BS is denoted by $\mathcal{F}_{BS_i}$, where $i \in \{1, \ldots, M\}$. We assume femto-cells adopt a dynamic no-mutual-interference approach when initializing their available resource set prior to the coalition formation process. More details on practical viability of this and other main assumptions are provided later in the chapter.

A block fading channel model is assumed for all sub-channels, where the fading distributions for all $f_i \in \mathcal{F}$, as monitored by each user, are i.i.d. The CSI between each UE and the corresponding BS/fBS is assumed to be known in both the transmitter and the receiver side. This assumption is feasible in LTE networks and, neglecting the associated signalling overhead, provides an upper bound on the performance of the network.

8.3.4 Effective Network Throughput

Various network resource allocation problems can be formulated as a constrained optimization problem with some utility function serving as a measure to quantify the level of customer and/or service provider satisfaction (Palomar & Chiang, 2006). Different applications require different utility functions. For instance, in best effort applications, where no QoS requirements are imposed, utility functions with respect to throughput are considered. On the other hand, in delay sensitive applications utility functions with respect to delay are considered (Song & Li, 2005). There are two reasons justifying why throughput is widely used in the literature as the utility function. First, being logarithmic, throughput is a concave function; thus, maximizing it will be efficiently solved using convex optimization techniques (if the constraints form a convex set). Second, sum throughput of all links directly captures the capacity of the network, maximizing which is a desirable property.

In the LTE context, due to the UFR property, the effective network throughput can be considered as the sum of cell throughputs, defined as the sum throughput of all links in a specific cell. In this chapter, aiming at approaching the maximal LTE network throughput, the utility function is defined as the revenue pertaining to the cell throughput reached by a given sub-channel allocation in that cell. Formally,

$$\mathcal{U}\left(\mathcal{C}_{BS_i}\right) = \mathbb{C}\left(\sum_{k \in \mathcal{C}_{BS_i}} \log\left(1 + \frac{P_{BS_i}(k) \cdot g_{BS_i,\Im(k)}(k)}{\sigma^2_{BS_i,\Im(k)}}\right)\right), \tag{13}$$

Here, $g_{BS_i,\Im(k)}(k)$ and $\sigma^2_{BS_i,\Im(k)}$ denote the channel gain for user k in sub-channel $\Im(k)$ when it is connected to BS_i and the noise power in that sub-channel, respectively. In addition, $\mathbb{C}(x)$ returns the revenue, as a function of the achieved throughput, and is assumed to be strictly increasing in x. Therefore, maximizing cell revenue will result in maximizing the cell throughput. This revenue value needs to be indirectly translated to a monetary regime, for instance, as the dollar equivalent of end users satisfaction proportional to its received service from BS_i. In its simplest form, $\mathbb{C}(x)$ can be considered as a linear function, i.e., $\mathbb{C}(x) = C \cdot x$, where $C > 0$ and the dimension for the proportionality constant is \$/Nats·s·Hz. Assuming $P_{BS_i}(k) = \bar{P}_{BS_i}$ be fixed and given the tuple $(\mathbf{x}_{BS_i}, \mathbf{y}_{BS_i})$, the utility function simply returns the cell revenue proportional to the total achievable throughput in the ith femto-cell.

Upon joining a coalition, user k increases the total throughput of that coalition and, in return, will receive a payoff, denoted by p_k. This payoff is defined as the *surplus* of the coalition utility, determined as the difference between the utility value before and after joining the coalition, i.e.,

$$p_k = \mathcal{U}\left(\mathcal{C}_{BS_i} \cup \{k\}\right) - \mathcal{U}\left(\mathcal{C}_{BS_i}\right). \tag{14}$$

Assuming $\mathbb{C}(x) = C \cdot x$, the proposed utility function will exhibit an additive characteristic. Hence, the surplus payoff allocated to a newly joining user k is simply proportional to its feasible throughput in coalition $\mathcal{C}_{BS_i}$.

An alternative utility function can be considered by allowing a uniform distribution of cell power amongst the associated links:

$$\bar{P}_{BS_i} = \frac{P_{max}}{\left|\mathcal{C}_{BS_i}\right|}. \tag{15}$$

A newly joining user k contributes to the total throughput in $\mathcal{C}_{BS_i}$ (by the added term to the sum in (13)) and, at the same time, decreases it due to reducing $\bar{P}_{BS_i}$ by a factor $|\mathcal{C}_{BS_i}|/(|\mathcal{C}_{BS_i}|+1)$ for all $l \in \mathcal{C}_{BS_i} \setminus \{k\}$. Here, the utility function still returns the cell revenue (proportional to the total cell throughput), whereas the payoffs are the revenue proportional to the added throughput to the cell by the newly joining users. It is clear that k will join $\mathcal{C}_{BS_i}$ only if there exists an unoccupied $f_j \in \mathcal{F}_{BS_i}$ such that

$$\mathcal{U}\left(\mathcal{C}_{BS_i} \cup \{k\}\right) - \mathcal{U}\left(\mathcal{C}_{BS_i}\right) > 0 \tag{16}$$

8.3.5 Practical Considerations

In this section, practical considerations and assumptions pertinent to the model considered in this chapter will be elaborated.

1. **Isolated LTE Cell with Macro-Cell BS and fBSs Operating Over the Same Spectral Ranges:** The choice of UFR is in line with the current trend of maximizing the utilization of licensed bands. Given the high-bandwidth spectrum requirement of UMTS LTE and the cost of acquiring licensed bands, it is logical to assume operators prefer to use UFR rather than higher frequency reuse factors in the cell planning phase. Extending the proposed solution to multi-cell scenarios, by assuming the UFR model, is straightforward. As long as neighbouring cells avoid mutual interference (See (b) below), existence of other cells only adds to the number of coalitions in the network. The dimensionality of multi-cell problem is non-prohibitive given that: (i) the proposed solution is distributed in nature, and (ii) the transmission range of femto-cells (mostly confined to the corresponding macro-cell) and macro-cells (diminishing effect beyond two macro-cell tiers) are limited.
2. **No-Mutual-Interference Policy is Adopted Within the Cell:** A critical assumption in the proposed model, captured in the utility function (13) and the set of resources for each coalition $\mathcal{F}_{BS_i}$, is that competing coalitions will not impose interference on each other. As discussed in (a) above, no-interference assumption implies a level of coordination between macro-BS and fBSs within one cell, as well as among neighbouring macro-BSs. There exist two solutions to realize this assumption in practice in a distributed fashion. First, the UMTS LTE architecture provisions a direct BS-to-BS networking interface, namely X2, which facilitates coordination of resource allocation amongst neighbouring cells, e.g., possibility of coordinated multi-point transmission (CoMP) mode in LTE networks. Therefore, macro-BSs can exchange frequency channel occupancy matrices amongst themselves via X2 so as to avoid CCI. Clearly, this is a dynamic approach facilitating flexibility of resource allocation from one period to the next, compared with static frequency reuse schemes. Given that femto-BSs also belong to the same operator, it is possible to envision the extension of this coordination approach to femto-cell scale too. However, if direct NNI interaction for femto-cells is not feasible, an alternative technical solution in the form of spectrum sensing can be adopted. Equipping femto-BSs with Cognitive Radio (CR) capability has recently gained a lot of attention in the literature which can be adapted to the scenario considered in this chapter (Attar, Krishnamurthy, & Namvar Gharehshiran, 2011; Li, Macuha, Sousa, Sato, & Nanri, 2009). It is imperative that if spectrum sensing errors cause mutual interference in one resource allocation period amongst some (f) BSs, the realized payoff of the affected coalitions will reduce. As a result, some users will disjoin such coalitions over the interfered channels, thereby realizing a degree of robustness (in the form of self-correction capability over time) for the distributed algorithm proposed later in this chapter.
3. **Users Join/Disjoin a Coalition Based on the Target Coalition Throughput Before and After Joining:** The issue of signalling overhead is a major concern in distributed networking solutions. In the proposed scenario, each user needs to calculate its feasible payoff over different coalitions in order to decide which coalition to join. The payoff

of the user will be the surplus of the utility of coalition after the user joins the coalition. However, the user does not need to explicitly know the channel state information (CSI) of all other users in order to calculate the coalition utility. Upon enquiring to join a coalition, each player can receive its potential payoff from the corresponding coalition head, i.e., (f)BS, which requires minimal signalling exchange. Therefore, the proposed distributed approach is scalable from signalling perspective.

4. **No Transmit Power Control Strategy is Assumed:** As detailed in Section β.4, transmission power is assumed to be fixed in (13). Alternatively, we proposed a uniform distribution of total coalition power among coalition subscribers. While it is possible to further enhance the proposed approach by developing an adaptive transmission power approach (such as water-filling schemes), we believe addition of this extra dimension of complexity will not change the spirit of the proposed algorithm and will not significantly enhance our results (Yu & Cioffi, 2006). Thus, transmit power control schemes are not considered in this chapter.

8.4 LTE NETWORK RESOURCE ALLOCATION PROBLEM

Given that the resources available in all cells belong to the same license holder, it is to the best interest of the service provider to maximally exploit the licensed spectrum by maximizing the achievable throughput in the network. Hence, as presented in Exhibit 2, the resource allocation problem in the LTE cell can be formulated as:

Here, $\mathbf{p} = \left(p_1, \cdots, p_K\right)$ denotes the *payoff allocation vector*. This vector represents the payoffs proportional to the users' (added) throughput. The payoff function $\mathcal{U}$ is also defined in (13). To make the problem mathematically tractable, the set of paoff values that could be allocated to the users is confined to a finite set. Suppose Δ ($) is the smallest payoff unit. Each user's demand is then restricted to the integer multiples of Δ in the closed interval $\left[0, \max_{BS_i \in B}\left(\max_{\mathcal{C}_{BS_i}} \mathcal{U}(\mathcal{C}_{BS_i})\right)\right]$, denoted by the set $\mathcal{X}$. Finally, $\lfloor x \rfloor_\Delta$ denotes the greatest "integer multiple" of Δ in x.

In (17), the objective function is the average payoff (throughput) allocated to the users in the LTE cell. Assuming C(x) = $C \cdot x$, solving (17) equivalently maximizes the total network throughput. The constraints (C1) guarantee that the payoff allocations in a particular coalition do not exceed the total achievable payoff in that coalition. Note that the solution to (17) satisfies:

$$\sum_{k \in \mathcal{C}_{BS_i}} p_k = \left\lfloor \mathcal{U}\left(\mathcal{C}_{BS_i}\right)\right\rfloor_\Delta \qquad \text{for all} \quad \mathcal{C}_{BS_i} \in \mathcal{S}. \tag{18}$$

The constraints in (C2) also ensure that the each user is allowed access to the fBS/BS to which it connects. Finally, due to the signalling overhead required for cooperation, it is assumed that each user k is interested in cooperation only if its payoff is greater than $\mathfrak{p}_k$, , where $\mathfrak{p}_k > 0$. Throughout, $\mathfrak{p}_k$ is referred to as *cooperation payoff threshold*.

Furthermore, a fairness criterion is defined on the allocation of resources to the users. Suppose users have formed a coalition structure $\mathcal{S}$ with the payoff allocation vector $\mathbf{p}$. The resource allocation is called *fair* if no group of users can improve their allocated payoffs by forming a new coalition $\mathcal{C}'_{BS_i} \notin \mathcal{S}$. This implies all users are satisfied with their current payoffs and the resources are divided in a fair fashion among users such that no user can gain higher payoff by exploiting resources in other coalitions as presented in Exhibit 3. Formally, for all $BS_i \in \mathcal{B}$,

Exhibit 2.

$$\begin{aligned} &\underset{\substack{\mathfrak{p}\in\mathcal{X}^K \\ \mathcal{S}\in\mathcal{P}}}{\text{maximize}} && \sum_{\mathcal{C}\in\mathcal{S}}\left(\sum_{k\in\mathcal{C}} p_k\right)/K && && \text{(P)} \\ &\text{subject to} && \sum_{k\in\mathcal{C}_{BS_i}} p_k \le \left\lfloor \mathcal{U}\left(\mathcal{C}_{BS_i}\right)\right\rfloor_{\Delta}, && \forall BS_i \in \mathcal{B}, && \text{(C1)} \\ & && k \in \mathcal{A}_{BS_i}, && \forall k \in \mathcal{C}_{BS_i}, && \text{(C2)} \\ & && p_k \ge \mathfrak{p}_k, && \forall k \in \bigcup_{BS_i\in\mathcal{B}} \mathcal{C}_{BS_i}. && \text{(C3)} \end{aligned} \tag{17}$$

In other words, there exists no other possible sub-channel allocation in the LTE cell which provides surplus payoff for any group of users, compared to the payoffs in the current allocation. Here, we denote the new coalitions by $\mathcal{C}'_{BS_i}$ to differentiate with the currently formed coalitions $\mathcal{C}_{BS_i}$ in $\mathcal{S}$.

By solving (17) subject to (19), the optimal resource allocation is determined such that the overall network throughput is maximized. Simultaneously, resources in each cell are allocated in a fair fashion to the users in that cell. Since channel gains change in consecutive resource allocation periods, the optimal coalition structure and resource allocations evolve over time. As a result, the above combinatorial optimization has to be solved separately in each period. Nevertheless, there exists no trivial relaxation to the above problem such that one can exploit existing methodologies in the combinatorial optimization literature. One natural approach would be the brute-force search on the set of all possible coalition structures and sub-channel allocations. This, however, incurs an immense computational overhead and has to be accomplished in a centralized manner. It is thus not well-suited for the framework in the LTE network due to the distributed nature of scheduling mechanisms required.

Perspective of the Approach

In the rest of this chapter, the resource allocation problem (17) is formulated as a coalition formation game. Users $\mathcal{K}$ constitute the set of players in the game. The characteristic function[3] $v(\mathcal{C})$, defined by (13), returns the total payoff achievable by a particular coalition $\mathcal{C}$. We devise a distributed coalition formation algorithm in which, at each iteration, users (as myopic optimizers) choose among existing coalitions to greedily maximize their expected payoffs. Formally, at each iteration, each user's strategy can be expressed as

Exhibit 3.

$$\sum_{k\in\mathcal{C}'_{BS_i}} p_k \ge \mathcal{U}\left(\mathcal{C}'_{BS_i}\right), \quad \forall \mathcal{C}'_{BS_i} \in 2^{\mathcal{K}} \setminus \varnothing, \mathcal{C}'_{BS_i} \notin \mathcal{S}, \mathcal{C}_{BS_i} \subseteq \mathcal{A}_{BS_i}. \tag{19}$$

$$\mathbb{C}_k(\omega) = \begin{cases} \arg\max_{BS_i \in \mathcal{B}} \left\{ v\left(\mathcal{C}_{BS_i} \cup \{k\}\right) - \sum_{l \in \mathcal{C}_{BS_i} \setminus \{k\}} p_l \right\}, & \text{w.p. } 1-\epsilon_k, \\ \text{Uniform}(\mathcal{P}), & \text{w.p. } \epsilon_k. \end{cases} \quad (20)$$

Here, $\omega = (\mathcal{S}, \mathbf{p})$ and $\text{Uniform}(\cdot)$ denote the state of the network and discrete uniform distribution, respectively. In addition, $\epsilon_k = \epsilon > 0$ *only* when there exists a coalition $\mathcal{C}'_{BS_i}$ comprising user k for which $\sum_{k \in \mathcal{C}'_{BS_i}} p_k < v(\mathcal{C}'_{BS_i})$. As will be explained later, the randomization amongst existing coalitions, which happens with probability ϵ, prevents users to get stuck in non-optimal coalition structures. It will be proved that if each user follows (20), iterations of the above algorithm eventually converge to the solution to (17). This approach brings about two main advantages: (i) it is performed in a distributed fashion among the users and eliminates the need for a central decision-making device, and (ii) in each iteration, unlike the NP-hard problem in (17), users solve a non-combinatorial optimization problem, for which the computational cost is linear in the number of base stations to which they are allowed access (Namvar Gharehshiran & Krishnamurthy, 2010).

8.5 COALITION FORMATION GAME IN THE LTE NETWORK

In this section, the resource allocation problem will be formulated as coalition formation game. The advantage of such interpretation is that one can utilize coalition formation algorithms to reach the solution. We utilize the modified core (Owen, 1995) as the solution concept for this game.

8.5.1 Cooperative Game and the Core

Since the optimal coalition structure in the LTE setting always comprises several disjoint coalitions of users (each exploiting the resources in a particular cell), it is natural to adopt the coalition formation game formulation. The resource allocation problem, formulated in (17), can be interpreted as a *non-superadditive cooperative game* (Owen, 1995) defined by the tuple $(\mathcal{K}, \mathcal{U})$. Here, the characteristic function $\mathcal{U} : 2^{\mathcal{K}} \setminus \varnothing \to \mathbb{R}$ formulates the reward for users collaboration in exploiting available resources in the LTE network efficiently and is given by (13). The payoff for each user k is a share p_k from $\mathcal{U}(\mathcal{C})$ that it claims from coalition $\mathcal{C}$ and tries to maximize it. Assuming $\mathbb{C}(x) = C \cdot x$, users' payoffs are proportional to their throughput. Specifically, assuming $C = 1$, each user's payoff represents its achievable throughput in the allocated sub-channel.

Each user k is encouraged to join a non-singleton coalition if it obtains a payoff $p_k > \mathcal{U}(\{k\})$. Considering constraints (C3), we set $\mathcal{U}(\{k\}) = \mathfrak{p}_k$. By this assumption, each user's payoff in non-singleton coalitions is restricted to the integers in the interval $\left[\mathfrak{p}_i, \max_{BS_i \in \mathcal{B}} \left(\max_{\mathcal{C}_{BS_i}} \mathcal{U}(\mathcal{C}_{BS_i})\right)\right]$. Throughout, this set is denoted by $\mathcal{X}_k$. It is worth mentioning that setting $\mathcal{U}(\{k\}) = \mathfrak{p}_k$ is just a virtual modification in the formulation of the game. The *reservation payoff* $\mathfrak{p}_k$ can, in fact, be considered as the *threshold* payoff for user k to collaborate with other users in a cell to optimally exploit resources in that cell.

Let $\mathbf{p}$ and $\mathbf{p}'$ denote two payoff vectors in $\mathbb{R}^K$. Then, $\mathbf{p} \leq \mathbf{p}'$ if and only if $p_k \leq p'_k$ for all $k \in \mathcal{K}$ This ordering is called the *product ordering*. Suppose an allocation $\mathbf{p}$ has been proposed by the users. If a group of users can form a coalition $\mathcal{C}'_{BS_i}$ where the payoff vector $\mathbf{p}'$ satisfies $\mathbf{p}' \geq \mathbf{p}$, this coalition will block the proposal. Formally, a coalition $\mathcal{C}'_{BS_i}$ will *block* an allocation $\mathbf{p}$ if:

$$\sum_{k\in \mathcal{C}'_{BS_i}} p_k < \mathcal{U}(\mathcal{C}'_{BS_i}). \quad (21)$$

Note that if (21) holds, allocations to the users $k \in \mathcal{C}'_{BS_i}$ can be improved by forming the new coalition $\mathcal{C}'_{BS_i}$ and dividing the surplus $\mathcal{U}(\mathcal{C}') - \sum_{k\in\mathcal{C}'} p_k$. By Definition 8, an allocation $\mathbf{p}$ is in the core if it is both *efficient* and *non-blocking*. The following definition extends the definition of the core to the coalition formation game pertinent to the resource allocation problem.

Definition 9: *An allocation* $\mathbf{p} \in \mathbb{R}^K$ *is called a core allocation if it satisfies the following two conditions:*

$$\sum_{k\in\mathcal{K}} p_k = \max_{S\in\mathcal{P}} \sum_{\mathcal{C}_{BS_i}\in S} \mathcal{U}(\mathcal{C}_{BS_i}), \quad p_k \in \mathcal{X}_k \quad (22)$$

$$\sum_{k\in \mathcal{C}'_{BS_i}} p_k \geq \mathcal{U}(\mathcal{C}'_{BS_i}), \quad \forall \mathcal{C}_{BS_i}{}' \in 2^{\mathcal{K}} \setminus \varnothing,\ BS_i \in \mathcal{B} \quad (23)$$

The modified core can be considered as an *equilibrium point* in the game in that reaching a core allocation and the coalition structure corresponding to it, no user can achieve higher payoffs by deviating from it.

The above definition ensures:

$$\sum_{k\in \mathcal{C}_{BS_i}} p_k = \mathcal{U}\left(\mathcal{C}_{BS_i}\right), \quad \text{for all} \quad \mathcal{C}_{BS_i} \in \mathcal{B}, \quad (24)$$

in the core. This is because having $\sum_{k\in \mathcal{C}_{BS_i}} p_k < \mathcal{U}\left(\mathcal{C}_{BS_i}\right)$, the non-blocking condition (23) is not met and users' allocations can be increased until (24) is satisfied. Comparing Definition 9 with (17) verifies that the modified core for the coalition formation game $(\mathcal{K},\mathcal{U})$ is the solution to the combinatorial optimization problem (17).

8.5.2 Properties of the Coalition Formation Game in the LTE Network

Suppose the number of users that are allowed access to cell i is less than the number of available resources, i.e., $\left|\mathcal{A}_{BS_i}\right| < N$. Then, assuming $\mathbb{C}(x) = C \cdot x$ in (13) and due to the additive property of the utility function, any resource allocation scheme for arbitrary disjoint coalitions $\mathcal{C}_1, \mathcal{C}_2 \subset \mathcal{A}_{BS_i}$ ensues

$$\mathcal{U}\left(\mathcal{C}_1 \cup \mathcal{C}_2\right) = \mathcal{U}\left(\mathcal{C}_1\right) + \mathcal{U}\left(\mathcal{C}_2\right).$$

Furthermore, assuming $\mathbb{C}(x)$ to be a convex function intersecting origin, i.e., $\mathbb{C}(0) = 0$, any resource allocation for disjoint coalitions $\mathcal{C}_1, \mathcal{C}_2 \subset \mathcal{A}_{BS_i}$ results in

$$\mathcal{U}\left(\mathcal{C}_1 \cup \mathcal{C}_2\right) \geq \mathcal{U}\left(\mathcal{C}_1\right) + \mathcal{U}\left(\mathcal{C}_2\right).$$

Thus, in any particular cell, it is always beneficial for all users to form a single coalition, rather than forming separate disjoint coalitions. Hence, one might think of the game $(\mathcal{K},\mathcal{U})$ as a superadditive game, where the grand coalition forms the equilibrium state. However, there are certain properties, as detailed below, that makes the cooperative game among users in the LTE network fit into the framework of non-superadditive coalition formation games.

First, even by assuming $K < N$, there exist local constraints that prevent all users to access resources in all cells. Second, although all users have access to resources in the macro-cell, some users might prefer to join femto-cells due to a better quality of service (higher throughput). In the game formulation, this corresponds to the case that the proposal allocation is blocked by the coalition formed in the femto-cell. Finally, cells have limited amount of resources. The number of users allowed access to a given cell could be larger than the available resources in that cell. This may particularly be the case for the macro-cell to which all users are allowed access. Therefore, users have to join different cells so that all users can be guaranteed a sub-channel. Hence, the resource allocation problem in LTE network best suits in the framework of non-superadditive coalition formation games.

8.5.3 Additional Constraints

In the context of cooperative games in characteristic form (Von Neumann J., Morgenstern, Rubinstein, & Kuhn, 2007), the basic assumption is that, given a fixed characteristic function, the coalition value only depends on coalition members (Aumann & Hart, 1994), i.e.,

$$v(\mathcal{C}_1) = v(\mathcal{C}_2) \quad \text{if} \quad \mathcal{C}_1 = \mathcal{C}_2. \tag{25}$$

In our formulation, although the characteristic function is fixed, the channel gains vary as the users attempt to access sub-channels in different cells. Hence, the characteristic function depends on the BS to which each coalition of users is connected. In order to avoid this inconsistency, it is assumed that BSs are included in the coalitions as players of the game achieving zero payoffs. Hereafter, to denote this convention, each coalition is denoted as $\{BS_i, \mathcal{C}_{BS_i}\}$ where BS_i and $\mathcal{C}_{BS_i}$ denote the i-th BS and the coalition of users in that cell, respectively. Singleton coalitions are also denoted by $\{\varnothing, \{k\}\}$, where $k \in \mathcal{K} \setminus \cup_{BS_i \in \mathcal{B}} \mathcal{C}_{BS_i}$. By this convention, the coalition value for each cell will be uniquely determined by the users in that coalition (Namvar Gharehshiran & Krishnamurthy, 2010).

As will be explained later in this chapter, users are allowed to leave coalitions or jump between coalitions until the LTE network reaches the core state. Therefore, for the formulation to be well posed, we need to disallow the BSs to leave or jump between cells. This requires the following assumptions on the characteristic function:

To prevent BSs virtually leave coalitions, the utility of a coalition without the corresponding BS is set to:

$$\mathcal{U}(\mathcal{C}_{BS_i}) = 0 \quad \text{for all} \quad BS_i \in \mathcal{B}. \tag{26}$$

To prevent BSs virtually jump between cells, we set:

$$\mathcal{U}(\{BS_i, BS_j, \mathcal{C}_{f_i}\}) = 0 \quad \text{for all} \quad BS_i, BS_j \in \mathcal{B}. \tag{27}$$

To allocate each sub-channel in each cell exclusively to one user, we set:

$$\mathcal{U}(\{BS_i, \mathcal{C}_{BS_i} \cup \{l\}\}) = 0 \text{ if } \exists k \in \mathcal{C}_{BS_i} \text{ such that } \Im(k) = \Im(l) = f_j. \tag{28}$$

Finally, the utility for singleton coalitions is set to:

$$\mathcal{U}(\{\varnothing, \{k\}\}) = 0, \tag{29}$$

and to prevent each singleton coalition join other singleton coalitions, we set:

$$\mathcal{U}(\{\varnothing, \mathcal{C}_\varnothing\}) = 0, \quad \text{where} \quad \mathcal{C}_\varnothing \subseteq \mathcal{K} \setminus \cup_{BS_i \in \mathcal{B}} \mathcal{C}_{BS_i}. \tag{30}$$

It is worth emphasizing that the above constraints are imposed only due to keeping consistency in the game formulation and are unrelated to the constraints (C1)-(C3) in (17).

Exhibit 4.

$$p_k^{n+1}\left(\omega^n\right) = \max_{BS_i \in \mathcal{B} \cup \varnothing} \left[\max_{\Im(k) \in \mathcal{Y}_{BS_i}^n} \mathcal{U}\left(\left\{BS_i, \mathcal{C}_{BS_i}^n \cup \{k\}\right\}\right) - \sum\nolimits_{l \in \mathcal{C}_{BS_i}^n \setminus \{k\}} p_l^n \right], \tag{31}$$

$$\mathbb{C}_k^{n+1}\left(\omega^n\right) \in \left\{ \arg\max_{BS_i \in \mathcal{B} \cup \varnothing} \left[\max_{\Im(k) \in \mathcal{Y}_{BS_i}^n} \mathcal{U}\left(\left\{BS_i, \mathcal{C}_{BS_i}^n \cup \{k\}\right\}\right) - \sum\nolimits_{l \in \mathcal{C}_{BS_i}^n \setminus \{k\}} p_l^n \right] \right\}, \tag{32}$$

Exhibit 5.

$$\mathbb{I}_k^{n+1}\left(\omega^n\right) \in \left\{ \arg\max_{\Im(k) \in \mathcal{Y}_{BS_i}^n} \mathcal{U}\left(\{BS_i, \mathcal{C}_{BS_i} \cup \{k\}\}\right) - \sum\nolimits_{l \in \mathcal{C}_{BS_i}^n \setminus \{k\}} p_l^n \; ; \; BS_i \in \mathbb{C}_k^{n+1}\left(\omega^n\right) \right\}. \tag{33}$$

8.6 DISTRIBUTED COALITION FORMATION FOR MAXIMIZING NETWORK OVERALL UTILITY

Having formulated the resource allocation problem (17) as a coalition formation game, in this section a distributed coalition formation algorithm will be proposed. It will be proved that iterations of the proposed algorithm converge almost surely to the core of the coalition formation game, which represents the solution to (17). The main advantage of this algorithm is that the solution to an NP-hard problem is reached in a distributed fashion by each user following a simple best-reply rule. Note that traditional (centralized) scheduling schemes have a myopic view of the isolated cells which, as shown in Section 7, will fail to maximize the overall cells utilities.

8.6.1 Distributed Resource Allocation via Coalition Formation Algorithm

Let $\mathcal{B}^n$ denote the set of users whose allocations are blocked and $\mathcal{C}^n(k)$ denote the coalition comprising user *k* in iteration n. The following algorithm is executed independently by each user in the LTE network at the beginning of each resource allocation period.

Algorithm 1. Distributed coalition formation for resource allocation

Step 0. *Initialization:* At $n = 0,$, select initial coalition structure and initialize the payoff allocation vector. Set $\omega^0 = \left(\mathcal{S}^0, \mathbf{p}^0\right)$. In addition, fix $\xi_k, \epsilon_k \in \left(0,1\right)$ for all $k \in \mathcal{K}$ in the network.
The following steps are done independently by each user $k \in \mathcal{K}$:

Step 1. *Revision Strategy:* Take a random draw from the Bernoulli trial with probability ξ_k. If the outcome is *"keep strategy"*, set $p_k^{n+1} = p_k^n$ and $\mathcal{C}^{n+1}\left(k\right) = \mathcal{C}^n\left(k\right)$ and go to Step 5. Otherwise, go to Step 2.

Step 2. *Evaluation of the Best Strategy for the Next Period:* Compute (see Exhibit 4) where the sub-channels in which p_k^{n+1} can be achieved is given by (see Exhibit 5).
Here, $\mathcal{C}_{\varnothing}^n = \varnothing$ and $\mathcal{Y}_{BS_i}^n$ denotes the set of available sub-channels in cell i at time n.

Algorithm 2. Randomized exploration step

```
For s = 1 to ζ do:
  Choose a coalition x^s and a resource allocation y^s randomly, i.e.,
  C^s = (x^s, y^s).
    For BS_i ∈ B do:
      If {BS_i, C^s_{BS_i}} ∉ S^n, then
        If Σ_{k∈C^s_{BS_i}} p_k < U({BS_i, C^s_{BS_i}}), then
        B^n_rs → B^n_rs ∪ {k ∈ K; x^s_k = 1}
      End if
    End if
  End For
End for
```

Step 3. *Perturbation:* If $k \in \mathcal{B}^n$, take a random draw from the Bernoulli trial with probability ϵ_k. If the outcome is *"experiment"*, choose $BS_i \in \mathcal{B} \cup \varnothing$ and $f_j \in \mathcal{F}_{BS_i}$ with equal probability $1/(|\mathcal{B}|+1)$ and $1/\left|\mathcal{F}_{BS_i}\right|$, respectively. In addition, demand $p_k^{n+1} \in \mathcal{X}_k$ with equal probability $1/|\mathcal{X}_k|$. Go to Step 5. Else, go to Step 4.

Step 4. *Best-ReplyProcess:* Set $p_k^{n+1} = p_k^{n+1}\left(\omega^n\right)$ and choose $C^{n+1}\left(k\right) \in \mathbb{C}_k^{n+1}\left(\omega^n\right)$ with equal probability $1/|\mathbb{C}_k^{n+1}(\omega^n)|$.

Step 5. *Recursion:* Set $n \leftarrow n+1$ and go to Step 1.

Steps 2 to 4 above correspond to the greedy strategy formulated in (20). Algorithm 1 is accompanied by a procedure for detecting the *blocked users* $\mathcal{B}^n$. In this procedure, one should seek all coalitions $C'_{BS_i} \in 2^{\mathcal{K}} \setminus \varnothing$, where $BS_i \in \mathcal{B}$, and feasible resource allocation possibilities (considering users' access to the resources in the femto-cells determined by the sets $\mathcal{A}_{BS_i}$) to find those which satisfy:

$$\sum_{i \in C'_{BS_i}} p_k < \mathcal{U}\left(\left\{BS_i, C'_{BS_i}\right\}\right). \tag{34}$$

Throughout, these coalitions are called *blocking coalitions* and correspond to the coalitions which do not satisfy (23) in Definition 9. This method is referred to as *exhaustive exploration* for blocked users and requires checking $O\left(M2^K N!\right)$ different resource allocations possibilities among users. As the number of users increases, this number blows out exponentially. Moreover, to prevent examining a particular coalition and resource allocation C'_{BS_i} repeatedly, one needs to keep track of those for which (34) has already been checked. The above search scheme thus imposes an immense memory and computational overhead.

As a variation to the exhaustive exploration explained above, we propose to construct a sample set from the set of all possible coalitions and examine (34) only for this set considering a random resource allocation to the users in each sample. This method is referred to as *randomized exploration* for blocked users and the resulting set of blocked users is denoted by $\mathcal{B}^n_{rs}$. In general, $\mathcal{B}^n_{rs} \subseteq \mathcal{B}^n$. However, it will be proved that by replacing $\mathcal{B}^n$ with $\mathcal{B}^n_{rs}$, Algorithm 1 still converges almost surely to the core of the defined game (Namvar Gharehshiran & Krishnamurthy, 2010). In what follows, the randomized exploration scheme is presented in pseudo-code format. (see Algorithm 2)

Remark: Using Algorithm 2 for detecting blocked users, the memory requirement at each user is reduced to $O(NK)$ from $O(N2^K)$ in the exhaustive exploration method. This is due to the fact that we do not keep track of the coalitions $\mathcal{C}'_{BS_i}$ for which (34) has already been checked. In addition, the computational cost at each user can be improved to $O(M\zeta)$ from $O(M2^K N!)$. These improvements, however, results in slower convergence to the core which provides a degree of freedom to compromise between the memory and computational costs and the convergence rate of Algorithm 1 using the size of the sample set, denoted by ζ.

Finally, Algorithm 1 should be accompanied by a mechanism to update the state of the network as it requires ω^n in order to compute $p_k^{n+1}\left(\omega^n\right)$ and $\mathbb{C}_k^{n+1}\left(\omega^n\right)$ at period $n+1$. Having the state of the network, the set of available resources in each cell $\mathcal{Y}^n_{BS_i}$ can also be determined. A practical solution for updating the network state is via the central macro-cell BS. In this approach, femto-cell BSs play the role of *coalition heads* (CH) in each coalition, for which there exist dedicated signaling channel to the macro-cell BS.

Assume that each CH is capable of broadcasting data to other BSs by first reporting to macro-cell BS which in turn broadcasts the aggregated information to all CHs. The main computational overhead in Algorithm 1 is to detect the blocked users. This task is being accomplished by the CHs collaboratively. Each base station BS_i is responsible for detecting the blocked coalitions in cell i. Hence, the BSs will also be responsible to inform the blocked users of their potential for gaining larger payoffs (throughputs) in other coalitions. In addition, the BSs will be in charge for updating the state of the network through broadcasting their own data, i.e., $\mathcal{C}^n_{BS_i}$, $\mathcal{Y}^n_{BS_i}$ and p_k, where $k \in \left\{k' \in \mathcal{K}; x^n_{BS_i,k'} = 1\right\}$, and receiving the aggregated information from the macro-cell BS. Therefore, computing (31)-(33) can also be turned over to the BSs.

Remark: If any sub-channel is left unoccupied following the allocations in the core, users which form singleton coalitions (due to being allocated sub-channels which provide them with payoffs less than the cooperation payoff threshold, i.e., $p_k < \mathcal{U}\left(\{k\}\right)$), will be allocated the unoccupied resources opportunistically.

8.6.2 Further Discussion on Algorithm 1

The coalition formation algorithm is decentralized in the sense that each user makes a sequence of decisions autonomously without considering other users' decisions at the current period. In what follows, detailed explanation and intuition is provided for the approach underlying Algorithm 1.

8.6.2.1 Myopic Best-Reply Strategy

In cooperative game theory, the common assumption is that players are *rational*. Rationality means that players always try to maximize their expected utility taking into account the strategies of opponents. In the proposed approach, it is assumed that the users (as players of the game) are *bounded rational*. A user which is selected to change strategy, considering the feasibility constraints, tries to maximize its expected payoff *only* for the next period. Users evaluate their expected payoffs based on the coalition structure and payoff allocation in the previous period.

Each user's strategic variables are its choice of coalition, sub-channel in the corresponding cell, and the share of the total payoff gained by the whole coalition. Hence, given a specific coalition structure $\mathcal{S}^n$, strategies available to user k are:

$$\Xi_k(\mathcal{S}^n) = \left\{\left(\mathcal{C}_k^*, p_k^*\right): \mathcal{C}_k^* = \mathcal{C}_{BS_i}^n \cup \{k\}, \Im(k) \in \mathcal{Y}_{BS_i}^n, BS_i \in \mathcal{B} \cup \varnothing, p_k^* \in \mathcal{X}_k\right\}. \quad (35)$$

At each time step, a random subset of the users gets the chance to revise their strategies for the next period. Formally, each user's opportunity to change strategy is determined by a random draw from the *Bernoulli trial* with probability ξ_k and outcomes: "*revise strategy*" and "*keep strategy*" (Step 1). In the literature, this trial is referred to as *receiving the learn draw* (Nöldeke & Samuelson, 1993).

If the outcome of the trial is "revise strategy", the user decides whether to join any sub-channel in any of the existing coalitions $\mathcal{C}_{BS_i}^n \in \mathcal{S}^n$ and, at the same time, announces its demand for the next period p_k^{n+1}. These decisions are based on the current state of the network $\omega^n = \left(\mathcal{S}^n, \mathbf{p}^n\right)$ and are determined greedily by a *best-reply rule*: a user switches coalition only if its expected payoff in the new coalition is strictly greater than its current payoff and also it demands the most it can obtain considering the feasibility constraints (Arnold & Schwalbe, 2002). Formally, if the network is in state ω^n,, each user determines its maximum expected payoff and the coalition and sub-channel in which it can be achieved by (31)-(33). If the maximizing coalition and sub-channel in (32) is not unique, i.e., $|\mathbb{C}_k^{n+1}(\omega^n)| > 1$ and/or $|\mathbb{I}_k^{n+1}(\omega^n)| > 1$, the user randomizes between them with equal probabilities $1 / \left|\mathbb{C}_k^{n+1}\left(\omega^n\right)\right| \cdot \left|\mathbb{I}_k^{n+1}\left(\omega^n\right)\right|$ (Step 2).

8.6.2.2 Best-Reply Process with Perturbation

The myopic best-reply strategy described above defines the following finite state Markov chain:

$$\Omega = \left\{\omega = (\mathcal{S}, \mathbf{p}) \mid \mathcal{S} \in \mathcal{P}, \mathbf{p} \in \times_{k \in \mathcal{K}} \mathcal{X}_k\right\}, \quad (36)$$

$$\mathbb{P}_{\omega\omega'} = \begin{cases} \left(\prod_{k \in \mathcal{R}_{\omega\omega'}} \xi_k \pi_k\left(\omega' \mid \omega\right)\right)\left(1 - \xi_k\right)^{K - |\mathcal{R}_{\omega\omega'}|}, & \text{if } \omega' \neq \omega, \\ 1 - \sum_{\omega' \in \Omega, \omega' \neq \omega} \mathbb{P}_{\omega\omega'}, & \text{if } \omega' = \omega, \end{cases} \quad (37)$$

where $\mathcal{R}_{\omega\omega'}$ denotes the set of users which have to change strategy in order to go from state ω to ω'. Let $\mathcal{C}(k)$ and $\mathcal{C}'(k)$ denote the coalition to which user k belongs in state ω and ω', respectively. Furthermore, let $\Im\left(k\right)$ and $\Im'\left(k\right)$ denote the sub-channels which user k occupies in state ω and ω', respectively. Then, $\pi_k(\omega' \mid \omega)$ is the probability of user i choosing sub-channel $\Im'_k\left(\omega\right)$ in $\mathcal{C}'(k)$ in state ω' if it occupies $\Im\left(k\right)$ in $\mathcal{C}(k)$ in state ω and can be expressed as the equation presented in Exhibit 6where $\wedge$ and $\vee$ denote the *"logical and"* and *"logical or"*, respectively. As stated before, ξ_k denotes the probability that user k gets the chance to revise its strategy. Therefore, $\xi_k \pi_k\left(\omega' \mid \omega\right)$ is the probability of user k switching to $\left(\mathcal{C}'(k), p_k'\right)$ from $\left(\mathcal{C}(k), p_k\right)$ in state ω. This Markov chain is referred to as the *best-reply process*.

Standard results on finite state Markov chains prove the probability of the best-reply process

Exhibit 6.

$$\pi_k\left(\omega' \mid \omega\right)=\begin{cases}\dfrac{1}{\left|\mathbb{C}_k\left(\omega\right)\right|\left|\mathbb{I}_k\left(\omega\right)\right|}, & \text{if } \mathcal{C}'\left(k\right)\in\mathbb{C}_k\left(\omega\right)\wedge\Im'\left(k\right)\in\mathbb{I}_k\left(\omega\right)\wedge p_k'=p_k\left(\omega\right),\\ 0, & \text{if } \mathcal{C}'\left(k\right)\notin\mathbb{C}_k\left(\omega\right)\vee\Im'\left(k\right)\notin\mathbb{I}_k\left(\omega\right)\vee p_k'\neq p_k\left(\omega\right),\end{cases} \tag{38}$$

Exhibit 7.

$$\pi_k\left(\omega' \mid \omega\right)=\begin{cases}\dfrac{1-\epsilon_k}{\left|\mathbb{C}_k\left(\omega\right)\right|\left|\mathbb{I}_k\left(\omega\right)\right|}+\dfrac{\epsilon_k}{\mid\Xi_k(\mathcal{S})\mid}, & \text{if } \mathcal{C}'\left(k\right)\in\mathbb{C}_k\left(\omega\right)\wedge\Im'\left(k\right)\in\mathbb{I}_k\left(\omega\right)\wedge p_k'=p_k\left(\omega\right),\\ \dfrac{\epsilon_k}{\mid\Xi_k(\mathcal{S})\mid}, & \text{if } \mathcal{C}'\left(k\right)\notin\mathbb{C}_k\left(\omega\right)\vee\Im'\left(k\right)\notin\mathbb{I}_k\left(\omega\right)\vee p_k'\neq p_k\left(\omega\right),\end{cases} \tag{39}$$

reaching an ergodic state or set of states, after n steps, tends to one as n tends to infinity (Kemeny & Snell, 1976). The outcome that which of these ergodic sets (states) will eventually be reached is determined by the initial state. Under the best-reply process, absorbing states do not necessarily guarantee reaching core (Arnold & Schwalbe, 2002). The fact that absorbing states may comprise non-equilibrium states is also observed in the literature of evolutionary models of non-cooperative games (Kandori, Mailath, & Rob, 1993). The solution to this problem is to introduce *perturbations*. That is, to allow players choose sub-optimal strategies with a small probability.

Here, "perturbation" from the myopic best-reply strategy is formally defined as follows: in any state $\omega^n=\left(\mathcal{S}^n,\mathbf{p}^n\right)$, when there exist a coalition and sub-channel allocation $\mathcal{C}'_{BS_i}\notin\mathcal{S}^n$ such that $\sum_{k\in\mathcal{C}'_{BS_i}} p_k<\mathcal{U}\left(\left\{BS_i,\mathcal{C}'_{BS_i}\right\}\right)$, each user $k\in\mathcal{C}'_{BS_i}$ chooses the best-reply strategy with probability $1-\epsilon$ and randomizes among possible strategies $(\mathcal{C}_k^{n+1},p_k^{n+1})\in\Xi_k^{n+1}\left(\mathcal{S}^n\right)$ with probability $\epsilon/\mid\Xi_k^{n+1}\left(\mathcal{S}^n\right)\mid$ (Step 3). The best-reply process modified by this convention is called *best-reply process with perturbation* (Arnold & Schwalbe, 2002). Let $\mathcal{B}\left(\omega\right)$ denote the set of blocked users in state ω. The Markov chain underlying the best-reply process with perturbation can then be expressed as in (36) and (37) by only modifying (38) as presented in Exhibit 7where

$$\epsilon_k=\begin{cases}\epsilon>0, & \text{if k}\in\mathcal{B}\left(\omega\right),\\ 0, & \text{otherwise.}\end{cases} \tag{40}$$

Therefore, if $k\notin\mathcal{B}\left(\omega\right)$, user k randomizes between the maximizing coalitions $\mathbb{C}_k(\omega)$ and sub-channels $\mathbb{I}_k(\omega)$ with probability $1/\left|\mathbb{C}_k(\omega)\right|\left|\mathbb{I}_k(\omega)\right|$ and demands the feasible payoff in that coalition $p_k\left(\omega\right)$. If the maximizing coalition is unique, the user joins that coalition with probability one, i.e. $1/\left|\mathbb{C}_k(\omega)\right|\left|\mathbb{I}_k(\omega)\right|=1$. However, if $k\in\mathcal{B}\left(\omega\right)$, it will join $\mathbb{C}_k(\omega)$ and demands $p_k\left(\omega\right)$ with probability $(1-\epsilon)/\mid\mathbb{C}_k(\omega)\mid\left|\mathbb{I}_k(\omega)\right|$ if it does not experiment and with probability $\epsilon/\mid\Xi_k(\mathcal{S})\mid$ if it experiments, thus the total probability adds up to $\frac{1-\epsilon}{|\mathbb{C}_k(\omega)||\mathbb{I}_k(\omega)|}+\frac{\epsilon}{|\Xi_k(\mathcal{S})|}$.

Best-reply process with perturbation constructs a Markov chain sample path such that the limiting distribution only assigns probability 1 to the core. Having constructed such a Markov chain, we form a realization of the chain $\left\{\omega^{(0)},\omega^{(1)},\omega^{(2)},\cdots\right\}$ (Algorithm 1). Once the convergence is reached, the network remains in ω^c in the consecutive iterations.

8.6.3 Convergence Analysis of Algorithm 1

The following theorem proves that the best-reply process with perturbation converges to the core of the defined coalition formation game with probability one as time tends to infinity. Noting that the exhaustive exploration method is a special case of Algorithm 2 (where all blocked users are detected), proof of convergence follows the same approach as that for Theorem 1.

Theorem 1: *Let* $\omega^c=\left(\mathcal{S}^c,\mathbf{p}^c\right)$ *denote the core of the game and suppose the randomized exploration method (Algorithm 2) is employed to detect blocked users. The best-reply process with perturbation then converges almost surely to the core, i.e.,*

$$\Pr\left(\lim_{n\to\infty}\omega^n=\omega^c\right)=1,\qquad \forall\omega^0\in\Omega,\qquad (41)$$

if the core of the game is nonempty and if every user in the LTE network follows Algorithm 1.

Proof: Suppose γ blocked coalitions exist in state ω^n. If ζ samples are taken (we assume $\zeta\gg\gamma$) from $2^{\mathcal{N}}\setminus\varnothing$, the probability of detecting ϱ blocked coalitions in state ω^n is given by:

$$\frac{\binom{\gamma}{\varrho}\binom{(M+1)2^K N!-\gamma}{(M+1)\zeta-\varrho}}{\binom{(M+1)2^K N!}{(M+1)\zeta}}.\qquad (42)$$

Particularly, the probability of detecting all blocked coalitions is

$$\frac{\binom{(M+1)2^K N!-\gamma}{(M+1)\zeta-\gamma}}{\binom{(M+1)2^K N!}{(M+1)\zeta}}.\qquad (43)$$

Therefore, there is a positive probability to detect blocked coalitions with only checking the sample set. Note that any two blocked coalitions may comprise overlapping blocked users. Hence, the probability of detecting all blocked users p_d follows:

$$p_d\geq\frac{\binom{(M+1)2^K N!-\gamma}{(M+1)\zeta-\gamma}}{\binom{(M+1)2^K}{(M+1)\zeta}}.\qquad (44)$$

Detecting at least one blocked user will guarantee destabilizing non-equilibrium absorbing states with probability ϵ.

In (Arnold & Schwalbe, 2002: Theorem 2), it is proved that *the vector of sleep times, allocated in an absorbing state of the best-reply process with perturbation, coincides with the set of core allocations of the game.* Therefore, it is implied that if $\omega=(\mathcal{S},\mathbf{p})$ is an absorbing state, then $\mathbf{p}$ will be a core allocation for the game and $\mathcal{S}$ will be the coalition structure in which it can be achieved.

Finally, we prove that *the* best-reply process with perturbation in Algorithm 1 converges to an *absorbing state* with probability one as time tends toward infinity when Algorithm 2 is deployed for detecting blocked users. This is proved by showing that the process will not get stuck in ergodic sets other than the absorbing states. Suppose that there exists a non-singleton ergodic set $\Psi \subset \Omega$ such that $|\Psi| \geq 2$. In (Arnold & Schwalbe, 2002: Theorem 2), it is guaranteed that none of the states in Ψ involve a core allocation (absorbing states are singleton ergodic sets). As a result, for each $\omega \in \Psi$ there exists a coalition and sub-channel allocation $\{BS_i, \mathcal{C}'_{BS_i}\} \notin \mathcal{S}$ such that

$$\sum_{k \in \mathcal{C}'_{BS_i}} p_i < \mathcal{U}\left(\left\{BS_i, \mathcal{C}'_{BS_i}\right\}\right). \tag{45}$$

Therefore, some users have the incentive to experiment. There is a positive probability that all the users in the blocked coalitions (m users) are detected. In addition, these users can experiment and form singleton coalitions with some positive probability. Hence, $\mathcal{S}^{n+1}$, which comprises of: (i) singleton coalitions, and (ii) non-singleton coalitions which have no blocked users, can be reached in one step. Since ω^{n+1} can be reached from ω^n with some positive probability, we have $\omega^{n+1} \in \Psi$. Now, using the fact that the core is nonempty and starting with ω^{n+1}, an absorbing state $\omega^c = (\mathcal{S}^c, \mathbf{p}^c)$ can be reached in one step. All users which are in non-singleton coalitions in ω^{n+1} do not experiment and all users in singleton coalitions experiment. This occurs with probability $(\xi\epsilon)^m$. Now, for every $\mathcal{C}^c_{BS_i} \in \mathcal{S}^c$, where $BS_i \in \mathcal{B}$, we fix one user denoted by $k(\mathcal{C}^c_{BS_i})$. There is a positive probability $1/|\Xi_k(\mathcal{S}^{n+1})|$ that all the other users in $\mathcal{C}^c_{BS_i}$, which experiment in ω^{n+1}, join the corresponding sub-channel in $\mathcal{C}^c_{BS_i}$ and demand p^c_k. The resulting state is $\omega^{n+2} = \omega^c$. Therefore, starting from ω^{n+1}, there is a positive probability to reach an absorbing state. This contradicts the assumption that ω^{n+1} is an element of an ergodic set and completes the proof.

8.6.4 Mean Time before Absorption

Theorem 1 proves that, starting from any initial state ω^0, the network reaches the maximum total throughput (equivalently, maximum average payoff to the users) conditional on feasibility. However, it does not say anything about how fast the algorithm converges to ω^c. In the theory of reducible homogenous Markov chains, *mean time before absorption* is introduced as a measure to quantify how fast a Markov chain is being absorbed by a given recurrent class $\mathfrak{C}_j$ starting from an initial (transient) state $\omega^0 \in \mathfrak{T}$. This measure will be employed to study the trade-off between the convergence rate and the cost of running Algorithm 1.

Suppose the core of the coalition formation game is nonempty. Then, there exists at least one recurrent state in addition to the transient states in the best-reply process with perturbation. The underlying Markov chain thus constitutes an *absorbing* chain and the state space Ω can be decomposed as

$$\Omega = \mathcal{T} + \sum_i \mathcal{R}_i. \tag{46}$$

Here, $\mathcal{R}_i$'s denote the states containing cores and $\mathcal{T}$ denotes the set of all transient states. Subsequently, the transition probability matrix can also be block-partitioned as

$$\mathbb{P} = \begin{bmatrix} \mathrm{R}_1 & \cdots & 0 & 0 \\ \vdots & \ddots & \vdots & \vdots \\ 0 & \cdots & \mathrm{R}_n & 0 \\ \mathrm{TR}_1 & \cdots & \mathrm{TR}_n & \mathrm{Q} \end{bmatrix}. \tag{47}$$

Here, R_i 's denote the sub-matrices with transition probabilities within each recurrent class and Q denotes the sub-matrix with transition probabilities within the transient states. TR_i 's also contain the probabilities of going from each transient state to each state in the recurrent classes. In our setting, there exists only one state comprising the core in each recurrent class, i.e., $\mathrm{R}_i = 1$ for $i = 1, 2, \ldots, n$. Therefore, (47) reduces to

$$\mathbb{P} = \begin{bmatrix} \mathrm{I}_n & 0 \\ \mathrm{TR} & \mathrm{Q} \end{bmatrix}, \tag{48}$$

where I_n denotes the identity matrix with n equal to the cardinality of the set of cores in the game.

We seek to compute the mean time before absorption by a given recurrent class starting from a given transient state. For this purpose, we need the *fundamental matrix* S for absorbing chains, defined by

$$\mathrm{S} = \sum_{n=0}^{\infty} \mathrm{Q}^n. \tag{49}$$

Staring from (49) and multiplying both sides by $(\mathrm{I}_n - \mathrm{Q})$, it is straight forward to show that

$$\mathrm{S}(\mathrm{I}_n - \mathrm{Q}) = \mathrm{I}_n. \tag{50}$$

Furthermore, since $|\mathcal{T}|$ is finite,

$$\mathrm{S} = (\mathrm{I}_n - \mathrm{Q})^{-1}. \tag{51}$$

Now, the following theorem provides mean time before absorption by any of the recurrent classes $\mathcal{R}_j$ (core of the coalition formation game) in terms of the above defined fundamental matrix.

Theorem 2: *(Brémaud, 1999) Consider a homogeneous Markov chain with transition probability matrix $\mathbb{P}$ as in (48). The expected absorption time from each transient state i is given by the i -th element of*

$$\mathbb{E}(\mathrm{T}_{\mathcal{R}}) = \mathrm{S}\mathbf{1}. \tag{52}$$

where $\mathbf{1}$ represents a column vector of ones and $\mathrm{T}_{\mathcal{R}}$ denotes the first-time visit to one of the recurrent classes $\mathcal{R}_j$ after time 0.

This measure will be used in the following section to compare the convergence rate of Algorithm 1 when accompanied with the two different search methods for blocked users.

8.7 NUMERICAL STUDY

In this section a numerical example is provided to illustrate the behavior and performance of the proposed resource allocation algorithm in the LTE cell depicted in Figure 2. The results are further compared with three opportunistic scheduling schemes. The small size of the network gives insight on how users will be contented to lower throughput, by choosing sub-channels with smaller channel gains, in favor of the co-existing users in the cell with the aim of increasing cell throughput. Since the radius of coverage for the femto-cells are significantly smaller than the macro-cell, we only illustrate the femto-cell BSs, (located at the center of each femto-cell) in Figure 2. Approximate location of the users and their proximity to femto-cell BSs can be extracted from Table 1, e.g., see the magnified femto-cells for HNB2 and HNB5 in Figure 3.

Figure 2. Configuration of the LTE cell: eNB and HNB denote the macro-cell and femto-cell BS, respectively. UEs which can only access the macro-cell resources are depicted in the above figure. UEs with closed access to the femto-cells are summarized in Table 1.

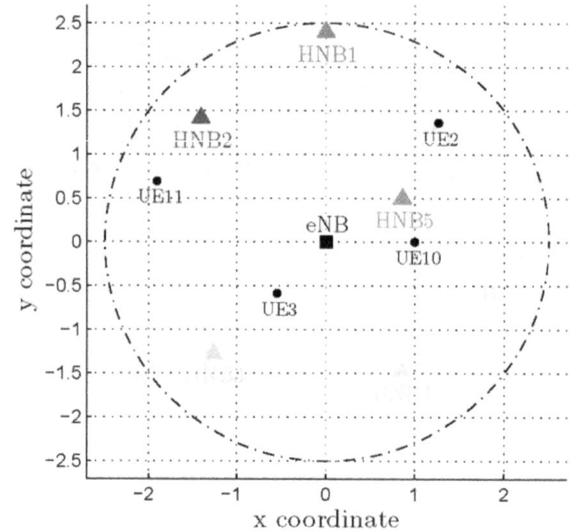

Table 1. Closed access structure for the LTE cell in Figure 2

	eNB	HNB1	HNB2	HNB3	HNB4	HNB5
$\mathcal{A}_{BS}$	$\mathcal{K}$	$\{6\}$	$\{8,12\}$	$\{4,9\}$	$\{7\}$	$\{1,5\}$

Figure 3. Magnified femto-cells associated to HNB1 and HNB5 and the corresponding UEs in the closed access scenario

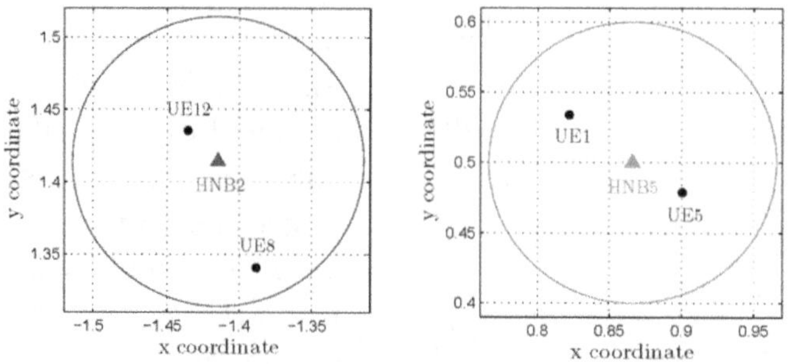

Table 2. Optimal coalition structure and resource allocation in the core using constant power allocation scheme

	f_1	f_2	f_3	f_4	f_5	f_6	f_7	f_8	f_9	f_{10}
eNB	11	0	0	2	1	0	0	6	3	10
HNB1	0	0	0	0	0	0	0	0	0	0
HNB2	0	0	8	0	0	0	12	0	0	0
HNH3	O	4	0	0	0	9	0	0	0	0
HNB4	0	0	0	0	0	7	0	0	0	0
HNB5	0	5	0	0	0	0	0	0	0	0

Table 3.

	UE1	UE2	UE3	UE4	UE5	UE6	UE7	UE8	UE9	UE10	UE11	UE12
Constant	6.2	10.8	5.4	2.9	3.7	4.3	6.1	3.9	4.1	5.8	10.5	2.7
Adaptive	6.7	13.1	5.2	3.8	6	5	8	6.1	6.4	5.9	11.4	3.5

In the simulation setup, it is assumed that the macro-cell BS (denoted by eNB in Figure 2) has a coverage radius of 2.5 Km, whereas femto-cell BSs (denoted by HNB) cover a radius of 100 m. The total transmit power of eNB and HNBs, denoted by P_{max}, are assumed 10 W and 100 mW, respectively. The channel propagation follows the COST 231 path-loss model, with a carrier frequency of 2140 MHz. There are 10 sub-channels of 180 KHz bandwidth available in the network. We assume 8 dB lognormal shadowing a unit-mean Rayleigh fading. The eNB antenna height is assumed 50 m, whereas user handset (denoted by UE) antenna height is assumed 2 m. As for the utility function, defined in (13), it is assumed $\mathbb{C} = 1$ \$/(Nats/s·Hz). Therefore, the payoff values represent the user's feasible throughput (Nats/s·Hz). Finally, we assume $\Delta = 0.1$ and set the reservation payoff to $\mathcal{U}(\{k\}) = 2$ for all $k \in \mathcal{K}$.

Table 2 shows the optimal coalition structure and resource allocations in the core, e.g., UE5 occupies f_2 in HNB5. Table 3 also gives the payoffs allocated to the UEs in the core for both cases when $P_{BS_i}(k) = P_{max} / N$ and $P_{BS_i}(k) = P_{max} / \mid \mathcal{C}_{BS_i} \mid$ in (13). Throughout, we refer to these power allocation schemes as *constant power allocation* and *adaptive power allocation*, respectively. The latter simply divides the total transmit power equally amongst the active links. In what follows, three special cases are emphasized to show how UEs, in contrast to the opportunistic scheduling schemes, cooperate with the goal of maximizing the LTE cell throughput.

8.7.1 Case I

As can be seen in Table 2, among the UEs which have access to the resources in HNB5, UE1 and UE5 occupy f_5 and f_2 in eNB and HNB5, respectively, in the core. Table 4 provides channel gains for the top three sub-channels for UE1 and UE5 in both eNB and HNB5. Considering the channel gains for UE5, it can be seen that f_2 in HNB5 is the only sub-channel which provides a reasonable channel gain for UE5. However, there exist other "good" channels (except f_2 in HNB5) for UE1 in both eNB and HNB5. In fact, UE1 can

Table 4. Channel gains for case I

		Sub-channel: Channel Gain		
UE1	eNB	$f_1 : 3.7e-6$	$f_5 : 8.3e-9$	$f_9 : 2.2e-19$
	HNB5	$f_2 : 1.6e-7$	$f_4 : 5.2e-8$	$f_7 : 4.4e-21$
UE5	eNB	$f_6 : 5.7e-17$	$f_3 : 1.2e-22$	$f_{10} : 4.6e-36$
	HNB5	$f_2 : 6.2e-8$	$f_9 : 6.7e-23$	$f_3 : 1.3e-36$

benefit a higher throughput by joining f_1 in eNB. Therefore, UE1 cooperates with UE5 by letting f_2 in HNB5 be allocated to UE5. This is in contrast with opportunistic scheduling schemes where sub-channels are allocated to the users with the highest channel gain. Surprisingly, UE1 occupies f_5 (instead of f_1) in eNB in the core. The reason is very similar to the one explained above and will be elaborated in Case II.

The above discussion can be verified noting the characteristic function values in the game formulation as explained below. If f_2 in HNB5 was allocated to UE1, f_6 would be the best sub-channel for UE5. In this case, assuming the constant power allocation scheme, the coalition throughput is given by

$$\mathcal{U}\left(\{HNB5,\{f_2 \rightarrow 1, f_6 \rightarrow 5\}\}\right) = 4.66,$$

whereas the sum payoffs of UE1 and UE5 in the core is given by: $\mathcal{U}\left(\{eNB,\{f_5 \rightarrow 1\}\}\right) + \mathcal{U}\left(\{HNB5,\{f_2 \rightarrow 5\}\}\right) = 10.02.$ Here, for simplicity in notation, we avoid using the vector notation $\left(\mathbf{x}_{BS_i}, \mathbf{y}_{BS_i}\right)$ and denote the coalition of UEs and resource allocation in eNB as: $\{eNB,\{f_5 \rightarrow 1\}\}$, i.e., $\mathcal{C}_{eNB}$ comprises UE1 occupying sub-channel f_5. Therefore, network throughput will be increased by collaboration amongst users.

8.7.2 Case II

This case provides an example on how UEs collaborate in the eNB to increase the cell throughput. As can be seen in Table 2, UE3 and UE11 occupy f_9 and f_1, respectively, in the core. Table 5 provides channel gains for the top three sub-channels for UE3 and UE11 in the eNB. Considering the channel gains for UE11, f_1 is the only sub-channel that provides a reasonable channel gain for this user. As can be seen, to increase the added payoff of both users (and the overall network throughput), UE3 chooses f_9. This also explains why UE1 chooses f_5 in Case I. This can be verified noting the characteristic function values in the game formulation. If f_1 was occupied by UE1, the trivial sub-channel allocation for UE3 and UE11 would be f_9 and f_8, respectively, and $\mathcal{U}\left(\{eNB,\{f_1 \rightarrow 1, f_9 \rightarrow 3, f_8 \rightarrow 11, f_6 \rightarrow 5\}\}\right) = 17.80.$ Note that, in this case, UE5 would also prefer f_6 in eNB as explained in Case I. Alternatively, if f_1 was occupied by UE3, the throughput for eNB and HNB5 would be $\mathcal{U}\left(\{eNB,\{f_1 \rightarrow 3, f_8 \rightarrow 11, f_5 \rightarrow 1\}\}\right) = 19.41$ and $\mathcal{U}\left(\{HNB5,\{f_2 \rightarrow 5\}\}\right) = 3.73$, respectively. Now, comparing the above values with the throughput of eNB and HNB5 in the core, $\mathcal{U}\left(\{eNB,\{f_1 \rightarrow 11, f_9 \rightarrow 3, f_5 \rightarrow 1\}\}\right) = 22.23$

Table 5. Gains for case II

	Sub-channel: Channel Gain		
UE3	$f_1 : 7.6e-6$	$f_9 : 3.4e-9$	$f_{10} : 5.2e-14$
UE11	$f_1 : 5.7e-7$	$f_8 : 1.0e-18$	$f_4 : 2.5e-28$

and $\mathcal{U}\left(\{HNB5, \{f_2 \rightarrow 5\}\}\right) = 3.73$, verifies optimality of resource allocation in the core.

8.7.3 Case III

This case provides an example on how the optimal resource allocation evolves when the adaptive power allocation scheme is selected. As can be seen in Table 2, UE6 occupies f_8 in eNB in the core when the constant power allocation scheme is in force. Table 6 provides channel gains for the top three sub-channels for UE6 in both eNB and HNB1. As can be seen, even though f_8 in HNB1 provides a higher channel gain, UE6 prefers eNB due to the higher transmission power in the macro-cell that guarantees a higher throughput to the user. This can be verified noting the user throughputs when occupying the same sub-channel in eNB and HNB1 which are $p_6 = 10.8$ and 7.3, respectively. However, employing the adaptive power allocation scheme, the payoffs will change to

$p_6 =$
$\mathcal{U}\left(\{eNB, \{1,2,3,6,10,11\}\}\right) - \mathcal{U}\left(\{eNB, \{1,2,3,10,11\}\}\right)$
$= 8.9$ and $\mathcal{U}\left(\{HNB1, \{6\}\}\right) = 11.9$ in eNB and HNB1, respectively. Therefore, f_8 in HNB1 will be allocated to UE6 in the core. In scenarios with large number of outdoor users, the adaptive power allocation exhibits a load balancing characteristic whereby users are encouraged to exploit resources in femto-cells.

Table 7 gives the expected time before absorption for the scenario considered above, starting from $\omega^0 = \left(\mathcal{S}^0, \mathbf{p}^0\right)$, where all UEs form singleton coalitions in $\mathcal{S}$, i.e., $\mathcal{S} = \bigcup_{k \in \mathcal{K}} \{k\}$, and $\mathbf{p}^0 = \mathbf{0}$, for both exhaustive and randomized exploration methods. As can be seen, decreasing the probability that the set of blocked UEs get the chance to experiment, the expected time before absorption will increase for both exploration methods. The trade off mentioned between the size of the sample set and the expected time before absorption can also be observed in Table 7.

To demonstrate performance of Algorithm 1, we compare the average payoff (throughput) achievable by the users in the core with three different opportunistic scheduling schemes. In practice, opportunistic scheduling is achieved as a two-stage process. First, users detect the pilot channel from nearby BSs, compare the received signal strength (RSS) and decide to attach to the BS with the highest RSS. Each BS will then allocate each sub-channel to the user with the highest channel gain. This scheduling policy is referred to as *scheduling scheme 1* in Figure 4 and Figure 5. We assume a separate pilot channel for each BS with an i.i.d. channel gain. An Alternative approach, named opportunistic *scheduling scheme 2*, allows each user being scheduled in a given sub-channel f_i to compare the channel gain with the identical sub-channel in the alternative (f)BSs. The user then joins the (f)BS and the sub-channel with the highest payoff value. To provide further insight, we have also simulated an additional op-

Table 6. Channel gains for case III

	Sub-channel: Channel Gain		
eNB	$f_8 : 1.12e-9$	$f_6 : 2.4e-10$	$f_1 : 5.2e-24$
HNB1	$f_8 : 2.3e-8$	$f_{10} : 9.8e-14$	$f_4 : 8.5e-19$

Table 7. Expected time before absorption: Exhaustive exploration method vs. randomized exploration method

	ξ	ϵ	ς	$\mathbb{E}(T_{\Re})$
Exhaustive Exploration	0.5	0.75	–	89.7
	0.5	0.5	–	116.4
Randomized Exploration	0.5	0.75	10000	221.5
	0.5	0.5	10000	342.4

portunistic scheduling scheme, named *scheduling scheme 3*, which is similar to scheme 2, but allows user to compare the channel gain in the scheduled sub-channel with the best available sub-channel in the alternative (f)BSs. The same utility function, as in (13), will be used to quantify performance in the opportunistic scheduling schemes.

In Figure 4 the average individual UE payoff, when Algorithm 1 reaches the core, is compared with the above explained opportunistic scheduling schemes as the LTE cell becomes more congested. Each point of the graph is an average over 1000 i.i.d. realizations of the LTE cell depicted in Figure 2. Relative configuration of the BSs is assumed to be fixed, whereas UEs are generated randomly according to the following scheme: in each realization, a random number of UEs (between 0 and 4) is generated with a random position within femto-cells. The remaining UEs, from the total number of UEs assumed in the LTE cell, are distributed uniformly in the macro-cell. This random placement scheme provides more realistic situations, as compared to the uniform distribution of the UEs in the LTE cell, and develops a better setup to compare the performance of Algorithm 1. This is due the fact that UEs will be positioned in the coverage area of the femto-cells with higher probabilities, hence, more UEs (in average) have the chance to whether choose the resources in the macro-cell or the associated femto-cell.

As the number of UEs increases (particularly outdoor users), they will become less flexible in choosing resources due to the conflict of interest with other UEs in the LTE cell. This is also because the occupied sub-channels in the macro-cell cannot be re-allocated in femto-cells due to high level of interference. Therefore, as can be seen in Figure 4, the average payoff of UEs decrease in all schemes. Since the number of resources is limited and fixed, the average payoff drops more rapidly as the number of UEs passes the number of available sub-channels. This is because some users will not be allocated any channels in that specific resource allocation period. Nevertheless, the distributed coalition formation approach demonstrates a significant improvement in the average throughput compared to the opportunistic

Figure 4. Average individual UE payoff vs. number of UEs in the LTE cell

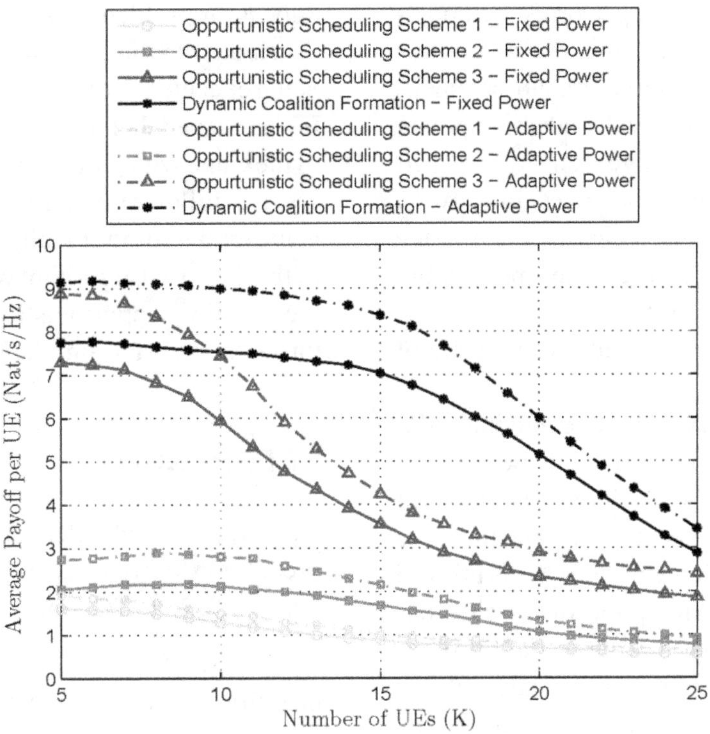

Figure 5. Average individual UE payoff vs. number of available resources

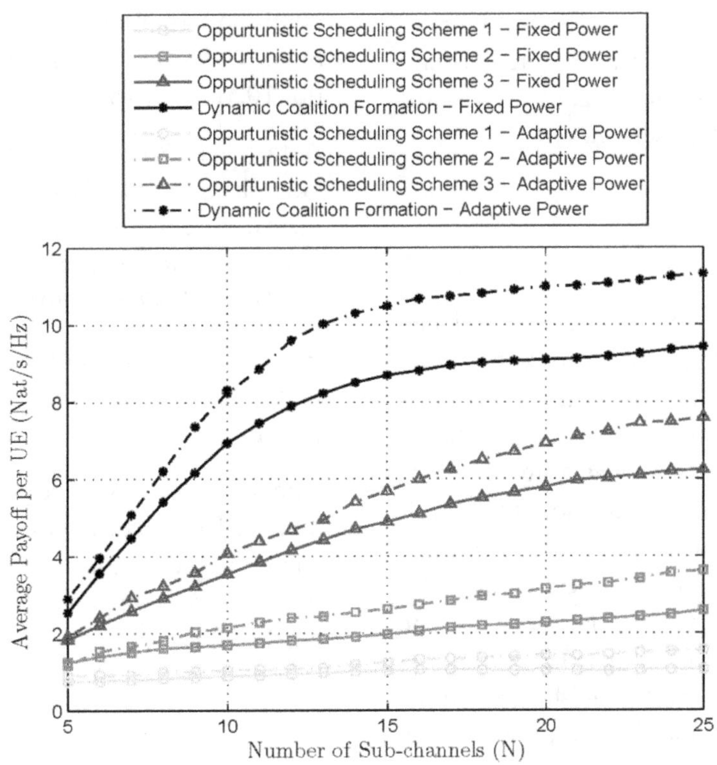

scheduling schemes. As can be seen in Figure 4, Algorithm 1 particularly outperforms the opportunistic scheduler when the number of users is slightly larger than the number of sub-channels. Figure 4 also illustrates how the adaptive power allocation scheme will improve the average payoff (throughput) of users in all schemes. Intuitively, when the number of UEs goes far beyond the number of available resources, the probability of emptiness of the core increases. In such cases, the core is replaced with the absorbing state of (36)-(38).

Figure 5 also illustrates the average individual UE payoff as the number of available resources increases in the LTE cell. Here, the number of UEs is assumed to be fixed $K = 15$ and the same random placement scheme for the UEs is utilized as explained above. As the number of sub-channels increases, more resources will be available to each UE and the probability that each user has multiple high quality channels increases. Therefore, UEs will be more flexible in choosing sub-channels, due to less conflict of interest with other UEs, and the average individual UE payoff increases in all schemes. However, as can be seen in Figure 5, the proposed coalition formation algorithm demonstrates a better performance compared to the opportunistic scheduling schemes.

8.8 CONCLUSION

UMTS LTE promises a truly mobile broadband connection with a better support for both indoor and outdoor users. To facilitate efficient coexistence of macro-cell BS and fBSs in LTE-based networks, intelligent self-organizing solutions pertinent to resource allocation should be developed. This chapter considered such an optimal resource allocation amongst users in an LTE setup. The problem was formulated as a coalition formation game with the modified core expressing the equilibrium point. It was shown that the coalition structure and sub-channel allocations in the core correspond to the solution to the resource allocation problem (17). We further proposed a distributed coalition formation algorithm which is being executed in each resource allocation period. It was proved that this algorithm reaches the core of the game with probability one provided that every user follows this algorithm and the core of the game is non-empty. Finally, superior performance of the proposed algorithm was demonstrated over the existing opportunistic scheduling schemes through extensive simulations.

REFERENCES

Agastya, M. (1999). Perturbed adaptive dynamics in coalition form games. *Journal of Economic Theory, 89*(2), 207–233. doi:10.1006/jeth.1999.2574

Apt, K. R., & Witzel, A. (2009). A generic approach to coalition formation. *International Game Theory Review, 11*(3), 347–367. doi:10.1142/S0219198909002352

Arnold, T., & Schwalbe, U. (2002). Dynamic coalition formation and the core. *Journal of Economic Behavior & Organization, 49*(3), 363–380. doi:10.1016/S0167-2681(02)00015-X

Attar, A., Krishnamurthy, V., & Namvar Gharehshiran, O. (2011). Interference management using cognitive base-stations for UMTS LTE. *IEEE Communications Magazine, 49*, 152–159. doi:10.1109/MCOM.2011.5978429

Aumann, R. J., & Hart, S. (1994). *Handbook of game theory with economic applications*. North Holland.

Aumann, R. J., & Peleg, B. (1960). Von Neumann-Morgenstern solutions to cooperative games without side payments. *Bulletin of the American Mathematical Society, 66*(3), 173–179. doi:10.1090/S0002-9904-1960-10418-1

Başar, T., & Jan Olsder, G. (1999). *Dynamic noncooperative game theory*. Philadelphia, PA: SIAM: Classics in Applied Mathematics.

Brémaud, P. (1999). *Markov chains: Gibbs fields, Monte Carlo simulation, and queues*. Springer-Verlag.

Buddhikot, M. M. (2010). Cognitive radio, DSA and self-X: Towards next transformation in cellular networks (Extended Abstract). *2010 IEEE Symposium on New Frontiers in Dynamic Spectrum*, (pp. 1-5).

Calin, D., Claussen, H., & Uzunalioglu, H. (2010). On femto deployment architectures and macrocell offloading benefits in joint macro-femto deployments. *IEEE Communications Magazine*, *48*(1), 26–32. doi:10.1109/MCOM.2010.5394026

Chandrasekhar, V., & Andrews, J. G. (2009). Spectrum allocation in tiered cellular networks. *IEEE Transactions on Communications*, *57*(10), 3059–3068. doi:10.1109/TCOMM.2009.10.080529

Harris, M. J., Hirst, J. L., & Mossinghoff, M. J. (2008). *Combinatorics and graph theory*. Springer Verlag.

He, G., Debbah, M., & Altman, E. (2010, January). A Bayesian game-theoretic approach for distributed resource allocation in fading multiple access channels. *EURASIP Journal on Wireless Communications and Networking - Special Issue on Femtocell Networks, 8*.

Jackson, M. O., & Wolinsky, A. (1996). A strategic model of social and economic networks. *Journal of Economic Theory*, *71*(1), 44–74. doi:10.1006/jeth.1996.0108

Kahan, J. P., & Rapoport, A. (1984). *Theories of coalition formation*. L. Erlbaum Associates.

Kandori, M., Mailath, G. J., & Rob, R. (1993). Learning, mutation, and long run equilibria in games. *Econometrica: Journal of the Econometric Society*, *61*(1), 29–56. doi:10.2307/2951777

Kemeny, J. G., & Snell, J. L. (1976). *Finite Markov chains*. New York, NY: Springer-Verlag.

Knisely, D. N., Yoshizawa, T., & Favichi, F. (2009). Standardization of femtocells in 3GPP. *IEEE Communications Magazine*, *47*(4), 68–75. doi:10.1109/MCOM.2009.5277458

Konishi, H., & Ray, D. (2003). Coalition formation as a dynamic process. *Journal of Economic Theory*, *110*(1), 1–41. doi:10.1016/S0022-0531(03)00004-8

Li, Y.-Y., Macuha, M., Sousa, E. S., Sato, T., & Nanri, M. (2009). Cognitive interference management in 3G femtocells. *IEEE 20th International Symposium on Personal, Indoor and Mobile Radio Communications*, (pp. 1118-1122).

Lien, S.-Y., Tseng, C.-C., Chen, K.-C., & Su, C.-W. (2010). Cognitive radio resource management for QoS guarantees in autonomous femtocell networks. *IEEE International Conference on Communications (ICC)*, (pp. 1-6).

Mathur, S., Sankar, L., & Mandayam, N. B. (2008). Coalitions in cooperative wireless networks. *IEEE Journal on Selected Areas in Communications*, *26*(7), 1104–1115. doi:10.1109/JSAC.2008.080908

Myerson, R. B. (1977). Graphs and cooperation in games. *Mathematics of Operations Research*, *2*(3), 225–229. doi:10.1287/moor.2.3.225

Myerson, R. B. (1997). *Game theory: Analysis of conflict*. Harvard University Press.

Namvar Gharehshiran, O., & Krishnamurthy, V. (2010). Coalition formation for bearings-only localization in sensor networks. *IEEE Transactions on Signal Processing*, *58*(8), 4322–4338. doi:10.1109/TSP.2010.2049201

Nöldeke, G., & Samuelson, L. (1993). An evolutionary analysis of backward and forward induction. *Games and Economic Behavior*, *5*(3), 425–454. doi:10.1006/game.1993.1024

Owen, G. (1995). *Game theory*. San Diego, CA: Academic Press.

Palomar, D. P., & Chiang, M. (2006). A tutorial on decomposition methods for network utility maximization. *IEEE Journal on Selected Areas in Communications, 24*(8), 1439–1451. doi:10.1109/JSAC.2006.879350

Ray, D. (2007). *A game-theoretic perspective on coalition formation*. Oxford University Press. doi:10.1093/acprof:oso/9780199207954.001.0001

3rd Generation Partnership Project. (2008). *Evolved universal terrestrial radio access (E-UTRA) and evolved universal terrestrial radio access network (E-UTRAN)*. Overall description. Tech. Rep. TS 36.300 V8.7.0.

3rd Generation Partnership Project. (2009). *Technical specification group radio access network: Evolved universal terrestrial radio access (E-UTRA) and evolved universal terrestrial radio access network (E-UTRAN)*. Overall description; Stage 2. Tech. Rep. TS 36.300 V9.0.0.

Saad, W., Han, Z., Debbah, M., Hjørungnes, A., & Basar, T. (2009). A game-based self-organizing uplink tree for VoIP services in IEEE 802.16j networks. *IEEE International Conference on Communications*, (pp. 1-5).

Saad, W., Han, Z., Debbah, M., Hjørungnes, A., & Basar, T. (2009). Coalitional game theory for communication networks. *IEEE Signal Processing Magazine, 26*(5), 77–97. doi:10.1109/MSP.2009.000000

Saad, W., Han, Z., Debbah, M., Hjørungnes, A., & Basar, T. (2009). *Coalitional games for distributed collaborative spectrum sensing in cognitive radio networks* (pp. 2114–2122). IEEE INFOCOM.

Sandholm, T., Larson, K., Andersson, M., Shehory, O., & Tohmé, F. (1999). Coalition structure generation with worst case guarantees. *Artificial Intelligence, 111*, 209–238. doi:10.1016/S0004-3702(99)00036-3

Shellhammer, S. J. (2009). A comparison of geo-location and spectrum sensing in cognitive radio. *18th International Conference on Computer Communications and Networks (ICCCN)*, (pp. 1-6).

Song, G., & Li, Y. (2005). Utility-based resource allocation and scheduling in OFDM-based wireless broadband networks. *IEEE Communications Magazine, 43*(12), 127–134. doi:10.1109/MCOM.2005.1561930

Thrall, R. M., & Lucas, W. F. (1963). N-person games in partition function form. *Naval Research Logistics Quarterly, 10*(1), 281–298. doi:10.1002/nav.3800100126

Von Neumann, J., & Morgenstern, O. (1944). *Theory of games and economic behavior*. Princeton University Press.

Von Neumann, J., Morgenstern, O., Rubinstein, A., & Kuhn, H. W. (2007). *Theory of games and economic behavior*. Princeton University Press.

Yu, W., & Ciof□, J. M. (2006). Constant-power waterfilling: Performance bound and low-complexity implementation. *IEEE Transactions on Communications, 54*(1), 23–28. doi:10.1109/TCOMM.2005.861678

Yucek, T., & Arslan, H. (2009). A survey of spectrum sensing algorithms for cognitive radio applications. *IEEE Communications Surveys & Tutorials, 11*(1), 116–130. doi:10.1109/SURV.2009.090109

ENDNOTES

1. Clearly, due to the allocation of different licensed bands to different operators, there is no CCI problem amongst service providers.
2. To this end, various solutions including database query and spectrum sensing have been proposed in the literature (Yucek & Arslan, 2009; Shellhammer, 2009), details of which are out of the scope of this chapter.
3. The term characteristic function is as used in cooperative games and is completely unrelated to characteristic functions in probability theory.

Chapter 9
QoS Scheduling with Opportunistic Spectrum Access for Multimedia

Pavol Polacek
National Central University, Taiwan

Chih-Wei Huang
National Central University, Taiwan

ABSTRACT

Thanks to the advances of multimedia application, mobile computing platform, and wireless communication technology, the research area has attracted serious attention in order to seamlessly provide interactive and ubiquitous user experience. To make it happen, the pursuit of higher system capacity in resource limited wireless networks is never-ending. Cognitive radio (CR) represents an exciting new communication paradigm with advantages on spectrum management so as to heighten channel utilization and capacity. The bandwidth demanding multimedia applications are excellent candidates to fully exploit the potential of CR. However, the research effort has been focused mainly on spectrum access while the application specific performance has been much less touched. The research considering both spectrum access and application data scheduling is emerging for maximal user experience. In this chapter, the authors first discuss advances in opportunistic spectrum access (OSA) strategies as well as multimedia QoS scheduling schemes, and then introduce the research trend on joint access and scheduling frameworks.

9.1 INTRODUCTION

Multimedia communication is exploding with growth. Thanks to the advances of multimedia application, mobile computing platform and wireless communication technology, the research area has attracted serious attention in order to seamlessly provide interactive and ubiquitous user experiences. To make it happen, the pursuit of higher system capacity in resource limited wireless networks is critical and never-ending.

DOI: 10.4018/978-1-4666-2005-6.ch009

Among emerging communication technology, CR represents an exciting new communication paradigm with advantages on spectrum management so as to improve channel utilization and overall available bandwidth (Sadler & Zhao, 2007). The bandwidth demanding multimedia applications are excellent candidates to fully exploit the potential of CR. However, research efforts have been focused mainly on spectrum access while the application specific performance issues have been relatively less touched. A novel aspect of view jointly considering spectrum access and application level quality of service (QoS) is desired to achieve maximal user experience.

The so called multimedia content is ranging from text, still images, audio, to video. With bandwidth hogging and interactive nature, delivering video is the most prominent topic in communication research. In comparison to non-real time types of multimedia applications, video streaming has stricter timing requirements while to other real time information, the need for bandwidth is considerably higher. As mobile data enabled handheld devices, such as smartphones and tablets, becoming rapidly ubiquitous, a growing group of users is expected to access popular video channels through wireless networks in the near future. Our focus is therefore on video over CR to demonstrate various design concepts under corresponding quality requirements.

Firstly, we define the background of video transmission through CR networks. Secondly the conditions under which this objective has to be achieved are discussed and problems, which might arise, are pointed out. Third, possible solutions are presented together with an outlook at the research directions. Overall, the chapter provides essential knowledge of transmitting multimedia in cognitive radio environment, most important problems, state-of-the-art solutions, and future perspectives.

9.2 QOS FOR VIDEO OVER COGNITIVE RADIO

In this section, we address related video coding technology, CR environment and QoS enabling mechanisms.

9.2.1 Video Coding Technology

Since the first introduction of digitally recorded video, the research and progress in this field had come a long way (Hwang, 2009). Due to the considerable data rate of raw video, effort towards effective video coding for enabling an effective transmission has been pursued early on with many variants being available at the current time. The evolution of video codecs brought better transmission efficiency with the same resulting quality and conversely better video quality at the same bit-rate. With scalable video coding extension of H.264/AVC called H.264/SVC presenting the state-of-the-art solution, the coding efficiency is comparable to that of the original H.264/AVC standard.

The video encoding scalability, indeed, is one of key factors to support high quality media over heterogeneous wireless medium (Schierl, Stockhammer, & Wiegand, 2007). Conventionally, source rate adaptation is realized through end-to-end (client to server) message exchange based on receiver side real-time bandwidth inference feedback. Scalable bit streams, in contrast, enable the capability of adaptation not only at end nodes, but also in intermediate entities. More robust rate adaptation corresponding to rapid fluctuating wireless channel condition can therefore be achieved. With H.264/SVC, one video stream is divided into a base layer providing basic video quality and one or more enhancement layers, which can be incrementally decoded by the receiver to achieve higher quality video playback. Users with heterogeneous channel throughput can therefore receive different amount of video data corresponding to their time varying receiving link condition. The capability to react to fluctuating bandwidth availability is a property making H.264/SVC video especially suitable for transmission using CR systems. Together with the basic loss resistance of encoded video due to error concealment techniques, using H.264/SVC video enables a formerly unprecedented system-wide quality video delivery.

9.2.2 Cognitive Radio Environment

CR presents a new paradigm, which changes the way we allocate wireless spectrum. Traditionally, licensed users (primary users, PU) access their own spectrum without cross-network competition, while unlicensed users access unlicensed spectrum with possible competition from other unlicensed users. The demand for spectrum and its scarcity have stimulated new development and over-thinking of the currently implemented fixed bandwidth access policies. According to Sadler and Zhao (2007), from a variety of solutions enabling dynamic spectrum access, the approaches showing the most promise, due to the best compatibility with current spectrum management policies and legacy wireless systems, are hierarchical access models: opportunistic spectrum access (OSA) and ultra wide band (UWB). For this reason, from now on we will use the term CR for a system implementing one of the hierarchical access models. In CR, unlicensed secondary users (SU) access spectrum that is being simultaneously, but not necessarily continuously, used by PU and need to share the same spectrum with other SUs. CR systems are aimed at exploiting available transmission opportunities as much as possible and at the same time forced by regulations to limit the amount of interference to the PU.

From the point of view of PU, the CR presents a source of interference. Because the redesign of already fielded technologies (e.g. TV, cell phone networks) is undesirable, the CR systems are responsible for enabling the coexistence. For a SU, the CR environment is very demanding, with not only the physical wireless channel conditions influencing the transmission and, but also the primary user (PU) activity having an impact. Physical channel and network related problems, like fading, shadowing, queue management etc., have been long studied in connection with other wireless technologies and a variety of solutions is available at the present day. Channel access in the presence of PUs is specific to CR with the following two alternatives available. In OSA, the CR system identifies spectrum white spaces through channel sensing and accesses idle channels, therefore SU and PU transmissions should ideally be orthogonal in time. If both, PU and SU access the same channel at the same time, a collision occurs with negative effect on both transmissions. In general, constrains are given for interference (e.g. allowed number of collisions or overlapping transmissions time) as not to hinder the normal PU system functioning. UWB on the other hand, does not depend on identification of whitespaces. UWB systems transmit with relatively low power, where the maximum is limited by the noise floor of the PU, thus enabling a simultaneous transmission of SUs and PUs in one frequency band, as long as the SU respects the power constrains. This approach requires more complex hardware compared to OSA systems and suffers from reduced reach due to power limitations. It is also possible, to combine OSA and UWB in one system, where higher transmission power can be used while the PU is idle and a low power when PU is active. A combined OSA and UWB system can achieve an overall higher throughput compared to a pure OSA or UWB solution.

In the simplest case there is only one pair of CR systems communicating in a PU network, where access of identified spectrum opportunities is rather simple. Mesh networks, infrastructure based networks or an arbitrary combination of such networks present scenarios that are more complicated and require spectrum management mechanisms, with the goal of achieving fairness and maximize the network-wide throughput. According to Akyildiz, Vuran, and Mohanty (2008) and Akyildiz, Lee, and Chowdhury (2009), spectrum access in a CR network (CRN) can be categorized based on four main criteria:

- Centralized and decentralized
- Cooperative and non-cooperative
- Underlay and overlay spectrum sharing
- Intranetworking and internetworking spectrum sharing

As our focus is on QoS related scheduling, we consider intranetwork sharing CRNs either centralized (infrastructure based or with a spectrum scheduler/broker) or cooperative decentralized. In a centralized and cooperative approach, it is both possible to prefer certain users or groups of users in channel access, to improve overall QoS. For non-cooperative decentralized approach, it is selfish in nature and lacking of organized scheduling of spectrum access.

9.2.3 Related QoS Mechanisms

A joint design has to be able to handle issues from multiple layers of the communication stack. CR has to change its behavior and parameters in reaction to the outer and inner state. Even though the hardware design is unchangeable in real time on an already fielded system, numerous options are available to a CR system. On the PHY layer, common wireless channel issues and the resulting channel quality is handled by changing the transmission parameters with adaptive modulation and coding (AMC) and using multiple input multiple output (MIMO) antenna systems. Constrains PU presence imposes on the CR require a spectrum sensor, which related parameters can be changed during CR operation to fit current situation. CR system can exploit the distribution of primary users in space by directing its transmission power into space unoccupied by the PUs or adjust the transmission power to limit interference to the PUs. On the MAC layer, the sensing strategy is aimed at maximizing the gain sensing information brings to the radio, preparing a favorable situation for access decision. Access strategy is the way CRNs decide to access the spectrum, ruled by two contradicting requirements, limitations on impairing PU functioning while maximizing the available resources. There are two kinds of scheduling on the MAC layer: one being data scheduling and the second user spectrum access scheduling. Data scheduling handles different data based on their priority, with QoS and fairness in mind. User spectrum access scheduling aims at fairness and satisfying user requirements, by distributing available resources to individual users or user groups.

For a joint design to work efficiently, information and their exchange is crucial. Bhat (2011) emphasizes the importance of information in a wireless systems, ranging from channel state information on the PHY layer, up to the source signal in application layer. He stipulates that incorporating available information into the decision process can gain a performance advantage over methods with less information available. Akyildiz et al. (2009), Akyildiz et al (2008) and Baldo and Zorzi (2008) present overviews of CR cross layer frameworks with different function block and information flowing between them. In fact, joint design and the high level of information flow between different layers, as already presented in Mitola and Maguire (1999) and were identified to be the main design features differentiating CR from other radio technologies.

Under the term QoS, a given minimum amount of bandwidth can be provided to the SU even in rapidly changing CR environment with active PUs (Willkomm & Wolisz, 2010). For multimedia applications, this translated into a minimum amount of data that will be transmitted on time to the end user and satisfying his expectations or predefined receiving quality. To provide users with quality service in CR, it is necessary to implement QoS management mechanisms adjusted to the CR environment.

Figure 1. Two types of scheduling can be implemented, packet scheduling, and spectrum scheduling

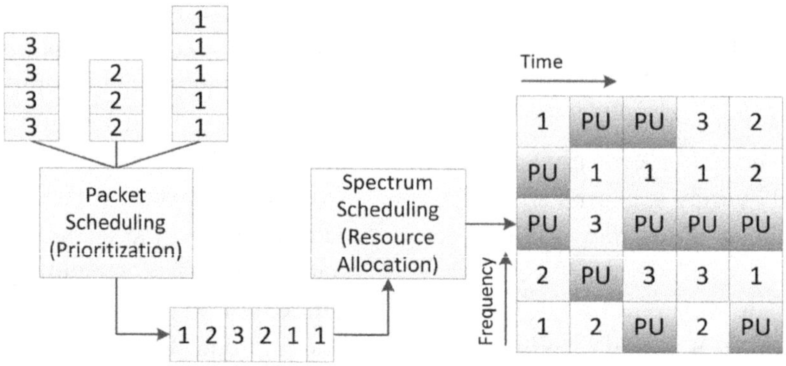

Scheduling algorithms are the core of QoS management, with the capability to provide guarantees of QoS parameters like throughput, loss, delay and jitter. In CRN, we can encounter two kinds of scheduling: 1) scheduling as a part of the spectrum sharing playing a crucial role in the network performance and 2) packet scheduling scheme deciding the order of packet transmission from different users/streams. An integrated algorithm plays an important role in achieving high resource utilization and system throughput (Wang, 2009). By selecting packets to be transmitted based on their priority in situations, we can achieve a satisfying system performance even when resources are scarce or identical handling of queued data would not be suitable. A scheduling algorithm distributes resources among different users (user groups or application sessions), according to predefined rules with specific objectives in mind.

For a scheduler to be effectively able to fulfill its role in the QoS management, it needs to possess a diverse set of features. A list of such selected desired features of scheduling algorithms follows (reduced list based on Fattah & Leung, 2002):

- Efficient channel utilization is important in order to minimize resource wasting.
- Ability to guarantee delay bounds, which is especially important for real-time applications.
- Fairness in resource distribution in order to satisfy a higher number of users.
- Maintaining at least the minimal necessary throughput for a given user which will provide the minimal required QoS level.
- Graceful service degradation in a quickly changing environment.
- Scalability making the algorithm efficient even with growing number of user sharing the same channel.

Figure 1 shows a simplified scheme of a packet and spectrum scheduler working in conjunction in a CR system. The packet scheduler firstly schedules the packets according to their priority and the predefined QoS. Secondly, the spectrum scheduler assigns resources, identified by the spectrum access mechanism, to each packet for transmission according to set rules. Such a combined system can provide better service quality to users and improve the system performance.

In wireless networks the links have time-varying channel capacities with errors occurring in bursts, due to time and location dependent signal attenuation, fading, interference and noise. In CRN, we also have to consider the presence of PUs, which furthermore complicates the situation. Under such demanding circumstances, a packet differentiation helps to selectively improve the probability of fulfilling the required quality parameters. To be able to make well informed scheduling decisions, a wide range of timely information is necessary in general, from internal queue states and application requirements to outer channel quality, regulatory policies and PU activity.

The use of forward error control (FEC) and automatic repeat request (ARQ) is a standard way of dealing with unfavorable wireless channel conditions and is therefore also part of many CR multimedia delivery frameworks (e.g. Hu, Mao, Hou, and Reed, 2010 and Polacek, Yang, and Huang, 2011), working in conjunction with scheduling to improve QoS. Both schemes remedy transmission loss by introducing redundancy but differ in their target scenarios for real-time multimedia applications with stringent timing requirements.

In addition, video is being distributed using unicast transmission to single users or multicast transmission for more efficiency, to a group of users. In a unicast scenario, the scheduler has to consider the requirements and capabilities of only one receiving user, thus being able to better adapt to the situation of the individual user. With a growing number of users in a multicast group, a considerable amount of bandwidth can be saved throughout the network. Multicast efficiency comes at a price however, as a nontrivial situation emerges, with different wireless users having in general heterogeneous channel conditions and different receiving equipment. A major performance factor in multicasting is the rate chosen by the sender and additional functionality has to be incorporated into the scheduler design to accommodate the heterogeneity of receiving rates. For users capable of receiving a high rate, a low rate will underutilize their link and provide them with inadequate QoS. For users with low link capacity, a high transmission rate can cause serious link congestion and loss. A conservative approach is to tailor the transmission rate to the needs of the bottleneck user and provide the lower quality stream to all users. Scheduling in conjunction with scalable video presents itself as a possible solution to this problem (Hwang, 2009) by being able to flexibly adjust the receiving rate and independently satisfy heterogeneous users. Examples of algorithm utilizing different kinds of layered video multicasting can be found in the next section.

9.3 JOINT SPECTRUM ACCESS AND SCHEDULING

An end-to-end transmission with CR as part of the transmission path can encounter a series of phenomena with possible negative effects, from queuing loss and delay, to CR specific problems, where the simultaneous usage of the same frequency space with the PU and bandwidth limited by PU activity are the foremost. PU transmissions have a twofold effect on CR. First, a CR cannot interfere with the PU functioning, therefore the resources available to CR can be scarce at times. The resulting bottlenecks can cause loss and delay. Second, PU transmission can directly influence the CR transmission if both occur simultaneously causing transmission errors. With such properties, the channel availability is very dynamic.

At the same time, the real time nature of multimedia strongly dictates their QoS requirements, putting delay, jitter and throughput at the forefront while, contrary to data transmissions, a limited level of loss being tolerable. Excessive delay degrades the multimedia experience, especially for real-time video data received after a certain threshold cannot be used for playback and are discarded. For this reason, delay has to be kept low to preserve the real-time properties and relatively constant to minimize the effects of jitter, which further complicates multimedia playback if it is too excessive. For these ends, a minimum amount of bandwidth has to be available at all times. Under these channel and application requirements, to implement a QoS enabled CR system is indeed a challenging task.

9.3.1 Motivation behind Joint Design Concepts

OSA, scheduling, AMC and other discussed mechanisms are solutions to specific problems that a CR system can encounter and the implementation of each of them improves the overall characteristics. To further improve the system performance, collaboration between different mechanisms is necessary, which calls for a joint design incorporating different partial concepts into a unifying solution.

We will illustrate this principle on the combination of OSA and scheduling. The OSA mechanism decides which parts of the spectrum to access, trying to maximize available resources and minimize PU collisions. The selected channels are then made available to the scheduler (in this case combining the functionality of the packet and the spectrum scheduler), which then distributes these resources to individual users. An OSA mechanism, which is not able to exploit all spectrum opportunities, will prevent the scheduler from fully utilizing the available resources and achieving the full potential for QoS delivery. Excessive probability of PU presence misdetection, causing frequent collisions again lowers the QoS level by raising the transmission errors probability. A correctly working OSA implementation will enable the scheduler to achieve full resource utilization, but to further improve performance, a joint OSA and scheduler design is necessary. An OSA mechanism that is capable of pushing information about the channels (e.g. probability of a channel being free of PU activity) to the scheduler, will improve the performance of the scheduler by enabling channel differentiation, facilitating lower transmission error for data of higher importance. A scheduler capable of incorporating the channel information into the scheduling process will fare better than a scheduler unable to understand such information. As this example shows, a joint design improves performance by enabling information flow and cooperation between different partial mechanisms.

The basic joint design in CRNs is combining optimization on lower protocol stack layers, with the aim of improving general application performance, but without direct incorporating of information about the applications itself into the design. The usual mechanisms being targeted for optimization are AMC, spectrum sensing, sensing strategy and access strategy. The improvement of system performance is mainly on throughput, with general beneficial effects to arbitrary applications. Another group of cross layers designs ultimately targets the application performance, compared to the general system performance of the already mentioned basic joint designs. The goal of improving application performance requires incorporating application layer information into the algorithm. The additional source of information enables a better and more specific adjustment of system behavior and parameters to suit the application needs. Figure 2 shows a schematic representation of a joint design, providing an brief overview of possibilities in term of layers, their functionalities and exchanged information that can be integrated into such a design.

9.3.2 Joint General Purpose System Designs

The next part presents examples of joint designs intended for CR based systems and networks. These examples highlight some mechanisms, considered criteria, and different approaches available in developing a joint design.

We start by describing designs intended to improve general system performance, which have a general design objective to improve operational parameters, mainly throughput and delay, and maintaining fairness. The first two designs are intended for ad hoc/mesh networks, where no central entity with a control function is present. The next four designs consider an infrastructure-based

Figure 2. A schematic representation of a joint design. Protocol layers, with selected functionalities, types of information, and information flow are also presented.

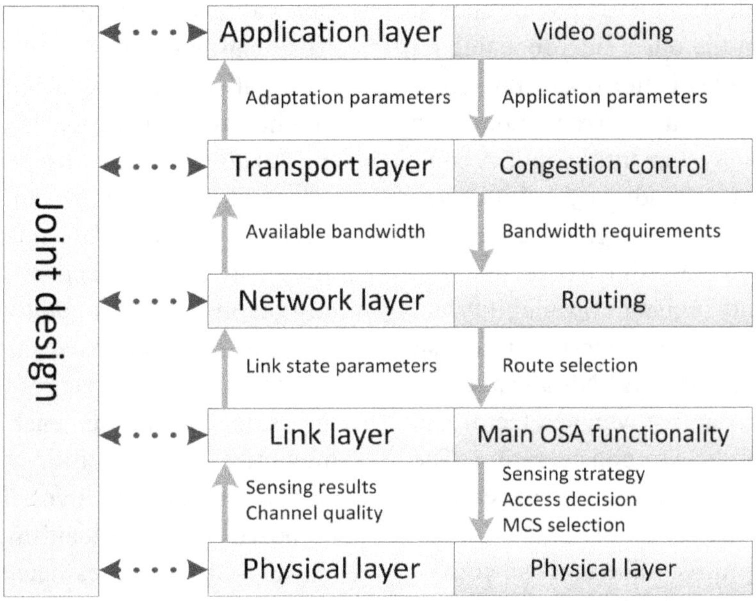

network; with a base-station present to facilitate centralized control.

An assisted multicast scheduling in wireless cognitive mesh networks is proposed in Almasaeid and Kamal, 2010. The cell of a mesh CRN consists of several mesh routers, which enable outside connectivity through one or more gateway nodes. Each of these mesh routers manages a number of mesh clients. In the unassisted multicast case, only mesh routers transmit data. The stipulated problem is potential heterogeneity in channel availability for members of a multicast group and the need to transmit data over multiple channels, effectively prolonging the multicast time. Assisted multicast scheduling mitigates the prolonging of multicasting time and covers two distinct activities, first being the forwarding of received data by the mesh clients inside of one multicast group and second forwarding data in the form of bitwise XORed packets to members of different multicast groups. Assisted multicast considerably reduces multicast time, which in turn increases the system throughput.

In Wang, Jiang, Li, and Liu (2010), the authors focus on the achievable throughput of an ad hoc CRN by means of spectrum access scheduling. Both the primary and secondary networks are TDMA based and it is interesting to note, that the authors assume the existence of a centralized scheduler responsible for scheduling all transmissions in a given network. A metric called throughput decrement ratio (TDR) is introduces, to measure the ratio of throughput of PUs in the presence of SUs. The PU have the right to determine the threshold of TDR, giving them control of the throughput reduction caused by SU activity. By using the schemes proposed by the authors, when density of PUs is sparser than the density of SUs and the densities ratio achieves a certain value, the primary and secondary ad-hoc networks can achieve simultaneously the same order of the aggregated multicast throughput as if they were stand-alone.

In his thesis Wang (2009) investigates the implementation of scheduling in single cell infrastructure based CRN to achieve higher resource utilization, system throughput and maintain fairness among users. He stipulates that fairness is at risk, if only users with the highest SNR are chosen for transmission. On the other side, focusing on fair scheduling provides better opportunities to users with lower SNRs, causing a reduction of the maximum achievable system throughput. A compromise between these contradictory requirements has to be found. An adaptive weighted scheduling is proposed to achieve the outlined objectives. To counter the uncertainty in channel availability and SNR values, an adaptive scheduler is devised, that takes the ratio of instantaneous SNR and average SNR in the form of adaptive weighed factor and the channel availability into consideration. The resulting scheme achieves balance between fairness and system throughput.

A design of a cognitive radio that can coexist with multiple parallel WLAN channels while abiding by an interference constraint is investigated in Geirhofer, Tong, and Sadler (2008). The authors study by experiment the impact of CR on the WLAN and derive a cognitive medium access (CMA) scheme. By channel measurement and predicting the WLAN behavior with the use of a continuous-time Markov chain model, the CMA enables coexistence in one spectrum space. CMA accesses the channel in a slotted manner and uses idle periods between bursty packet transmissions of the PU and frequency hopping, thus limiting interference with PU transmission. This design is based on a frequency hopping system and could be viewed as a "smart" Bluetooth device. The authors derive access schemes for fully observable scenario, when all channels can be observed simultaneously and partially observable scenario, when only one channel can be sensed at a time. Simulations show that the partially observable performing similarly as the fully observable. Performance evaluation shows improvement over plain Bluetooth/WLAN, which proves advantages of cognitive based channel access.

In Huang and Krishnamurthy (2009), the authors investigate a game theory based approach for resource allocation in a network with a central controller. The controller decisions are based on state values reported to the controller by the users. A user can abuse this mechanism by reporting false information to gain an advantage at the expenses of other users; rising his own utility while lowering the utility of other users and the system-wide value. This presents an undesired scenario and a real possibility with emerging users that are more sophisticated and can exploit cooperative designs in a non-cooperative manner. A pricing mechanism is proposed that binds OSA with a mechanism that prevents users from misbehavior, thus safeguarding that each rational selfish user maximizes his own utility function, at the same time optimizing the overall system utility. The proposed pricing mechanism is based on the classic Vickrey-Clark-Groves mechanism and had several desirable economic properties. Nash equilibrium is obtained by implementing a learning algorithm and is shown to be system optimal.

The authors of Jin, Xu, and Li (2010) present a multicast scheduling protocol for CRN. Instead of relying on a single-hop approach requiring globally available channels, a multi-hop approach is chosen, which requires only locally available channels mitigating multicast loss and delay effects. The algorithm jointly considers design factors as power control, relay assignment, buffer management, dynamic spectrum access, primary user protection, and fairness. Two scheduling algorithms, greedy centralized multicast scheduling and distributed online multicast scheduling protocol based on stochastic Lyapunov optimization are proposed, with the second being able to provide explicit over-time performance guarantees in realistic systems. Simulations undertaken by the authors show that the multicast performance in CRN can be significantly improved with the effective use of scarce wireless spectrum, by applying power control, cooperative communication and network coding.

9.3.3 Joint QoS System Designs

Based on general CR systems, QoS systems have final objective to improve the QoS of the user. By defining a measurable and objective quantity, an algorithm can not only target on throughput or latency improvement, but specifically aim at refining application performance. Such a measure in video transmission related designs can be video utility, PSNR or distortion. We will present three designs devised for ad hoc and two designs developed for infrastructure based networks.

In Luo, Ci, Wu, and Tang, (2010), the authors present a framework for real time video transmission in multi-hop networks. The end-to-end system design objective is the optimization of the user-perceived video quality. This objective is achieved under delay constrains in a wireless CRN. The system optimizes encoding parameters at the application layer, scheduling at the MAC layer and modulation and coding schemes (MCS) at the PHY layer, to determine the coding and transmission of the video packet. By mathematically expressing the objective (video distortion) and the constrains (delay bounds), the authors were able to formulate the proposed problem as a MIN-MAX problem, which can be solved with the means of dynamic programming. Simulations show the improved PSNR of the received video, outperforming schemes that optimize only PHY layer parameters or PHY and APP layer parameters. The performance gain is especially evident in more demanding circumstances with lower SNR or tighter delay demands.

The authors of Bocus, Coon, Canagarajah, McGeehan, Armour, and Doufexi (2011) present a method for resource allocation in OFDMA based multiuser CRN, specifically designed for transmission of fine grain scalability (FSG) video. A minimum amount of bandwidth necessary for base-layer transmission is defined together with the maximum channel width equal to the maximum total video rate, as not to waste resources. In these boundaries, they design a near-optimal resource allocation scheme for video transmission. To improve video quality, optimization for bit and power allocation of subcarrier, is achieved through means of linear programming. The proposed scheme never exceeds the imposed interference power limit, while maximizing the sum rate over all users and achieving fairness among multiple transmissions. Adjusting to accommodate coarse grain scalable (CGS) or medium grain scalable (MGS) video transmission is also possible.

Authors of Hu and Mao (2010) propose a system for delivery of FGS and MGS videos, to enable video transmission in an environment with heterogeneous channel availability and fast changing network condition. The algorithm is conceived with these objectives: to maximize the overall video quality and to achieve fairness among the concurrent video sessions, subject to bounded interference to PUs. A mesh network is considered, where individual users pose as relay stations for video data streamed by other users belonging to the same CRN. This network shares the spectrum with one or more fixed primary networks (PN). The authors assume the existence of a common control channel for the exchange of management information among individual users, which non-intrusively utilize the spectrum for unicast video communications. As the data have to be potentially relayed through several users, an amplify-and-forward approach is implemented to model and guarantee end-to-end video performance. This prevents cumulative processing delay from seriously degrading video quality, by minimizing delay and jitter and maximizing the available bandwidth. A mixed integer nonlinear programming (MINLP) problem, which is further decomposed into a channel scheduling problem and a path selection problem, is formed accordingly. A fast greedy algorithm is also derived to solve the channel-scheduling problem for each user and a distributed algorithm is presented for solving the path selection problem. Simulations of the presented solution suggest that it performs close to the upper bound in the examined cases.

An interesting observation made by the authors is the performance gap between MGS and FGS video, with MGS performing better due to the rate-distortion efficiency of MGS.

For the infrastructure type of systems, the authors of Hu et al. (2010), investigate fine grained scalability (FSG) video multicasting application over it. The goal of this design is optimizing the overall received video quality, achieve proportional fairness among multicast users and to keep the interference to PU in the predefined limits. The problem is defined as a mixed integer nonlinear programming (MINLP) problem, for which solving a sequential fixing algorithm and a greedy algorithms are devised. The algorithms are able to set parameters of MCS and FSG video layers. ARQ is used to counter transmission loss and guarantee video data delivery. The channels are accessed under limitations on the maximum allowed collision probability, with the objective of maximizing the system-wide utility function. For each video, a base layer and an enhancement layer are created. The algorithm then assigns resources to enhancement layers to determine the most efficient data rate for each of them, thus maximizing the overall system utility. The greedy algorithm is also shown to be performing well in low complexity implementation.

The design of Polacek et al. (2011) presents a joint opportunistic spectrum access and scheduling for layered video multicasting in infrastructure based CRNs. Besides adopting H.264/SVC video layering, quality based ranking in OSA and opportunistic layered multicasting (OLM) (Huang et al., 2011) are added into the scheduling equation. The resource optimization scheme firstly serves the mandatory video layers to all subscribed users using minimum amount of resource and the offers additional enhancement layers to some selected users by optimizing the total utility.

The OSA implemented enables more available bandwidth and simultaneously keeps the PU interference at a predefined level. OLM, which means a more aggressive selecting of MCS, uses the time variation of channel conditions to serve users with momentary relative good channel quality and FEC to recover intended data loss. Thus the amount of successfully decoded video data can be significantly increased. Also the channel 2 step ranking process that incorporates information about the channel occupancy estimation and physical channel quality into the scheduling is a major novelty. For scheduling purposes, primary channels (PC), containing several secondary channels (SC), are ranked according to their estimated availability probability, with better quality PCs being selected first, in order to minimize the impact of collisions with the PU on more important video data. SCs inside of one PC are also sorted according to the MCS dependent transmission rate to reduce the overall number of used SCs. This information flow improves the algorithm performance in accordance with the formerly stated paradigm of the importance of information flow in CR systems.

For comparison, 3 other schemes were implemented and evaluated in Polacek et al. (2011). Scheme 2 differs from the basic design in a more conservative channel access scheme that lowers the interference to the PU but simultaneously lowers the available bandwidth. Schemes 3 and 4 both use the same access scheme as scheme 1, but select MCS as to fulfill the requirements of the users with worst channel quality. Scheme 3 does not implement any FEC protection and scheme 4 only a basic amount to counter the expected PU collisions.

We demonstrate system performance measurements, which were improved by the joint scheduling framework. The normalized utility is the designated performance measure and is used for final comparison. System performance is shown in Table 2. and Table 3. and points to the advantage these designs have in delivering quality video. As Table 2. shows, the throughput is higher for schemes 1 and 2, thanks to OLM like scheduling, with more aggressive MCS selection and using situation appropriate FEC protection. Schemes 3 and 4 suffer from lower throughput that deterio-

Table 1. Overview of compared schemes. Scheme 1 is the Joint Opportunistic Spectrum Access and Scheduling algorithm that uses OSA and opportunistic multicasting. Scheme 2 uses more conservative channel access and schemes 3 and 4 user non opportunistic multicasting scheduling and no or only minimal amount of FEC protection.

Scheme ID	1	2	3	4
Channel Access	OSA	NOSA	OSA	OSA
Receiving rate tracking	Yes	Yes	No	No
FEC Protection	Dynamic	Dynamic	No	Basic amount
MCS choice	OLM	OLM	NOLM	NOLM
Channel ranking	2 step	2 step	Simple	Simple

rates faster with the rising numbers of multicast group members, when compared with schemes 1 and 2. The decoding error rate (DEC) is shown in Table 3. The advantage provided by the dynamic FEC can be seen, as the DEC for schemes 1 and 2 are stayed at zero regardless of the number of users in the multicast group. Scheme 3 suffers from low DEC under 1% thanks to the minimal implemented FEC protection. Scheme 4 is not protected by FEC and suffers from 8.9 to 9.4% of DEC, which strongly influences the video utility.

Final comparison can be seen in Figure 3, with scheme 1 performing the best, outperforming scheme 2 only due to more aggressive channel access. Scheme 3 shows the worst results, due to lower throughput and poor error resilience. Scheme 4 is somewhat better in base-layer delivery, but lacks considerably behind schemes 1 and 2 in enhancement layers delivery.

9.4 FUTURE RESEARCH DIRECTIONS

In the future, we reckon with development towards more efficient and flexible designs able to process a great amount of diverse information fueled by progress in CR technology. With the CR technology maturing over time, the still unsolved user centric issues are expected to receive serious attention.

As the CR environment is very dynamic, we can expect transmission errors and this induced loss is detrimental to multimedia QoS. Approaches like hybrid-ARQ and data re-fragmentation can be

Table 2. System wide throughput achieved by different schemes

Users	Throughput for different schemes (Mbps)			
	1	2	3	4
10	12.5	10.8	9.6	9.1
25	11.4	9.9	6.9	6.6
40	10.8	9.4	6	5.8
55	10.2	8.9	5.5	5.4
70	10.2	8.9	5.3	5.3
85	10.1	8.8	5.2	5.2
100	10.1	8.8	5.2	5.1

Table 3. Decoding error rate for different schemes

Users	Decoding error rate for different schemes (%)			
	1	2	3	4
10	0	0	8.9	0.5
25	0	0	9.2	0.3
40	0	0	9.3	0.4
55	0	0	9.2	0.6
70	0	0	9.3	0.5
85	0	0	9.2	0.4
100	0	0	9.4	0.6

considered for future implementations; in order to find a better balance between protection and the induced overhead. Spreading information into different primary channels, as not to suffer from QoS degradation if a collision occurs (Willkomm & Wolisz, 2010) is another way of data protection, which mitigates the negative impact of errors localized on one channel.

To improve application performance, specialized algorithm designs with improved internal information flow and build to fulfill user QoE requirements are necessary. Which information to exchange across layers and how to facilitate this exchange is indeed an interesting academic research area that promises interesting outcomes (Steenkiste, Sicker, Mindn, & Raychaudhuri, 2009). Channel ranking based on their quality (SNR value, collision probability, interference, etc.) is one way of improving information exchange and enabling a better informed decision

Figure 3. Normalized system utility for different schemes. Scheme 1 outperforms all other schemes, with scheme 2 still showing satisfying results. Utility achieved by scheme 4 is lacking behind schemes 1 and 2 while scheme 3 shows a very low overall utility.

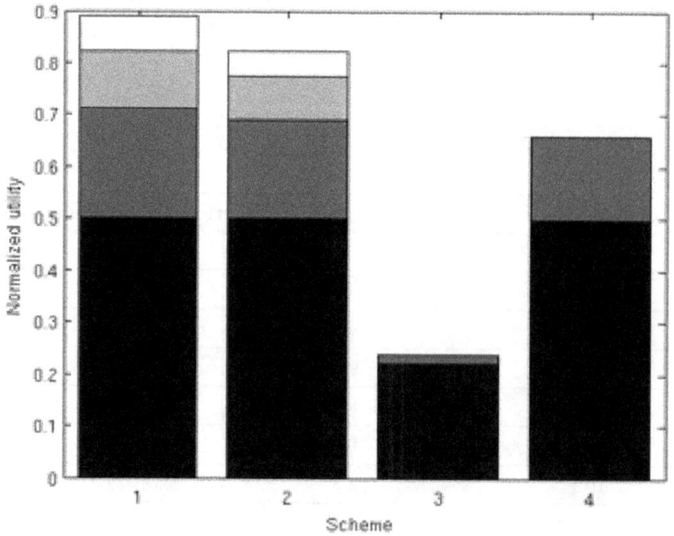

when selecting channels for transmission. This way, higher priority data can be transmitted over a better quality channel with less error probability. In multiuser scenarios, game theoretical approaches can be applied to better respond to needs of individual users and facilitate their coexistence.

Source-channel resource management is another area that deserves attention. Except the CRN responding to the needs of the multimedia source by adapting its parameters, it is possible to change the source-channel parameters in response to the current situation. A flexible and fully integrated adaptation starting from the source requires fewer interventions at the lower layers. For example reducing the output video rate in response to less available bandwidth reduces the negative impact on the resulting video quality.

9.5 CONCLUSION

In this chapter, we have covered the topics concerning QoS multimedia scheduling and opportunistic spectrum access. We described video, the background of video transmission through CR networks and the problems and issues that might arise in this context. Examples of possible solutions were presented, together with an outlook at the research directions. Overall, the chapter provides essential knowledge of transmitting multimedia in cognitive radio environment, most important problems, state-of-the-art solutions, and future perspectives.

REFERENCES

Akyildiz, I., Lee, W. Y., & Chowdhury, K. (2009). Spectrum management in cognitive radio ad hoc networks. *IEEE Network*, *23*(4), 6–12. doi:10.1109/MNET.2009.5191140

Akyildiz, I. F., Vuran, M. C., & Mohanty, S. (2008). A survey on spectrum management in cognitive radio networks. *IEEE Communications Magazine*, *46*(4), 40–48. doi:10.1109/MCOM.2008.4481339

Almasaeid, H. M., & Kamal, A. E. (2010). Assisted-multicast scheduling in wireless cognitive mesh networks. *2010 IEEE International Conference on Communications (ICC)*, (pp. 1–5).

Baldo, N., & Zorzi, M. (2008). Fuzzy logic for cross-layer optimization in cognitive radio networks. *IEEE Communications Magazine*, *46*(4), 64–71. doi:10.1109/MCOM.2008.4481342

Bhat, C. T. P. S. (2011). *Opportunistic scheduling, cooperative relaying and multicast in wireless networks.* Unpublished doctoral dissertation Arizona State University, Phoenix Metropolitan Area, Arizona, USA

Bocus, M. Z., Coon, J. P., Canagarajah, C. N., McGeehan, J. P., Armour, S. M. D., & Doufexi, A. (2011). Resource allocation for OFDMA-based cognitive radio networks with application to H.264 scalable video transmission. *EURASIP Journal on Wireless Communications and Networking*, *2011*, 1–10. doi:10.1155/2011/245673

Chen, Y., Zhao, Q., & Swami, A. (2008). Joint design and separation principle for opportunistic spectrum access in the presence of sensing errors. *IEEE Transactions on Information Theory*, *54*(5), 2053–2071. doi:10.1109/TIT.2008.920248

Erlebach, T., & Grant, T. (2010). *Scheduling multicast transmissions under SINR constraints. algorithms for sensor systems* (pp. 47–61). Berlin, Germany: Springer.

Fattah, H., & Leung, C. (2002). An overview of scheduling algorithms in wireless multimedia networks. *IEEE Wireless Communications*, *9*(5), 76–83. doi:10.1109/MWC.2002.1043857

Geirhofer, S., Tong, L., & Sadler, B. (2008). Cognitive medium access: constraining interference based on experimental models. *IEEE Journal on Selected Areas in Communications*, *26*(1), 95–105. doi:10.1109/JSAC.2008.080109

Geirhofer, S., Tong, L., & Sadler, B. M. (2006). Dynamic spectrum access in WLAN channels: Empirical model and its stochastic analysis. *Proceedings of the First International Workshop on Technology and Policy for Accessing Spectrum.*

Harrison, K., Mishra, S. M., & Sahai, A. (2010). How much white-space capacity is there? *IEEE Symposium on New Frontiers in Dynamic Spectrum Access Networks (DySPAN 2010)*, Singapore, (pp. 6–9).

Hincapie, R., Tang, J., Xue, G., & Bustamante, R. (2008). QoS routing in wireless mesh networks with cognitive radios. *IEEE Global Telecommunications Conference*, (pp. 1–5).

Hu, D., & Mao, S. (2010). Streaming scalable videos over multi-hop cognitive radio networks. *IEEE Transactions on Wireless Communications*, *9*(11), 3501–3511. doi:10.1109/TWC.2010.092810.100098

Hu, D., Mao, S., Hou, Y. T., & Reed, J. H. (2010). Scalable video multicast in cognitive radio networks. *IEEE Journal on Selected Areas in Communications*, *28*(3), 334–344. doi:10.1109/JSAC.2010.100414

Huang, C.-W., Huang, S.-M., Wu, P.-H., Lin, S.-J., & Hwang, J.-N. (2011). OLM: Opportunistic layered multicasting for scalable IPTV over mobile WiMAX. *IEEE Transactions on Mobile Computing*, *11*(3), 1–9.

Huang, J., & Krishnamurthy, V. (2010). Transmission control in cognitive radio as a Markovian dynamic game: Structural result on randomized threshold policies. *IEEE Transactions on Communications*, *58*(1), 301–310. doi:10.1109/TCOMM.2010.01.080157

Huang, J. W., & Krishnamurthy, V. (2009). Truth reveling opportunistic scheduling in cognitive radio systems. *IEEE 10th Workshop on Signal Processing Advances in Wireless Communications*, (pp. 91–95).

Huang, S., Liu, X., & Ding, Z. (2008). Opportunistic spectrum access in cognitive radio networks. *2008 IEEE INFOCOM- The 27th Conference on Computer Communications*, (pp. 1427-1435).

Hwang, J. (2009). *Multimedia networking: From theory to practice*. Leiden, The Netherlands: Cambridge University Press. Retrieved from http://cdsweb.cern.ch/record/1253368

Jin, J., Xu, H., & Li, B. (2010). Multicast scheduling with cooperation and network coding in cognitive radio networks. *2010. Proceedings - IEEE INFOCOM*, 1–9. doi:10.1109/INFCOM.2010.5461917

Kumar, A., Wang, J., Challapali, K., & Shin, K. G. (2010). Analysis of QoS provisioning in cognitive radio networks: a case study. *Wireless Personal Communications*, *57*(1), 53–71. doi:10.1007/s11277-010-0006-8

Li, S. (2010). *Channel allocation for smooth video delivery over cognitive radio networks: Perspective*. University of Waterloo. Waterloo, Ontario, Canada. Retrieved from http://hdl.handle.net/10012/5416

Low, T.-P., Pun, M.-O., Hong, Y.-W., & Kuo, C.-C. (2010). Optimized opportunistic multicast scheduling (OMS) over wireless cellular networks. *IEEE Transactions on Wireless Communications*, *9*(2), 791–801. doi:10.1109/TWC.2010.02.090387

Low, T. P., Pun, M. O., & Kuo, C. C. J. (2008). Optimized opportunistic multicast scheduling over cellular networks. *IEEE Global Telecommunications Conference*, (pp. 1–5).

Lu, J., & Ma, M. (2011). Cognitive radio-based framework and self-optimizing temporal-spectrum block scheduling for QoS provisioning in WiMAX. *Computer Networks*, *55*(6), 1291–1309. doi:10.1016/j.comnet.2010.12.003

Luo, H., Ci, S., Wu, D., & Tang, H. (2010). Cross-layer design for real-time video transmission in cognitive wireless networks. *2010 INFOCOM IEEE Conference on Computer Communications Workshops*, (pp. 1-6).

Mitola, J. III, & Maguire, G. Q. Jr. (1999). Cognitive radio: making software radios more personal. *IEEE Personal Communications*, *6*(4), 13–18. doi:10.1109/98.788210

Ngo, D. T., Tellambura, C., & Nguyen, H. H. (2009). Resource allocation for OFDM-based cognitive radio multicast networks. *IEEE Wireless Communications and Networking Conference*, (pp. 1–6).

Niyato, D., Hossain, E., & Han, Z. (2009). Dynamic spectrum access in IEEE 802.22-based cognitive wireless networks: A game theoretic model for competitive spectrum bidding and pricing. *IEEE Wireless Communications*, *16*(2), 16–23. doi:10.1109/MWC.2009.4907555

Polacek, P., Yang, T.-Y., & Huang, C.-W. (2011). *Joint opportunistic spectrum access and scheduling for layered multicasting over cognitive radio networks*. IEEE International Workshop on Multimedia Signal Processing, 2011.

Razavilar, J., Liu, K. J. R., & Marcus, S. I. (2002). Jointly optimized bit-rate/delay control policy for wireless packet networks with fading channels. *IEEE Transactions on Communications*, *50*(3), 484–494. doi:10.1109/26.990910

Sadler, M. B., & Zhao, Q. (2007). A survey of dynamic spectrum access. *IEEE Signal Processing Magazine*, *24*(3), 79–89. doi:10.1109/MSP.2007.361604

Shan, Y. (2005). Cross-layer techniques for adaptive video streaming over wireless networks. *EURASIP Journal on Advances in Signal Processing*, *2*, 220–228. doi:10.1155/ASP.2005.220

Steenkiste, P., Sicker, D., Minden, G., & Raychaudhuri, D. (2009). Future directions in cognitive radio network research. *Network*, 1–40. Retrieved from www.cs.cmu.edu/~prs/NSF_CRN_Report_Final.pdf

Stevenson, C., Chouinard, G., Shellhammer, S., & Caldwell, W. (2009). IEEE 802.22: The first cognitive radio wireless regional area network standard. *IEEE Communications Magazine*, *47*(1), 130–138. doi:10.1109/MCOM.2009.4752688

Su, H., & Zhang, X. (2008). Cross-layer based opportunistic MAC protocols for QoS provisionings over cognitive radio wireless networks. *IEEE Journal on Selected Areas in Communications*, *26*(1), 118–129. doi:10.1109/JSAC.2008.080111

Tian, C., & Yuan, D. (2008). Cross layer opportunistic scheduling for multiclass users in cognitive radio networks. *2008 4th International Conference on Wireless Communications, Networking and Mobile Computing*, (pp. 1-4).

Tsai, J. T., & Cruz, R. L. (2011). Opportunistic multicast scheduling for information streaming in cellular networks. *IEEE Transactions on Wireless Communications*, *10*(6), 1776–1785. doi:10.1109/TWC.2011.040411.100517

Wang, C., Jiang, C., Li, X.-Y., & Liu, Y. (2010). Multicast throughput for large scale cognitive networks. *Wireless Networks*, *16*(7), 1945–1960. doi:10.1007/s11276-010-0237-3

Wang, F. (2009). *Adaptive weighted scheduling in cognitive radio networks.* Waterloo, Canada: University of Waterloo. Retrieved from http://hdl.handle.net/10012/4452

Willkomm, D., & Wolisz, A. (2010). Efficient QoS support for secondary users in cognitive radio systems. *IEEE Wireless Communications*, *17*(4), 16–23. doi:10.1109/MWC.2010.5547917

Won, H., Cai, H., Eun, D., Guo, K., Netravali, A., Rhee, I., & Sabnani, K. (2009). Multicast scheduling in cellular data networks. *IEEE Transactions on Wireless Communications*, *8*(9), 4540–4549. doi:10.1109/TWC.2009.080330

Yu, F. R., Sun, B., Krishnamurthy, V., & Ali, S. (2010). Application layer QoS optimization for multimedia transmission over cognitive radio networks. *Wireless Networks*, *17*(2), 371–383. doi:10.1007/s11276-010-0285-8

Yucek, T., & Arslan, H. (2009). A survey of spectrum sensing algorithms for cognitive radio applications. *IEEE Communications Surveys & Tutorials*, *11*(1), 116–130. doi:10.1109/SURV.2009.090109

Zhang, L., Zeng, K., & Mohapatra, P. (n.d.). Opportunistic spectrum scheduling for mobile cognitive radio networks in white space. *IEEE Wireless Communications and Networking Conference (WCNC)*, (pp. 844–849).

Zhao, Q., Geirhofer, S., Tong, L., & Sadler, B. M. (2008). Opportunistic spectrum access via periodic channel sensing. *IEEE Transactions on Signal Processing*, *56*(2), 785–796. doi:10.1109/TSP.2007.907867

Chapter 10
Cognitive Cooperation in Wireless Networks

Eng Hwee Ong
Nokia Research Center, Finland

Jamil Y. Khan
The University of Newcastle, Australia

ABSTRACT

In recent years, vertical handover (VHO) has been identified as the primary vehicle to provision seamless mobility and quality of service (QoS) transparency for end-user in composite network. This allows end-user to enjoy ubiquitous connectivity in the most efficient way, irrespective of time and place, commonly known as always best connected. In this chapter, the authors introduce the notion of cognitive cooperation as a means to provide optimized VHO opportunistically in order to exploit the inherent heterogeneity that exists within such composite network to improve radio resource usage. Through the cognitive cooperation, the chapter proposes a distributed load adaptation strategy (LAS) framework which exploits the benefits of joint optimization, particularly between link adaptation and load adaptation on-demand. The authors advocate that such synergetic interactions between the physical layer (PHY) and medium access control (MAC) layer have advantages over the PHY approach based only on link adaptation. Comprehensive performance analyses show that the LAS framework arbitrates a QoS-balanced system in which statistical QoS guarantee for multimedia traffic can be provisioned and overall system capacity can be maximized.

10.1 INTRODUCTION

Recent advancements in wireless network technologies have resulted in several new concepts and terms to support advanced radio system technologies. For instance, the composite network architecture has been defined in the IEEE 1900.1 standard (IEEE Std 1900.1-2008, 2008) as a type of wireless communication network that comprises of multiple radio access technology (multi-RAT) under a single or multiple network management control to support efficient commu-

DOI: 10.4018/978-1-4666-2005-6.ch010

nication. Note that the key difference between a composite network and heterogeneous network, according to the IEEE 1900.1 standard, is that the former focuses on the sharing of resources while the latter does not. In similar context, the term 'cognition' central to cognitive radio has been dilated to cognitive network lately. Such cognitive network postulates the ability to optimize both user and network performances, as well as bring about better utilization of radio resources by adopting cognitive functionality, e.g., in the radio to exploit spectrum holes.

In fact, the IEEE 1900.1 standard states that cognitive functionalities may be applied to a composite network so that radio can select the best available option for communication. This is propelled by the fact that although cognitive radio can identify these spectrum holes, a parallel cognitive functionality is still required in the higher layer to harness any effective use of such additional spectrum. Given that the frequency spectrum is typically divided into multiple channels, the ramification of dynamic spectrum access supported by cognitive radio will manifest as the main challenge of managing these dynamically available heterogeneous channels which emanate from diverse parts of the spectrum with different propagation characteristics.

On the other hand, the fundamental problem of a composite network is imputed to the coordination of VHO between different radio access technologies (RATs). Recently, the research community is motivating the unification of cooperative and cognitive principles to exploit their highly complementary characteristics. According to (Fitzek & Katz, 2007), cooperation and cognition in wireless networks have significant inter-dependencies which intensify with the increasing heterogeneity of networks, terminals, and services. Although significant progress has been achieved in cooperative (Fitzek & Katz, 2006) and cognitive (Thomas, 2007) networks, their advancements are mainly autonomous. Therefore, it will be beneficial to explore the *cross-fertilization* of cooperation and cognition, which will become important as the heterogeneity of access networks, terminals, and services escalates. This elicits that cognition depends on the data acquisition of performance metrics through cooperation while cooperation depends on the awareness of surroundings through cognition. In particular, this chapter motivates that cooperation will become pertinent when there exists a mutual goal that requires the coordinated effort of more than one RAT. In this case, the mutual goal of a composite network is to harness overall system capacity and QoS improvements through optimized VHO fueled by cognitive cooperation.

For example, link adaptation or adaptive modulation and coding has been widely implemented across standards such as the 3GPP long term evolution (LTE) and IEEE 802.11xx wireless local area network (WLAN) to provide spectrally efficient and flexible data rate access while adhering to a target error performance over wireless channels. In general, it is known that the link adaptation of LTE system is more robust than WLAN. To be more specific, LTE system comprises of inner loop link adaptation which acts on channel quality information reported by the user equipment and (optional) outer loop link adaptation which relies on acknowledgment (ACK)/negative ACK (NACK) from hybrid adaptive repeat and request (HARQ) to correct systematic errors (Holma & Toskala, 2009). In contrast, most of WLAN's link adaptation are typically implemented based on the automatic rate fallback mechanism (Kamerman & Monteban, 1997), (Pang, Leung, & Liew, 2008).

Although link adaptation in LTE and WLAN systems has the ability to achieve optimum throughput by matching transmission parameters to the time-varying wireless channel conditions of a realistic wireless environment, it introduces an upper bound on the maximum achievable throughput due to the reduction of transmission rate in reality. This would inevitably affect users with bandwidth intensive and delay sensitive multimedia applications where QoS guarantee is extremely important. Such impact is especially pronounced in contention-based networks such as WLAN since link adaptation is known to give rise to the rate anomaly phenomenon (Heusse, Rousseau, Berger-Sabbatel, & Duda, 2003), (Tan & Guttag, 2004). Furthermore, it is worth noting that the link adaptation and handover control in LTE and WLAN have been designed to operate independently. However, we advocate that the exploitation of link adaptation and load adaptation through optimized VHO by leveraging on cognitive cooperation will bring about a QoS-balanced system with several advantages. In this chapter, we demonstrate this concept by using a multirate, multiple access point (multi-AP) WLAN and show that the rate anomaly problem of WLAN can also be mitigated. Without loss of generality, such cognitive cooperation concept could also be applied to other wireless system such as LTE or composite network such as LTE/WLAN to achieve optimized VHO.

The remainder of this chapter is organized as follows. Section 2 identifies some of the key challenges of multirate WLAN-based composite network which will arise from the rate anomaly phenomenon and reviews some pertinent works. Section 3 introduces the novel concept of cognitive cooperation. Section 4 presents the LAS framework that leverages on cognitive cooperation and discusses how this can be used to mitigate rate anomaly from a QoS perspective. Section 5 gives a detailed account of the simulation models and preliminaries. Section 6 presents a comprehensive performance evaluation of the LAS framework. Section 7 concludes this chapter.

10.2 CHALLENGES OF MULTIRATE WLAN-BASED COMPOSITE NETWORK

Within the research community, WLAN is envisioned as one of the introductory de-facto wireless access networks in future composite network, thanks to its pervasive deployments over many diverse environments. According to the Federal Communications Commission (FCC 05-57, 2005) and IEEE 1900.1 standard (IEEE Std 1900.1-2008, 2008), WLAN could be considered as a network with cognitive functionalities since it operates with listen-before-talk access protocol together with dynamically changing transmission powers and data rates to allow more efficient spectrum use. However, the delivery of QoS-demanding multimedia applications over WLAN is not trivial. Particularly, in the presence of the well-known rate anomaly problem, which is an inherent issue associated with a multirate WLAN-based composite network, arising from dynamically varying wireless channel conditions as a result of non-line-of-sight (NLOS) transmissions and dynamic spectrum access.

The rate anomaly of distributed coordination function (DCF) can occur when contending stations (STAs) in the same basic service set (BSS), with similar wireless channel conditions, transmit packets of same size but with different data rates. Under such conditions, the DCF preserves throughput fairness such that each STA would yield approximately the same throughput, regardless of its own data rate as illustrated in Figure 1(a). The rate anomaly of DCF in uplink transmissions is a direct consequence of throughput fairness which dictates an equal probability of channel access. In other words, slower rate STAs will disadvantage higher rate STAs by occupying more channel time to acquire approximately the same number of transmission opportunities. In fact, Figure 1(b) also illustrates that the aggregate throughput of multirate transmissions by two STAs using 11 Mbps and 1 Mbps is much lesser than the average of the total throughput that could be achieved when both STAs transmit using either 11 Mbps or 1 Mbps.

While the DCF will affect channel bandwidth allocation and cause rate anomaly in the uplink, the access point (AP) queueing mechanism will affect channel bandwidth allocation and similarly induce rate anomaly in the downlink for an

Figure 1. Impact of the rate anomaly of DCF in uplink transmissions. (a) STA's throughput. (b) Aggregate throughput

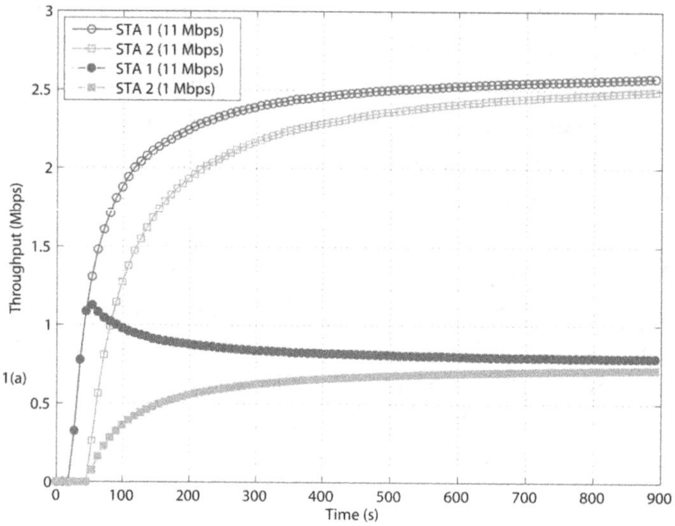

Figure 1(a)

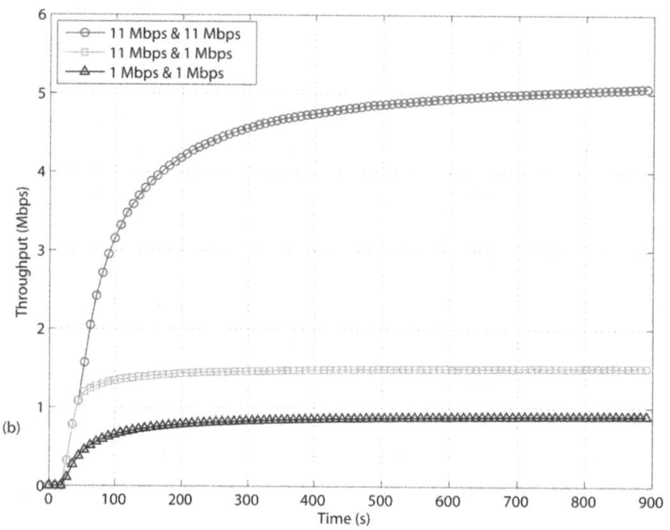

Figure 1(b)

infrastructure BSS. The rate anomaly of AP queueing mechanism in downlink transmissions is attributed to varying wireless channel conditions encountered at different STAs. For the case of a single first in first out (FIFO) queue in a typical IEEE 802.11 DCF-based AP, the head-of-line (HOL) blocking will occur as soon as the wireless channel quality toward the STA of the HOL packet is degraded due to burst error (Bhagwat, Bhattacharya, Krishna, & Tripathi, 1996) since the AP first performs retransmissions and eventually link adaptation to transmit that packet at a lower rate. Such prolonged HOL blocking will cause the unfair allocations of channel bandwidth and severe packet losses due to buffer overflow once the arrival rate exceeds the service rate. To overcome this form of unfairness due to location-dependent errors, many fair queueing mechanisms have been developed for wireless environments, e.g., wireless fair service, idealized wireless fair

Figure 2. Impact of the rate anomaly of AP queueing mechanism in downlink transmissions

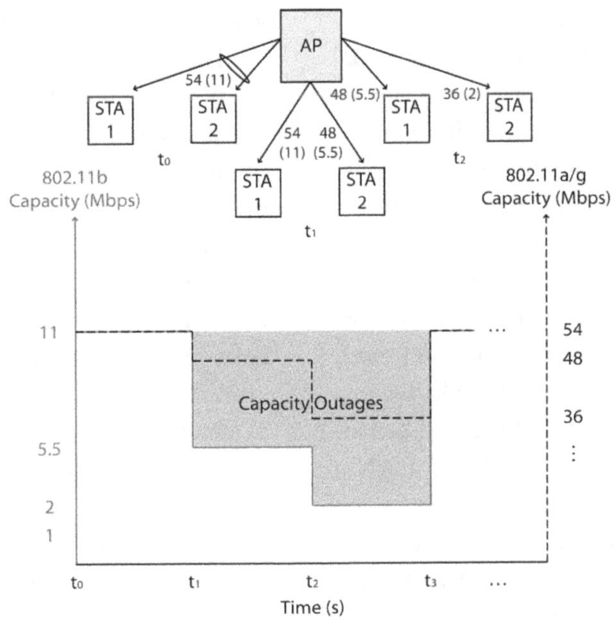

queueing algorithm, and channel-condition independent fair queueing (Bharghavan, Lu, & Nandagopal, 1999). These algorithms are primarily designed to consider only *downlink* scheduling based on a *single* rate server to provide throughput fairness. In reality, however, the IEEE 802.11 networks are half-duplexed and subjected to multirate operations as a consequence of link adaptation used to combat varying wireless channel conditions at the STAs. Hence, when these fair queueing mechanisms are deployed in the AP of multirate WLAN, the rate anomaly of AP queueing mechanism will occur as exemplified in Figure 2 and described in the following.

At time t_0, both STAs associated to the AP experience similar wireless channel conditions and receive packets from the AP at the maximum rate of 54 (11) Mbps. At time t_1, STA 2 experiences poor wireless channel conditions and receives packets from the AP at a reduced rate of 48 (5.5) Mbps due to link adaptation. By the throughput fairness property of AP queueing mechanism, STA 1 will be disadvantaged and eventually receives packets at a rate similar to STA 2. When this happens, the effective capacity of the system will be reduced to 48 (5.5) Mbps. At time t_2, STA 2 continues to suffer wireless channel quality degradation and the effective capacity of the system will be further reduced to 36 (2) Mbps. Clearly, the rate anomaly of AP queueing mechanism will create severe *capacity outages* in terms of packet loss when the offered load exceeds the downgraded system capacity, particularly, when the AP performs link adaptation to combat varying wireless channel conditions experienced by the STAs. Notice that such capacity outages will be even more significant for the IEEE 802.11b WLAN, as indicated by the data rates in parentheses, due to its coarser link adaptation steps.

Apparently, rate anomaly will be unavoidable in a multirate WLAN-based composite network since link adaptation is widely employed to combat varying wireless channel conditions which could occur as a consequence of: (i) channel impairments in hotspot and indoor environments arising from frequent NLOS transmissions caused by structures and obstacles; and (ii) dynamic spectrum access supported by cognitive radio, resulting in heterogeneous channels operation. The impact of rate anomaly will escalate for WLAN-based composite network with dynamic spectrum access capability since cognitive radio can opportunistically access diverse channels which are subjected to very different amount of frequency-dependent path loss, multipath effects, and attenuations. Specifically, the negative impact of rate anomaly will be significant when two or more STAs contend with diverse data rates *and* during heavy network load situations. Moreover, this impact would be magnified in the coexistence of the IEEE 802.11b and IEEE 802.11g WLANs as the former has coarser link adaptation steps. It is also worth to mention that the achievable throughput is not easily predictable if different data rates vary both on uplink and downlink, even though all STAs use the DCF and the AP employs a fair queueing mechanism (Tan & Guttag, 2004). Hence, a serious implication of rate anomaly is that it will *dilute* the benefits derived from any link adaptation techniques to exploit the tradeoff between date rate and bit error rate under varying wireless channel conditions.

10.2.1 Related Work

Numerous solutions have been proposed to address rate anomaly prevalent in multirate environment. Some notable works in (Tan & Guttag, 2004), (Tinnirello & Choi, 2005), and (Joshi, Mukherjee, Yoo, & Agrawal, 2008) address this problem from a temporal fairness perspective in order to overcome the throughput fairness of DCF. Tan and Guttag (Tan & Guttag, 2004) are the first to tackle the rate anomaly problem by advocating the concept of temporal fairness where each contending STA receives an equal share of the channel occupancy time. They have highlighted that such time-based fairness has an important baseline throughput property, which is not available from the throughput fairness of the DCF, and this could be used to mitigate the rate anomaly of multirate WLAN.

Tinnirello and Choi (Tinnirello & Choi, 2005) have shown that the enhanced distributed channel access (EDCA) channel access mechanism in the forthcoming IEEE 802.11e standard can also provide temporal fairness by using transmission opportunity (TXOP) to maintain equal channel occupancy time for all contending STAs. However, the authors show that there is a tradeoff between achieving temporal fairness and overall system bandwidth efficiency as the equalization of channel occupancy time requires fragmentation which introduces significant overheads. Most importantly, the study also reveals that temporal fairness can be guaranteed only under uniform TXOP settings. However, this would be difficult to realize in practice as STAs can select different fragmentation sizes without a fixed threshold, as opposed to the DCF fragmentation rule. Hence, the challenges in QoS provisioning for multimedia traffic delivery over the DCF remain as an open research issue.

Recently, Joshi *et al.* (Joshi, et al., 2008) present a distributed time fair carrier sense multiple access scheme to mitigate the rate anomaly problem of DCF. The authors use the baseline property suggested in (Tan & Guttag, 2004) to estimate a target throughput for each contending STA. The key idea is that STAs would adjust their minimum contention window to meet the estimated baseline throughput which has the intrinsic property of providing temporal fairness. However, they make several strong assumptions that limit pragmatic implementation. First, the authors assume that every STA has a priori knowledge of the number of contending STAs which they concede is non-trivial, especially, in multirate WLAN. Second, in their proposal, each STA assumes that transmission rate, packet size, and packet error rate experienced by all other STAs are the same as itself. However, this scheme will fail to guarantee airtime fairness when traffic of different packet sizes exists, in reality, due to the fact that the baseline throughput is dependent on packet sizes (Tan & Guttag, 2004).

This chapter takes a novel approach to mitigate rate anomaly from a QoS perspective which is first proposed in (Ong & Khan, 2009b) for a single rate WLAN and later extended in (Ong & Khan, 2009a) for a multirate WLAN. The key idea is to exploit heterogeneity within such multi-AP composite network, whenever possible, to promote a QoS-balanced system by redistributing load through optimized VHO in an opportunistic manner. By introducing cognitive cooperation to induce synergetic interactions between link adaptation and load adaptation, the QoS requirements of multimedia STAs can be better satisfied and will not be penalized unnecessarily due to rate anomaly. Figure 3 illustrates that STA with degrading QoS can be identified and transferred to a better quality AP in an effort to maintain a QoS-balanced system through load adaptation. As soon as the STA with degrading QoS is relocated, link adaptation will increase the transmission rates of both the STA and AP, which in turn recover the QoS and overall system capacity, respectively. In fact, such synergetic interactions between link adaptation and load adaptation on-demand is an important aspect of optimized handover that is also raised in the recent IEEE 802.21 standard (IEEE Std 802.21-2008).

To the best of the authors' knowledge, this is the first attempt to address rate anomaly in a multirate WLAN-based composite network by using the concept of cognitive cooperation from a single unifying QoS framework. This chapter substantially extends our previous works (Ong & Khan, 2009b), (Ong & Khan, 2009a) in five significant aspects: (i) an exhaustive discussion on the key challenges of multirate WLAN-based composite network with examples illustrating how rate anomaly can occur in both uplink and downlink transmissions; (ii) a formal introduction of the novel cognitive cooperation concept to enable optimized VHO through the usage of cooperative load metric (CLM), cooperative exchange of QoS context information, and joint optimization; (iii) a detailed exposition of the distributed LAS framework including the optimization functions and its schematic which highlights the thematic connections between the prime network and terminal entities; (iv) a complete account of the PHY channel models, network configurations, and key performance indicators (KPIs); and (v) additional performance comparison and discussions on the QoS fairness between APs, QoS fairness between STAs of different service classes, various sources of packet loss, throughput fairness between STAs, retransmission attempts, and MAC delay.

10.3 COGNITIVE COOPERATION

In general, VHO decision relies on the selection of the 'best' access network that could meet the QoS requirements of the end-user, allowing them to enjoy ubiquitous connectivity in the most efficient way, irrespective of time and place. As a matter of fact, an optimal network selection technique (Ong & Khan, 2010b) is the core component of cognitive cooperation. To be more specific, it is important for directing informed VHO to exploit heterogeneity within a composite network opportunistically and optimize radio resource usage. Clearly, such optimized VHO will be imperative for supporting seamless mobility and QoS transparency. Seamless mobility refers to handover which is both smooth and fast so that the end-user is unaware of this process. Specifically, smooth handover minimizes packet loss whilst fast handover minimizes packet delay (PD) during the actual handover process. According to the IEEE 802.21 standard, seamless mobility also implies the ability to assess and adapt to prevailing network conditions such that the handover decision and execution phases can be effectively optimized. On the other hand, QoS transparency refers to the QoS support that is required to meet end-user's expectation in different scenarios, regardless of RATs. Accord-

Figure 3. Cognitive cooperation: A QoS perspective. Solid lines depict the implication of rate anomaly arising from link adaptation. Dashed lines depict synergetic interactions between link adaptation and load adaptation to mitigate rate anomaly and maintain a QoS-balanced system.

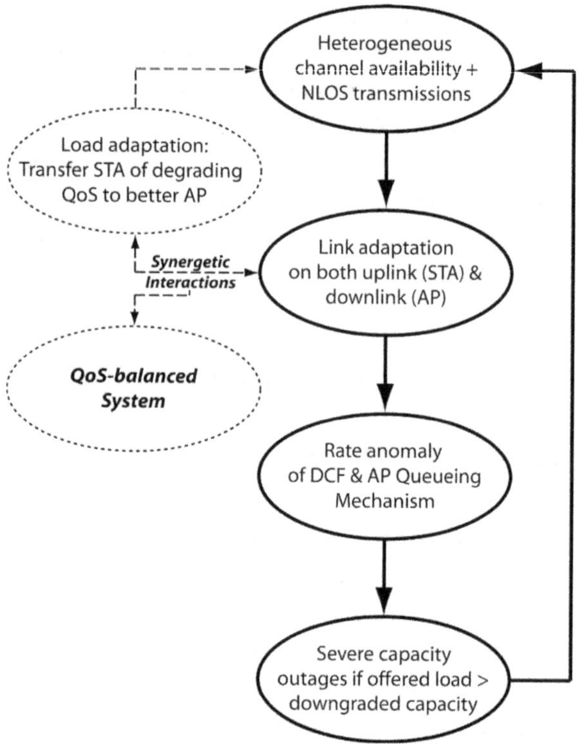

ingly, the chief advantages of optimized VHO such as: (i) permitting QoS-based handover or 'better alternative' handover; (ii) increasing the QoS satisfaction level of end-users; (iii) enabling network operators to integrate several RATs into a multi-RAT environment which will improve their coverage and QoS; and (iv) improving trunking efficiency, hence, overall system capacity of networks through dynamic load distribution could be realized through the notion of cognitive cooperation.

To illustrate this, cognitive cooperation, which involves four domains of cooperation, viz., inter-network, inter-entity, intra-layer, and lastly inter-layer cooperation, is implemented within the distributed load adaptation strategy (LAS) framework as shown in Figure 4. Note that both inter-network cooperation and inter-entity cooperation form the baseline design of cognitive cooperation. In other words, these are the two minimum cooperation domains required to achieve cognitive cooperation. On one hand, the distributed terminal-oriented network-assisted (TONA) handover architecture enables inter-network cooperation to facilitate the cooperative exchange of QoS context information within a composite network (Ong & Khan, 2010a). On the other hand, the generic dynamic access network selection algorithm enables inter-entity cooperation between network-terminal entities to make an informed initial access or VHO decision (Ong & Khan, 2010b).

In order to offer a more comprehensive QoS guarantee for delay-sensitive real-time services in the presence of channel impairments, an additional intra-layer cooperation is introduced within the MAC layer to optimize load distribution according to: (i) different services using the network-QoS entity; and (ii) dynamic network conditions through the connection-QoS entity. The network-QoS entity consists of service prioritization and admission control to deal with different user service profiles while the connection-QoS entity consists of network selection and handover control to deal with dynamic network conditions associated with network congestions, channel impairments, and dynamic spectrum access. Furthermore, an additional inter-layer co-operation is incorporated between the PHY and MAC layer to enable synergies between: (i) link adaptation and connection-QoS entity; and (ii) link adaptation and load adaptation such that their benefits can be exploited on-demand.

Through these four domains of cooperation, an efficient reactive QoS balancing (RQB) algorithm can be devised by incorporating fast handover in conjunction with soft admission control (Ong, 2010). The novel concept of RQB will harmonize optimized VHO in composite wireless networks

Figure 4. Cognitive cooperation within the distributed LAS framework

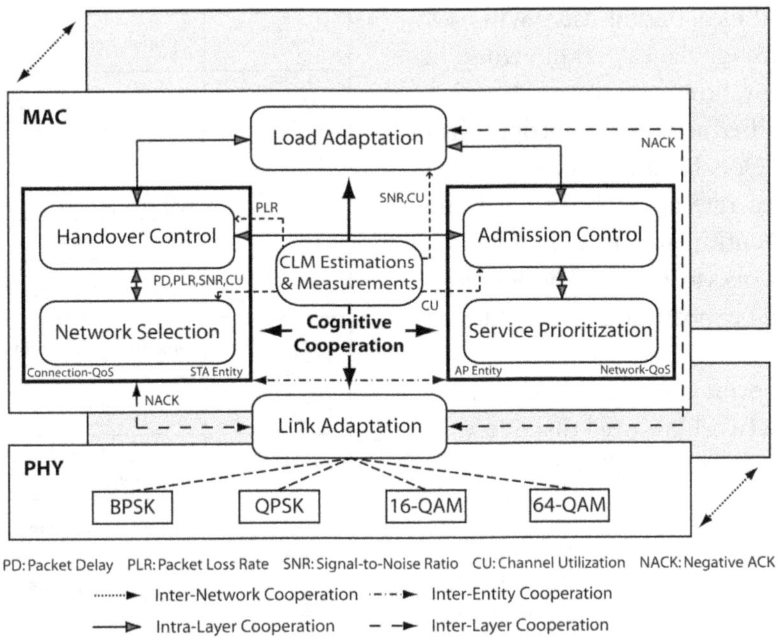

so that the end-to-end goal of maintaining a QoS-balanced system can be accomplished. Under the notion of RQB, optimized VHO will be triggered *only* if: (i) the QoS requirements of STAs cannot be sustained (i.e., when the PLR of source AP exceeds 1%); (ii) a better quality AP exists (i.e., the optimal target AP has been selected); and (iii) the requested VHO will not disadvantaged the existing connections of the target AP (i.e., when subjected to soft admission control). Note that the opportunistic yet altruistic exploitation is achieved when all the above conditions are met. Readers are referred to (Ong, Khan, & Mahata, 2011) for a comparison of reactive and predictive approaches in radio resource management of composite wireless networks.

It is now clear that link adaptation aims to achieve optimum throughput by matching transmission parameters to dynamic wireless channel conditions, whereas load adaptation aims to achieve optimal resource allocations by transferring load dynamically to maintain a QoS-balanced system. However, as explained in Section 2, the rate anomaly problem tends to dilute any link adaptation gains. Moreover, link adaptation introduces an upper bound on the maximum achievable throughput due to the reduction of transmission rate. Hence, it is conjectured that RQB can effectively minimized the capacity outage time described in Section 2 by maintaining a QoS-balanced system. Thus, cognitive cooperation provides an attractive avenue to complement link adaptation so that both system capacity and QoS can be further improved. To be more specific, load adaptation through optimized VHO will be invoked when opportunistic yet altruistic exploitation is possible, otherwise link adaptation will be invoked. This conjecture will be validated in Section 6.

10.4 LOAD ADAPTATION STRATEGY FRAMEWORK

The ability of cognitive cooperation, which entails multi-domain cooperation, to exploit heterogeneity as an enabler for improving the overall system capacity and QoS of end-users will now be demonstrated through the LAS framework. In what follows, we show how cognitive cooperation benefits from the unified actions of joint optimization and results in a QoS-balanced system, which effectively mitigates rate anomaly by enabling different functional entities to form synergies and multiple access networks to interact. The notations used in the following exposition are summarized in Table 1.

The underlying principle of the LAS framework is based on network-assisted discovery, in which the source AP broadcasts the measurement report of neighboring APs together with its own, over the TONA handover architecture. To be more specific, the TONA handover architecture consists of an entity known as the access point controller (APC) to facilitate the cooperative exchange of QoS context information within a composite network. The key idea is to leverage on the internet protocol core network for enabling the integration of different RATs and providing a flat, common platform to support access network heterogeneity. Readers are referred to (Ong, 2010) for the details of the TONA handover architecture.

The QoS broadcast (Ong, 2010) contains the QoS context information of PD, packet loss rate (PLR), channel utilization (CU), and signal-to-noise ratio (SNR), which is global to both AP and STAs. Although it has been shown in (Ong & Khan, 2009b) that the QoS context information of PD and PLR can capture wireless channel variations implicitly, the SNR is employed here so that it can be used to make explicit decisions concerning the wireless channel conditions proactively. For instance, STA could select an AP according to its SNR requirements on top of its QoS requirements and the AP load conditions. In addition, the AP could make use of the SNR information to effectuate load distribution more proactively when dealing with multirate WLAN. For the case of a multi-RAT environment with heterogeneous access networks, the SNR can be

Table 1. List of notations

N_k^{opt}	Optimal network selection outcome at time *k*
CF_k^i	Cost function of network *i* at time *k*
$\Upsilon_k^{i,j}$	Network anomaly identifier of network *i* at time *k*
$\Psi_k^{i,j}$	Bayes estimate of *j*th QoS parameter of network *i* at time *k*
τ_j	Required QoS threshold of *j*th QoS parameter
$NACK_i^n$	Negative ACK of *i*th STA in source AP *n*
CU_j^n	Channel utilization of *j*th flow in source AP *n*
CU_i^n	Channel utilization of *i* th STA (i.e., uplink and downlink flows) in source AP *n*
CU_{ave}	Average channel utilization of all APs
CU_{total}^m	Total channel utilization of *m* target APs
CU_Δ^m	Available channel utilization in target AP *m*
$T_{s_j}^n$	Average successful packet transmission time of *j*th flow in source AP *n*
T_{BO}^n	Average backoff time in source AP *n*
T_{SLOT}	Slot time
T_{SIFS}	Short interframe space time
T_{DIFS}	DCF interframe space time
T_{ACK}	Acknowledgment time
δ	Propagation delay
CW_{min}	Minimum contention window size
$QSF_{i,j}$	QoS satisfaction factor for *i*th service class of *j*th STA
$\overline{QSF}$	Average QoS satisfaction factor of all STAs
$TFI_{i,j}$	Throughput fairness index for *i*th service class of *j*th STA
$\overline{TFI}$	Average throughput fairness index of all STAs
$QBI(x)$	QoS balance index of metric *x* between APs/STAs

Figure 5. Thematic connections of the LAS framework to support cognitive cooperation

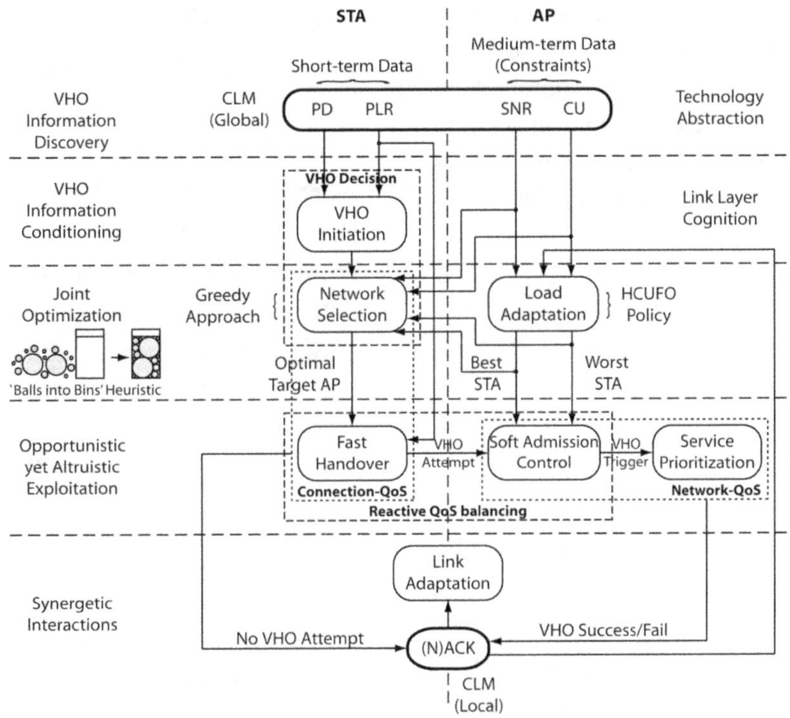

normalized by using the Shannon capacity formula to relate the maximum achievable data rate from different technologies as in (Yang, Gondal, Qiu, & Dooley, 2007).

An additional NACK, which is a local information to both AP and STAs, is included to the QoS context information of PD, PLR, SNR, and CU to form the CLM. The CLM is then exploited for adaptation to the dynamic network conditions prevalent in a multirate WLAN-based composite network. The NACK is used as input by: (i) the AP to detect STAs with degrading QoS; and (ii) the respective AP and STAs to invoke link adaptation procedures. The idea of NACK, in general, is similar to missed ACK (MACK), except for the fact that MACK could also occur due to collisions. It has been pointed out in (Pang, et al., 2008) that link adaptation procedures such as automatic rate fallback (Kamerman & Monteban, 1997) which depend on MACK will suffer throughput degradation and prolonged collision time, determined by the slower STA, when transmission failure is due to collisions rather than link errors. Hence, the idea of NACK is used instead for distinguishing between link errors and collisions. This is an enhancement over MACK for two reasons: (i) link adaptation procedure will not be falsely triggered to reduce data rate; and (ii) STA will not be unnecessarily identified as the 'worst' STA to perform load adaptation during collisions in which exponential backoff is more appropriate.

Figure 5 illustrates the thematic connections of the LAS framework. Accordingly, the PD and PLR are short-term data used by STA to make VHO decisions, whereas the additional SNR information together with the CU are medium-term data that serve as constraints for network selection and load adaptation. The VHO information discovery and VHO information conditioning by using the technology agnostic approach and Bayesian learning to provide link layer cognition, respectively can be found in (Ong, 2010). The measurement-based network selection relies on the greedy approach in which the optimal AP will be selected based on the short-term data of PD and PLR. Particularly, the optimal network selection for *j* QoS parameters, satisfying both user preferences and prevailing network conditions at time *k*, can be expressed as an optimization function conditionally by

$$N_k^{opt} = \begin{cases} \max CF_k^i \quad s.t. \quad \Upsilon_k^{i,j} \neq 0, \ \forall j, \ \ if \exists i : \Upsilon_k^{i,j} = 0 \\ N_{k-1}^{opt}, \ \ otherwise \end{cases} \tag{1}$$

where

$$\Upsilon_k^{i,j} = H\left[\Pr\left(\Psi_k^{i,j} \le \tau_j\right) - \frac{1}{2}\right], \quad H\left[n\right] = \begin{cases} 0, & n < 0 \\ 1, & n \ge 0 \end{cases}. \tag{2}$$

CF_k^i is the simple additive weighting (SAW) cost function to rank network *i* at time *k* as the weighted w_j sum of *j*th QoS parameter. $\Upsilon_k^{i,j}$ is the network anomaly identifier that indicates whether the network quality is sufficient in which $\Upsilon_k^{i,j} = 1$ if network quality is sufficient and $\Upsilon_k^{i,j} = 0$ otherwise. $H\left[n\right]$ is the unit step function used to implement $\Upsilon_k^{i,j}$. Note that VHO is triggered *only* if the source network quality is insufficient, i.e., $\Pr\left(\Psi_k^{i,j} \le \tau_j\right) < \frac{1}{2} \Rightarrow \Upsilon_k^{i,j} = 0$.

On the other hand, the load adaptation plays a central role by serving two main functions. First, it identifies the 'worst' STA with highest NACKs among STAs with degrading QoS for unplanned handover. Second, it attempts to perform RQB by selecting the 'best' STA for planned handover. Note that the 'worst' STA has priority over the 'best' STA in order to mitigate rate anomaly by reducing the capacity outage time. Specifically, when poor wireless channel condition is detected, the source AP would nominate a 'worst' STA as

$$\begin{aligned} &\max \quad NACK_i^n \\ &s.t. \quad \left\{i \in STAs : i = \min\left(r_i \in R_{DATA}\right)\right\} \end{aligned} \tag{3}$$

where $n \in APs$ and R_{DATA} is the PHY data rate. For the case of multirate WLAN, the constraint ensures that the STA that is receiving data frames with the lowest R_{DATA} and has the highest NACK is transferred to a better quality AP first. Otherwise, the source AP would nominate a 'best' STA as

$$\begin{aligned} &\min \quad CU_{ave} - CU_{total}^m - CU_i^n \\ &s.t. \quad CU_{ave} - CU_{total}^m - CU_i^n \ge 0, \quad m \ne n \end{aligned} \tag{4}$$

where $i \in STAs$, *m*, $n \in APs$, and CU_{ave} is the average CU of all APs. CU_i^n is the per-STA CU which includes both uplink and downlink flows owing to the fact that all flows in an infrastructure BSS are relayed via the AP. The constraint prevents load distribution to an overloaded AP and overloading the target AP. Accordingly, the nominated STA will then perform VHO to the optimal target AP. The CU corresponds to the bandwidth requirement of each flow, which is simply the product of its packet arrival rate and the average time required to transmit a packet successfully given in (Garg & Kappes, 2003) as

$$CU_j^n = \frac{R_j}{L_j} \times T_{s_j}^n, \tag{5a}$$

$$T_{s_j}^n = T_{DIFS} + T_{BO}^n + 2T_{PHY} + T_{DATA_j}^n + T_{SIFS} + T_{ACK} + 2\delta, \tag{5b}$$

$$T_{BO}^n = \frac{CW_{\min}}{2} \times T_{SLOT} \tag{5c}$$

where R_j is the average data rate, L is the packet length of *j*th flow, $T^n_{s_j}$ is the average time required for a successful packet transmission, T^n_{BO} is the average backoff time of *n*th AP, and δ is the propagation delay. T_{DIFS}, T_{SIFS}, T_{SLOT}, and $CW_{\min}$ are defined in the IEEE 802.11 standard (IEEE Std 802.11-2007). Note that the CU accounts only for channel occupation time during successful transmissions while channel busyness ratio proposed in (Zhai, Chen, & Fang, 2005) accounts for channel occupation time during both successful transmissions and collisions. However, the work of (Zhai, et al., 2005) has shown that the CU is almost the same as the channel busyness ratio when WLAN is operating in non-saturation mode as the probability of collision is very small. Hence, the channel occupation time during collisions can be disregarded as admission control can accurately regulate input traffic with the CU.

It is important to note that both 'best' and 'worst' STAs correspond to the *most aggressive* traffic source associated with the source AP. The 'best' STA will be the traffic source with the highest CU that can fit into the available capacity of the target AP. The 'worst' STA will be the traffic source with the highest NACKs, which also corresponds to the traffic source with the highest CU since the probability of NACKs increases with higher traffic arrival rate. This *highest channel utilization first out (HCUFO)* policy, also based on the greedy approach, is attractive as it can essentially reduce the number of unnecessary VHOs and result in the long-term uniform distribution of aggressive traffic sources over the multi-AP composite network, which is beneficial from a load distribution perspective. Such greedy approaches are adopted due to the fact that obtaining an optimal allocation of STAs to the available APs such that the allocation maximizes the overall system capacity is a combinatorial problem which is NP-hard (Gazis, Houssos, Alonistioti, & Merakos, 2003). Collectively, these greedy approaches contrive the classical 'balls into bins' heuristic which provide an intuitively simple way to jointly optimize overall system capacity while mitigating rate anomaly and providing statistical QoS guarantee. The benefits of such joint optimization to orchestrate optimized VHO can be achieved by embracing cognitive cooperation introduced in Section 3. The pseudo codes describing the algorithm of the LAS framework based on cognitive cooperation in both AP and STA can be found in (Ong & Khan, 2009a).

10.5 SIMULATION MODELS AND PRELIMINARIES

To demonstrate the effectiveness of the LAS framework, built on the basis of cognitive cooperation, simulation models are developed by using OPNET™ Modeler® 14.0 with Wireless Module. Modifications to the DCF models are performed in order to provide QoS support with service prioritization, admission control, network selection, fast handover, link adaptation, and load adaptation which are the focus of this study. The OPNET PHY model is also enhanced to include shadow fading, Rayleigh flat fading, and the capability to simulate NLOS transmissions by varying the path loss exponent. These enhancements allow different propagation characteristics as a result of NLOS transmissions and/or dynamic spectrum access to be captured.

10.5.1 Unified Analytical Model

A unified analytical model has been developed in (Ong, Khan, & Mahata, 2010) as illustrated in Figure 6 to validate the baseline DCF models of OPNET. The random backoff process is first modeled by using a discrete time Markov chain which is solved numerically to obtain the transmission and its failure probabilities. Subsequently, closed-form expressions for the average MAC

Figure 6. Unified analytical model: Markov chain model in conjunction with finite queueing model

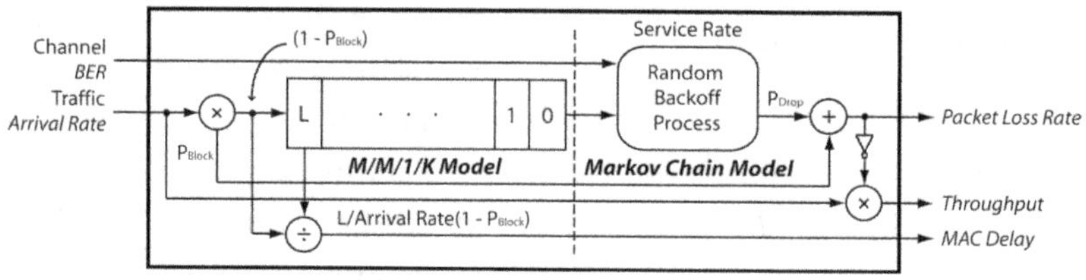

service time are derived from the random backoff process based on the transmission failure probability. This average MAC service time, i.e., $1/\mu$ is then used in conjunction with the $M / M / 1 / K$ model to obtain the average MAC delay, PLR, and throughput efficiency for each STA and the AP. The results within (Ong, et al., 2010) shows that there is a good agreement between the analyses and simulations for various traffic sources, PHYs, and data rates, which confirms the accuracy of our modeling assumptions and validates the OPNET simulation models.

10.5.2 Log-Normal Path Loss Model

An error-prone channel is considered by including shadow fading and variable path loss exponent to capture different propagation environments which may contain obstacles to cause NLOS transmissions. Accordingly, the log-normal path loss model (Rappaport, 2001) is

$$L_{path}(dB) = 10\log\left[\frac{(4\pi)^2 d^n}{\lambda^2}\right] + X_\sigma \qquad (6)$$

where λ is the wavelength associated with the carrier frequency, d is the distance between the transmitters and receivers, n is the path loss exponent which determines the rate of loss, and X_σ is a zero-mean Gaussian distributed random variable with standard deviation σ (in dB). NLOS transmissions due to obstructions that are prevalent in hotspots and indoor environments are simulated by varying n uniformly $U[3,4]$ and shadow fading is accounted by varying σ uniformly $U[6,9]$. Additionally, background noise is modeled as

$$N = FkT_0B \qquad (7)$$

where k is the Boltzmann's constant, T_0 is the ambient room temperature of $290K$, B is the equivalent bandwidth of the receiver, and F is the noise figure of the receiver simulated with a normal distribution $N(10,0.1)$.

10.5.3 Exponential Channel Model

The exponential channel model (O'Hara & Petrick, 1999), adopted by the IEEE 802.11 Task Group-b for its simplicity and accuracy, is modeled as a finite impulse response filter where k_{max} is the maximum tap length. The taps are independent complex Gaussian distributed random variables with an exponentially decaying power delay profile in which the coefficient of kth tap is given by

$$h_k = N\left(0,\frac{1}{2}\sigma_k^2\right) + jN\left(0,\frac{1}{2}\sigma_k^2\right), \quad k = 0,\ldots,k_{max}, \qquad (8a)$$

$$k_{max} = \left\lceil 10\frac{T_{RMS}}{T_s}\right\rceil, \quad \beta = e^{T_s/T_{RMS}}, \quad \sigma_k^2 = \frac{(1-\beta)\beta^k}{1-\beta^{k_{max}+1}} \qquad (8b)$$

Figure 7. Simulation model of a multirate multi-AP WLAN-based composite network with multimedia traffic subjected to wireless channel variations

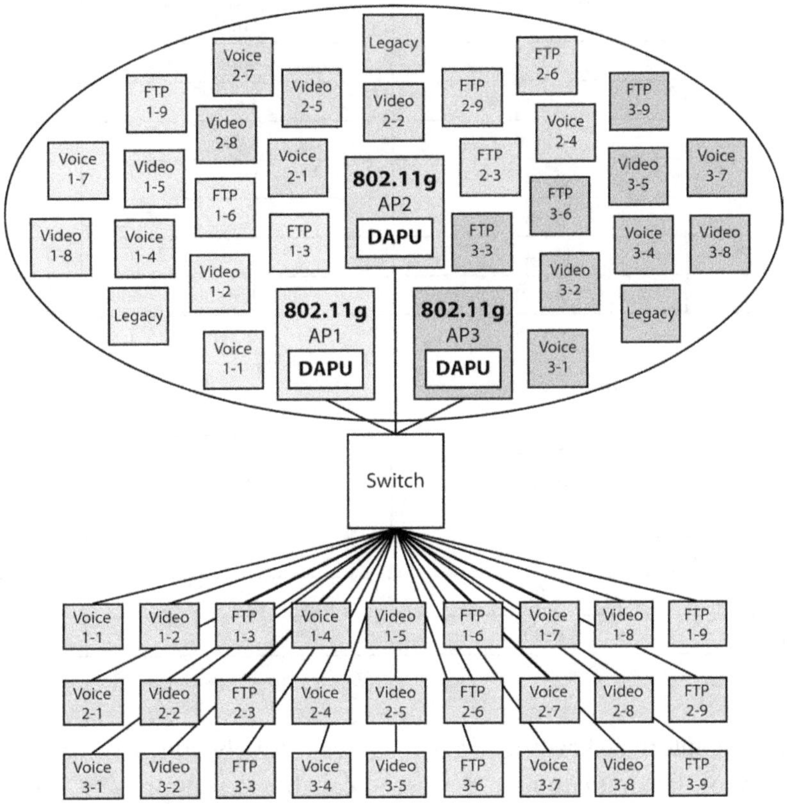

where $N\left(0, \frac{1}{2}\sigma_k^2\right)$ is a zero-mean Gaussian distributed random variable with variance $\frac{1}{2}\sigma_k^2$. T_s is the sampling period and T_{RMS} is the root mean square (RMS) delay spread of the channel. Thus, Rayleigh flat fading is simulated as a special case of the exponential channel model by considering a single tap channel with zero RMS delay spread, i.e., setting $k_{\max} = 0 \Rightarrow \sigma_k^2 = 1$ such that

$$h_0 = N\left(0, \frac{1}{2}\right) + jN\left(0, \frac{1}{2}\right). \tag{9}$$

10.5.4 Network Configurations

A typical hotspot which consists of a multirate multi-AP WLAN-based composite network with three IEEE 802.11g APs operating at an initial data rate 54 Mbps is simulated as shown in Figure 7. It is assumed that at least one legacy STA is associated with the IEEE 802.11g AP since it is unlikely to operate in the IEEE 802.11g only mode given the vast IEEE 802.11b deployments. However, the legacy STA does not transmit any traffic so that all the system resources are available for the IEEE 802.11g STAs. Additionally, the CTS-to-Self protection mechanism is used by the IEEE 802.11g STAs. The network interface parameters are modeled according to the Cisco Aironet 1130AG series AP and CB21AG series

Table 2. Wireless channel variations

State	AP1	SNR AP2	AP3
0	High	High	High
1	High	High	Low
2	High	Low	High
3	High	Low	Low
4	Low	High	High
5	Low	High	Low
6	Low	Low	High
7	Low	Low	Low

Table 3. Traffic generation parameters

Traffic Type	Packet Size (Bytes)	Inter-arrival (ms)	Avg. Data Rate (Kbps)
Voice-CBR (G.711)	80	10	64
Video-CBR	1000	125	64
Data-FTP (UL)	450	120	30
Data-FTP (DL)	1500	40	300

wireless client reference interface specifications to improve the realism of the simulations.

In this simulation, a balanced load of three voice, three video, and three FTP STAs in each BSS is considered. The wireless channel variations in each BSS are simulated by introducing NLOS transmissions and the effects of dynamic spectrum access. Specifically, the wireless channel variations are introduced according to Table 2 where binary representations of 0 to 7 are used to simulate 2^3 possible states, resulting in either 'high' or 'low' SNR in each AP per 100 s interval. Shadow fading and multipath are included for the entire simulation duration, and multimedia traffic sources are simulated according to Table 3. The MAC service data unit (MSDU) lifetime limit mechanism is also incorporated to discard MSDUs from the transmitter queue if they exceed the MSDU lifetime before successful transmission. The MSDU lifetime for voice, video, and data packets are chosen as 50 ms, 150 ms, and 1 s, respectively. Furthermore, all STAs are roaming capable to support VHO events.

10.5.5 Performance Indicators

The performance of the LAS framework is examined using the common performance indicators such as the MAC delay, retransmission attempts, aggregate throughput, and packet loss in a multi-AP composite network. In addition, the QoS performance of the LAS framework is evaluated from three critical aspects using a set of KPIs defined in the following.

First, the QoS performance of STA or end-user is quantified as a function of two QoS metrics. Each QoS element is the ratio of the required QoS metric threshold and measured QoS value. Accordingly, QoS satisfaction factor (QSF) is defined as the minimum between the two QoS elements given by

$$QSF_{i,j} = \min_{k \in Links} \left(\frac{PD_i^t}{PD_{i,j}^{m,k}}, \frac{PLR_i^t}{PLR_{i,j}^{m,k}} \right), \tag{10a}$$

$$\overline{QSF} = \frac{1}{cn} \sum_{i=1}^{c} \sum_{j=1}^{n} QSF_{i,j} \tag{10b}$$

where PD_i^t and PLR_i^t are the PD threshold and PLR threshold, respectively. $PD_{i,j}^{m,k}$ and $PLR_{i,j}^{m,k}$ are the measured PD and measured PLR for *i*th service class of *j*th STA and *k* links (i.e., both uplink and downlink), respectively. $\overline{QSF}$ is the average QSF of all STAs. QSF < 1 when the QoS requirements of STA cannot be met. For the purpose of QSF computation, PD thresholds for voice and video are 50 ms while data is 300 ms, and PLR thresholds for all service classes are 2%. It is important to note that for infrastructure BSS, particularly, in the presence of many two-way communications such as voice over internet protocol (VoIP) connections, downlink becomes the capacity bottleneck since each VoIP connection has duplex traffic which will eventually result in higher downlink load and asymmetric traffic load on both links (Ong & Khan, 2010a). Therefore, unless otherwise stated, the PD of interest is taken as the MAC delay experienced by the AP. On the other hand, the PLR is a function of data dropped due to buffer overflow, MSDU lifetime expiry, and retry limit exceeded.

Second, the effect of load distribution on the achievable throughput of STA or end-user is quantified by throughput fairness index (TFI) which is defined as

$$TFI_{i,j} = \frac{\left| \overline{S}_{i,j}^{m} - S_i^t \right|}{S_i^t}, \tag{11a}$$

$$\overline{TFI} = \frac{1}{cn} \sum_{i=1}^{c} \sum_{j=1}^{n} TFI_{i,j} \tag{11b}$$

where $\overline{S}_{i,j}^{m}$ and S_i^t are the measured average throughput and target (maximum) throughput for *i*th service class of *j*th STA, respectively. $\overline{TFI}$ is the average TFI of all STAs. $TFI \in [0,1]$ reflects the amount of deviation from target throughput and serves as a normalized measure of throughput fairness in the network. It has a value of 0 when the throughput of STA is fair and a value of 1 when the throughput of STA is tremendously unfair.

Third, the Jain's fairness index is adopted to quantify the effect of load distribution on the QoS fairness among the APs (STAs). Suppose x_i is the QSF of AP (STA) *i*, then QoS balance index (QBI) can be defined as

$$QBI(x) = \left(\sum_{i=1}^{n} x_i \right)^2 \Big/ n \left(\sum_{i=1}^{n} x_i^2 \right) \tag{12}$$

where *n* is the number of APs (STAs.) over which the load, i.e., STAs will be redistributed. $QBI \in [0,1]$ is a continuous function which is independent of scale. It has a value of 1 when all APs (STAs) have exactly the same QSF and a value of $1/n$ when the QSF of each AP (STA) is extremely unbalanced, which is 0 in the limit as $n \rightarrow \infty$.

10.6 PERFORMANCE EVALUATION

A comparative study between the performance of the LAS framework and the typical load balancing mechanism (LBM) implementation in (Velayos, Aleo, & Karlsson, 2004), which does not consider wireless channel variations and utilizes CU as the only load metric, is also investigated. Figure 8 illustrates an overview on the implementation of the LBM and LAS framework. In this case, an additional SNR metric is explicitly employed in the LAS framework together with the CU to function

Figure 8. Overview on the implementation of the LBM vs. LAS framework

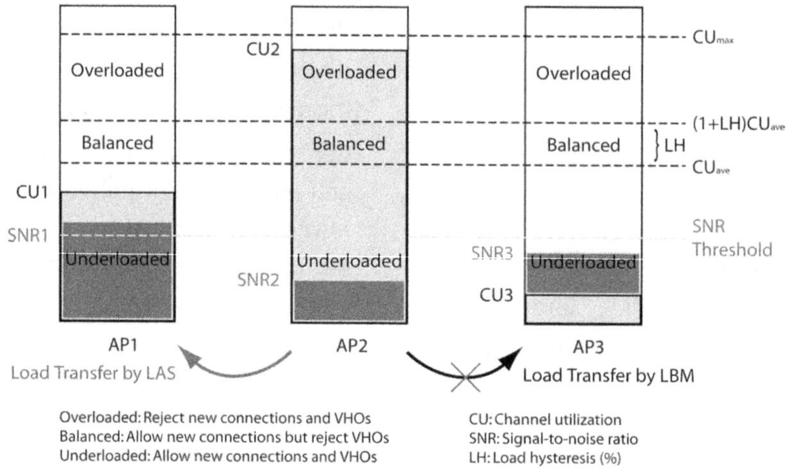

as constraints for load distribution as discussed in Section 4 (cf. Figure 5). Ideally, according to the definitions of the three KPIs in (10) – (12), the QSF should be greater than 1, the TFI should be close to 0, and the QBI should be close to 1 so as to offer QoS guarantee, throughput fairness, and QoS fairness, respectively. The results are analyzed in eight states, which correspond to the simulated scenario, starting from 100 s (0 – 100 s is the warm-up period).

From Figure 9, LAS has an average QSF of greater than 1 except for state seven. LBM also has an average QSF of greater than 1 except for state seven and at the beginning of states four and six. In general, LBM has higher average QSF than LAS except for state seven when all the APs have low SNR. This counterintuitive result can be explained from the design philosophy of RQB to avoid unnecessary VHOs when the QoS requirements of STAs can be supported (cf. Section 3 and Figure 5). Since the traffic load distribution is the same with 68% load per AP, LBM does not trigger any VHOs as the load hysteresis δ is set to 10% and the CU of all three APs is deemed balanced. However, note that when the load hysteresis is zero, LBM has similar results in terms of average QSF and aggregate throughput but a massive 40% increase in VHOs as compared to LAS. Although the load hysteresis could prevent unnecessary VHOs, it is static, and hence it is not responsive to wireless channel variations. Evidently, the major pitfall of using the CU as a single load metric in LBM is the inability to manage wireless channel variations which could arise due to NLOS transmissions and/or dynamic spectrum access in a WLAN-based composite network.

On the other hand, LAS enables the VHO of STAs with degrading QoS to a better quality AP in an opportunistic yet altruistic manner. Accordingly, the load of a good quality AP (i.e., one which maximizes the SAW cost function while fulfilling the constraints of SNR and CU as explained in Section 4) may be filled up to 87% of its maximum capacity. Thus, the average QSF of LAS under such conditions will be lower than LBM. However, it is important to emphasize that the QoS requirements of STAs are not compromised as the average QSF of LAS is still greater than 1, thanks to its altruistic design. On the contrary, LAS has higher average QSF than LBM during state seven even when all APs have low SNR since one of the APs has only 33% load while the remaining two APs have about 85% load each. It follows that the AP which is lightly loaded is not affected by low SNR. Hence, all the

Figure 9. Average QSF of STAs

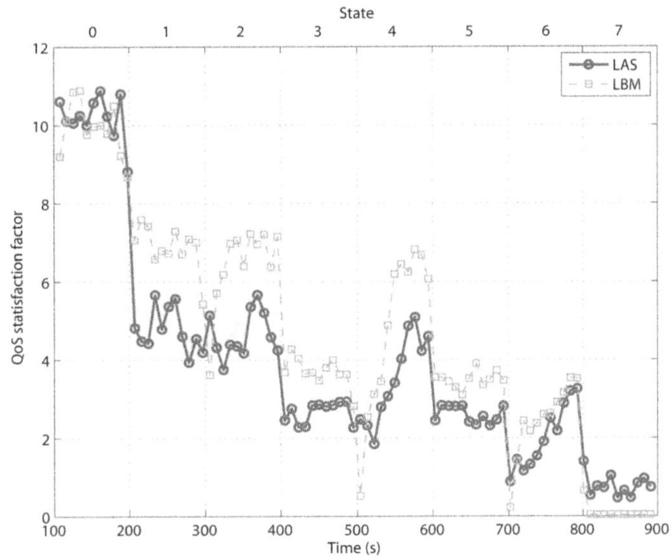

associated STAs can still meet their QoS requirements, which result in higher average QSF.

From Figure 10(a), LAS achieves higher QBI than LBM for all states except for state seven where LBM has higher QBI as the QSFs of all STAs are equally *bad.* Similar trend is found in Figure 10(b) which depicts the QBI of APs' QSF. These observations are in direct contrast with that of Figure 9, which suggest that tradeoffs exist between the average QSF and QBI. For every increase in the QBI achieved by LAS, there is a corresponding decrease in the average QSF as compared to LBM. In other words, LAS *trades* the QSF for the QBI in order to maintain a QoS-balanced system, thanks to its joint optimization powered by cognitive cooperation. Although LBM has higher average QSF, it fails to provide any QoS fairness which means that there is a huge disparity in the QoS or throughput between STAs of the same service class. Moreover, Figure 11 and Figure 12 illustrates that the QBI of STAs' QSF per service class also shows similar trends to Figure 10(a). Specifically, it shows that voice STAs yield the best improvement in the QBI over each state duration, followed by video STAs and

Figure 10. Average QBI of STAs' and APs' QSF: (a) Average QBI of STAs' QSF; (b) Average QBI of APs' QSF

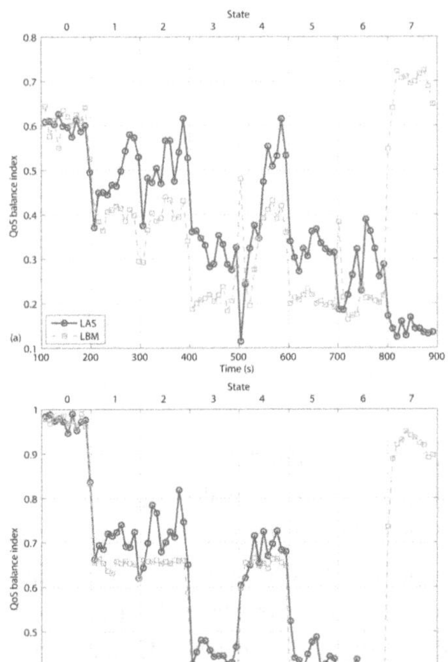

Figure 11. Average QBI of STAs' QSF for different service classes: (a) Voice STAs; (b) Video STAs;

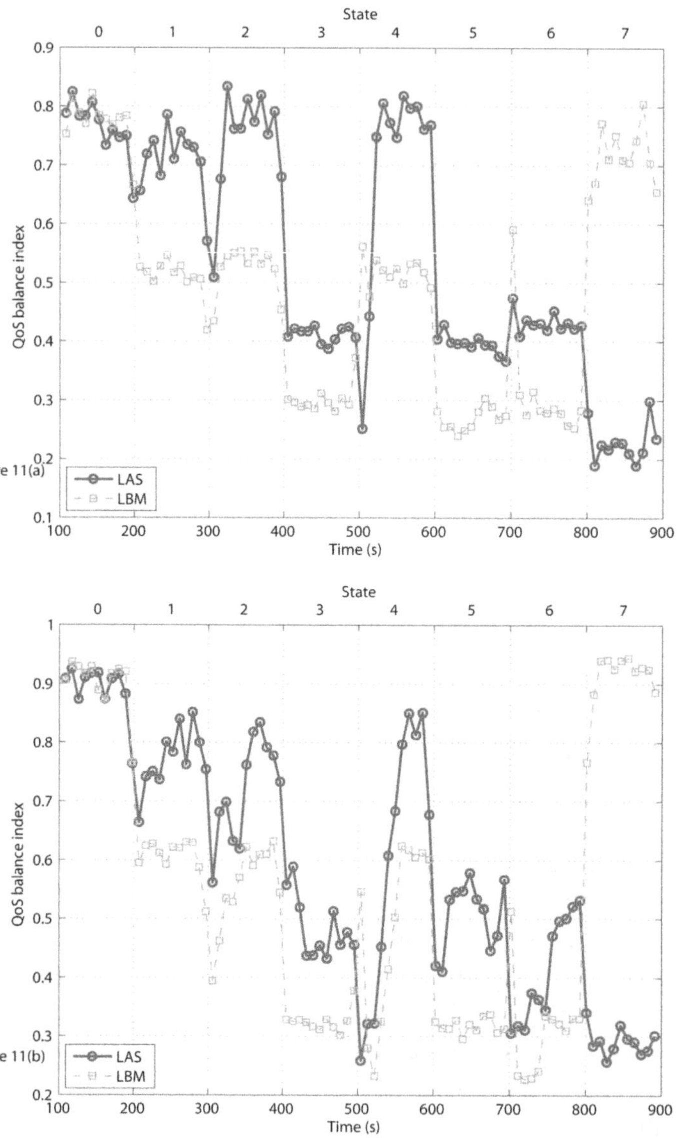

Figure 11(a)

Figure 11(b)

then data STAs. Such strict QoS prioritization is a desirable outcome of employing the strict priority queueing technique (Yu & Choi, 2006) to provide service prioritization in the DCF, which the EDCA cannot guarantee as explained in (Wang, Jiang, & Zhuang, 2006) due to its statistical prioritization.

From Figure 13(a), LAS attains the highest aggregate throughput as compared to LBM and DCF across all states. Note that the DCF serves as a baseline for this comparative study since both LAS and LBM are essentially DCF-based. Apparently, both LBM and DCF suffer marked throughput degradation as a consequence of rate anomaly in the downlink which causes HOL blocking at the AP's FIFO queue. This subsequently extends to buffer overflow and excessive real-time packets being dropped due to MSDU lifetime expiry for the case of LBM as depicted in Figure 13(b). Note that the MSDU lifetime expiry mechanism is not implemented in the DCF

Figure 12. Average QBI of Data STAs

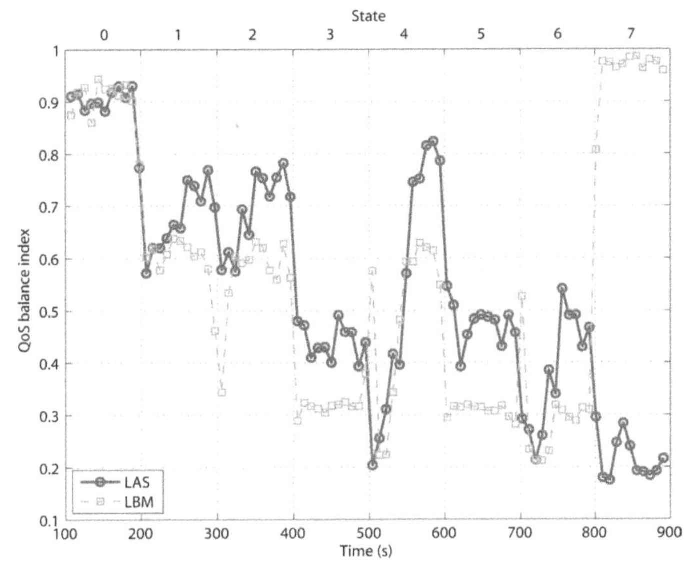

as specified by the IEEE 802.11 standard, thus no packets are lost due to the expiry of MSDU lifetime. In this simulation, rate anomaly in the downlink has a more dominant effect than the uplink since an infrastructure BSS WLAN is downlink limited and the DCF has only a single FIFO queue in the MAC layer. Thus, it can be concluded that although LBM (with link adaptation) yields better throughput than DCF (without link adaptation) under varying wireless channel conditions, the throughput enhancement is *rate-limited* by link adaptation. In fact, Figure 13(b) also illustrates that although link adaptation reduces the amount of packet loss in the PHY due to retry limit exceeded, it results in significant packet loss due to buffer overflow and MSDU lifetime expiry in the MAC layer. It is worth noting that this dilution of link adaptation gains will become more substantial with the IEEE 802.11b or mixed IEEE 802.11b/g WLANs as explained in Section 2 (cf. Figure 2).

Another interesting observation in Figure 13(a) is that LAS has a throughput gain of 20% over LBM in states one, two, and four where 2/3 of the APs have high SNR. On the other hand, LAS has a throughput gain of 12.5% over LBM in states three, five, and six where only 1/3 of the APs have high SNR. Similarly, Figure 13 illustrates that LAS obtains the highest improvement of throughput fairness in states one, two, and four followed by states three, five, and six. It is worth noting that LBM with higher TFI on average implies that some STAs suffer larger deviations from their target throughput, and hence LBM is unlikely to provide any throughput fairness. As a matter of fact, these observations are consistent with the QBI but inversely related to the QSF as LAS trades the QSF for the QBI. Obviously, these suggest that the throughput and QoS performance gains of LAS will scale with the heterogeneity of increasing number of APs as the probability of all APs with low SNR at the same time is remote. Given that multi-AP (multi-RAT) deployments in excess of three APs (RATs) will become a reality in future wireless networks, we advocate that heterogeneity should be exploited by using cognitive cooperation, whenever possible, to harness higher overall system capacity.

Besides, Figure 15(a) also shows that LAS has the least retransmission attempts in states one, two, and four while it has similar retransmission attempts as LBM in states three, five, and six. The

Figure 13. Average aggregate throughput and packet loss in a multi-AP composite network: (a) Average aggregate throughput; (b) Average packet loss due to (i) buffer overflow, (ii) MSDU lifetime expiry, and (iii) retry limit exceeded

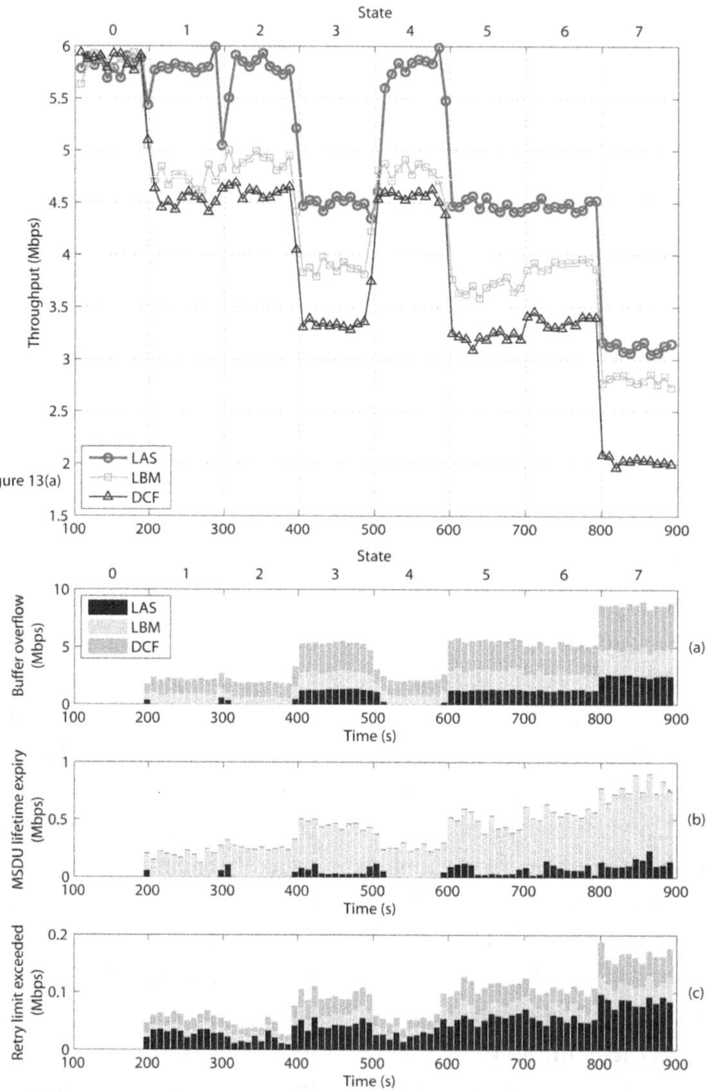

Figure 13(a)

Figure 13(b)

reduction of retransmission attempts implies that more radio resources can be used to transmit useful traffic and boost system capacity. The only exception is found in state seven where LAS has similar retransmission attempts as DCF. This is due to the fact that it will be beneficial to have a balanced load as in the case of LBM when SNR of all APs are equally low. However, such a scenario will again be unlikely. Finally, Figure 15(b) shows that LAS has the lowest MAC delay as compared to both LBM and DCF across all states, whereas DCF suffers from the longest MAC delay which renders it unable to meet any QoS requirements for real-time traffic.

Clearly, these results corroborate that by maintaining a QoS-balanced system, LAS provides

Figure 14. Average TFI of STAs

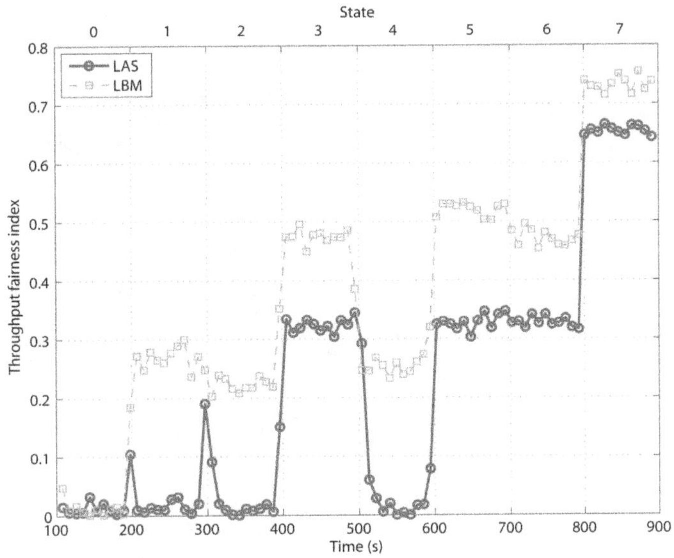

better utilization of radio resources through optimized VHO fueled by cognitive cooperation. In fact, RQB has exhibited the intrinsic properties of mitigating rate anomaly in multirate environments, providing statistical QoS guarantee, and maximizing overall system capacity by maintaining QoS and throughput fairness. These validate the conjecture in Section 3 that the capacity outage time can be minimized by maintaining a QoS-balanced system.

The importance of cognitive cognition in composite network has been demonstrated. However, any derived benefits come at a cost to both network and terminal which we would briefly discuss. On the network side, there will be some storage and communication overheads associated with the exchange of network context information incurred by the APs. However, such overheads will not be significant as discussed in (Ong & Khan, 2010a). On the other hand, the proposed algorithm has a linear time complexity, and it is distributed between both APC and STAs. To be more specific, the load adaptation function in the APC involves the computation of the total and average channel utilization across its connected APs, as well as the nomination of the 'best' and 'worst' STAs. On the terminal side, the network selection algorithm in the STAs involves the ranking of candidate networks in order to select an optimal network. In fact, these algorithms in both APC and STAs are known to have linear time complexity of O (n) where n is the number of 'reachable' APs within a geo-localized area. It is worth noting that n will be typically upper bounded in practice due to received signal strength requirements. The specifics of these deployment issues are outside the scope of the chapter which focuses on the novel approach to mitigate rate anomaly from a QoS perspective. Further cost benefit analysis will be conducted as part of our future work.

10.7 CONCLUSION

This chapter has introduced the novel concept of *cognitive cooperation* to enable optimized VHO as a harmonized solution for a multirate multi-AP WLAN-based composite network by exploiting link adaptation and load adaptation on-demand.

Figure 15. Average retransmission attempts and MAC delay: (a) Average retransmission attempts; (b) Average MAC delay

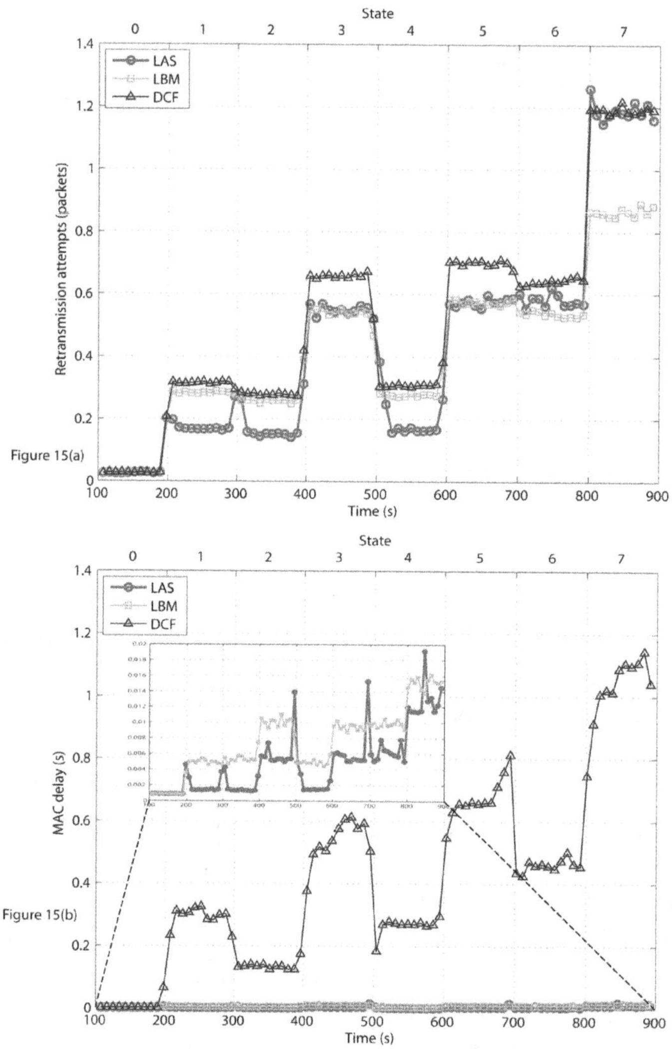

Through cognitive cooperation, we have demonstrated that the LAS framework can provide adaptation to dynamic network conditions in a self-adjusting manner through optimized VHO. Particularly, the LAS framework maintains a *QoS-balanced* system by introducing the CLM, facilitating the cooperative exchange of QoS context information, and performing joint optimization. Additionally, we have shown that rate anomaly arising from link adaptation, which is used to combat varying wireless channel conditions, can also be mitigated by maintaining a QoS-balanced system without any elaborated modifications. It is worth mentioning that the LAS framework is compatible with both DCF and EDCA channel access mechanisms. In fact, the possibility to augment the legacy DCF in providing a unifying QoS solution has significant advantages over the EDCA owing to its prevalence. Furthermore, our results have indicated that the heterogeneity of multi-AP (multi-RAT) composite network should be exploited by using cognitive cooperation as a scalable means to orchestrate better utilization of radio resources, which will lead to overall system capacity and QoS improvements.

REFERENCES

Bhagwat, P., Bhattacharya, P., Krishna, A., & Tripathi, S. K. (1996). *Enhancing throughput over wireless LANs using channel state dependent packet scheduling*. IEEE 15th Annual Joint Conference of the IEEE Computer Societies.

Bharghavan, V., Lu, S., & Nandagopal, T. (1999). Fair queuing in wireless networks: Issues and approaches. *IEEE Personal Communications*, *6*(1), 44–53. doi:10.1109/98.752787

FCC 05-57. (2005). *Facilitating opportunities for flexible, efficient, and reliable spectrum use employing cognitive radio technologies*. ET docket no. 03-108, Federal Communications Commission.

Fitzek, F. H. P., & Katz, M. D. (Eds.). (2006). *Cooperation in wireless networks: Principles and applications: real egoistic behavior is to cooperate!* Dordrecht, The Netherlands: Springer. doi:10.1007/1-4020-4711-8

Fitzek, F. H. P., & Katz, M. D. (Eds.). (2007). *Cognitive wireless networks: Concepts, methodologies and visions inspiring the age of enlightenment of wireless communications*. Dordrecht, The Netherlands: Springer.

Garg, S., & Kappes, M. (2003). *Admission control for VoIP traffic in IEEE 802.11 networks*. IEEE Global Telecommunications Conference.

Gazis, V., Houssos, N., Alonistioti, N., & Merakos, L. (2003). *On the complexity of "always best connected" in 4G mobile networks*. IEEE 58th Vehicular Technology Conference.

Heusse, M., Rousseau, F., Berger-Sabbatel, G., & Duda, A. (2003). *Performance anomaly of 802.11b*. IEEE 22nd Annual Joint Conference of the IEEE Computer and Communications Societies.

Holma, H., & Toskala, A. (Eds.). (2009). *LTE for UMTS: OFDMA and SC-FDMA based radio access*. West Sussex, UK: John Wiley & Sons, Ltd.

Joshi, T., Mukherjee, A., Yoo, Y., & Agrawal, D. P. (2008). Airtime fairness for IEEE 802.11 multirate networks. *IEEE Transactions on Mobile Computing*, *7*(4), 513–527. doi:10.1109/TMC.2007.70740

Kamerman, A., & Monteban, L. (1997). WaveLAN-II: A high-performance wireless {LAN} for the unlicensed band. *Bell Labs Technical Journal*, *2*(3), 118–133. doi:10.1002/bltj.2069

O'Hara, B., & Petrick, A. (1999). *The IEEE 802.11 handbook: A designer's companion*. New York, NY: Standards Information Network, IEEE Press.

Ong, E. H. (2010). *A generalized cooperative and cognitive radio resoure management architecture for future wireless networks. Ph.D.* NSW, Australia: The University of Newcastle.

Ong, E. H., & Khan, J. Y. (2009a). *On load adaptation for multirate multi-AP multimedia WLAN-based cognitive networks*. IFIP Wireless Days Conference.

Ong, E. H., & Khan, J. Y. (2009b). *A unified QoS-inspired load optimization framework for multiple access points based wireless LANs*. IEEE Wireless Communications and Networking Conference.

Ong, E. H., & Khan, J. Y. (2010a). Cooperative radio resource management framework for future IP-based multiple radio access technologies environment. *Computer Networks*, *54*(7), 1083–1107. doi:10.1016/j.comnet.2009.11.001

Ong, E. H., & Khan, J. Y. (2010b). On optimal network selection in a dynamic multi-RAT environment. *IEEE Communications Letters*, *14*(3), 217–219. doi:10.1109/LCOMM.2010.03.092378

Ong, E. H., Khan, J. Y., & Mahata, K. (2010). *A simple model for non-homogeneous and non-saturated IEEE 802.11 DCF infrastructure BSS*. IEEE 12th International Conference on Communication Systems.

Ong, E. H., Khan, J. Y., & Mahata, K. (2011). *Radio resource management of composite wireless networks: Predicitve and reactive approaches*. IEEE Transactions on Mobile Computing.

Pang, Q., Leung, V. C. M., & Liew, S. C. (2008). An enhanced autorate algorithm for wireless local area networks employing loss differentiation. *IEEE Transactions on Vehicular Technology*, *57*(1), 521–531. doi:10.1109/TVT.2007.905418

Rappaport, T. (2001). *Wireless communications: Principles and practice* (2nd ed.). Upper Saddle River, NJ: Prentice Hall PTR.

IEEE Std 802.11-2007. (2007). *IEEE standard for local and metropolitan area networks part 11: Wireless LAN medium access control (MAC) and physical layer (PHY) specifications*.

IEEE Std 1900.1-2008. (2008). *IEEE standard definitions and concepts for dynamic spectrum access: Terminology relating to emerging wireless networks, system functionality, and spectrum management*.

IEEE Std 802.21-2008. (2009). *IEEE standard for local and metropolitan area networks part 21: Media independent handover services*.

Tan, G., & Guttag, J. (2004). *Time-based fairness improves performance in multi-rate WLANs*. USENIX Annual Technical Conference.

Thomas, R. W. (2007). *Cognitive networks. Ph.D.* Blacksburg, VA, USA: Virginia Polytechnic Institute and State University.

Tinnirello, I., & Choi, S. (2005). *Temporal fairness provisioning in multi-rate contention-based 802.11e WLANs*. IEEE 6th International Symposium on a World of Wireless Mobile and Multimedia Networks.

Velayos, H., Aleo, V., & Karlsson, G. (2004). *Load balancing in overlapping wireless LAN cells*. IEEE International Conference on Communications.

Wang, P., Jiang, H., & Zhuang, W. (2006). IEEE 802.11e enhancement for voice service. *IEEE Wireless Communications*, *13*(1), 30–35. doi:10.1109/MWC.2006.1593522

Yang, K., Gondal, I., Qiu, B., & Dooley, L. S. (2007). *Combined SINR based vertical handoff algorithm for next generation heterogeneous wireless networks*. IEEE Global Telecommunications Conference.

Yu, J., & Choi, S. (2006). Comparison of modified dual queue and EDCA for VoIP over IEEE 802.11 WLAN. *European Transactions on Telecommunications*, *17*(3), 371–382. doi:10.1002/ett.1128

Zhai, H., Chen, X., & Fang, Y. (2005). How well can the IEEE 802.11 wireless LAN support quality of service? *IEEE Transactions on Wireless Communications*, *4*(6), 3084–3094. doi:10.1109/TWC.2005.857994

Chapter 11
Fuzzy Systems for Spectrum Access, Mobility and Management for Cognitive Radios

Prabhjot Kaur
ITM University, India

Moin Uddin
Delhi Technological University, India

Arun Khosla
National Institute of Technology India

ABSTRACT

This chapter addresses the issues in air interface designs for Cognitive Radios. Fuzzy logic system is used as one of the soft computing techniques to learn to sub optimality and vagueness. Many good, simple, and quick fuzzy based solutions have been developed since few decades in many diverse domains. Through this chapter, the authors first discuss the significance and need of soft computing techniques in designing such solutions and then present fuzzy based solutions for spectrum access, mobility, and management. Hierarchical fuzzy systems have been used to get over to the problem of curse of dimensionality. The proposed solutions consider an architecture, similar to the one proposed by IEEE 802.22 working group, for spectrum sharing and management. Models have been designed using fuzzy logic toolbox in MATLAB, and the system performance is checked using SIMULINK.

11.1 INTRODUCTION

Lack of spectra has emerged as a unanimous problem among all countries. Researchers throughout the world are focusing to increase the spectrum efficiency to mitigate spectrum scarcity. Cognitive radio (CR) is looked as one of the promising proposals for improving spectrum utilization. Many efforts are being put up in standardizing the architecture and air interface parameters for CR. To optimize resource use, next generation networks require smart devices like CR to be able to model their location, their users, networks and

DOI: 10.4018/978-1-4666-2005-6.ch011

the larger environment. Based on monitored set of these parameters, CR can adapt to appropriate frequency bands, protocols and interfaces.

11.2 SOFT COMPUTING AND COGNITIVE CYCLE

CR was first coined by Mitola (1999) who represented its major functions to adapt the transmission parameters in changing environments through a cognitive cycle. This six stage cognition cycle is briefed as follows.

- **Observe**: Know the information of operating environment through sensing and signaling mechanisms.
- **Orient**: Evaluate this information to determine its significance and relevance.
- **Plan**: Based on this evaluation, the radio determines its options or alternatives for resource optimization.
- **Decide**: An alternative is chosen that evaluates more favorably than other options, including the current ongoing action.
- **Act**: The radio implements the decisions taken for resource optimizations. These changes are then reflected in the interference profile presented by the cognitive radio in the outside world.
- **Learn**: Throughout the process, the radio uses its observations and decisions to improve its own operation, creating new modeling states and alternatives.

Cognitive cycle as briefed above, forms the heart of a Cognitive Radio. Out of the six stages of cognitive cycle, the stages *"decide"* and "act*"* is important functionalities that may work efficiently using soft computing techniques. Thus, artificial intelligence and soft computing techniques form important enabling techniques for the success of CR. Some of the key functionalities (Mitola 1999), (Linda 2009) supported by these techniques are as follows:

1. **Optimization of Resources:**
    - CR to make best use of use of resources it has been given or it has identified for itself while protecting the rights of others
    - Auto reconfigurability of transmission profile and operating parameters and redirecting resources around the network e.g. *"who gets what"*

   Thus the purpose is to evaluate all choices or solutions to find the one with highest rank and adapting to the "best" parameters. Thus, this amounts to the cross layer optimization largely focussing on physical (PHY) and medium access control (MAC) layers.
2. **Signal Shaping:** CR needs to represent, organise, store and analyse the collected knowledge so that appropriate optimization routines can be evoked. Some uniform mechanism for knowledge representation to be selected and then a suitable form of reasoning could be used. The operating parameters that need manageable decisions are: waveform type and antenna usage. The decisions are to be taken from the choices available as shown in Figure 5.1. Regulatory policies like any other knowledge need to be represented in a machine readable manner and the CR needs to reason about the policies and make decision.
3. **Knowledge Gain Through Learning:** Learning is crucial when dealing with unknown and unplanned scenarios and thus can be of great benefit to a CR in improving the overall performance. There are many learning techniques ranging from very simple memorising techniques to complex ones e.g. artificial intelligence and machine learning.

Figure 1. Structure of a fuzzy logic system.

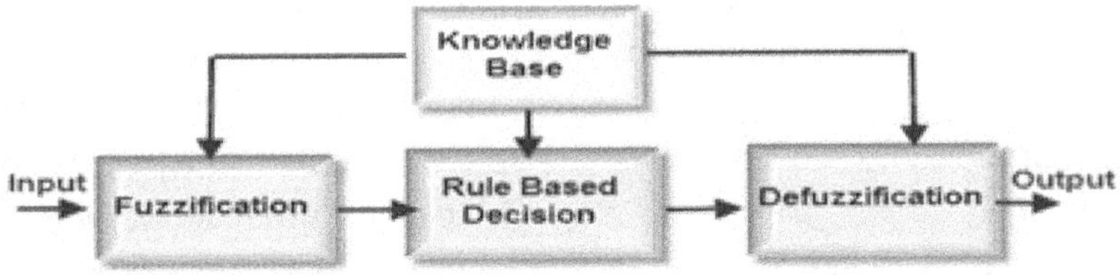

Good solutions have been proposed to cater to the above requirements in past few years by exploiting soft computing techniques. Some of these techniques include neural networks, genetic algorithms, game theory, expert systems, intelligent agents and fuzzy systems. These techniques have been used alone or a mix of two or more of these has been used for designing solutions to different aspects of CR. This chapter is focused to discuss some air interfacing solutions using fuzzy systems.

11.3 AIR INTERFACE SOLUTIONS USING FUZZY SYSTEMS

Though not much of the work is reported for fuzzy systems in the field of CR, some interesting research is available to address the air interface issues. These are detailed in the following section.

Before discussing these solutions for CR, a background on fuzzy systems is presented to make the readers familiar with this technique.

11.3.1 Fuzzy Logic Systems

As its name suggests, it is the logic underlying modes of reasoning which are approximate rather than exact. The importance of fuzzy logic derives from the fact that human reasoning and especially common sense reasoning are approximate in nature. Fuzzy logic has been extended to handle the concept of partial truth, where the truth value may range between completely true and completely false. Thus, it is tolerant to sub optimality and vagueness. Fuzzy logic was first proposed by L.A. Zadeh (1965). In Fuzzy, exact reasoning is viewed as a limiting case of approximate reasoning and everything is a matter of degree. It is stated that any logical system can be fuzzified which makes boolean logic as a subset of fuzzy logic. He presented fuzzy probability and also generalized it to what is called possibility theory which is beyond the scope of this chapter.

Fuzzy logic system (FLS) is a suitable intelligence system for implementation of incomplete knowledge representation and qualitative reasoning. The control actions of this system are determined using fuzzy logic reasoning. The structure of an FLS is depicted in Figure 1. The input to the controller is often hard in nature thus requires preprocessing before it enters the system. This preprocessor can be a quantizer, differentiator and filter etc.

Knowledge Base

Knowledge base represents the characteristics of its universe, the set of linguistic attributes and the membership functions. The universe is complete domain over which a variable can occupy its value, linguistic attributes are the labels and each label is defined by its membership function.

Fuzzification

The controller receives a crisp input and the process of Fuzzification translates this input to a degree of membership function for linguistic terms of fuzzy sets. The membership function is used to associate a grade to each linguistic term.

Rule Based Decision

Rule base is the heart of FLS. A set of rules is formed to relate the possible combination of inputs with the output using IF-THEN statements. The decision rules are formed based on expert experience, learning, operator's actions or fuzzy model.

Consider as an example *S, R* as the input variables, each representing linguistic variables good and poor. Let the output be *P* representing any of the variables as change or freeze. We can have following four decision rules:

Rule 1: IF *S* is good and *R* is good THEN *P* is freeze
Rule 2: IF *S* is poor and *R* is good THEN *P* is change
Rule 3: IF *S* is good and *R* is poor THEN *P* is freeze
Rule 4: IF *S* is poor and *R* is poor THEN *P* is change

If we have x input variables and y membership functions, then number of rules *R,* needed to construct a complete single layer fuzzy controller is y^x. As x increases, the rule base will make the fuzzy controller difficult to implement. The system or the memory will be overloaded. This is called "curse of dimensionality". To deal with this problem, hierarchical fuzzy system (HFS) is designed which is discussed in next section.

Defuzzification

The defuzzification process converts resulting fuzzy output to crisp value in the usable form in real world. There are several defuzzification methods but only three of these are in common use. These are centre of gravity (COG), Maximum defuzzification and weighted average technique. Here, COG is used in the schemes proposed as part of this chapter.

Associated transparency of fuzzy systems enables interpretation and analysis of many complex systems. Thus, fuzzy systems are not just black box models and are considered universal approximators. However, when fuzzy systems are applied to more complicated and high dimensional systems, the *"curse of dimensionality"* becomes increasingly apparent. To avoid this bottleneck to wider applications, hierarchical fuzzy systems were proposed.

Hierarchical Fuzzy Systems

If we increase the number of input variable, complexity of the fuzzy system increase and its memory is overloaded. This increase in complexity of fuzzy systems with increase in number of input variables is referred to as *curse of dimensionality.* To overcome the problems arising out of curse of dimensionality, HFS was introduced. HFS was first introduced by Raju, Zhou and Kisner (1993) who proposed that instead of using a high dimensional standard fuzzy system, an HFS consists of a hierarchy of number of low dimensional sub-fuzzy systems called fuzzy logic modules contributing to the final solution. The outputs of lower level fuzzy module are fed as an input to the higher level module. The upper fuzzy sub system may also take the input as a combination of lower level output and an original input variable. Thus, an HFS may have multiple levels and multiple sub systems at each level.

In HFS, total number of rules increase only linearly with the increase in number of input variables as against the exponential growth of rules in standard fuzzy system.

11.3.2 Air Interface: Background and Solutions

The air interface for CR is based on four main procedures: Spectrum sensing, Spectrum management, mobility and spectrum sharing. (C. Cordeiro et al., 2006). This section briefs these procedures and highlights the use of fuzzy systems designed by various researchers for solving these issues. In regards to CR, these solutions largely address the cross layer functionalities.

11.3.2.1 Spectrum Sensing

Spectrum sensing refers to the detection of spectrum holes with the help of spectrum sensing techniques as transmitter/energy detection, interference based detection, matched filters and cooperative detection. The ability of a CR network to coexist with other existing networks is predicated upon being able to sense the existence of other networks.

A collaborative spectrum sensing scheme in a CR network is proposed by Wendong Yang et al. (2007). Spectrum sensing is performed at individual SU level and these results are compiled at access point (AP). The AP takes the decision on sensed spectrum and evaluates the credibility of SU using fuzzy system. The collected sensing data and the creditability of each SU help make the sensing decision. Here the credibility is evaluated during the training stages using fuzzy inference. The work presented by Wendong Yang et al. (2007) is based on the concepts of Q. Peng et al., (2006) for credibility evaluation.

11.3.2.2 Spectrum Access: SU Credibility Evaluation

In order to confirm "who" should be given "what" spectrum, it is important to evaluate the credibility of SU which is different for every user due to scenarios like fading, hidden terminal problem and shadowing etc. The above sensing works consider probability of detection, false alarm and missed detection for evaluating the credibility of an SU. However, applying fuzzy inference rules on parameters which are computed as probability may not provide good approximations.

Another scheme for evaluating SU credibility is presented by Hong-Sam T. Le1 and Hung D. Ly2 (2008). Here, the credibility is evaluated on the basis of three antecedents as spectrum utilization efficiency of the secondary user, its degree of mobility, and its distance to the primary user. Following section details this scheme and additionally studies an impact of fuzzy based distance evaluation between a transmitter and the receiver. Thus, these works do not consider probabilistic parameters for credibility evaluation of unlicensed users.

11.3.2.3 Power Control

Fuzzy based power control mechanism is designed for secondary user (SU) to coexist with the primary user (PU/s) for better spectrum utilization. (Le, H.-S.T. Qilian Liang, 2007). This work presented FLS based DSA to obtain lesser transmission power consumption as compared to the fixed power control. This has been done based on three antecedents as the interference level caused by the SU to the reclaiming PU, distance from the SU to the reclaiming PU and the received power difference of the SU at BS.

However, these antecedents considered in this work seem to be related to each other and convey redundant information. Thus, as part of this dissertation work, power control is optimized with only two antecedents sufficient to decide the power control. These antecedents are PU power received at SU and the actual PU transmitted power.

11.3.2.4 Spectrum Mobility

Spectrum mobility or handover refers to the change of operating frequency or band. Mobility occurs when CR changes its frequency band upon detection of PU signal. CR needs to switch to another frequency, maintaining seamless communication requirements during the transition to better spectrum.

Once an SU occupies a channel for transmission and the PU activity is detected, a decision on spectrum mobility needs be made. Spectrum handover is initiated either when the interference at PU exceeds certain threshold level or if QoS at SU is not satisfactory. This handover strategy is designed using fuzzy logic by L. Giupponi, Ana I. Pérez-Neira, (2008). Factors deciding if handover is required or not are signal strength received at the SU from the PU, PU SNR, the SU transmission power and the bit rate of SU. The work presented in following section presents a re-arranged spectrum mobility strategy based on the works of L. Giupponi, Ana I. Pérez-Neira, (2008) and proposes power modification before a decision on handover may be made rather than considering it at the same level while deciding the handover.

11.3.2.5 Cross Layered Optimizations

For any communication system, groups of certain set of functionalities together form communication layers. For developing successful CR platforms, it is required to design optimized cross layered solutions as independent optimized layered solutions may provide only suboptimal performance. Baldo, Nicola and Zorzi Michele (2008) proposed a cross layer optimization architecture between medium access control (MAC) and transport layer for knowledge representation and control implementation. This work developed a fuzzy TCP for error and congestion control and is proved to outperform the traditional TCP protocol.

Thus, fuzzy systems have proved to be an important enabling technique for solving a number of CR air interface problems.

11.4 SPECTRUM ACCESS POSSIBILITY

This section proposes a fuzzy based approach to define spectrum access possibility of an unlicensed user in a context characterized by uncertain and incomplete information. A hierarchical Fuzzy System of two fuzzy controllers is presented. One of these Controllers is used to estimate distance between unlicensed and licensed user and the second controller receives the output of this controller as input along with two other antecedents to find the possibility of spectrum access for an unlicensed user. Thus here we evaluate the effect of distance fuzzification on access possibility of SU proposed by Hong-Sam T. Le1 and Hung D. Ly2 (2008).

This model comprises of two FLS in hierarchical form as shown in Figure 2. Selection policy of both FLS is based on the *min, max, min, max and centroid* methods for the fuzzy *and, or, implication, aggregation and defuzzification* operators respectively. The flowchart of this scheme is as shown in Figure 3.

11.4.1 Distance Estimation between Primary and Secondary Users

Before a base station (BS) decides to assign free available frequency to the desired SU, it must know that SU will not cause interference with the PU. Interference between two users is directly proportional to the distance between them. The power received at SU from PU is generally considered enough to determine the distance between the two, without realizing the strength of signal power transmitted by the PU. Thus, in the scheme presented here, we compute the distance between PU and SU using two antecedents namely power transmitted by PU (TPower) and power received by SU (RPower). Assuming free space propagation model, received power and transmitted power are related using well known Friis free space formula in as in (1)

$$P_r(d) = \frac{P_t G_t G_r \lambda^2}{(4\pi)^2 d^2 L} \tag{1}$$

Figure 2. System model for spectrum access scheme

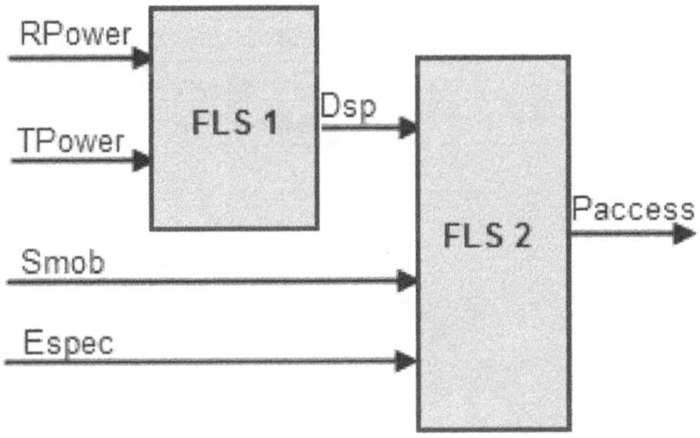

where $P_r(d)$ is the received power (RPower) as function of transmitter and receiver separation distance d in meters, P_t is the transmitted power (TPower), G_t and G_r are transmitter and receiver antenna gain respectively, λ is the wavelength in meters and L is the system loss factor not related to propagation ($L \geq 1$). It is also clear from (1) that the received power falls off as the square of separation between the transmitter and receiver. This implies that the received power decays with distance at a rate of 20dB/decade which further emphasises the need to estimate distance between SU and PU before SU is given spectrum access by BS.

Figure 3. Flowchart for spectrum access possibility with distance estimation between PU and SU

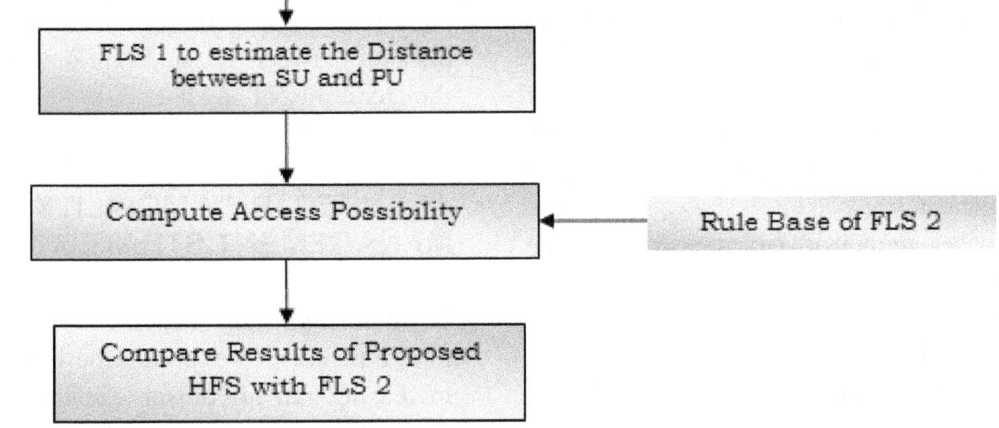

Figure 4. Comparison of spectrum assignment solution of proposed scheme with standalone fuzzy system for evaluation of access probability

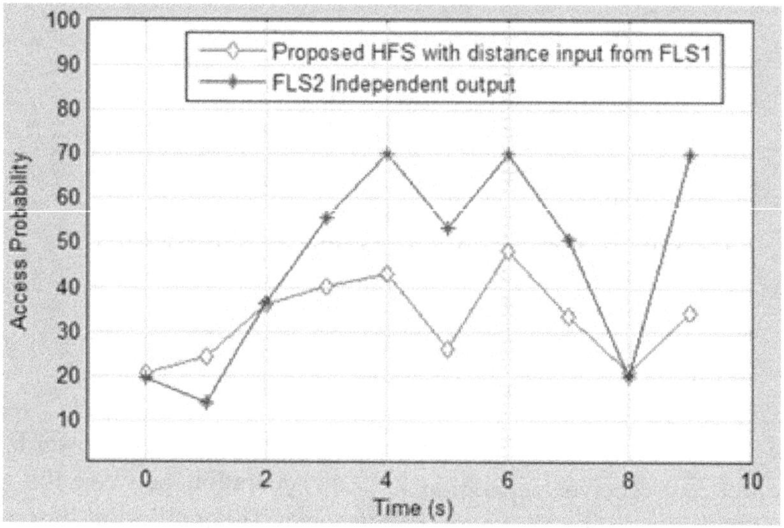

11.4.2 Simulation Setup and Results

FLS1 and FLS 2 are designed using Fuzzy controller toolboxes in MATLAB and the complete system is simulated using SimuLink. Transmission and reception ranges of subscriber station (SS) of IEEE 802.16 Standard (2009) are considered here to define the membership function ranges of T_{Power} and R_{Power} respectively. For FLS1, we consider a random generation of Tpower and RPower in the range of -84 dBm to 43 dBm and -123 dBm to -30 dBm respectively for an SU. In FLS2, the spectrum efficiency and SU mobility is made to vary randomly between 0 to 17 b/s/Hz and 1 to 10 m/s respectively. The results obtained are an average of 20 simulations with simulation time of 10 s each. For every simulation, a new seed of random generators is selected. Access possibility of SU to be determined by BS to select best SU in its network using this scheme is shown in Figure 4.

Two cases are considered here to obtain the access possibility.

1. **Case a:** Proposed HFS
2. **Case b:** Simulation of FLS2 as an independent entity.

In case b, Dsp is varied in a range of 1 to 100 m/s randomly.

The maximum possibility so obtained for *case a* is 49% and that of case b is 70%. This signifies the distance Fuzzification process in evaluating the credibility of SU as more close to reality. Thus a better judgment of SU selection can be made at the *Head SU*.

Thus, fuzzified distance estimation between SU and PU tends to vary the credibility in contrast to only estimating it without distance measurement which will change the credibility of every contending user. Thus, the selection for best candidate for spectrum allocation will change. Hence, it becomes crucial to measure/fuzzify the distance between SU and PU prior to determining the credibility of any user.

11.5 SPECTRUM MOBILITY MANAGEMENT STRATEGY

To obtain high signal to noise ratio for better data rate, transmit power should be high. However, increasing transmit power levels has an undesirable effect of increasing interference. Thus, it is not possible to represent the overall system performance with single index of performance. Rather tradeoffs for decision to control transmit power or switch to another frequency needs to be adopted. This model proposes to tackle this issue in computationally traceable fashion using fuzzy logic. A hierarchical model with similar structure as that of Figure 2 is designed where two FLS are used, one to control the transmit power and another to decide if handover to another frequency is required or not (Prabhjot Kaur et al., 2009). This work proposes power modification before a decision on handover may be made rather than considering it at the same level while deciding the handover unlike the works by L. Giupponi and Ana I. Pérez-Neira (2008). This model considers only two antecedents for deciding power control and thus the number of handovers may be reduced by controlling the power before taking the decision on spectrum handover. This strategy focuses on the idea that instead of returning the band to PU whenever it wants and switching to another available band, CR should try to modify transmit power in order to avoid interference with PU. As illustrated in flowchart given in Figure 5, SU at first estimates the power it receives from PU and compares it with the power being transmitted by the PU.

This comparison will help estimate the power modification required in the transmit power of SU. However, if power transmitted becomes very low, signal to noise ratio (SNR) will drop and also the rate of transmission. In effect, QoS will be deteriorated. Spectrum handover will be the only choice here. Thus, first it is tried to reduce the interference between SU and PU and depending upon various QoS parameters, we make SU decide whether or not spectrum handover should take place. Power modification is determined with the help of FLS 1 and spectrum handover decision is taken using FLS 2.

11.5.1 Power Control

We compute the percentage change in power required for SU transmission in order to avoid interference with PU. The two antecedents used here are:

Antecedent 1: R_{Power}.as power received by SU from PU.

Antecedent 2: T_{Power} as actual power transmitted by PU.

Both these antecedents and the consequence are characterized by three fuzzy sets *T(x)* where *x* represents antecedent and consequence variable as shown in (2)

$$T(RPower) = T(TPower) = T(P\,\mathrm{mod}\,ified) = \{Low, Medium, High\} \tag{2}$$

We compare R_{Power} with T_{Power} in order to access the interference level, which in this case helps estimate the modification required in transmit power of SU.

11.5.2 Spectrum Handover Probability

Second FLS is designed to find the probability of handover with two antecedents and one consequence. One of the inputs R_{Power} is same as that of FLS1. Consequence of FLS1 is another input to FLS2. The third antecedent is bit rate of SU, Rsu considered as an indicator of QoS for SU. The bit rate chosen is 96 Mb/s with a roll off factor of 0.25, channel size of 25 MHz, 64-QAM modulation scheme corresponding to IEEE 802.16 standard (2009). Rsu is characterized with a term set of three fuzzy sets as shown in (3) and the output linguistic variable $P_{handover}$ is characterized into four fuzzy sets as shown in (4).

Figure 5. Flowchart for spectrum mobility management strategy

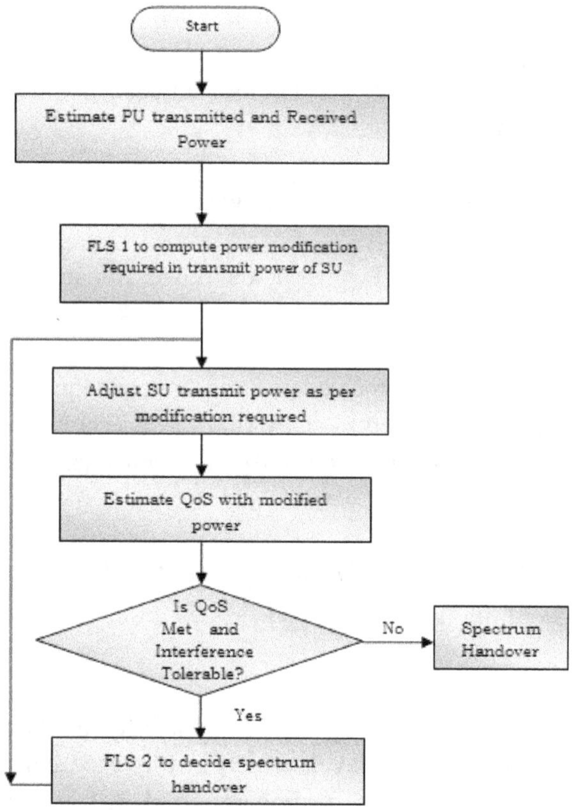

$$T(Rsu) = \{Low, Medium, High\} \tag{3}$$

$$T(PHandover) = \{No, PNo, PYes, Yes\} \tag{4}$$

where P_{No} and P_{Yes} are probably no and probably yes respectively.

11.5.3 Simulation Setup and Results

We assume that data rate of PU is deterministic for SU which helps deducing SNR of PU and consequently the T_{Power} of PU. Transmission and reception ranges of subscriber station (SS) of IEEE 802.16 (2009) are considered here to define the membership function ranges of T_{Power} and R_{Power} respectively. Membership function for R_{Power} is defined with the help of received signal strength indicator (RSSI) measured at SS. From a succession of RSSI measurements, the SS derives and updates the estimates of RSSI mean and standard deviation. The range over of these measurements extend 3 dB on each side beyond the –40 dBm to –123 dBm limits for the final averaged statistics that are reported [IEEE 802.16 Standard, 2009]. Also, SS receiver is capable of decoding a maximum on-channel signal of –30 dBm. Thus, we consider a random generation of T_{Power} and R_{Power} in the range of -84 dBm to 43 dBm and -123 dBm to -30 dBm respectively for an SU.

Different parameters considered for simulations are based on IEEE 802.16 standard with their values same as discussed in previous section. Percentage modification made in transmit power of SU with variation in power transmitted by the PU and that of received at SU is shown in Figure 4. The results show modification level of SU transmits power with R_{Power} which increases with increase in R_{Power}. Thus our strategy helps SU to modify its transmit power gradually rather than in absolute fixed steps. This helps CR to smoothly adapt to channel conditions. The modification level in SU transmit power helps it decide if handover is required or not. The determined possibility of handover with variation in R_{Power}, $P_{modified}$ and data rate of SU is as shown in Figure 5. A reduction in handover frequency by 50% is obtained with a variation of Rsu. Thus using our strategy, a handover takes place when either QoS is not met or the interference limit is above tolerable limit.

11.6 FUTURE RESEARCH DIRECTIONS

In the models presented in this chapter, it is assumed that data rate of PU is deterministic for SU which helps deducing SNR of PU and consequently the T_{Power} of PU. In such scenarios, the proposed scheme seems fit for point-to-point PU communication scenarios but probably not for the systems with many undetectable receivers such

Figure 6. Percentage modifications in transmit power of SU with varying PU power received at SU

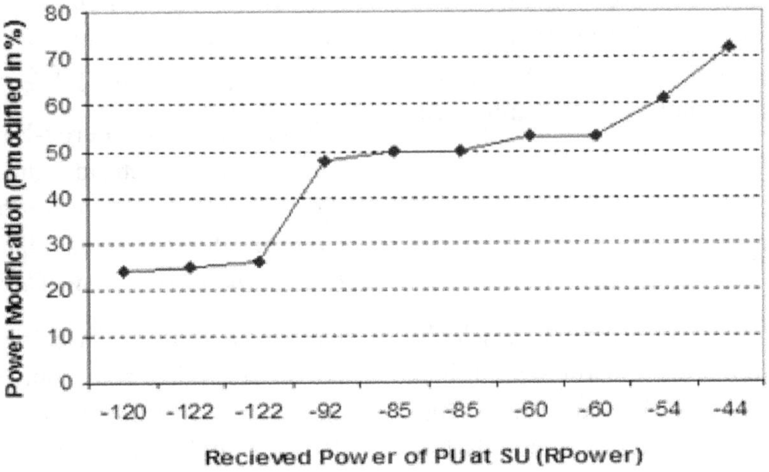

Figure 7. Handover probability achieved due to variation in (a) power modification done in transmit power of SU as per output received using FLS1 (b) data rate of SU

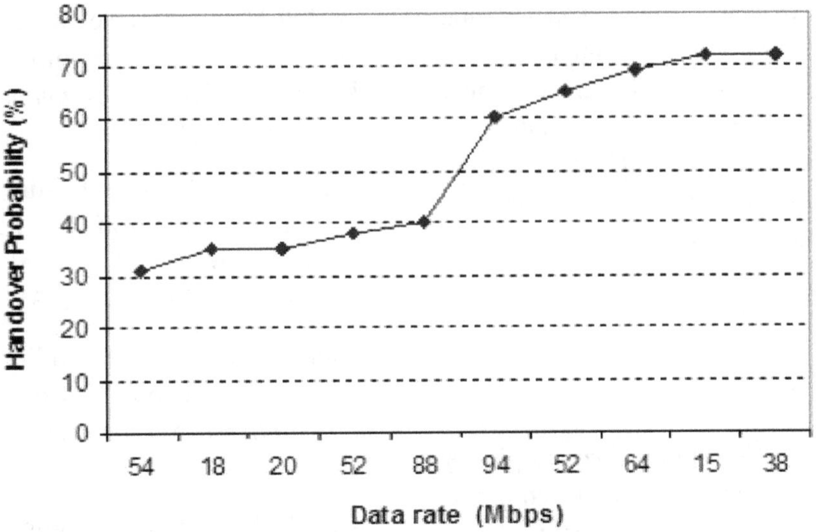

as TV broadcast etc. Thus, integrated systems for broadcast services may also be considered.

Additionally, these being one of the initial proposals of using fuzzy logics in CR, it does not encompass learning approaches to optimize the proposed FLS, and the fuzzy rule base is selected based on the logical reasoning. Thus optimized solutions may tend to improve the performance of CR intelligence for its cognition cycle. As an example, a combination of neural networks, game theory and fuzzy control may be used for optimization of resources and spectrum access decisions at SU. However it depends on the network architecture and spectrum utilization of licensed users that helps select the protocols and intelligence techniques for optimizing the resources.

This work may further be extended for QoS evaluation in real time scenarios with a network of SU and PU comprising different traffic types. Cross layered integrated solutions can be designed for effective implementation of air interface solutions.

11.7 CONCLUSION

CR will be intelligent devices in the market which will require a combination of soft computing techniques for design effective cognition. This chapter proposed some fuzzy based models for spectrum access, spectrum mobility and spectrum management. The strategies have been discussed to present the effectiveness in air interface designs even by using simple hierarchical fuzzy systems. The first model analyzed the impact of distance measurement between an SU and PU on deciding the strongest contender for spectrum allocation. Second strategy was designed to lower the number of handovers in a given network by controlling the interference power levels of SU. Finally, some limitations are discussed and future directions are briefed for interested readers to pursue their research.

REFERENCES

Baldo, N., & Zorzi, M. (2008). Fuzzy logic for cross-layer optimization in cognitive radio networks. *IEEE Communications Magazine*, *46*(4), 64–71. doi:10.1109/MCOM.2008.4481342

Cordeiro, C., Challapali, K., Birru, D., & Shankar, S. (2006). IEEE 802.22: An introduction to the first worldwide wireless standard based on cognitive radios. *The Journal of Communication*, *1*(1), 38–47.

Doyle, L. E. (2009). *Essentials of cognitive radio*. Cambridge University Press. doi:10.1017/CBO9780511576577

Doyle, L. E., Nolan, K. E., Forde, T. K., et al. (2006). *A platform for dynamic spectrum experimentation*. 1st International Workshop on Technology and Policy for Accessing Spectrum.

Giupponi, L., & Pérez-Neira, A. I. (2008). Fuzzy-based spectrum handoff in cognitive radio networks. *3rd International Conference on Cognitive Radio Oriented Wireless Networks and Communications, CrownCom 2008*, (pp. 1-6).

Haykin, S. (2005). Cognitive radio: Brain-empowered wireless communication. *IEEE Journal on Selected Areas in Communications*, *23*(2), 201–220. doi:10.1109/JSAC.2004.839380

Jing, X., Mau, S. C., Raychaudri, D., & Matyas, R. (2005). *Reactive cognitive radio algorithms for coexistence between 802.11b and 802.16a networks* (pp. 2465–2469). IEEE GLOBECOM.

Jondral, F. K. (2005). Software-defined radio–Basics and evolution to cognitive radio. *EURASIP Journal on Wireless Communications and Networking*, 275–283.

Kaur, P., Khosla, A., & Uddin, M. (2010). Adaptive bandwidth allocation scheme for cognitive radios. *International Journal of Advancements in Computing Technology*, 35-41.

Kaur, P., Khosla, A., & Uddin, M. (2010). *Fuzzy based adaptive bandwidth allocation scheme in cognitive radio networks* (pp. 41–45). IEEE ICT-KE.

Kaur, P., Khosla, A., & Uddin, M. (2011). Cognitive radios: Need, capabilities, standards, applications and research challenges. *International Journal of Computers and Applications*, *30*(1), 31–38.

Kaur, P., Khosla, A., & Uddin, M. (2011). Markovian queuing model for dynamic spectrum allocation in centralized architecture for cognitive radios. *International Journal of Engineering and Technology*, *3*(1), 96–101.

Kaur, P., Uddin, M., & Khosla, A. (2009). An efficient spectrum mobility management strategy in cognitive radio networks. *IEEE International Workshop on Cognitive Wireless Systems*, (pp. 1-6).

Kaur, P., Uddin, M., & Khosla, A. (2009). Cognitive radio as common information exchange reference model for interoperability between heterogeneous networks. *IEEE Workshop on Cognitive Wireless Systems*, (pp. 1-4).

Kim, K. W., Park, J., Cho, J., Lim, K., Razzell, C. J., & Kim, K. … Laskar, L. (2007). Interference analysis and sensing threshold of detect and avoid (DAA) for UWB coexistence with WiMax. *Vehicular Technology Conference*, (pp. 1731-1735).

Klir, G., & Yuan, B. (1995). *Fuzzy sets and fuzzy logic: theory and applications*. Upper Saddle River, NJ: Prentice-Hall PTR.

Lansford, J. (2004). UWB coexistence and cognitive radio. *International Workshop on Ultra Wideband Systems joint with Conference on Ultra-wideband Systems and Technologies*, (pp. 35-39).

Le, H.-S. T., & Liang, Q. (2007). An efficient power control scheme for cognitive radios. *IEEE Conference on Wireless Communications and Networking*, (pp. 2559-2563).

Le, H. S. T., & Ly, H. D. (2008). Opportunistic spectrum access using fuzzy logic for cognitive radio networks. *Second International Conference on Communications and Electronics*, (pp. 240–245).

Lee, M.-L., Chung, H.-Y., & Yu, F.-M. (2003). Modeling of hierarchical fuzzy systems. *Elsevier Fuzzy Sets and Systems*, *138*(2), 343–361. doi:10.1016/S0165-0114(02)00517-1

Mitola, J. (2001). Cognitive radio for flexible mobile multimedia communications. *Springer Mobile Networks and Applications*, *6*(5), 435–441. doi:10.1023/A:1011426600077

Mitola, J. III, & Maguire, G. Q. Jr. (1999). Cognitive radio: Making software radios more personal. *IEEE Personal Communications*, *6*(4), 13–18. doi:10.1109/98.788210

Mohamadian, M., & Stonier, R. J. (2000). Hierarchical fuzzy logic control. *IPMU Conference*, (pp. 622-629).

Neel, J. O. (2006). *Analysis and design of cognitive radio networks and distributed radio resource management algorithms*. Ph.D dissertation, Virginia Polytechnic Institute and State University.

Nolan, K. E., Rondeau, T. W., Sutton, P. D., Bostian, C. W., & Doyle, L. E. (2006). *A framework for implementing cognitive functionality.* 51st SDR Forum General Meeting and 2006 Technical Conference.

Peng, Q., Zeng, K., Wang, J., & Li, S. (2006). A distributed spectrum sensing scheme based on credibility and evidence theory in cognitive radio context. *IEEE 17th International Symposium on Personal, Indoor and Mobile Radio Communications*, (pp. 1-5).

Raju, G. V. S., & Zhou, J. (1993). Adaptive hierarchical fuzzy controller. *IEEE Transactions on System and Cybernetics*, *23*(4), 973–980. doi:10.1109/21.247882

Rieser, C. J., Rondeau, T. W., Bostian, C. W., & Gallagher, T. M. (2004). Cognitive radio testbed: Further details and testing of a distributed genetic algorithm based cognitive engine for programmable radios. *Military Communications Conference*, Vol. 3, (pp. 1437-1443).

Scaperoth, D., Le, B., Rondeau, T., Maldonado, D., Bostian, C. W., & Harrison, S. (2006). Cognitive radio platform development for interoperability. *IEEE Military Communications Conference. MILCOM* 2006, (pp. 1-6).

Spectrum Policy Task Force. (2002). *Spectrum policy taskforce report*. Federal Communications Commission ET Docket 02-135.

IEEE Standard 802. (2009). *IEEE standard for local and metropolitan area networks part 16: air interface for broadband wireless access systems*.

Takagi, T., & Sugeno, M. (1985). Fuzzy identification of systems and its applications to modelling and control. *IEEE Transactions on Systems, Man, and Cybernetics*, *15*, 116–132.

Tsekourasa, G., Sarimveisb, H., & George, E. K. (2005). A hierarchical fuzzy-clustering approach to fuzzy modelling. *Fuzzy Sets and Systems*, *150*, 245–266. doi:10.1016/j.fss.2004.04.013

Valenta, V., Maršálek, R., Baudoin, G., Villegas, M., Suarez, M., & Robert, F. (2010). *Survey on spectrum utilization in Europe: measurements, analyses and observations. Cognitive Radio Oriented Wireless Networks & Communications* (pp. 1–5). CROWNCOM.

Wang, D., Zeng, X. J., & Keane, J. A. (2006). A survey of hierarchical fuzzy systems. *International Journal of Computational Cognition*, *4*(1), 18–29.

Wang, L.-X. (1999). Analysis and design of hierarchical fuzzy systems. *IEEE Transactions on Fuzzy Systems*, *7*(5), 617–624. doi:10.1109/91.797984

Yang, W., Cai, Y., & Xu, Y. (2007). A fuzzy collaborative spectrum sensing scheme in cognitive radio. *International Symposium on Intelligent Signal Processing and Communication System*, (pp. 566-569).

Zadeh, L. A. (1965). Fuzzy sets. *Information and Control*, *8*(3), 338–353. doi:10.1016/S0019-9958(65)90241-X

Zhao, Q., & Sadler, B. M. (2007). A survey of dynamic spectrum access. *IEEE Signal Processing Magazine*, *24*(3), 79–89. doi:10.1109/MSP.2007.361604

Chapter 12
Cross-Layer Design for Cognitive Radio Networks

M. Ayyash
Chicago State University, USA

Y. Al-Sbou
Mu'tah University, Jordan

ABSTRACT

Nowadays, due to the tremendous growth of wireless communications technologies and multimedia applications, the radio spectrum is starting to be crowded and scarce to meet the continuous growth of frequency requirements. Additionally, interference management is one of the key issues in wireless networks. Therefore, network solutions and evolutions have crucial challenges to overcome the inefficiency in configuring and managing network resources. To optimize wireless network operations and spectrum scarcity, a new networking paradigm, known as cognitive radio networks (CRNs), has been introduced. Due to the limited capabilities of the conventional layered protocol, CRNs adjust layer parameters adaptively according to the spectrum environment and Quality of Service (QoS) requirements. Hence, cross-layer design (CLD) solutions were necessary to allow for improving and optimizing CRNs performance. This chapter provides an extensive and exclusive overview of cognitive networks, CLD methodologies and properties, and cross-layer optimization (CLO) schemes among different layers. Moreover, it presents possible research solutions for cognitive networking. Finally, indispensable highlights of future work research directions are provided.

12.1 INTRODUCTION

Cognitive Radio Networks (CRNs) have gained importance in recent years. Addressing the issues of CRNs is a challenging task. The purpose of this chapter is to provide an overview of the key issues that face CRNs and the suggested solutions by researchers. Additionally, this chapter presents possible research avenues that can be explored by researcher in the area of CRNs.

The Cognitive Radio (CR) technology addresses a very important fact in wireless commu-

DOI: 10.4018/978-1-4666-2005-6.ch012

nications which is the scarcity of the frequency spectrum. The huge demand for frequency bands and the expected growth in wireless technologies required attention to how to utilize the limited frequency spectrum. The CR technology presents a promising solution for spectrum scarcity issues as it provides the capability to share the wireless spectrum with licensed users in an opportunistic manner (Akyildiz; Lee; Vuran; & Shantidev, 2008). Therefore, the CR technology concentrates on what's called "spectrum holes" utilization without interfering with *primary users* (PUs) signals (Reddy & Bullmaster, 2008). These "holes" are parts of the frequency spectrum that are underutilized. The CR mode searches for these "spectrum holes" and selects the best channel to use for communicating with other CR nodes. While the CRN nodes are responsible for the coordination among each other on how to access the selected communication channel, they must pay close attention to PUs' needs and presence. Once the PUs are detected, CRNs must free the selected channel.

Designing efficient CRNs is challenging due to the fact that the secondary users (SUs) are required not to interfere with the PUs. Hence, interference management is a key issue in CRN design. Additionally, whenever multimedia services are part of the traffic, *Quality of Service* (QoS) guarantees are challenging issues (Liu & Zhou, 2009). Since CRNs are wireless by nature, the conventional layered protocol architectures, such as the Open system Interconnection (OSI) and TCP/IP, don't address all the design aspects of the wireless medium of communication. Layered architectures were initially designed for wired networking environments; therefore, they are not completely ready to combat all the characteristics of the wireless environment.

To avoid the rigidity of the layered architectures, cross-layered architectures have become the suitable frameworks for CRNs (Reddy & Bullmaster, 2008; Liu & Zhou, 2009; Foukalas, Gazis, & Alonistioti, 2008; Shakkottai, Rappaport, & Karlsson, 2003). Cross-layer design (CLD) provides an adaptive framework which is aware of the working conditions of wireless networks. Consequently, the CLD provides a new and promising design paradigm which is responsive to wireless network variations (Foukalas et al., 2008; Shakkottai et al., 2003). The CLD is meant to improve the performance of wireless networks, in general, CRNs in particular.

The thrust of this chapter is to provide an overview of the usage of CLD in CRNs. The chapter is divided into six sections. The second section presents the background information. The third section discusses the focus topics in details. The fourth section is the solutions section. The fifth section presents future research directions. Finally, the chapter is concluded.

12.2 BACKGROUND

The recent development in radio communications and mobile computing provides several technologies for wireless connectivity where, different wireless network standards have appeared such as *wireless local area networks* (IEEE 802.11b or HiperLAN), *wireless wide area networks* (WiMax and UMTS) (Baldo & Zorzi, 2008a). The deployment of wireless networks can be in two modes: *infrastructure-based mode* in which access point links the nodes to the wired network and *infrastructureless-based* or *ad hoc* mode. *Mobile ad hoc networks* (MANETs) consist of wireless mobile nodes without reliance on fixed base stations or wired infrastructures (Ayyash, Ucci, & Alzoubi, 2010).

A wireless ad hoc network is a complex distributed system which consists of a collection of wireless mobile or static nodes that can dynamically self-organize into an arbitrary and temporary topology to form a network without necessarily using any pre-existing infrastructure (Al Hanbali, Altman, & Nain, 2005). Therefore, wireless ad hoc networks are self-creating, self-organizing, and self-administrating (Chakrabarti & Mishra, 2001).

The main advantages of ad hoc networks are in their flexibility, low cost, mobility, and robustness (Rubinstein, Moraes, Campista, Costa, & Duarte, 2006). These characteristics make ad hoc networks well suited for military activities, emergency operations, and disaster recoveries, meetings or conventions in which users wish to quickly share information, and data acquisition operations in hostile terrain. In ad hoc networks, nodes may communicate directly with each other or through intermediate nodes. This means, all the nodes communicate with each other's in a wireless multi-hop routing style. Every ad hoc node acts as a router in the network. Because the routers are mobile and can switch on/off, the network topology is changed frequently due to these events among mobile nodes (Ayyash, 2005; Chang & Hsu, 2000).

The wireless ad hoc network design and operation are constrained by some traditional difficult problems. Generally, these problems are related to the limitations of the free-space channel (i.e. relatively low bandwidth, electromagnetic interference, fading...etc), scarce resources (i.e. bandwidth, battery power, processing and storage limitations, etc.), lack of any established backbone infrastructure, and high error rates (Dupcinov & Krco, 2002). More specifically, these problems can be classified as physical medium (i.e., high bit error rates, noise, limited range), MAC sub-layer (i.e., the difficulty of collision detection, the hidden and the exposed terminal problems), network layer (i.e., routing protocols), and transport layer problems (i.e., TCP performance).

Given the above limitations, however, it is a major challenge to guarantee QoS in a wireless environment. Providing QoS in ad hoc networks means guaranteed end-to-end delay, limited packet losses and minimum bandwidth to specific flows especially over multihop communications (Rubinstein et al., 2006). These constraints arise with delay sensitive applications such as video and audio multimedia applications. To guarantee QoS, the whole protocol stack layers must be aware of the QoS requirements. This can be achieved using some structures that aim to define procedures to assure QoS in ad hoc networks using cross-layer models (Xiao, Seah, Lo, & Chua, 2000; Ahn, Campbell, Veres, & Sun, 2002).

Nowadays, with the fast development of wireless communications, the limited unlicensed spectrum bands cannot meet the continuous growth of requirement (Yong, Lifeng, & Quan, 2009). Actually, these spectrum bands are statically allocated and centrally managed. Therefore, the assigned unused spectrum by some specific users cannot be used under these strategies. To make use of these wasted bands, a new technology is needed to improve the spectrum utilization where the wireless network needs to be intelligent enough to automatically sense the conditions of the present environment and then to act and optimize its performance. This new technology is called CRN which plays a key in improving spectrum utilization efficiency under the current static spectrum allocation policy (Yan, Ma, Cao, Li, & Wang, 2008; Kim & Shin, 2008).

The primary reason behind performance inefficiency and degradation is the intricacy and difficulty in configuring and managing the wireless networks. Therefore, self-awareness, self-management, and self-healing are characteristics that have been proposed in order to optimize network operation, reconfiguration, and management (Kliazovich, Granelli, & Fonseca, 2009). Due to this, improvement in data transfer performance will be achieved by bringing "intelligence" into the network, thereby creating a new paradigm known as cognitive networking (Market, 2006). Therefore, a cognitive network is a network architecture that is based on CR.

CR is a brand-new technology that is supposed to be the future technology for wireless communications and is expected to solve the spectrum scarcity problem. In cognitive wireless networks, new features that are not present

Figure 1. Spectrum holes

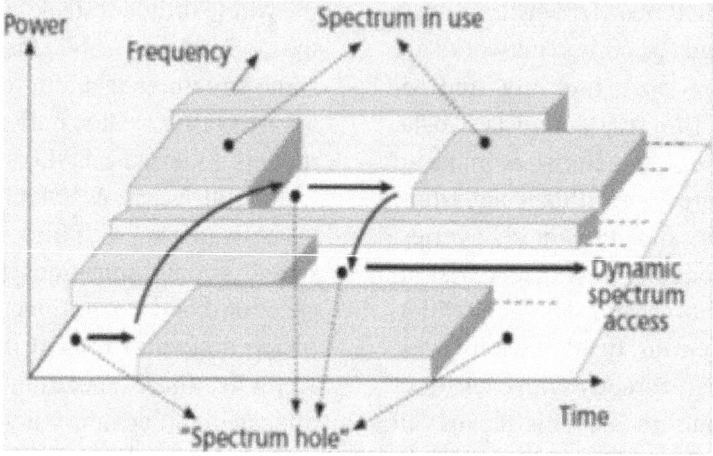

in traditional design approach are faced. These features are the *spectrum sensing* (SS) and the *opportunistic spectrum access* (OSA). The SS is a feature used to detect the channel status: idle or busy. The OSA feature is used for effective utilization of the bandwidth of the channel that may be dynamically changing depending on the activity of the PU. PUs are users who have the right to access the channel and are authorized to use the radio spectrum while SUs are not. However, and based on the SS feature, it is allowed for SUs to search the idle channels and sense whether a portion of the spectrum (holes) is being used or not. Based on the sensing outcomes, SUs will decide whether or not to access the sensed channels as illustrated in Figure 1. Therefore, improvement of SUs' performance is a fundamental factor that determines the success of CR technologies.

There are different definitions on what CRN concept means:

- The term CR for wireless devices and networks has been defined by correlating it with computational intelligence about radio resources and related communications, in order to detect user communication needs and provide most appropriate radio resources and wireless services (Mitola, 2000).
- Cognitive networking is defined as: "the knowledge plane: representation, learning, and reasoning that allow the knowledge plane to be "aware" of the network and its actions in the network." (Clark, Partridge, Ramming, & J. T. Wroclawski, 2003).
- The cognitive network can be defined as: a network with a cognitive process that can perceive current network conditions, and then plan, decide and act on those conditions. The network can learn from these adaptations and use them to make future decisions, all while taking into account end-to-end goals (Thomas R., 2007).
- CR can be described as an intelligent wireless communication system that is aware of its surrounding environment, and uses the methodology of understanding-by-building to learn from the environment and adapt its internal states to statistical variations in the incoming RF stimuli by making corresponding changes in certain operating parameters in real-time, with two primary objectives in mind: highly reliable communications whenever and wherever needed;

Figure 2. Cognition radio cycle

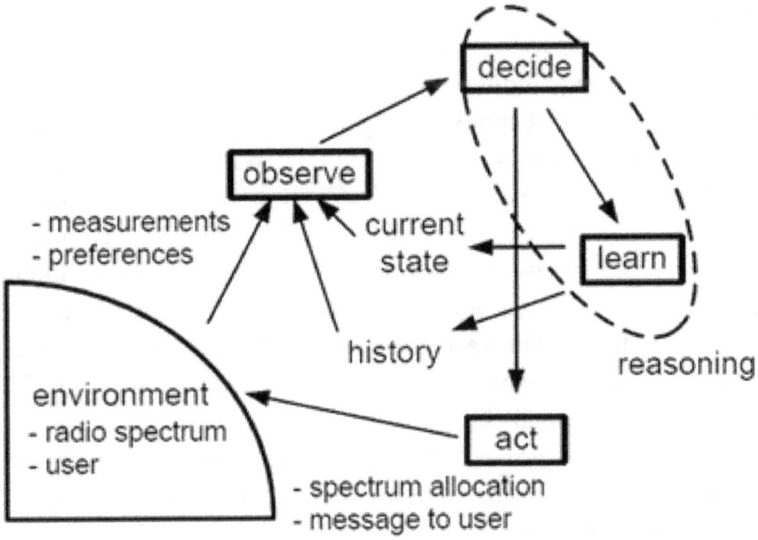

efficient utilization of the radio spectrum (Haykin, 2005).

It is difficult to come up with a unique definition that combines all the different CR perspectives; however, the following characteristics are frequently attributed to a CR (Bogatinovski & Gavrilovska, 2008):

- **Environment Awareness:** The CR can observe the communication environment it is operating in by measuring some parameters that affect its performance, like the packet-error-rate, signal to noise ratio, etc.
- **Reconfigurability:** It can reconfigure some network parameters at different layers of the protocol stack aiming at an efficient spectrum usage.
- **Decision Making Based on Optimization Principles:** Node can make decisions based on available knowledge on environmental parameters, current system configuration, and measurements of the extent of attaining certain requirements.
- **Learning:** The intelligent part of the CR can learn based on past experiences by first recognizing recurrent contexts and patterns.

Resulting from the above features of the CR and based on observing the environment, the CR usual operation implies a decision making based on learning from observation history and the consideration of the actual state of the environment. This process can be represented as a *cognition cycle* as depicted in Figure 2.

Wireless ad hoc networks need to be adaptive to the changes of wireless environment and service requirements and needs. Therefore, ad hoc network should have cognitive capabilities in order to adjust network parameters and protocols to improve its performance. One of the promising approaches to achieve performance amelioration is to employ CLD architecture which allows direct interaction and information sharing between adjacent and nonadjacent layers. Based on cross-layer and cognitive capabilities, it is expected that ad hoc networks will provide QoS and achieve highly-improved performance for real-time multimedia transmissions.

The traditional design of data and communications systems is based on a layered architecture. This 7-layer reference model abstracts features that are common to all data communications systems, and organize them into layers such that every layer only considers the layer directly above it and the one directly below it. This layered network stack design has been very successful and its principles have been widely adopted throughout various implementations and applications worldwide (Aune, 2004). However, the traditional layered module has come under serious censure in wireless networks. This is because radio transmissions are subject to interference therefore; performance degradation and varying channel conditions lead to serious inefficiency with the original and traditional OSI protocol design (Shakkottai et al., 2003). In addition, this implies that a change in conditions on one link can induce changes in the conditions of all links surrounding it.

The traditional layered network stack design has always been to treat the different OSI layers as separate entities. However, one of the key attributes of modern wireless data communications systems is the parameterized operation at different layers of the protocol stack. This attribute aims at providing them with the ability of adapting to the rapidly changing traffic, channel and system conditions and finally enjoying improved performance (Triantafyllopoulou, Passas, & Kaloxylos, 2007). An interesting approach is the combination of the adaptation mechanisms in every individual layer into a joint-layer interaction design that can maximize their efficiencies. *Joint-optimization* or CLD is defined to modify the pure layered approach by direct communication between nonadjacent layers or sharing of internal information between layers so as to enhance the system performance. This design is the most promising alternative to inefficient traditional layered protocol architectures especially for wireless networks. Therefore, for this new design to be efficient, some design principles should be taken into account like (Bogatinovski & Gavrilovska, 2008):

- Which layers to select to exchange parameters between them?
- Which parameters will be chosen to be exchanged between layer?
- How will the interaction between layers be? (i.e., direct interactions between layers, shared database across the layers or through new abstractions which may be added (Yong et al., 2009).

The central idea behind CLD is to improve the overall performance of wireless networks by exchanging information between the different layers of the protocol stack. Recent research on CLD showed a substantial improvement in the routing efficiency, throughput, and fairness and delay variance among different applications (V & M., 2005; Liu, Zhou, & Giannakis, May 2005). Generally, CLD has been thought of as one of the most effective and efficient ways to provide QoS over wireless networks. Figure 3 shows some parameters that can be provided by individual layers to be used for CLD and optimization. A simple example of cross-layer system is the cooperation between the physical layer and MAC layer, where the data rate in the physical layer is dynamically changed based on the packet loss rate in the MAC layer.

The most fundamental challenge in any network reference design is how to allocate the available resources among the different network users. However, CRNs adjust the parameters and protocol adaptively according to the awareness of spectrum environment and user requirement which need the combination of different network layers. This means, an efficient protocol stack is needed for such cognitive networks that can react to network conditions and service requirements. To achieve such efficient protocol, information sharing among physical, MAC, network, transport

Figure 3. Examples of OSI layers parameters for cross-layer optimaization

Application Layer (encoding , packetization, compression...etc)
Transport Layer (congestion windows ize, Timeout clock,...etc)
Network Layer (routing protocol, Routing lifetime, ... etc)
Data Link Layer/MAC (Minimum contention window size, Maximum back off stage, Link bandwidth ARQ...)
Physical Layer (modulation, transmitting power, SNR information ...)

and application layers is also needed. All this can be achieved by CLD.

As the original purpose of the CRNs is to obtain efficient radio spectrum utilization, this new standard has to devolve to include also general objectives of network usage optimization and/or satisfying user requirements based on application preferences (Bogatinovski & Gavrilovska, 2008). These are difficult tasks to be achieved as they require cooperation of the radio system itself with new cross-layer cognitively concepts, and artificial intelligence, as well as optimization algorithms in the conditions of changing characteristics of the wireless channel. Therefore, powerful optimization methods combined with machine learning techniques are promising candidates for optimization methodologies (Riihijärvi, Wellens, & Mähönen, 2006). Optimization of each OSI layer separately is likely to yield *suboptimal* performance optimization rather than an *overall system* optimization. Therefore, prior to implementing cognitive networks, it is important to understand how varying parameters at different layers can affect the performance and reliability of a wireless system. Understanding these performance changes allows us to consider how a system might be optimized to better suit the needs of an application or set of applications over the network (Weingart, Sicker, & Grunwald, 2006).

12.3 FOCUS

This section covers three key topics: cross-layer optimization, interference management in cognitive networks, and CLD challenges and issues.

12.3.1 Cross-Layer Optimization (CLO)

As mentioned earlier, there is no doubt that a dramatic change is happening in the management of the frequency spectrum. This is due to the fact that data transmission quality and network performance have degraded due to an ever increasing number of heterogeneous wireless devices and applications are being deployed. This is true in both licensed and un-licensed frequency bands. Therefore, the heterogeneous nature of the wireless environment makes wireless communications very prone to interference issues.

To address the need for acceptable network performance, guaranteed QoS, and interference mitigation, researchers are working on devising a dynamic and robust framework for network protocols and services. The static layered and stratified protocol reference models have become limited and not able to address all wireless networking issues. It is noteworthy to remind here that wireless networks have unique characteristics, which are different than wired networks.

Wireless communications channels are time-varying and space-varying due to either mobility of the wireless devices or changes in the physical environment around these devices. Such variations can possibly lead to communication errors (Shakkottai et al., 2003). The built-in time- and space-varying nature poses a challenge that must be carefully addressed. Thus, designing wireless protocols while considering a specific protocols stack layer is not practical and the design paradigm is being modified to include more collaborative and cooperative cross-layer techniques. The CLD relies on the fact that the conventional static reference models can be revised in many ways to fulfill the QoS, network performance, and interference mitigation needs.

Again, the CR is defined as a radio that "can change its parameters based on interaction with its environment" (FCC, 2003). The CR technology was originally devised in part a response to the fact that large amount of frequency bands are underutilized by the PUs while the frequency spectrum is very scarce. Thereby, spectrum management is the focus of the CR technology. Spectrum management deals with the following issues and challenges (Akyildiz, Lee, Vuran, & Shantidev, 2008): (1) *spectrum sensing;* (2) *spectrum decision;* (3) *spectrum sharing;* and (4) *spectrum mobility*.

Spectrum management in CRNs requires a CLD approach where various layers can be considered in achieving the optimal design and implementation results. Following are some CLD examples. A simple example is the need for multi-user scheduling design in wireless networks. Another example is the mapping of the interference level in a Wi-Fi network to the network layer parameters such as end-user throughput to provide better QoS (Shakkottai et al., 2003). In most cases, CLD jointly adjusts a number of lower layers' parameters such as channel states information to upper-layers functions like transport and network (Georgiadis, Neely, & Tassiulas, 2006).

It is clear from the above simple examples that CLD would improve network performance. The CLD has a great potential to improve the performance aspects of current wireless network. It is quite intuitive that separate optimization of each layer of the protocol stack is likely to produce *sub-optimal* performance when compared, in theory, with joint optimization of all layers (Baldo & Zorzi, 2008b). Therefore, researchers around the globe have concentrated their efforts on CLD solutions (Foukalas et al., 2008). Current wireless protocols are designed while CLD is in mind.

Stratification or what is referred to as the black-box approach, renders each protocol layer isolated from the functionality embedded within other protocol layers (Foukalas et al., 2008). Once the stratified reference model is violated, this is considered a CLD. As shown in (Liu & Zhou, 2009), different kinds of CLD (OSI or TCP/IP violations) can be proposed among the layers of the stratified model. The CLD allows for interactions between adjacent and non-adjacent layers of stratified reference models. Hence and for best results, the spectrum management aspects of CRNs should be addressed while CLD is clearly kept in mind.

The ovals shown in Figure 4 are meant to solve specific problems in wireless mobile networks. The dashed ovals depict the flexibility of CLDs. Each CLD technique focuses coordinating similar functionalities amongst protocols layer, which in turn would lead to more desirable results for wireless networks needs such wireless efficiency. The example of *power efficiency* is an excellent example of the needed coordination among stratified protocol layers. Another example is QoS provisioning. QoS provisioning requires the collaboration among non-adjacent layers in order to achieve the best results. Some researchers refer to the above "ovals" as the *coordination planes*. Generally, such planes are meant to include security, mobility, QoS, and the adaptation of the wireless link (Shakkottai et al., 2003). Another excellent example that requires CLD coordination amongst layers is the wireless link adaptation. Wireless links properties are not constant all the time. For instance, the bit-error-rate (BER), channel fading, transmission delay, etc., are some properties that require upper layers participation in order to avoid performance degradation (Kliazovich et al., 2009).

Overall, CLD introduces additional management entities which are needed to manage cross-layer interactions. Such management entities are

Figure 4. Spectrum management with CLD

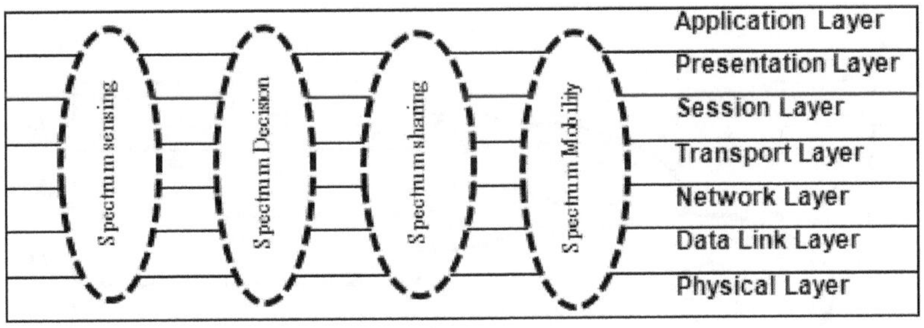

either internal within the protocol stack or external using external nodes. In (Shakkottai et al., 2003), authors present using excellent examples the details of the internal and external cross-layer entities. Additionally, they review the possible impact of such entities on network performance and overhead.

CRNs are also wireless networks that share some reasons for utilizing what we mentioned about CLD techniques. As we stated earlier, designing CRNs while CLD is in mind is very challenging due to complexity, modularity, information imperfection/imprecision, and interoperability issues (Baldo & Zorzi, 2008b). However, CLD optimization is likely to be employed in CRNs. Nevertheless, the current state of the art in wireless communication systems is still far from having ultimate cognitive cross-layer architecture (Baldo & Zorzi, 2008b).

Most of CLD architectures and solutions for cognitive networks are generally meant to address specific scenarios. These solutions concentrate on improving the performance of a particular multimedia application or transport protocol over a given radio link. Therefore, such solutions are not general enough to be reused in different wireless technologies (Baldo & Zorzi, 2008b).

Improving the performance of wireless networks based on CLD is an area that has received significant attention (known as cross-layer optimization). However, a strict layered design is not flexible enough to cope with the dynamics of ad hoc environments, and will thus prevent performance optimizations (Goldsmith & Wicker, 2002). Therefore, it is widely believed that combination of cross-layer approach with cognitive designs in wireless networks is one of the most effective and efficient proposals of future CRN real-world implementations. This combination must be optimized in order to provide the required service and the best performance for all transmitted applications.

One of the significant features of CRNs is their capability to facilitate the simultaneous optimization of multiple operating parameters across the different OSI layers of a wireless network node. A general layout of CLO architecture is shown in Figure 5 (Chen & Wyglinski, 2009). Each node collects environment information from local sensing and from other nodes via overhead channels and performs a CLO to enhance both system and overall network performance. This is executed by an optimization engine implemented in every wireless node. This engine decides which operating parameters across the different layers of the node in order to content with other nodes and to satisfy application requirements with as little resource as possible. Every node should select the operating parameters in a way that keep the contention between nodes as minimum as possible by cooperation and optimization across network nodes. Adjacent nodes exchange information about their current operating parameters (e.g., modulation, power, bandwidth, center frequency) and their environmental conditions. The optimization engine interfaces with the different layers and

Figure 5. Framework of a CLO architecture of a cognitive wireless node

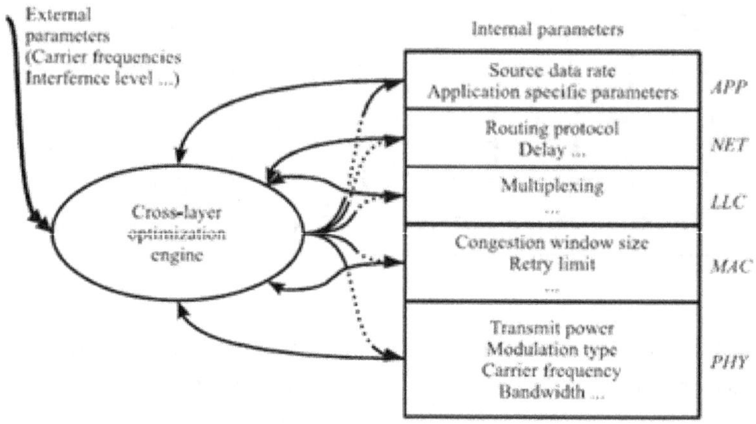

can act as shared database. This engine is adopted to utilize the cognitive information from all layers. Thereby; implement performance optimization of cognitive ad hoc networks (Yong et al., 2009).

The overall end-to-end performance of an application is determined by the combination of the performance of each single layer in the protocol stack; in other words, in order to improve end-to-end performance the configuration parameters of each layer should be tuned. The problem of optimization in cognitive networks is not very straight forward and it is mainly a numerical problem that can be solved using machine learning techniques (Riihijärvi et al., 2006). The optimization problem in case of cognitive wireless radio with CLD is complex, due to (Bogatinovski & Gavrilovska, 2008):

- Existence of many parameters that can be tuned in a CR, which makes the search (configuration) space very large.
- Current network state information has to be known at all transmitters.
- Deriving analytical expressions for the objective function of the channel conditions is very difficult because these functions are nondeterministic and non-linear (Schaar & N, 2005; Baynast, Mahonen, & Petrova, 2008).
- Wireless channel conditions might be changing, so the optimization process should also support a dynamic QoS management scheme depending on the available network resources.
- Real cognitive wireless networks require performing multi-objective optimization, often with conflicting goals (Baynast et al., 2008; Petrova & Mähönen, 2007).

The majority of CLO CLD techniques found in the literature today can be generally classified into four classes (Aragorn, 2008):

1. **Evolutionary Versus Revolutionary:** Based on the changes in the protocol stack, the CLO can be classified as evolutionary or revolutionary techniques (Aune, 2004; Gavrilovska & Prasad, 2006). The evolutionary approaches only use a vertical shortcut through the protocol layers keeping the existing protocol stack in the background. They aim at providing only on system compatibility rather than performance. While the revolutionary CLO techniques are not constrained with any existing layered protocols, they concentrate on maximum performance without any consideration of the compatibility.

2. **User Centric Versus Network Centric:** The CLO approaches can be classified as user centric or network centric (Fu & Schaar, 2007). This depends on the optimization process target: the user or the network performance. The user centric optimizations are commonly used rather than network centric. This due to the fact that the user centric approaches are used to adapt the user behavior to the actual network conditions, aiming at providing QoS guarantees, whereas the network centric focus on Network Utility Maximization (NUM) (Chiang, Low, Caldbank, & Doyle, 2007).
3. **Targeted Versus Joint Optimization:** Based on the number of protocol stack layers in focus, the CLO schemes can be viewed as being targeted or joint (Khan, Peng, Steinbach, Sgroi, & Kellerer, 2006). Targeted CLO methods are used to improve performance based on a particular layer or a subset of protocol layers. Therefore, even when all the layers are involved in the CLO design, the major benefit is experienced only by a restricted part of the protocol stack. On the other hand, the joint CLO designs permit all protocol stack layers to be shared in the performance improvement.
4. **Centralized Versus Layered:** The CLO techniques can also be classified as centralized or layered based on the practical implementation of the CLO method (Fu & Schaar, 2007). In centralized approaches, a central cross-layering optimizer is employed to cooperate with all protocol stack layers to provide a common cross layer solution. On the contrary, the layered CLO approaches are structured as a distributed cross layer. This will produce an optimizer of independent re-configurability at each layer.

By observing the current conditions, CRNs perform inherently and indirectly CLO. That is because it shares information that is not available in the stratified layered architecture (Thomas, Frlend, DaSilva, & MacKenzie, 2006). Therefore, in order to attain best end-to-end performance, a CRN must perform tradeoffs between multiple goals by executing *multi-objective* optimizations using CLO techniques. There are various techniques have been proposed for CLO in CR based on the *Artificial Intelligence* (AI) in the literature. *Genetic Algorithms*, *Neural Networks*, and *Fuzzy Logic* are AI techniques that have been the most heavily used models due to their capability for fulfilling specific optimization tasks.

Artificial Neural Networks (ANNs) approach has been used to solve the link activation problem in CRNs (Barnhart, Wieselthier, & Ephremides, 1995). ANNs were applied to solve routing without interference in multihop radio networks (Wieselthier, Barnhart, & Ephremides, 1994), and more recently, for signal classification (Like, Chakravarthy, & Wu, 2007). Another work (Baldo & Zorzi, 2008a) proposed the usage of *Multi-layer Feed Forward Neural Networks* (MFNNs) technique for real-time characterization of the communication performance of CRs components emphasizing the modularity of this approach, making it feasible to be applied in several layers. All of these works are problem-specific solutions which may not be possible to be reused as a general-purpose CLO design.

Genetic algorithms (GAs) are very suitable to be employed in cognitive networks because it is well-suited to deal with large sets of heterogeneous variables, and possibly carrying out CLO over heterogeneous wireless interfaces, protocols and applications (Rieser, Rondeau, Bostian, & Gallagher, 2004). Rieser (2004) suggested applying a *Hidden Markov Model* (HMM) which is trained by a genetic algorithm to model the channel response in CR. In addition, GAs were invoked for iteratively finding the optimum parameters based on the acknowledgment (ACK) signal only. The set of accurate *objective functions*, which were derived in the GA implementation control the optimization process at the system level toward any QoS (Baynast et al., 2008). Also, an adaptive component using Gas was presented to evolve a radio defined by a chromosome to represent the adjustable parameters in a given radio, and by genetically manipulating the chromosomes, the GA could find a set of parameters that optimize the radio for the user's current needs (Rondeau, C. J. Rieser, & Bostian, 2004). Siddique (2010) proposed a GA technique for spectrum allocation solution for CR networks to provide an optimum solution among the available solution set, assuring that the QoS requirements of the application are met. Although genetic algorithms are very good optimizing techniques, a very long learning processes are usually needed in order to give acceptable solutions, which may or may not be acceptable according to the requirements of the applications.

Fuzzy logic is a tool that has been used as a convenient knowledge representation scheme for CLO. Fuzzy logic was employed to implement specific cross-layer scenarios without generalizing the proposed approaches to be used as general-purpose CLO designs (Xia & Liang, 2005; Otal & Alonso, 2004). A fuzzy logic system was used to coordinate physical layer, data-link layer and application layer for CLD for adjusting factors for the *adaptive modulation and coding* (AMC), transmission power, retransmission times and rate control decision (Xia & Liang, 2005). A PHY-MAC cross-laver energy-saving mechanism based on a distributed queuing MAC protocol and fuzzy logic theory was introduced (Otal & Alonso, 2004). On the other hand, Baldo & Zorzi (2007) used fuzzy logic as a special effort towards a general-purpose CLO. Because of its superior performance, CLO is expected to be extensively used in CRNs. The cognitive engine at the heart of the CR will leverage cross-layer information exchange and interactions to exploit different wireless interfaces and protocol stack configurations in order to achieve the best service quality for all communications (Baldo & Zorzi, 2008b).

12.3.2 Interference Management in CRNs

Again, as wireless applications with high data rate have become more widely used, the demands for efficient radio spectrum resources have increased significantly. Consequently and as discussed earlier, there has been a terrific interest in spectrum sharing technology in order to recycle assigned frequency spectrum to improve the utilization of spectrum radio resources. Federal Communications Commission (FCC) found that while the available spectrum becomes increasingly scarce, the assigned spectrum is significantly underutilized (Kolodzy, 2002). Most importantly, CR is a spectrum sharing technology that has the capabilities of solving this scarcity by taking advantage of the free spectrum (Garcia, 2010). This is because CRN is a wireless network that can adapt its transmission and reception parameters in a way to reduce or avoid interference with licensed or unlicensed users based on monitoring of several factors, like radio frequency spectrum, user behavior, and network state. Based on the basic characteristics of the CRNs, non-interference basis to the PUs, SUs can use available spectrum to improve spectrum utilization (Kim, 2011). A non-interference basis means that SUs are dynamically able to sense, access and use available spectrum only if PUs are not seriously interfered and affected by them. Therefore, for the radio spectrum sharing to be efficient, interference management is considered as a central part of the CRNs.

12.3.3 CLD Challenges and Issues

Unlike cross layer issues in traditional wireless networks, the CLD in CRNs has number of challenges due to broad range of available spectrum. Recently, cross layer proposals and designs for CRNs have witnessed an increased interest that relies on interactions between different layers of the protocol stack in order for achieving performance improvements. This is still a very hot and attractive research topic, therefore many research problems and challenges still need to be identified and solved. This section focuses on some issues, cautions, and challenges that face the deployment of cross layer architectures in cognitive networks scenarios.

Generally, effective CLD is challenged by *modularity, interpretability, imprecision, scalability and complexity* issues (Baldo & Zorzi, 2008b). In addition, interaction with all layers of protocol stack and signaling are other key challenges.

In examining and evaluating CLDs, the trade-off between performance and architecture of these designs need to be fundamentally considered because there is always a fundamental tension between performance and architecture (Kawadia & Kumar, 2005). The above mentioned issues and other challenges are detailed below:

1. **Modularity:** Conventionally, the legacy OSI protocol stack has always been effective in allowing independent implementation of all layers. Modularity should be present in the CLD in order to allow components with cross-layer capabilities to be designed independently and used interchangeably (Kawadia & Kumar, 2005). This means if it is possible for a layer to obtain measurement information from other layers in an abstract way, this design is modular and can operate with different technologies. Otherwise, the design will be technology-specific and not useful for other technologies. Abstracting from the underlying technology is a key prerequisite to this concept of modularity.
2. **Information Understanding (Interpretability):** Interchanged information between layers of a CLD should be correctly interpreted. But this is not always possible because these measurements may be technology-specific and not generic. As an example, if the *Signal-to-Noise Ratio* (SNR) is used as generic *link layer information* shared by all wireless technologies, the interpretation of SNR measurements by other layers is not always true and would be misleading. It is impossible to infer the goodness of a wireless link from SNR without knowing some information like the type of modulation used in that wireless technology.
3. **Faulty Measurement Information (Imprecision):** Most information interchanged among different layers of a CLD is usually obtained from measurements. These measurements are affected by errors in precision and accuracy, which may be misinterpreted due to these errors. Suppose that upper layers (transport and/or application) possess the technology-specific knowledge and a correct interpretation of SNR measurements is needed to be used for CLO. In most wireless technologies, SNR measurements are very inaccurate and have a very sharp transition from excellent to bad performance. As a consequence, if a SNR measurement is considered as an exact value, a bad channel might be misinterpreted as good and vice versa.
4. **Computation-related Problems (Complexity):** An effective CRN is needed to observe all available resources in order to find the optimal solution to meet the user's requirements. CLO is often a computationally intensive process. Therefore, CLDs may suffer from computation-related problems in terms of decision and learning processes (Pawełczak, 2008). This computational load increases as the number of layers, information nature, and parameters exported by each layer involved in the optimization process increase. Therefore both the decision algorithm and the learning algorithm represent challenges for cognitive CLD due to the fact that measurements are related to how to perform them and to complexities of designing optimization algorithms. In addition, wireless channel conditions might be changing which will result in cognitive

wireless networks to perform multi-objective optimization often with conflicting goals (Bogatinovski & Gavrilovska, 2008; Petrova & Mähönen, 2007).

5. **Interaction with all Layers of Protocol Stack:** Based on CLDs, a number of issues arises and are related to the interaction between layers. That is because the nature of interactions can affect not only the layers concerned, but also other parts of the system (Kawadia & Kumar, 2005). Examples of such issues are: which layers need to be interacted, which parameters in these layers are chosen to be interchanged and will be of the most influential (Bogatinovski & Gavrilovska, 2008), how will be the cross-layer interactions performed, what if several cross-layer interactions are implemented, etc. The cost of interaction in CLD can be the overhead on the system due to cross-layer communication. These issues could have a drastic negative impact on the overall performance.
6. **Signaling:** Signaling is a challenging issue for CRNs (Pawełczak, 2008). In order to provide efficient operation, cognitive network nodes must have cross-layer interfaces suitable for adaptation and optimization. Therefore, specific signaling designs are needed for internal or network-wide exchange of information. In addition, while CLD has gained maturity even with conflicting ideas (Kawadia & Kumar, 2005). It is valuable to address signaling architecture to support cross layer and optimization issues and solutions.
7. **QoS Provisioning:** Once more, the growing demand for wireless network resources and the exponential increase in the number of users have led to a spectrum scarcity problem. This high demand increases the requirement of QoS architecture for the various applications that use the CR. To date, several research products have been dedicated to designing QoS architectures for wireless ad hoc networks; however, none of them can be directly applied to cognitive networks (Yau, Komisarczuk, & Teal, 2008). QoS support in cognitive networks is a key challenge as the capacity of the wireless channel that the SUs are operating is always subject to change dependent on spectrum utilization from PUs. The various reasons for the need of a proper QoS system over CRN are listed below (Sivanantha, Ehsan, Dhamodaran, & Gatimu, 2010):
    - Constrained resources such as spectrum bandwidth, transmission power, and data rate.
    - Dynamic and time-varying network conditions due to interference, shadowing, and multipath fading.
    - Diverse applications lead to diverse design requirements. For example, video is encoded with different profiles and priorities. Loss of certain important frames may degrade the video quality for the receiver significantly. Another example, delay generated from sensing and channel switching activities may cause some video frames to miss the deadline.
    - Upon arrival of PUs, SUs with multimedia traffic are more vulnerable than typical data traffic.

Generally, CLD, as the most effective and efficient approach, is expected to provide QoS over wireless networks and especially over CRNs. Before provisioning, QoS over these networks needs to be assessed and monitored. This involves devising QoS measurement techniques that measure individual QoS parameters or combine and summarize these parameters in a single value (Al-Sbou, 2006). For some multimedia applications that many parameters affect their QoS, it is better to use QoS assessment systems that provide a single QoS value especially when using CLDs (Al-Sbou, Saatchi; Al-Khayatt; Strachan; Ayyash;

Saraireh, 2008); (Al-Sbou, 2006). This value represents the QoS level provided to the applications based upon the network conditions compared to the QoS level expected for such applications. A good example on such assessment approaches is presented in (Al-Sbou, 2010), in which an intelligent fuzzy logic assessment system was devised. However, this may be a challenging issue due to the existence of different multimedia applications. Each of these applications has its own QoS constraints. Even more, end-to-end performance optimization of the cognitive network must meet the different QoS requirements of users, and must adapt to the conditions of the wireless channel, network environment, and resources. So, multi-objective decision is usually needed in a cognitive ad hoc network.

Therefore, any proposal of cognitive cross-layer architecture must consider these key characteristics. This will produce a CLD which permits efficient sharing of information across all protocol layers, will be independent of any specific family of protocols, and enable better system performance through cognition.

12.4 SOLUTION

Clearly, the number and nature of interference sources are continually increasing and also becoming more complex to manage their negative effects. Cross-layer framework for interference mitigation and management represents a viable solution for problems and challenges mentioned in the previous section. Hence, interference is becoming a significant factor to be considered in the design and evaluation of networking nodes. At the same time, cross-layer techniques should plan an important role in enhancing the network performance and capacity. In CRNs, coupling these emerging insights should dramatically improve the spectral utilization and therefore the efficiency of wireless networks. Wireless interference can be mitigated using a variety of methods in the physical layer, from changing the modulation method and manipulating error control coding, to using smart antennas, power control, handover margin, or other means to adjust the coverage of the computing devices in a wireless access network. These methods are not fully exploited in existing standard protocols in the unlicensed band. Consequently, algorithms and protocols should be created to allow for cross-layer communication so that both upper and lower layers of the protocol stack become "interference-aware" and can acclimate to changing interference environments. We are confident that it is a research feat to develop discrete interference mitigation algorithms and protocols as they are complex undertakings due to the variability of wireless networks' parameters and characteristics.

In summary, the solution should focus on the investigation of how CRNs can be implemented to be interference-aware using reliable CLD and CLO techniques. Obviously, there is no solution that fits all. That is, there is no one cross-layer framework that can effectively combat the needs of CR devices and networks. Hence, several open and interesting issues are still under consideration by the research community.

12.5 FUTURE RESEARCH DIRECTIONS

An excellent overview of future research directions in CRNs is published in (Steenkiste, Sicker, Minden, & Raychaudhuri, 2009). As discussed in (Steenkiste et al., 2009), significant and novel research is needed to address CRNs challenges and issues such as spectrum management issues, dynamic spectrum allocation (DSA) methods, CRNs security, CRNs architectures and protocols, etc. Additionally, testbeds are necessary to evaluate CRNs at various stages of their development.

As mentioned in previous sections, researchers are expected to concentrate on CLDs and CLOs techniques in order to devise new, efficient, and effective interference mitigation protocols for CRNs. Therefore, future research directions are expected to lead to better solutions for CRNs while CLD is the focus of such solutions.

12.6 CONCLUSION

Cognitive radio networks represent a logical solution for frequency spectrum scarcity in wireless network. The scarcity nature of frequency bands required special attention from wireless networks researchers. Primarily, it is of great importance to come up with a suitable research approaches by which the unique characteristics and issues of CRNs could be addressed. This chapter concentrated on the newly and promising research tactics to overcome the challenges and inefficiency problems of CRNs. Currently, cross-layer design (CLD) and cross-layer optimization (CLO) are the promising research tactics to address CRNs issues such as QoS provisioning and interference management. This chapter, for the most part, presents a detailed overview of the CLD and CLO concepts and related topics. Additionally, the chapter includes some suggested solutions to address CRNs' interference management needs. Finally, a summary of future research directions is discussed. Actually, this chapter can be used as a starting point for interested researchers in the area of interference management and QoS provisioning in CRNs.

REFERENCES

Ahn, G. S., Campbell, A. T., Veres, A., & Sun, L. H. (2002). SWAN: Service differentiation in stateless wireless ad hoc networks. *IEEE Conference on Computer Communications (INFOCOM)*, (pp. 457–466).

Akyildiz, I. F., Lee, W.-Y., & Vuran, M. C., & Shantidev. (2008). A survey on spectrum management in cognitive radio networks. *IEEE Communications Magazine*, *46*(4). doi:10.1109/MCOM.2008.4481339

Al Hanbali, A., Altman, E., & Nain, P. (2005). A survey of TCP over ad hoc networks. *IEEE Communications Surveys & Tutorials*, *7*(3), 22–36. doi:10.1109/COMST.2005.1610548

Al-Sbou, Y. (2006). *Quality of service assessment and analysis of wireless multimedia networks*. Sheffield, UK: Sheffield Hallam University.

Al-Sbou, Y. A. (2010). Fuzzy logic estimation system of quality of service for multimedia transmission. *International Journal of Quality of Service Issues in Networking*, *1*(1).

Al-Sbou, Y. A., Saatchi, R., Al-Khayatt, S., Strachan, R., Ayyash, M., & Saraireh, M. (2008). A novel quality of service assessment of multimedia traffic over wireless ad hoc networks. *The 1st IEEE International Workshop on Future Multimedia Networking (IEEE FMN'08) in The Second International Conference on Next Generation Mobile Applications, Services, and Technologies (IEEE NGMAST'08)*, (pp. 479-484).

Aragorn, F. P. (2008). *Adaptive reconfigurable access and generic interfaces for optimization in radio networks. Document Number D2.1: State-of-the-art*. Retrieved from Public Deliverable.

Aune, F. (2004). *Cross layer design tutorial*. Norwegian University of Science and Technology: Published under Creative Commons License.

Ayyash, M. (2005). *A new robust quality of service routing protocol for wireless mobile ad hoc networks*. USA: Illinois Insitute of Technology.

Ayyash, M., Ucci, D., & Alzoubi, K. (2010). A proactively maintained quality of service infrastructure for wireless mobile ad hoc networks. *International Journal of Communication Networks and Information Security*, *2*(2).

Baldo, N., & Zorzi, M. (2007). *Fuzzy logic for cross layer optimization in cognitive radio networks*. IEEE Consumer Communications and Networking Conference (CCNC'07).

Baldo, N., & Zorzi, M. (2008a). Learning and adaptation in cognitive radios using neural networks. *IEEE CNNC 2008 Proceedings*.

Baldo, N., & Zorzi, M. (2008b). Fuzzy Logic for cross layer optimization in cognitive radio networks. *IEEE Communications Magazine*, *46*(4), 64–71. doi:10.1109/MCOM.2008.4481342

Barnhart, C., Wieselthier, J., & Ephremides, A. (1995). A neural network approach to solving the link activation problem in multihop radio networks. *IEEE Transactions on Communications*, *43*(234), 1277–1283. doi:10.1109/26.380169

Baynast, A. D., Mahonen, P., & Petrova, M. (2008). ARQ-based cross-Layer optimization for wireless multicarrier transmission on cognitive radio networks. *Computer Networks*, *52*(4), 778–794. doi:10.1016/j.comnet.2007.11.009

Bogatinovski, M., & Gavrilovska, L. (2008). Overview of cross-layer optimization methodologies for cognitive radio. *TELFOR '08. ARAGORN (ICT-216856)*.

Chakrabarti, S., & Mishra, A. (2001). QoS issues in ad hoc wireless networks. *IEEE Communications Magazine*, *39*(2), 142–148. doi:10.1109/35.900643

Chang, Y.-L., & Hsu, C.-C. (2000). Routing in wireless/mobile ad-hoc networks via dynamic group construction. *Mobile Networks and Applications*, *5*(1), 27–37. doi:10.1023/A:1019175605040

Chen, S., & Wyglinski, A. M. (2009). Efficient spectrum utilization via cross-layer optimization in distributed cognitive radio networks. *Elsevier Computer Communications Journal, 32*(18).

Chiang, M., Low, S. H., Caldbank, A. R., & Doyle, J. C. (2007). Layering as optimization decomposition: A mathematical theory of network architectures. *Proceedings of the IEEE*, 95.

Clark, D. D., Partridge, C., Ramming, J. C., & Wroclawski, J. T. (2003). *A knowledge plane for the internet*. ACM SIGCOMM.

Dupcinov, M., & Krco, S. (2002). *Routing in ad hoc networks*. Ireland: Applied Research Lab, EEI, Ericsson.

FCC. (2003). *Notice of proposed rule making and order*. ET Docket No. 03-322.

Foukalas, F., Gazis, V., & Alonistioti, N. (2008). Cross-layer design proposals for wireless mobile networks: A survey and taxonomy. *IEEE Communications Surveys and Tutorials*, *10*(1), 70–85. doi:10.1109/COMST.2008.4483671

Fu, F., & Schaar, M. v. (2007). *A new theoretic foundation for cross layer optimization*. UCLA Technical Report.

Garcia, I. P. (2010). *Interference management in cognitive radio systems*. Master's Thesis, Aalto University.

Gavrilovska, L., & Prasad, R. (2006). *Ad hoc networking towards seamless communications*.

Georgiadis, L., Neely, M. J., & Tassiulas, L. (2006). *Resource allocation and cross-layer control in wireless networks*. Now Publishers Inc.

Goldsmith, A., & Wicker, S. (2002). Design challenges for energy-constrained ad hoc wireless networks. *IEEE Wireless Communications*, *9*(4), 8–27. doi:10.1109/MWC.2002.1028874

Haykin, S. (2005). Cognitive radio: Brain-empowered wireless communications. *IEEE Journal on Selected Areas in Communications*, *23*(2), 201–220. doi:10.1109/JSAC.2004.839380

Kawadia, V., & Kumar, P. R. (2005). A cautionary perspective on cross layer design. *IEEE Journal on Wireless Communications, 12*.

Khan, S., Peng, Y., Steinbach, E., Sgroi, M., & Kellerer, W. (2006). Application-driven cross layer optimization for video streaming over wireless networks. *IEEE Communications Magazine*, *44*(1), 122–130. doi:10.1109/MCOM.2006.1580942

Kim, H., & Shin, K. (2008). In-band spectrum sensing in cognitive radio networks: Energy detection or feature detection? *Proceedings of the 14th ACM International Conference on Mobile Computing and Networking*, (pp. 14-25).

Kim, M. (2011). *Project*. Retrieved from https://engineering.purdue.edu/~ee695n/projects/Misook_Kim.pdf

Kliazovich, D., Granelli, F., & Fonseca, N. L. (2009). *Handbook on sensor networks: architectures and cross-layer design for cognitive networks*. River Edge, NJ: World Scientific Publishing Co, Inc.

Kolodzy, P. (2002, November). *Spectrum policy task force report*. ET Docket No. 02-135.

Like, E., Chakravarthy, V., & Wu, Z. (2007). *Reliable modulation classification at low SNR using spectral correlation*. IEEE Consumer Communication and Networking Conference.

Liu, Q. W., Zhou, S. L., & Giannakis, G. B. (2005, May). Cross-layer scheduling with prescribed QoS guarantees for adaptive wireless networks. *IEEE Journal on Selected Areas in Communications*, *23*(5), 1056–1066. doi:10.1109/JSAC.2005.845430

Liu, Y., & Zhou, Q. (2009). State of the art in cross-layer design for cognitive radio wireless networks. *IEEE 2009 International Symposium on Intelligent Ubiquitous Computing and Education*, (pp. 366-369).

Market, M. (2006). *Clash of the titans — WiMAX and 4G: The battle for convergence is joined.*

Mitola, J. (2000). *Cognitive radio: An integrated agent architecture for software defined radio*. Ph.D. dissertation, Sweden, Royal Institute of Technology (KTH).

Otal, B., & Alonso, L. (2004). A cross layer energy saving mechanism for an enhancement of 802.11 WLAN systems. *IEEE 59th Vehicular Technology Conference (VTC 2004 Spring)*, (p. 3).

Pawełczak, P. (2008). *Technical challenges of cognitive radio-related systems*. First Competition and Regulation in Network Industries Conference, Brussels, Belgium.

Petrova, M., & Mähönen, P. (2007). Cognitive resource manager: A crosslayer architecture for implementing cognitive radio networks. In Fittzek, F., & Katz, M. (Eds.), *Cognitive wireless networks*.

Reddy, Y. B., & Bullmaster, C. (2008). *Cross-layer design in wireless cognitive networks*. Ninth International Conference on Parallel and Distributed Computing, Applications and Technologies.

Rieser, C. (2004). *Biologically inspired cognitive radio engine model utilizing distributed genetic algorithms for secure and robust wireless communications and networking*. Virginia Polytechnic Institute and State University.

Rieser, C., Rondeau, T., Bostian, C., & Gallagher, T. (2004). Cognitive radio testbed: Further details and testing of a distributed genetic algorithm based cognitive engine for programmable radios. *IEEE Military Communications Conference MILCOM 2004*, (p. 3).

Riihijärvi, J., Wellens, M., & Mähönen, P. (2006). Link-layer abstractions for utility-based optimization in cognitive wireless networks. *Proceedings of CrownCom 2006*.

Rondeau, T. W., & Rieser, C. J. B. L., & Bostian, C. W. (2004). Cognitive radios with genetic algorithms: Intelligent control of software defined radios. *Proceedings of SDR Forum Technical Conference and Product Exposition*, (pp. C-3–C-8). Phoenix, USA.

Rubinstein, M. G., Moraes, I. M., Campista, M. E., Costa, L. H., & Duarte, O. C. (2006). A survey on wireless ad hoc networks. *IEEE Conference on Mobile and Wireless Communication Networks*, (pp. 1-33).

Schaar, M. v., & N, S. S. (2005). Cross-layer wireless multimedia transmission: Challenges, principles and new paradigms. *IEEE Wireless Communications, 12*(4), 50-58.

Shakkottai, S., Rappaport, T., & Karlsson, P. (2003). Cross-layer design for wireless networks. *IEEE Communications Magazine*, *41*(10), 74–80. doi:10.1109/MCOM.2003.1235598

Siddique, T. (2010). *Spectrum optimization in cognitive radio networks using genetic algorithm*. MSc Thesis, Sweden, Blekinge Institute of Technology.

Sivanantha, A., Ehsan, S., Dhamodaran, A. N., & Gatimu, K. (2010). *A survey on quality of service (QoS) provisioning for multimedia transmission in cognitive radio networks (CRNs)*. Oregon State University.

Srivastava, V., & Motani, M. (2005, December). Cross-layer design: A survey and the road ahead. *IEEE Communications Magazine*, *43*(12), 112–119. doi:10.1109/MCOM.2005.1561928

Steenkiste, P., Sicker, D., Minden, G., & Raychaudhuri, D. (2009). *Future directions in cognitive radio network research*. NSF.

Thomas, R. (2007). *Cognitive networks*. PhD Dissertation. Virginia Tech.

Thomas, R. W., Frlend, D. H., DaSilva, L. A., & MacKenzie, A. B. (2006). Cognitive networks: Adaptation and learning to achieve end-to-end performance objectives. *IEEE Communications Magazine*, 51–57. doi:10.1109/MCOM.2006.273099

Triantafyllopoulou, D.-K., Passas, N., & Kaloxylos, A. (2007). *A cross-layer optimization mechanism for multimedia traffic over IEEE 802.16 networks*. Paris, France: European Wireless.

Weingart, T., Sicker, D., & Grunwald, D. (2006). Evaluation of cross-layer interactions for reconfigurable radio platforms. *First International Workshop on Technology and Policy for Accessing Spectrum: WiCon '06*.

Wieselthier, J., Barnhart, C., & Ephremides, A. (1994). Neural network approach to routing without interference in multihop radio networks. *IEEE Transactions on Communications*, *42*(1), 166–177. doi:10.1109/26.275311

Xia, X., & Liang, Q. (2005). Bottom-up cross layer optimization for mobile ad hoc networks. *IEEE Military Communications Conference (MILCOM)*, (pp. 1-7).

Xiao, H., Seah, W. K., Lo, A., & Chua, K. C. (2000). *A flexible quality of service model for mobile ad hoc networks* (pp. 445–449). IEEE VTC-Spring.

Yan, Z., Ma, Z., Cao, H., Li, G., & Wang, W. (2008). Spectrum sensing, access and coexistence testbed for cognitive radio using USRP. *IEEE International Conference on Circuits and Systems for Communications*, (pp. 270-274).

Yau, K.-L. A., Komisarczuk, P., & Teal, P. D. (2008). *C2net: A cross-layer quality of service (QoS) architecture for cognitive wireless ad hoc network*. IEEE Australasian Telecommunication Networks and Applications Conference (ATNAC'08).

Yong, Y., Lifeng, W., & Quan, Y. (2009). *Cross-layer architecture in cognitive ad hoc networks*. WRI International Conference on Communications and Mobile Computing.

Chapter 13
Security for Cognitive Radio Networks

Rajni Dubey
Shri Ram College of Engineering & Management, India

Sanjeev Sharma
SOIT, UIT, Rajiv Gandhi Proudyogiki Vishwavidyalaya, India

Lokesh Chouhan
ABV-Indian Institute of Information Technology and Management, India

ABSTRACT

Most of the frequency spectrum bands have already been licensed, and the licensed spectrum is not being utilized efficiently. Cognitive Radio Networks (CRNs) are the kind of full duplex radio that automatically altered its transmission or reception parameters, in such a way that the entire wireless communication network of which it is a node communicates efficiently, while avoiding interference with primary or secondary users. In this chapter, the authors introduce the concept of security threats that may pose a serious attack in CRN. Due to the unique characteristics of CRN, such network is highly vulnerable to security attacks compared to wireless network or infrastructure-based wireless network. The main objective of this chapter is to assist CR designers and the CR application engineers to consider the security factors in the early development stage of CR techniques. Challenges and various security issues are explored with respect to OSI (Open Systems Interconnection) reference model. Various possible and attacks are discussed broadly and respective solutions are also proposed by this chapter. Different architectures and models are also explained, and compared with the existing models.

13.1 INTRODUCTION

Today's wireless networks are characterized by a fixed spectrum assignment policy. However, a large portion of the assigned spectrum is used sporadically and geographical variations in the utilization of assigned spectrum ranges from 15% to 85% (FCC Std., 2003). The influence of CRNs functions on the performance of the upper layer protocols such as routing and transport are exigent and open research issues in these areas are also challenging issues in CRN (Q. Y. How & K.C.,

DOI: 10.4018/978-1-4666-2005-6.ch013

2011; E. Cesena & Cuomo, 2011). So there is always requirement of CR-MAC (Cognitive Radio Medium Access Control) protocol to provide efficient, fair, and seamless services to user. One of the factors which should be considered during design process of CR-MAC protocol is security of the network infrastructure and security of transmitted information. Without proper network security, attacker responsible for the disaster would be able to eavesdropping important information and utilize it for future attacks. Moreover, the network elements due to their poor security could become a target of attack itself (H. S. Lean, O., 2010; J. Burbank, 2008; G. X. Zhang, Y. 2008). Because cognitive radio constitutes a new approach for building wireless networks, it simultaneously opens a door for new methods of attacks on their physical infrastructure and architecture.

The recent years have been seen major and remarkable development in the field of CRN technologies. Cognitive radio is a revolutionary technology, they promise to improve the utilization of radio frequencies, making room for new and additional commercial data, emergency and military communication services. In this chapter we will introduce the concept of security threats that may pose a serious attack in Cognitive radio network. Due to the unique characteristics of cognitive radio network, such network is highly vulnerable to security attacks compared to wireless network or infrastructure based wireless network. We know that most of the frequency spectrum band has already been licensed and the licensed spectrum is not being utilized efficiently. Cognitive radio helps to efficiently utilize the spectrum band when the primary user is not using it. The main objective of this chapter is to assist CR designers and the CR application engineers to consider the security factors in the early development stage of CR techniques.

Cognitive radios facilitate in the secondary usage of the licensed spectrum (when not in use). When in use by a primary user, the secondary user cannot use it. Therefore, accurate spectrum occupancy information needs to be maintained by a secondary user. This minimizes the interference. A malicious user can try to falsify the spectrum occupancy information, which may cause interference.

This chapter has attempted to exemplify the multi-discipline nature of developing cognitive radio security models, and has attempted to outline analogies and comparisons to applicable research communities. This chapter has attempted to illustrate that the traits of a CR (awareness, adaptation, etc.) are human traits that we as designers are attempting to impose on machines. Frequency spectrum is a limited resource for wireless communications and may become congested owing to a need to accommodate the diverse types of air interface used in cognitive radio networks. However, since conventional wireless communications systems also utilize the frequency bands allocated by the Telecom Regulatory Authority of India (TRAI) and Federal Communications Commission (FCC) in a static manner, they lack adaptability (TRAI std., 2012). Also, several studies show that while some frequency bands in the spectrum are heavily used, other bands are largely unoccupied most of time. These latent vacant spectrums result in the under-utilization of available frequency bands. To defeat the overcrowding (Figure 1), different governing and non governing agencies and organization such as TRAI and Federal FCC have been investigating new ways to manage Radio Frequency (RF) resources (FCC Std., 2003). With advances in software and cognitive radio, practical ways of doing this are on the perspective. In 2003, the FCC released a memorandum seeking comment on the interference temperature model for controlling spectrum use (FCC Std., 2003). Spectrum Hole: In the literature (S. Haykin, 2005) Simon Haykin has defined (Figure 1), "A spectrum hole is a band of frequencies assigned to a primary user, but, at a particular time and specific geographic location, the band is not being utilized by that user (M. S. Tandra, R., 2009).

Figure 1. Cognitive radio concept

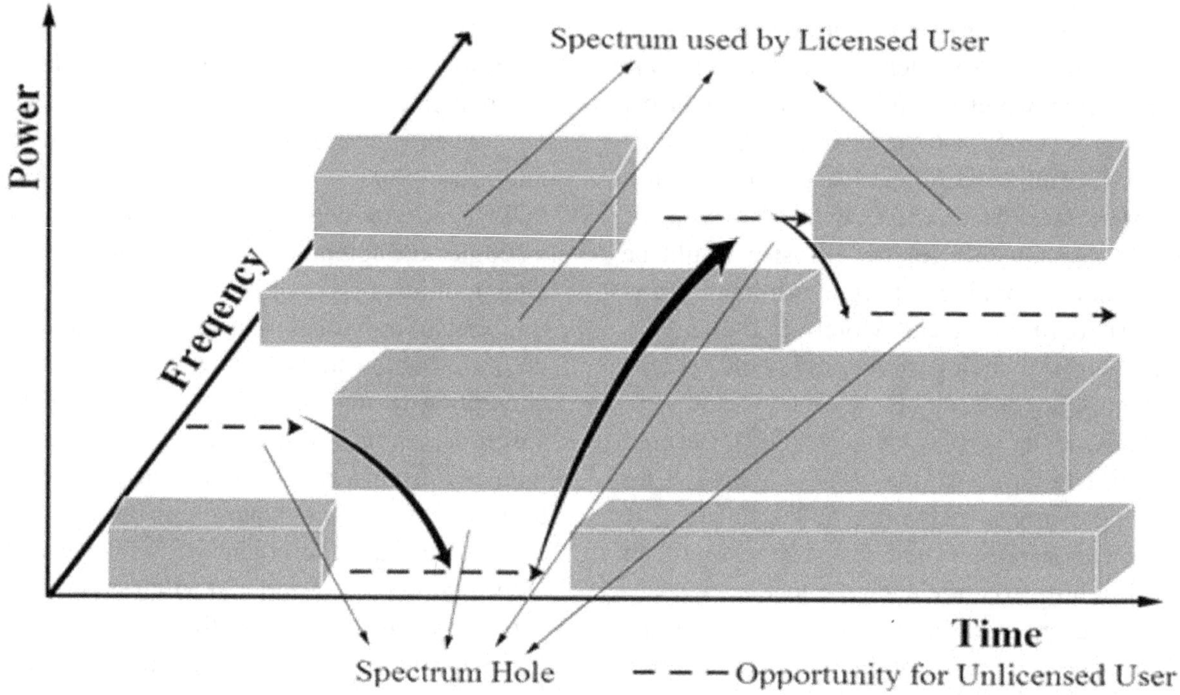

13.2 REQUIREMENT OF EFFICIENT SECURITY IN NETWORK

Cognitive radio is definitely a promising solution to achieve dynamic spectrum access and alleviate the inefficient spectrum utilization. In dynamic spectrum access networks, mutual sharing of the spectrum occupancy information helps accomplish the goal to use the spectrum efficiently and minimize the interference to the primary users. However, it is also important to detect and conclude the information is from the trustworthy user and the information is not compromised, so that the spectrum sharing objective is accomplished. So each aspect of security should be precisely pointed in any kind of CRN. It should require following four properties of security system (C. Murthy & B. Manoj, 2004):

- **Confidentiality:** The data sent by the sender (source node) must be comprehensible only to the intended receiver (destination node). Though an intruder might get hold of the data being sent, he/she must not be able to derive any useful information out of the data. One of the techniques used for ensuring confidentiality is data encryption.
- **Integrity:** The data sent by the source node should reach the destination node as it was sent-unaltered. In the other word, it should not be possible for any malicious node in the network to tamper with the data during transmission.
- **Availability:** The network remains operational all the time. It must be robust enough to tolerate link failure and also capable of surviving various attacks mounted on it. It should be able to provide the guaranteed service whenever authorized users require them.

- **Non-repudiation:** Non-repudiation is a mechanism to guarantee that the sender of a message cannot later deny having sent the message and that the recipient cannot deny having received the message.

13.3 PROBLEM STATEMENT ON SECURITY

As a limited natural resource, wireless spectrum becomes increasingly scarce and sparse with the ever increasing number of users, and the utilization ratio is a critical problem in wireless applications, then cognitive radio appears.There are two scenarios in spectrum sharing: Cooperative and Non-Cooperative scenarios. In the cooperative scenario, the primary users and secondary user may share the spectrum occupancy information thereby maximizing the spectrum utilization and minimizing the interference. However, the source of spectrum occupancy information needs to be verified to be from the primary user. In this case, malicious user can claim as a primary user and might cause interference with the primary user. So, trustworthiness between the server and client needs to be ensured so that the secondary users make sure that they are communicating with the primary user not with the malicious user.

13.4 ISSUES AND CHALLENGES IN SECURITY PROVISIONING

Similar to conventional wireless networks, CRNs also have various security issues and challenges. Some of the issues and challenges are explicitly to the wireless nature of CRNs, and few issues are specifically to the ad hoc nature of the cognitive radio networks, also known as cognitive radio ad hoc networks.

13.4.1 Issues Related to Wireless Nature (C. Murthy & B. Manoj, 2004)

13.4.1.1 Shared Broadcast Radio Channel

Unlike in wired network where a separate dedicated transmission line can be provided between pair end users, the radio channel used for communication in CRN network is broadcast in nature and is shared by all nodes in the network. Data transmitted by a node is received by all nodes within its direct transmission range. So a malicious node could easily obtain data being transmitted in the network.

13.4.1.2 Insecure Operational Environment

The operating environments where CRN networks are used may not always be secure. One important application of such networks is in battlefields. In such applications, nodes may move in and out of hostile and secure enemy territory, where they would be highly vulnerable to security attacks.

13.4.1.3 Limited Resource Availability

Resources such as bandwidth, battery power and computational power are scarce in CRN networks. Hence, it is difficult to implement complex cryptography-based security mechanism in such networks.

13.4.1.4 Physical Vulnerability

Nodes in these networks are usually compact and hand-held in nature. They could get damaged easily are also vulnerable to theft.

13.4.2 Issues Related to Cognitive Radio Ad Hoc Networks (CRAHNs)

13.4.2.1 Lack of Central Authority

In wired networks and infrastructure-based wireless networks, it would be possible to monitor the traffic on the network through central points and implement security mechanism at such points. Since CRN ad hoc network do not have any such central points, these mechanism cannot be applied in CRN ad hoc networks (C. Murthy & B. Manoj, 2004; L. J. S. J. & Lee, Y., 2007).

13.4.2.2 Lack of Association

Since these networks are dynamic in nature, a node can join or leave the network at any point of the time. If no proper authentication mechanism is used for associating node with a network, an intruder would be able to join into the network quite easily and carry out his/her attacks.

13.5 COGNITIVE RADIO NETWORK SECURITY ATTACKS

This subsection examines various attacks and vulnerabilities related to CRN. The security threats due to each characteristic are discussed in detail, besides some countermeasures and keys need to attention are mentioned. In General, we can classify various attacks is into two broad categories:

1. Active Attacks
2. Passive Attacks
    a. Attack on cognitive radio networks can be classified into two broad categories-namely passive and active attacks.
    b. A Passive attack does not disrupt the operation of the network like snooping. An Active attack attempts to alter or destroy the data being exchanged in the network. Active attack divides on the basis of cognitive radio layer architecture.

We can define attacks on CRNs as any activity that results in (a) unacceptable interference to the license primary users or (b) missed opportunities for secondary users (G. N. Clancy & C.T., 2009). An attack is considered strong, if it involves a minimal number of adversaries performing minimal operations but causing maximum damage/loss to the primary and or secondary users in the network (H. Z. Li, Z., 2008; X. Y. Hu, F., 2009). We have considered five layers in the protocol stack, namely, the physical layer, link layer, network layer, transport layer and application layer (Chen R. & Park J. M., 2008; L. C. Zhang, X., 2010; S. N.-D. Yu, H., 2010; A. A.L. Sodagari, S., 2010; H. S, Parvin, S., 2010; M. H. Zhu, L., 2011). In addition to these attacks that are specific to a given layer, we have also discusses some cross-layer attacks that can be applied at one layer to affect at another layer (V. Tepandi, J., 2010; S. Wang, W., 2010), see Figure 2.

13.5.1 Various Type of Attack

13.5.1.1 Denial–of–Service (DoS) Attack

Cognitive radio network is a wireless regional area network utilized unused spectrum opportunistically. This attack based on the inability of the secondary user to distinguish the primary user and the secondary user. In a denial-of-service (DoS) attack, an attacker attempts to prevent legitimate users from accessing information or services.

- **Denial of service attack in centralized DSA:** In this section considers a centralized DSA (Chetan N. Mathur & K. P. Subbalakshmi, 2007) network represented by a Figure 3. Here are five CR (Secondary User) $\boldsymbol{S_1}$ to $\boldsymbol{S_5}$, controlled by the Secondary Base Station ***B*** and Primary User ***P***, CR

Figure 2. Layered attacks

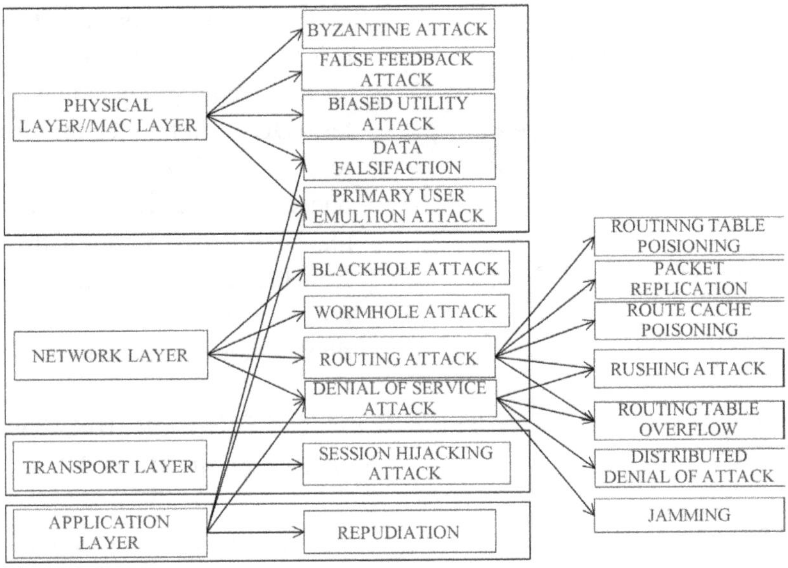

user access the band opportunistically in the same band as that of primary user transmission. Therefore when primary user ***P*** begins transmission, secondary user S_1, S_2 and S_5 can sense it and report it to the secondary base station ***B***. The secondary base station then orders all its associated secondary users to vacate the channel corresponding to the primary user and select the next available channel. Suppose that the secondary user S_2 was a malicious user. So S_2 perform denial attack, all that S_2 needs to do is to transmit during the sensing periods of other secondary user. Primary user detection scheme is energy based the other secondary users cannot make an accurate distinction between the primary and the malicious secondary user's transmission. Furthermore, the malicious user ***S2*** can repeat this attack in all the channels selected by the base station ***B*** and effectively starve all the secondary users form access to the transmission medium. Since the malicious user needs to transmit only during the quiet periods, this attack can be performed by an adversary with limited resources. Hence, the denial of service attack against centralized DSA networks is very simple. Note that this attack is unique to DSA networks due to the coexistence of two different classes of users, the licensed primary and the unlicensed secondary. Public key cryptography based primary user identification mechanism used for prevents malicious secondary users from masquerading as primary users.

- **Distributed Denial of Service Attack (DDOS):** This is the other form of the DOS attack, in the attack several malicious that are distributed throughout the network collude and prevent legitimate users from accessing the services offered by the network.

13.5.1.2 Incumbent Emulation Attack

When an incumbent is detected in a given band, all Secondary users avoided accessing that band. However, when a secondary is detected, other secondary user may choose to share that same band (R.

Figure 3. A scenario of denial of service attack

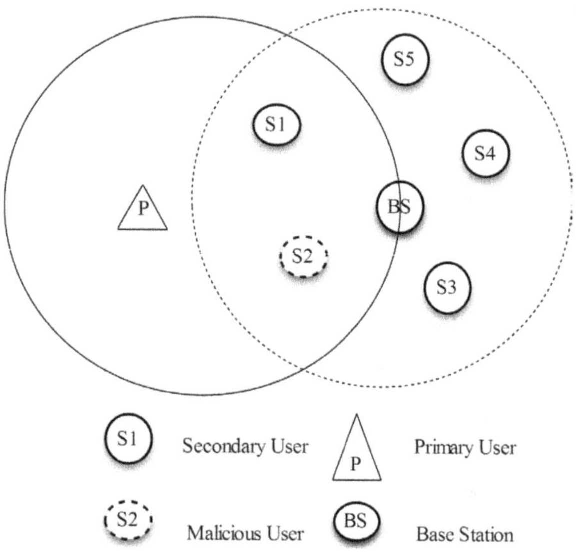

Chen et al., 2008). In other words, incumbents have higher priority than secondary user in accessing spectrum Resources. In an incumbent emulation (IE) attack, a malicious secondary tries to gain priority over other secondary user by transmitting signals that emulate the characteristics of an incumbent. The potential impact of an IE attack depends on the legitimate secondary's abilities to distinguish the attacker's signal from actual incumbent signals while conducting spectrum sensing.

There are various robust techniques against IE attack for verifying the authenticity of an incumbent signal. One approach for verifying incumbent transmitters is simply to embed signature in an incumbent signal, another approach is to employ an authentication protocol between an incumbent transmitter and a verifier. These approaches are inappropriate because no modification to an incumbent system should be required to accommodate opportunistic spectrum use by secondary user .other approach is location based verification of the transmitter but in the CRN, location of the transmitter is not fixed. Two techniques are presented in to address the problem. The first technique is called a distance ratio test (DRT), which uses received signal strength measurements obtained from a pair of location verifiers (LV) to verify the location of the transmitter. The second technique is called a distance difference test (DDT), in this techniques signal is transmitted from a single source to two LVs, a relative phase difference can be observed, when the signal reaches the two LVs due to their differing distances from the transmitter. Although, DDT does not suffer from the drawbacks of a DRT, DDT requires tight synchronization among the LVs.

13.5.1.3 Spectrum Sensing Data Falsification Attack

The second security threat to DSA is the transmission of false spectrum sensing data by malicious secondary's (R. Chen et al., 2008). An attacker may send false local spectrum sensing results to a data collector, causing the data collector to make a wrong spectrum-sensing decision. We use the term spectrum sensing data falsification (SSDF) attack to refer to such an attack.

To counter the SSDF attacks effectively, a two level defense is required, at the first level; all spectrum sensing result must be authenticated by the data collector. The purpose of this security measure is to prevent replay attacks or false data injection committed by entities outside the CR network. The second level of defense is the deployment of a data fusion schemes are vulnerable against SSDF attacks.

13.5.1.4 Wormhole Attack

In this attack (A. A.L. & Sodagari, S., 2010), malicious user (adversary node) receives a signal from the transmitter in the network and tunnels them to another location in the network. In this attack resent the signal into the network.

13.5.1.5 Black Hole Attack

In this type of attack (C. Murthy & B. Manoj, 2004), a malicious node falsely advertises good paths (shortest path or stable path) to the destination node during update messages. The intention of the malicious user could be to hinder the path finding process or to intercept all the transmission process.

13.5.1.6 Byzantine Attack

In this, attack a compromised intermediate node or set of compromised intermediate nodes works in collusion and carries out attacks such as creating routing loop, routing packets on non-optimal paths, and selectively dropping packets (Ankit Singh Rawat et al., 2010).

13.5.1.7 Jamming Attack

Due to the vulnerabilities of spectrum sensing noted in recent studies, a cognitive radio can be attacked during the sensing interval by an adversary who puts spoofing signals in unused bands. Further, once secondary users access unused bands, the adversary can use traditional jamming to interfere with them during transmission. That is, if we optimally spoof in the sensing duration and then optimally jams in the transmission slot. Then the attackers can utilize cognitive radio's fast channel switching capability to amplify their jamming impact across multiple channels using a single radio (Qihang peng et al., 2011).

13.5.1.8 Biased Utility Attack

Selfish user changes utility function parameters to get more bandwidths, so the bandwidths of other unlicensed users are decreased (Xueying Zhang & Cheng Li, 2009).

13.5.1.9 False Feedback Attack

Malicious user hides the truth about the occurrence of licensed users, and other nodes cannot sense the information due to signal fading or long distance (Xueying Zhang & Cheng Li, 2009).

13.5.1.10 Routing Attack

There are several types attacks mounted on the routing protocol which are aimed at disrupting the operation of the network, we can classify the various attack on the routing protocol are as follows (C. Murthy & B. Manoj, 2004; J. Burbank, 2008; E. Cesana & Cuomo, 2011):

- *Routing table overflow*
- *Routing table poisoning*
- *Packet replication*
- *Route cache poisoning*
- *Rushing attack*

13.5.1.11 Session Hijacking Attack

A malicious user takes control over a session between two nodes. Since most authentication processes are carried out only at the start of a session, once the session between two nodes get established, the malicious node masquerades as one of the end nodes of the session and hijacks the session (C. Murthy & B. Manoj, 2004).

13.5.1.12 Repudiation Attack

Repudiation refers to the denial or attempted denial by a node involved in a communication of having participated in all or part of the communication.

13.6 THREATS RELATED TO COGNITIVE RADIO

We can classify different attacks into two main categories, Artificial Intelligence behavior and

Dynamic spectrum access threats. Artificial Intelligence behavior threats contains policy threats, learning threats, and parameters threats whereas dynamic spectrum access threats includes spectrum sensing threats, spectrum management threats, spectrum mobility threats, and spectrum sharing threats. Both the types of threats directly affect CR applications (Sharma, S. & Dubey R., 2010; Zenon C. & Ruckshan W., 2010).

13.6.1 Artificial Intelligence Behavior Threats

- **Policy Threats:** In order to communicate more effectively in an intelligent way, a CR needs policies for reasoning different environment or from different conditions. Policy threats come from two aspects: lack of policy and failure when using policy.
- **Learning Threats:** AI involves reasoning and learning. Some CRs are designed with the capability of learning. These CRs can learn from the past experience or current Situations to predict future environment and select optimal operations, and they are Vulnerable because of learning capability.
- **Parameters Threats:** A CR control a large number of radio parameters. Both in policies and learning process, CR use parameters to control operations and estimate its performance. The functionalities of these parameters are variety. Some of these parameters are used to weight and estimate the performance of CR. Some of them are the conditions or the switching bases of policies. Altering these parameters can bases of policies. Altering these parameters can cause sub-optimal or wrong operations for a CR.

13.6.2 Dynamic Spectrum Access Threats

Dynamic Spectrum access is the process involved in getting a CR to detect and occupy a vacant Spectrum. It involves spectrum sensing, spectrum management and spectrum mobility threats.

- **Spectrum Sensing Threats:** In DSA environment, primary user has the license to use the certain frequency band whenever they want. When the Primary uses don't use their Spectrum, the Spectrum is idle, and secondary users could use the available spectrum opportunistically. Such secondary users need sensing algorithms to detect spectrum holes for communications, and CRs have the capability of detecting the Spectrum holes. In addition a CR has to vacate the channel when the primary users use it.
- **Spectrum Management Threats:** The threats here come from the possibility of false or fake spectrum characteristics parameters. The false or fake parameters impact the results of spectrum analysis, and then impact result of spectrum decision. So a CR may select the wrong band or the sub-optimal band, and the performance of communications may be impaired. For example, in Spectrum analysis, spectrum characterization is focused on the capacity estimation recently.
- **Spectrum Mobility Threats:** During Spectrum handoff, the security threats are seriously. Because a failed handoff may need a long time to resume the communication. An attacker can induce a failed spectrum handoff through ways of: compelling the CR vacating the current band by masking primary user, jamming to slower the process of selecting for a new available band or to cause a communication failure.

13.7 DEFENSE AGAINST ATTACK

(see Table 1)

13.8 SECURITY ARCHITECTURE OF COGNITIVE RADIO NETWORK

The proposed architecture of different heterogeneous networks and proposes Embedded Layered Security architecture for the future Cognitive Radio Network. The proposed architecture consists of two sub-layers. Figure 4 shows the Security Architecture of the cognitive Radio Network (Li Zhu & Huaqing Mao, 2011).

Sublayer 1 (Security of Physical layer) consists of the following modules.

13.8.1 Encryption Algorithms Layer

In encryption algorithms layer, symmetrical encryption is used for data encryption, asymmetrical encryption is used for key exchange and identity authentication. Message digest algorithms such as MD5 (Message Digest 5), SHA (Secure Hash Algorithm) (Kim I. G. & Choi J. Y., 2004; Rivest

Table 1. Defense against attack

Attack	Targeted Layer in the Protocol Stack	Proposed Solution
Denial of Service	Network Layer, Application Layer	Digital signature (Eastlake D. & Jones P., 2002), Issues of Certificates, Implementation of trust metrics (Chen R. & Park J. M., 2006; Haverinen H. & Salowey J., 2004), WPKI security in CRN (L. J & Lee, Y., 2007).
Incumbent Emulation	Physical/Mac Layer, Application Layer	Threshold Cryptography (Chen R. & Park J. M., 2006; Haverinen H. & Salowey J., 2004), End-to-End Authentication (Li Zhu & Huaqing Mao, 2010; Zenon C. & Ruckshan W., 2010; Eastlake D. & Jones P., 2002; Simpson W., 1994), Implementation of trust metrics, WPKI security in CRN (L. J & Lee, Y., 2007).
Spectrum Sensing Data Falsification	Physical/Mac Layer Application Layer	Misbehaving user detection (G. N. Clancy & C.T., 2009; Haverinen H. & Salowey J., 2004), Attack proof Collaborative spectrum sensing (Chen R. & Park J. M., 2008; Wang W. & Li H., Sun Y., 2009).
Byzantine Attack	Physical/Mac Layer	Misbehaving user detection (Chen R. & Park J. M., 2006; Haverinen H. & Salowey J., 2004) Attack proof Collaborative spectrum sensing (Chen R. & Park J. M., 2008; Wang W. & Li H., Sun Y., 2009).
False Feedback Attack	Physical/Mac Layer	Misbehaving user detection (Chen R. & Park J. M., 2006; Haverinen H. & Salowey J., 2004), Attack proof Collaborative spectrum sensing (Chen R. & Park J. M., 2008; Wang W. & Li H., Sun Y., 2009).
Biased Utility Based Attack	Physical/Mac Layer	Misbehaving user detection (Chen R. & Park J. M., 2006; Haverinen H. & Salowey J., 2004), Attack proof Collaborative spectrum sensing (Chen R. & Park J. M., 2008; Wang W. & Li H., Sun Y., 2009).
Wormhole Attack	Network Layer	Authenticated Routing for Cognitive radio network (E. Cesana & Cuomo, 2011; Qihang peng et al., 2011; Jakimoski G. & Subbalakshmi K. P., 2008; Li Zhu & Huaqing Mao, 2010).
Black hole Attack	Network Layer	Authenticated Routing for Cognitive radio network (E. Cesana & Cuomo, 2011; Qihang peng et al., 2011; Jakimoski G. & Subbalakshmi K. P., 2008; Li Zhu & Huaqing Mao, 2010).
Routing Attack	Network Layer	Authenticated Routing for Cognitive radio network (E. Cesana & Cuomo, 2011; Qihang peng et al., 2011; Jakimoski G.& Subbalakshmi K. P., 2008; Li Zhu & Huaqing Mao, 2010).
Session Hijacking	Transport Layer	Co-operative security mechanism (Chen R. & Park J. M., 2008; Haverinen H. & Salowey J., 2004; Li Zhu & Huaqing Mao, 2010).
Repudiation Attack	Application Layer	End-to-End Authentication (Jakimoski G. & Subbalakshmi K. P., 2008; Li Zhu & Huaqing Mao, 2010; Eastlake D. & Jones P., 2002).

Figure 4. Security architecture for cognitive radio network

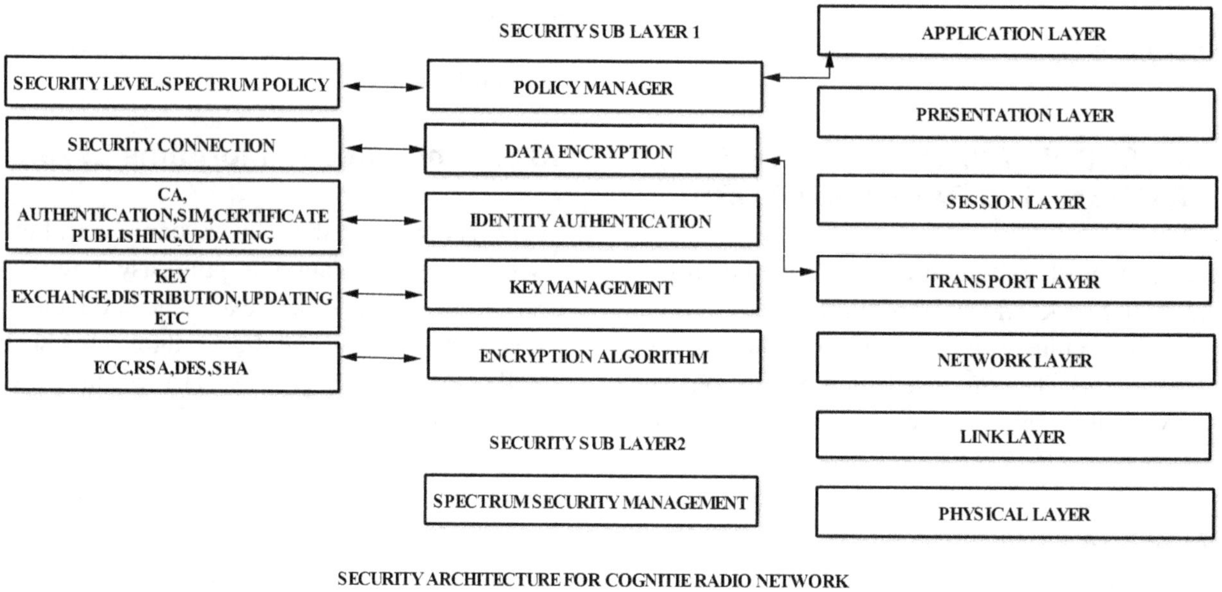

R., 1992; Wang X. & Yu H., 2005) are used for data security. Encryption algorithms are extensible, which not only supports current algorithms but also user-defined algorithms. Core algorithms provide interface to upper layers, algorithm extensible module provides user-defined interface to user. Core algorithms are supported by Random number, user record information and other random information. In cognitive radio networks, random number generated from location information and spectrum information (Chen R. & Park J. M., 2006). Cryptographic algorithm is the core concept of security architecture, which including all types of symmetrical/asymmetric algorithm and message digest algorithm.

13.8.2 Key Management Layer

Key management layer is responsible for key distribution, management, updating and destruction during communication. Keys exit security hidden danger including disclosure, duplication etc, so the regularly key updating is needed. Keys need to be destroyed when communication process ends. Key management center plays important role in whole key lifetime. Business process of key management layer includes normal business process and abnormal business process. For normal business process, asymmetrical encryption algorithms are used for key distribution; key management center updates keys regularly, and destroys keys when session ends. For abnormal business process, it is needed to update key generation algorithms if the keys face risks of disclosure and duplicating, Key Management layer is responsible for key redistribution, key transport, key arbitration and key agreement.

13.8.3 Identity Authentication Layer

Identity authentication layer includes mutual authentication between base station and terminal users. EAP-SIM (Extensible Authentication Protocol Method for GSM Subscriber Identity Module) method is used to certificate identity of terminal users, and each terminal has unique identification card. The third party authentication method is adopted for BS (Base Station) users by CA (Certifying Authority) issues certificate to BS. BS connects to core networks, identity of terminals are certificated by authentication unit of core networks certificates. Terminal authentication realize by IMSI (International Mobile Subscriber Identity), and identity of users by USIM (Universal Subscriber Identity Module). Third-party authentication has some advantages. Firstly, primary user emulation attack (Chen R. & Park J. M., 2008) is avoided by CA awards certificate to BS. When BS access to networks, CA can verifies if the BS is legal by comparing the awarding information; Secondly, for terminal users, certificate are written in device which assure legitimacy of terminal devices. CA is in core networks which adopts wired connections makes it faces less attack and will more security.

13.8.4 Data Encryption Layer

Data encryption layer main includes security connection data encryption and decryption. It is based on identity authentication layer. Communication data is encrypted on the premise that identity of both sides has been authenticated. According to requirement of application layer, different encryption methods are adopted for different data. The highest level encryption is adopted for system Signaling data. But for general data flow, stream cipher is used for encryption. For video and audio signal, it only needs distorting and convolution. During data transmission, data encryption layer resetting connection of networks timely to prevent forge connection and distort.

13.8.5 Policy Strategy Layer

Policy strategy layer includes security strategies of communication and spectrum policy. Future spectrum policy may soften condition for work frequency. Meanwhile, there exist problems such as work time limitation, power limitation etc. Policy layer overall considerate work frequency the network provides and requirement of users in order to choose suitable security strategies and spectrum strategies. Spectrum strategy management is to manage vacant spectrum and authorization spectrum. Emission power, received power and usable frequency of cognitive devices need to be limited. Emission power control is to guarantee the cognitive devices don't become disturbance sources for other network devices who are regularly working. Received power controlling can assure it can't monitor other distant devices.

13.8.6 Spectrum Security Management Layer

Spectrum security management layer is proposed for dynamic spectrum of future network. If work frequency is unfixed, the attackers will increase the power and make primary users in state of no frequency available. To solve the problem, special spectrum security management layer is needed. Spectrum strategy layer is used for managing available spectrum and spectrum etiquette. In cognitive networks, the cognitive devices have dynamical spectrum access (Chen R. & Park J. M., 2006) ability, but not all the frequency can be accessed. Spectrum strategy points out which frequency can access, when access etc. Besides that, access spectrum etiquette needs to be ruled. Spectrum sensing module is to solve spectrum sense (Wang W., Li H. & Sun Y., 2009) information fusion. For different terminal nodes, spectrum sense information is different, in order to get exact spectrum information, it is needed to eliminate false and anomaly information, fusion information by weight which assure information is accurate. Security of spectrum sensing assures the report information is true from attack nodes. Spectrum allocation makes the frequency band of the nodes within the pale and not demand-based which prevent attackers get available spectrum illegally and spectrum resources of other nodes exhaust. The aim is to assure spectrum allocation is effective, fair and security between cognitive users and primary users. Generation of random number is dependent on spectrum history information of terminal nodes and BS. Therefore,

Figure 5. Two layers security system

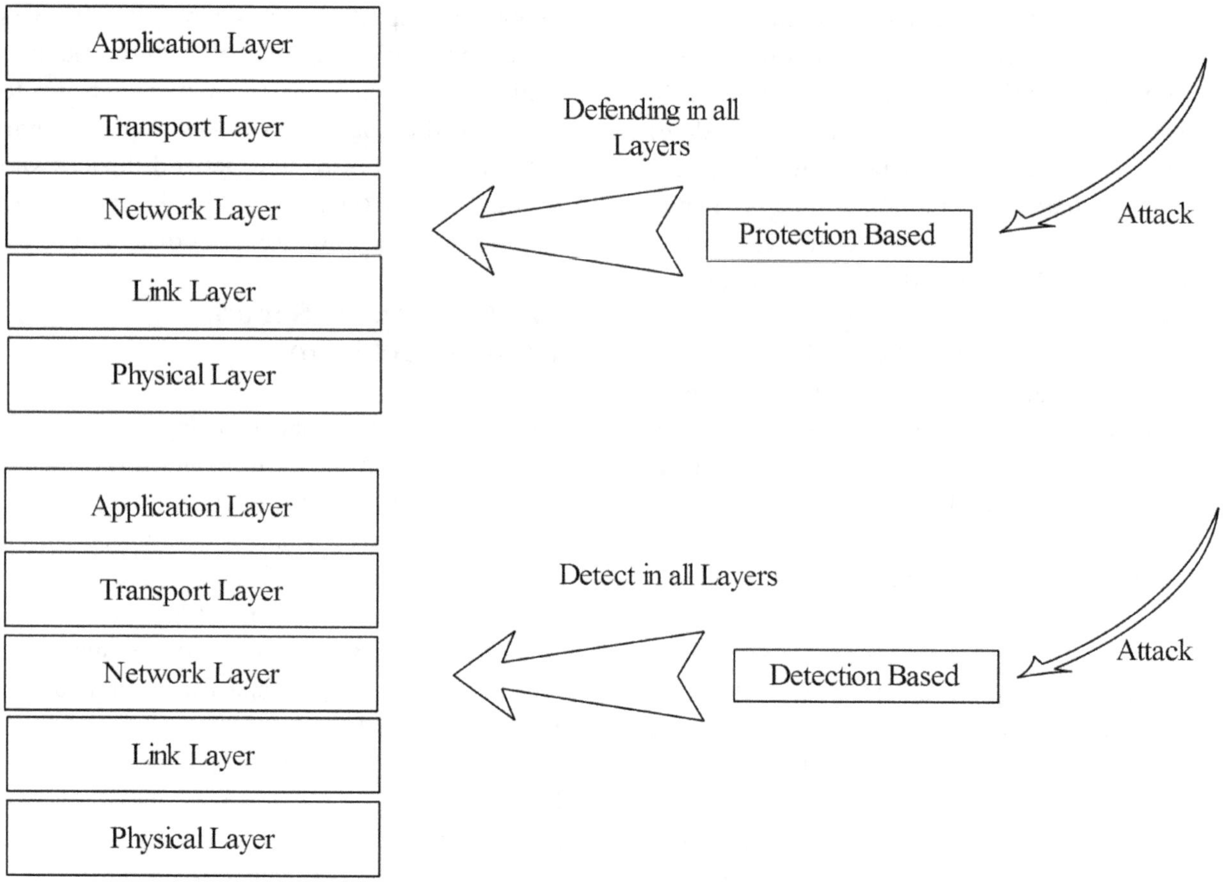

security of history information becomes key point of spectrum management. If attackers acquire history information of terminal nodes, they will guess random numbers of communication, and leads to a series of security problems. Security of spectrum history information adopts two methods: 1) Authorization mechanisms for terminal nodes; and 2) the history record needs to be synchronized and disturbed.

13.9 TWO LAYER SECURITY SYSTEM

Security in CRN can be divided into two categories- First is protection based technology in this type protects the network to the malicious user by using various type of encryption algorithms. Second is detection based technology in this type detect the malicious user. We maintain that security system in cognitive radio network should be constructed based on both protection and detection and each protocol layer should be considered due to the specific characteristics (Xueying Zhang & Cheng Li, 2009). See Figure 5.

13.10 WIRELESS PKI SECURITY IN COGNITIVE RADIO NETWORK

Wireless public key infrastructure technologies used in many security applications for digital signing and authentication. As this technology is new, service providers are concerned about its security.

Figure 6. Security algorithms for CRN

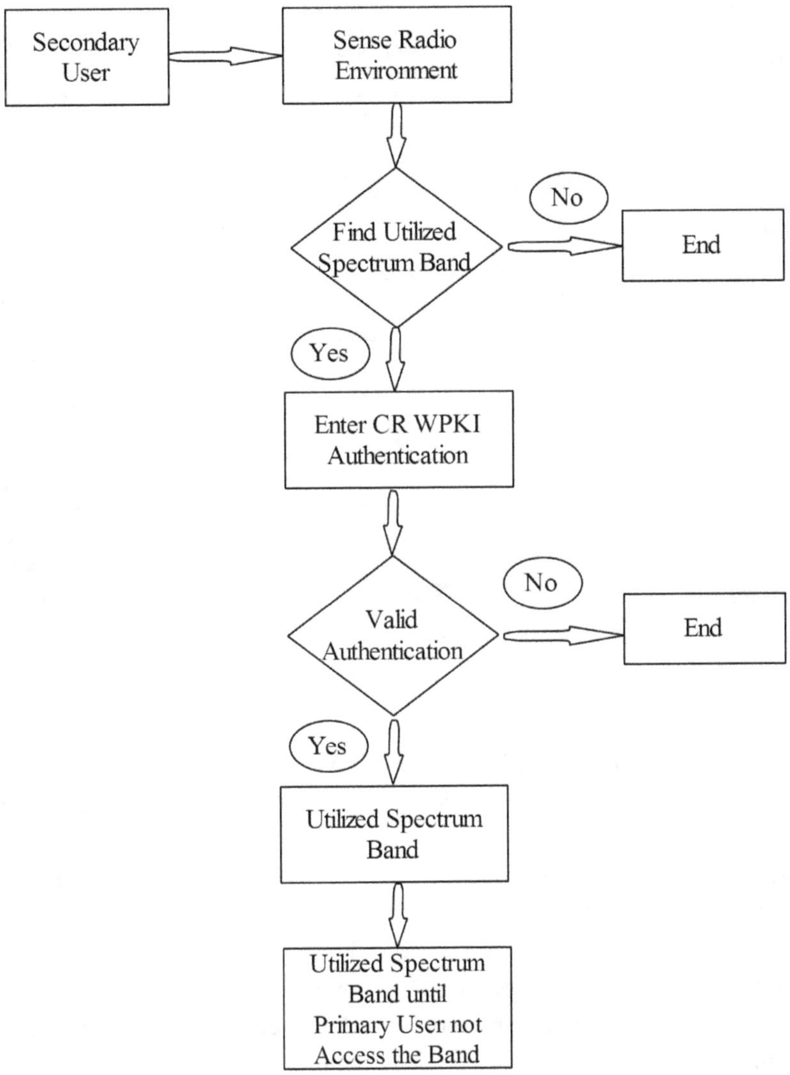

The list of threats is extensive, as WPKI (Wireless Public Key Infrastructure) used in connection with varying environments. The study considered three main types: general threat relate to legal issues, cryptography and software development. PKI binds user identities to their public keys by means of Certificate Authority (CA). PKI can support important security-critical functionality like digital signing. This assumes fulfillment of certain requirements by CA, the user registration process, and the PKI environment. To satisfy these requirements, nationally recognized CAs may issue qualified certificates stored on ID cards. A user USIM (universal subscriber identification module) card provides the usual SIM card functionality and also incorporates the private keys for authentication and digital signature. ID-card is based on emerging wireless PKI specifications. WPKI can be used to obtain client authentication and no repudiation.

Proposed security solution and algorithm is based on wireless public key infrastructure (WPKI). Wireless public key infrastructure (WPKI) is a two-factor authentication scheme

Figure 7. Proposed security frameworks and architecture

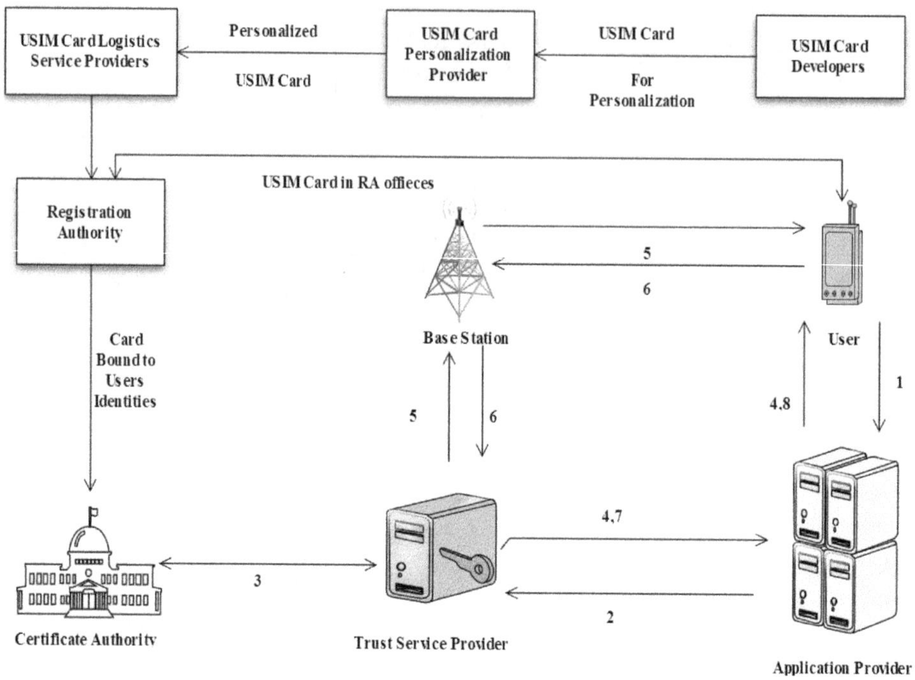

using mainly the mobile phone. It is mainly promoted by banks, mobile operators, and mobile network manufacturers L. J, Lee, Y. (2007), and V. Tepandi, J. (2010)].WPKI is useful concept when we dealt with CRNs. Figure 6, shows how the secondary user access the unutilized band of the licensed spectrum band, if unutilized band is find then check the validity of the particular band by using WPKI (Figure 7).

USIM card developers- the USIM card manufacturer, the USIM operating system developer, the WPKI application procurer, and the WPKI application developer and installer-make cards available for personalization.

WPKI management entities are in charge of user registration, certificate handling, and the main services. The RA register user, issues USIM cards, and provides customer car, often on the behalf of the CA. The CA manages certificate activation, suspension, revocation and storage. The Trust Service Provider (TSP) handles authentication and signing requests from application providers (APS). Figure 7 represents a scenario to check the authority and validity of the user.

13.10.1 Working Operation of WPKI

Figure 7 shows the basic WPKI authentication operation, it has eight steps for authentication of the cognitive radio user. User equipped with a computer and mobile phone with USIM card. The user communicates and accesses an application on server ***S***. The user identified by '***I***' her phone number or personal ID code. USIM card holds a public-private key for authentication and a PIN code for using the authentication private key, simplified authentication process are as follows

1. The user accesses a service on a particular spectrum band send a request, including ***I*** to ***S***.
2. S requests identity service for ***I***'s from the ***TSP***.
3. The ***TSP*** validates ***I***'s certificates with the ***CA***.

4. The ***TSP*** sends a verification code ***V*** to ***S***. ***S*** forwards ***V*** to the user's computer, explaining that the authentication request including code ***V*** will follow on a mobile phone and the user should check equivalence of both verification codes.
5. The ***TSP*** sends the same verification code ***V*** with hash code ***H*** to the base station and then transfer to the user's mobile phone and asks to checks code ***V*** and supplies the authentication PIN.
6. The user inputs the PIN on her mobile phone and signs ***H*** with the authentication private key.
7. The ***TSP*** verifies the user's signature and sends the result to the application provider.
8. The ***AP*** authenticates the user or rejects authentication, depending on result from the result.

13.11 CONLUSION

Cognitive radios sense spectrum activity and apply spectrum policies in order to make decisions on when and in what bands they may communicate. These activities go beyond what is done when traditional radios communicate. This chapter examines the denial of service vulnerabilities that are opened by these additional activities and explores potential protection remedies that can be applied. The security threats due to these special characteristics are mentioned in detail, besides some countermeasures and keys need to attention are mentioned. In order to follow the flexible and cognition characteristics of CR, new and robust architectures and techniques are required. In addition, corresponding countermeasures against these security threats are also required. Cognitive radio designers are encouraged to consider these assessments in the light of potential vulnerabilities and remedies as they continue to develop cognitive radios.

13.12 SUMMARY

- In this chapter, security requirements for cognitive radio network are highlighted. In the first section brief introduction about the cognitive radio, its behavior and requirement of security are discussed.
- In the second section four important properties confidentiality, integrity, availability, and non-repudiation are explained. Problem statement of security in the cognitive radio systems is described in third section.
- Issues and challenges are described with wireless nature and ad hoc nature of the CRN in the forth section. In fifth section, Attacks are defined with respect to the OSI model. Several attacks like incumbent emulation, false feedback attacks etc. are specific to layer and few are multilayer attacks like DoS and routing attacks. In sixth section, two types of threats, artificial intelligence threats and dynamic access threats are discussed.
- Defense and possible solutions against various attacks are depicted by the Table 1 with the help of three parameters, attack, targeted layer and possible solution, in the seventh section.
- In Eight sections Security architecture of CRN is explained with five layers, i.e., encryption algorithm, key management, identity authentication, data encryption, policy strategy and spectrum security management layers.
- In ninth section, two layer, protection based and detection based security system is represented.
- Our Proposed solution WPKI is explained in the tenth section. Working operation of the WPKI is also described in by the block diagram.In the last section, conclusion is summarized, which is evidently demanding of specific security requirement and

new framework architecture for successful deployment of cognitive radio networks in the future.

REFERENCES

Burbank, J. (2008). *Security in cognitive radio networks: The required evolution in approaches to wireless network security*. 3rd International Conference on Cognitive Radio Oriented Wireless Networks and Communications.

Cesana, M., Cuomo, F., & Ekici, E. (2011). Routing in cognitive radio networks: Challenges and solutions. *Ad Hoc Networks*, *9*(3), 228–248. doi:10.1016/j.adhoc.2010.06.009

Chen, R., & Park, J. M. (2006). Ensuring trustworthy spectrum sensing in cognitive radio networks. *1st IEEE Workshop on Networking Technologies for Software Defined Radio Networks*, (pp. 110-119).

Chen, R., & Park, J. M. (2008). Defense against primary user emulation attacks in cognitive radio networks. *IEEE Journal on Selected Areas in Communications*, *26*(1), 25–37. doi:10.1109/JSAC.2008.080104

Chen, R., Park, J.-M., Hou, Y. T., & Reed, J. H. (2008). Toward secure distributed spectrum sensing in cognitive radio networks. *IEEE Communications Magazine*, *46*(4), 50–55. doi:10.1109/MCOM.2008.4481340

Clancy, T. C., & Goergen, N. (2009). *Security in cognitive radio networks: threats and mitigation*. 3rd International Conference on Cognitive Radio Oriented Wireless Networks and Communications.

Eastlake, D., & Jones, P. (2002). *US secure hash algorithm 1* (SHA1). RFC 3174, September 2001: 2002.

Haas, Z. J., Deng, J., & Hall, F. R. (2002). Dual busy tone multiple accesses (DBTMA)-A multiple access control scheme for ad hoc networks. *IEEE Transactions on Communications*, *50*(6), 975–985. doi:10.1109/TCOMM.2002.1010617

Haverinen, H., & Salowey, J. (2004). (Manuscript submitted for publication). EAP SIM authentication. *Work (Reading, Mass.)*.

Haykin, S. (2005). Cognitive radio: Brain-empowered wireless communications. *IEEE Journal on Selected Areas in Communications*, *23*(2), 201–220. doi:10.1109/JSAC.2004.839380

How, K. C., Ma, M., & Qin, Y. (2011). Routing and QoS provisioning in cognitive radio networks. *Computer Networks*, *55*(1), 330–342. doi:10.1016/j.comnet.2010.09.008

Hu, F., Dong, D.-C., & Xiao, Y. (2009). Attacks and countermeasures in multi-hop cognitive radio networks. *International Journal of Security and Networks*, *4*(4), 263–271. doi:10.1504/IJSN.2009.028672

Jakimoski, G., & Subbalakshmi, K. P. (2008). Denial-of-service attacks on dynamic spectrum access networks. *IEEE International Conference on Communications Workshops*, (pp. 524-528).

Kim, I. G., & Choi, J. Y. (2004). *Formal verification of PAP and EAP-MD5 protocols in wireless networks: FDR model checking*. 18th International Conference on Advanced Information Networking and Applications.

Lee, Y., Lee, J., & Song, J.-S. (2007). Design and implementation of wireless PKI technology suitable for mobile phone in mobile-commerce. *Computer Communications*, *30*(4), 893–903. doi:10.1016/j.comcom.2006.10.014

Leon, O., Hernandez-Serrano, J., & Soriano, M. (2010). Securing cognitive radio networks. *International Journal of Communication Systems*, *23*(5), 633–652.

Mathur, C. N., & Subbalakshmi, K. P. (2007). Digital signature for centralized DSA networks. *4th IEEE conference on Consumer Communications and Networking Conference*, (pp. 1037-1041).

Murthy, C. S. R., & Manoj, B. S. (2004). *Ad hoc wireless networks: Architectures and protocols*. Prentice Hall PTR.

Parvin, S., Han, S., et al. (2010). Trust-based authentication for secure communication in cognitive radio networks. *2010 IEEE/IFIP International Conference on Embedded and Ubiquitous Computing*, (pp. 589-596).

Parvin, S., Han, S., Tian, B., & Xie, M. (2010). Authenticated spectrum sharing for secondary users in cognitive radio networks. *2nd IEEE International Conference on Network Infrastructure and Digital Content*, (pp. 509-513).

Peng, Q., Cosman, P. C., & Milstein, L. B. (2011). Spoofing or jamming: Performance analysis of a tactical cognitive radio adversary. *IEEE Communications*, *29*(4), 903–911.

Rawat, A. S., Anand, P., Chen, H., & Varshney, P. K. (2010). *Countering byzantine attacks in cognitive radio networks.* IEEE International Conference on Acoustics, Speech, and Signal Processing.

Rivest, R. (1992). *RFC1321: The MD5 message-digest algorithm*. RFC Editor United States.

Rivest, R. (2009). The MD4 message digest algorithm. *Advances in Cryptology, CRYPT0'90*, (pp. 303-311).

Sharma, S., & Dubey, R. (2010). *Distributed shared spectrum techniques for cognitive wireless radio networks.* The International Conference on Computational Intelligence and Communication Networks (CICN2010).

Simpson, W. (1994). *PPP challenge handshake authentication protocol* (CHAP). RFC Editor United States, 1996.

Sodagari, S., & Bil, S. G. (2010). *Denial of service attacks in cognitive radio networks through channel eviction triggering*. IEEE Global Telecommunications Conference.

Std, F. C. C. (2003). *Notice of proposed rulemaking and order*. ET Docket No 03-222, The Federal Communications Commission (FCC) Std., (2003). Retrieved from www.fcc.gov

Std, T. R. A. I. (2012). *Telecom regulatory authority of India* (TRAI). Retrieved from http://www.trai.gov.in

Std, W. P. C. (2011). *National frequency allocation plan*. Wireless Planning & Coordination (WPC) Std. Retrieved from www.wpc.dot.gov.in

Tandra, R., Sahai, A., & Mishra, S. M. (2009). What is a spectrum hole and what does it take to recognize one? *Proceedings of the IEEE*, *97*(5), 824–848. doi:10.1109/JPROC.2009.2015710

Tepandi, J., Andahhirov, I. T. X. C., & Vassiljev, S. (2010). Wireless PKI security and mobile voting. *Computer*, *43*(6), 54–60. doi:10.1109/MC.2009.418

Wang, W., Li, H., & Sun, Y. (2009). *Attack-proof collaborative spectrum sensing in cognitive radio networks*. IEEE 43rd Annual Conference on Information Sciences and Systems.

Wang, W., & Sun, Y. (2010). *Cross-layer attack and defense in cognitive radio networks*. IEEE Global Telecommunications Conference.

Wang, X., & Yu, H. (2005). How to break MD5 and other hash functions. *Advances in Cryptology–EUROCRYPT*, *2005*, 19–35.

Yu, H., Shen, Z., Miao, C., Leung, C., & Niyato, D. (2010). A survey of trust and reputation management systems in wireless communications. *Proceedings of the IEEE*, *98*(10), 1755–1772. doi:10.1109/JPROC.2010.2059690

Zenon, C., & Ruckshan, W. (2010). Security threats in cognitive radio applications. *2010 14th International Conference on Intelligent Engineering Systems (INES)*, (pp. 209-215).

Zhang, X., & Li, C. (2009). The security in cognitive radio networks: A survey. *International Wireless Communications and Mobile Computing Conference*, (pp. 309-313).

Zhang, X., & Li, C. (2010). Constructing secured cognitive wireless networks: Experiences and challenges. *Wireless Communications and Mobile Computing*, *10*(1), 50–69. doi:10.1002/wcm.878

Zhang, Y., Xu, G., & Geng, X. (2008). Security threats in cognitive radio networks. *10th IEEE International Conference on High Performance Computing and Communications*, (pp. 1036-1041).

Zhu, L., & Mao, H. (2010). *Research on authentication mechanism of cognitive radio networks based on certification authority*. International Conference on Computational Intelligence and Software Engineering.

Zhu, L., & Mao, H. (2010). *Research on security issues of cognitive radio networks*. The Conference on Web Based Business Management (WBM 2010).

Zhu, L., & Mao, H. (2011). *An efficient authentication mechanism for cognitive radio networks*. 2011 Asia-Pacific Power and Energy Engineering Conference.

Zhu, L., & Mao, H. (2011). *Unified layered security architecture for cognitive radio networks*. Conference on Power and Energy Engineering Conference.

Zhu, L., & Zhou, H. (2008). Two types of attacks against cognitive radio network MAC protocols. *International Conference on Computer Science and Software Engineering*, Vol. 4, (pp. 1110-1113).

Chapter 14
Dynamic Resource Management and Optimization in Heterogeneous Wireless Networks under IEEE1900.4 Framework

Chungang Yang
Xidian University, China

Jiandond Li
Xidian University, China

ABSTRACT

Next generation (xG) wireless networks will contain a number of radio access networks (RANs), which mainly depend on various radio access technologies (RATs). Meanwhile, the subscribers in the current networks will be equipped with multi-mode, reconfigurable, and cognitive equipments, with or without multi-homing abilities. This chapter looks at the definition, principle, architecture, optimization objective, specific approach, and optimization of resource for such a composite network /hybrid network context under IEEE1900.4 framework. First, the challenges of spectrum scarcity, and the requirements of a novel resource management framework in the limited resource scenario are summarized. Second, the authors focus on the techniques of resource management and optimization architecture. After a simple description of the basics of IEEE P1900.4, a novel double cognition cycle (DCC) design in the fashion of utility function theory is proposed in this chapter. The DCC contains two cognition cycles corresponding to the terminal and network side, respectively, and is uniformed in the frameworks of network utility maximization (NUM). Third, the authors capture the challenges of resource management protocols of the current developing heterogeneous networks and investigate the dynamic churning behaviors, where the subscribers are equipped with intelligent terminals with the aid of cognitive radios.

DOI: 10.4018/978-1-4666-2005-6.ch014

14.1 INTRODUCTION

A dynamically interactive resource management mechanism/protocol between multiple RANs and subscribers is presented via network utility maximization approach under the IEEE P1900.4 architecture framework. Specifically, from the network centric/user centric perspectives, we design the interactive protocol between the (Terminal Reconfiguration Manager) TRM and (Network Reconfiguration Manager) NRM entities and the corresponding distributed algorithms, e.g., the best link selection. Finally, simple tests verify the performance improvement of our proposed resource management approach and the feasibility of the presented protocol.

The rapid development of wireless services over the past decades has illuminated the growing demands for radio spectrum, which however is limited, valuable and increasingly congested. On the other hand, recent spectrum measurement campaigns indicate that most of the licensed spectrum has been under-utilized (Haykin, S., 2005). To deal with the seemingly contradictory problem between spectrum congestion and spectrum under-utilization, cognitive radio (CR) technology has been proposed and advocated (Akyildiz, I. F., Lee, W.-Y., Vuran & M. C., Mohanty, S., 2006), which can well improve the utilization of the spectrum. However, with the further development of wireless networks, the scarcity of spectrum resource is faced with more severe challenges. Specifically, future networks may comprise multiple heterogeneous wireless access networks, which are unified in a management framework with the aid of CR technology (Akyildiz, I. F., Lee, W.-Y., Vuran & M. C., Mohanty, S., 2006). Given the more and more applications of the intelligent terminal equipments, various dynamic behaviors, different kinds of service diversity and more harsh wireless context, there is no doubt that the shortage of resources is getting even worse.

As a promising development trend of future networks, heterogeneous wireless access networks have undergone many significant changes in profound, where a number of radio access networks (RANs) coexist. Moreover, these networks tend to use different radio access technologies (RATs), such as second generation or third-generation access technology: GSM, GPRS, EDGE, UMTS, HSDPA, 3GPP-LTE, IEEE 802.11x, etc.. As a result, it urgently calls for new generation of resource management system, architecture and technology to achieve seamless resource switching and dynamic spectrum access technology to achieve higher transmission rate (Marojevic, V., Salazar, J., Revés, X. & Gelonch, A., 2008). We expect a unified management structure, which can support a greater range of business types, such as online gaming, video conferences and, more importantly, further improve the spectrum efficiency. In addition to these developments of the current heterogeneous network, the reconfigurable multi-mode mobile terminals provide subscribers with the ability of choosing among multiple access networks. These terminals will be equipped with the cognitive module to achieve dynamic spectrum access, and to allow of different modes of communication between different systems for real-time dynamic spectrum sharing. Moreover, reconfigurable multi-mode mobile terminal is of a wide range of applications; as a result, user equipment may be of different operations. All these also lead to the urgent need of the novel management architecture to support a wide range of wireless interface (Buljore, S. Merat & V. Harada, H., 2008). As a result of these changes, there is a need to develop a standard that addresses the requirements and leverages the opportunities posed by such a versatile radio environment. To this end, IEEE P1900.4 aims to standardize the overall system architecture and information exchange between the network and mobile devices, which will allow these communication elements to be optimally choose from available radio resources. In other words, the standard facilitates the distributed dynamic optimization of the usage of spectrum offered by the heterogeneous wireless network, relying on a collaborative information exchange between networks and mobile devices, thereby acting as a common means to improve overall composite capacity and Quality of Service (QoS) for the served networks (Buljore, S. Merat & V. Harada, H., 2008).

There is no doubt that the development and the deployment of various radio access networks and the more and more intelligent terminal sets facilitate subscribers with more convenience, better QoS guarantee and higher-bit-rate. Also, this leads to some enormous challenges including 1) the dilemma of limited-resource, 2) the restriction of fixed spectrum management approach, 3) the deficiency of the traditional analysis tool, e.g., convex optimization (MacKenzie, A. B. & Dasilva, L. A., 2005; Chiang, M., Low & S. H., 2007) and 4) new emerging user churning behavior which depicts the behaviors that users jump from one service provider to another (Niyato, Dusit & Hossain, Ekram, 2009). These challenges are described in detail as follows: First, there are not hundreds of megabytes of bandwidth resource required by 4G (Akyildiz, I. F., Lee, W.-Y., Vuran & M. C., Mohanty, S., 2006). The only way is to exploit, employ and share the available spectrum which has been fixed allocated to these RANs using the CR technology (Haykin, S., 2005). Second, recent research and test reports reflect the physical reason of the spectrum shortage. The fixed spectrum allocation chart arouses wide attention and is with a lot of criticism. Therefore, there is an urgent need for an efficient, effective, flexible and dynamic spectrum allocation scheme and a more senior resource management structure. Also, here we should consider not only the technique aspects but the economical aspects of the spectrum (Niyato, Dusit & Hossain, Ekram, 2009), e.g., the pricing issue. Third, newly designed resource management will provide more dimensions to achieve efficient utilization of various resources, e.g., the radio, the network, the device and the computing resource (Buljore, S. Merat, V. Harada, H., 2008). Last but not the least, the traditional mathematical tools, e.g., convex optimization theory, encounter the unprecedented obstacles. Thus, we need more sophisticated analysis tools, here we investigate game theory (MacKenzie, A. B. & Dasilva, L. A., 2005) and network utility maximization (Chiang, M., Low & S. H., 2007).

As the new paradigm, cognitive radio technology can provide a powerful and practical way to implement the dynamic spectrum sharing and solve the above listed problems (Haykin, S., 2005; Akyildiz, I. F., Lee, W.-Y., Vuran & M. C., Mohanty, S., 2006). In this chapter, we investigate the resource management and optimization problems in such a scenario: mobile subscribers are equipped with reconfigurable and multi-mode mobile terminal; moreover, they may churn from one service provider to another, which is expected to become a common feature. This scenario investigated here is in line with the settings of IEEE P1900.4 (Buljore, S. Merat, V. Harada, H., 2008; Chiang, M., Low & S. H., 2007; Filin, S., Ishizu, K. & Murakami, H., 2008; Ishizu, K., Murakami, H. & Miyamoto, G., 2008), where we consider both the technical and the economical aspects of wireless network design from a network view and a wireless operator perspective (Niyato, Dusit & Hossain, Ekram, 2009). These are in line with the settings of IEEE P1900.4 (Buljore, S. Merat, V. Harada, H., 2008; Chiang, M., Low & S. H., 2007; Filin, S., Ishizu, K. & Murakami, H., 2008; Ishizu, K., Murakami, H. & Miyamoto, G., 2008) and impact both the technical and the economical aspects of wireless network design (Niyato, Dusit & Hossain, Ekram, 2009). How to provide seamless Quality of Service (QoS) provisioning and better satisfy the user's QoS requirement in this dynamic cognitive scenario are full of big challenges and worth exploring in depth.

For wireless communication systems, the concept of resources is very broad including wireless resources, network resources, equipment resources, and the management of resources. No matter from which angle, wireless communication networks are resource-constrained. Meanwhile,

Figure 1. Composition wireless networks (CWNs): a multi-network on the basis of cognitive radio technology. Here, an OSM, multiple NRM and their multi-network under their separate management and operation constitute one plane. The authors consider such a scenario that multiple operators co-exist in the same area, so there are multiple planes here to describe the complex cognitive context.

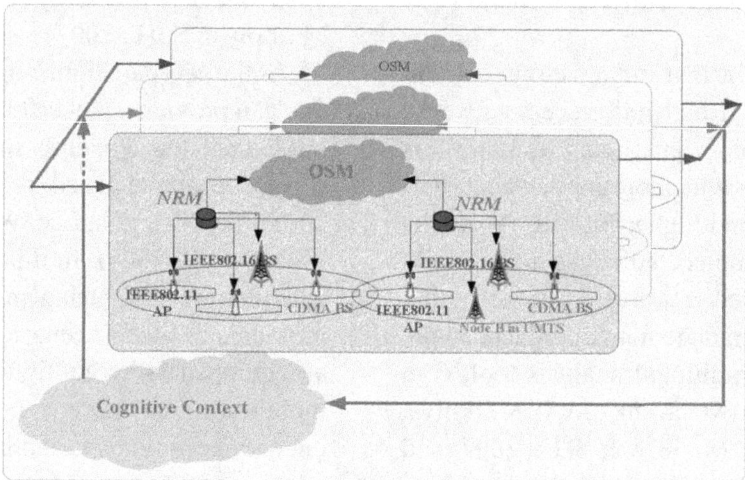

the widening range of business types, the growing number of subscribers, and the increasingly complex wireless environments all lead to the emergence of IEEE P1900.4. Motivated by the development of the heterogeneous networks and the innovation of terminal technique, there is an urgent need for a more senior management structure. Especially, we consider the multi-networks coexistence scenario shown as Figure 1.

Here such a heterogeneous wireless environment might include the following elements: multiple operator spectrum managers (OSMs), multiple radio access networks (RANs), multiple radio interfaces, and multiple terminals. In this chapter, such a network is denoted as composition wireless network. Specifically, one OSM may control and manage several Network Reconfiguration Managers (NRMs), and each NRM is the management center of the RAN side. Moreover, the Terminal Reconfiguration Manager (TRM) is used for the terminal side. Furthermore, these multiple RANs include IEEE 802.11x wireless networks, IEEE 802.16 networks, TD-SCDMA networks, and so on. The contributions of this proposal are summarized as follows:

1. This chapter summarizes the current development of wireless network, especially the development trend of heterogeneous network. And it also focuses on the impacts, factors and challenges of which the resource management may be subject to.
2. Under the current challenges as well as the urgent need for a new management structure, more attention should be paid to the IEEE P1900.4 and the main architecture and functional framework of it are summarized in this chapter.
3. Spectrum leasing and spectrum trading, a possible use case, are investigated as a new use case in the IEEE P1900.4, besides the dynamic spectrum allocation, dynamic spectrum sharing and the distributed optimization.
4. Meanwhile, a novel interactive protocol between the NRM and TRM is proposed, and the especially important thing is we propose a double leading cycle for the terminal and the RAN side, respectively.

14.2 BACKGROUND (RESOURCE DEFINITIONS AND RESOURCE MANAGEMENT OBJECTIVE, CONCEPT, AND USAGE)

The basic problems of communication networks are effectiveness and reliability of the data transmission. Both of them are related to how to utilize the existing resource effectively under the guarantee of a certain degree of QoS requirements of subscribers. The basic purpose of communication is to transmit the data, voice and multimedia in a given time delay from the source to the destination shown in Figure 2 (A). Further, the ultimate goal of communication is to achieve an effective transmission in any time, at any place and in any form of information format. In fact, achieving effective use of resources has always been a great matter of concern, and it is particular important for the developing heterogeneous network today, which will take unified network architecture in the future, that is, various radio access networks will co-exist in the same area under unified management architecture. Therefore, how to provide enough resource for the terminal who is trying to transmit data to their potential receivers is full of uncertainty factors due to the environmental variability, business diversity and interactivity property. As a result, there appears the need for good resource management and control schemes to provide an effective resource supporting, and such programs should have the self-adaptive, dynamic, and even intelligent prediction abilities. To cope with these increasing diversity and complexity challenges, mechanisms for self-optimization, self-organization, self-healing, self-configuration (self-x) are essential to guarantee efficient cost and high quality network operation. The resource management objective is to realize the best mapping from the service requests for the current user requirement to a variety of network resources, and to achieve effective resource utilization.

In cognitive network, heterogeneous access technologies, multi radios, multi channels, multi interfaces, multi homing, and multi streaming should be practically and efficiently used (Haykin, S., 2005). And this process may be carried out in the description of multiple dimensions, such as time, space and frequency, including radio spectrum, power, channel link, processing power, and capacity. This is the most traditional concept of resource definition. In a traditional way, we can define the communication network resource as: communication network = physical media + computing resources (capacity), such as: wireless transmission channel=radio frequency +modulation, Ethernet=twisted pair + modulation, and fiber-optic network = fiber + transmission standard.

As we known, the basic resource for wireless communications is the available radio frequency at the given space for the specified time duration. Due to the emerging wireless transmission and network technologies (e.g., MIMO, relay, software defined radio, adaptive transmission, self-optimization, self-organization, self-healing, self-configuration (self-x)), a variety of resources, such as antennas, power, air interfaces, access networks, storage and computing capacity, except frequency can now be used independently or combined to transmit and receive different kinds of services for a user. To make efficient use (collecting and allocating) of different kinds of resources, we need an abstract of those resources. For any given communication link, it can be described simply by one kind of physical media (e.g., radio frequency, twisted wire, optical fiber) plus a modem, which can be described by computing capability (e.g., MIPS). Specifically, for MIMO link, distributed antennas, relay and cooperative transmissions, resource scheduling and locations, varieties of signal processing algorithms are needed, which also can be abstracted as the requirements to computing capability. The transmit power and its control can also be included in the computing capability. While for the access points or base stations in access networks, their information storage and switching/routing capability can

Figure 2. Examples of a typical resource management process: (A) The resource distribution in the space; (B) The allocated set of resource vectors; (C) The resource allocation process

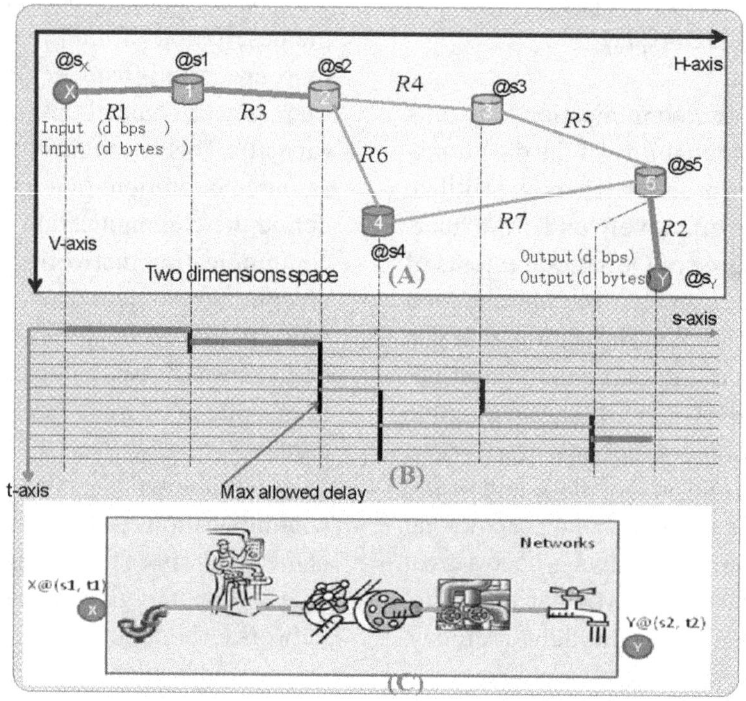

be described by the allowed delay. By using this delay, we can flexibly bridge the resources from the different networks. Furthermore, for each access network, it can use different air interfaces or transmission standards and belongs to a different operator. So the operator identification and corresponding standard is the import property of the resource. In summary, for the heterogeneous wireless access network, their resources can be expressed by the resource vector $R = \{f; net; c; \tau; id\}$ where*f*=(*startfrequency,end frequency*) in Hz, e.g., *f* = (825*MHz*,826*MHz*), *net*=network transmission capacity, e.g., 10Mbps or 1000Mbps by using Ethernet or other infrastructure; *c*= computing capability in MIPS (FLOPS) to perform a modem or other processing/controlling functions, e.g., 10MIPS; τ =allowed packet or service storage delay in a node, e.g., 100ms, which is determined by the storage and switching/routing capacity; *id*=the identifications of operator and standards, e.g., {China mobile, TD-SCDMA}.

As an example, we want to setup a path between a user *x* at space location s_x and another user *y* at space location s_y through a wireless access network between user *x* and node1, one backbone network between node 1 and node 2, another backbone network between node 2 and node 5, and another wireless access network between node 5 and user y as shown in Figure 2 (A). The resources used for links on the path are {*R*1,*R*2,…,*R*7}, which is called as the set of resource vectors. Here, take *R*1 and *R*3 as examples to show the meanings of the resource vectors. *R*1(1Mbps) = {(823MHz,824MHz), 10MIPS, 2*ms*, *OA*} means the link between user *x* and node1 can provide 1*Mbps* data rate by using radio frequency between 823MHz and 824MHz, 10MIPScomputing resource, and a maximum of 2*ms* allowed delay at node 1, and it belongs to operator OA. $R3$(1Mbps) = {(823MHz,824MHz), 100Mbps, 5MIPS, 6*ms*, *OB*} means the link between node 1 and node 2 can also provide 1Mbps data rate by using 100Mbps Ethernet, 5MIPS computing resource, and a maximum of 6*ms* allowed delay at node 2, and it belongs to operator OB. For the path section between node 2 and 5, two parallel cascaded links, each providing 0.5Mbps data rate, are used.

The finally allocated set of resource vectors for the path is shown in Figure 2 (B). Based on the above abstract of resources, the resource allocation process is very similar as the setup process of the water pipeline system as shown in Figure 2(C). We need to setup a cascaded pipeline to transmit a batch of bytes or $d$bps rate data at the space location s_x and in time t_1 from user x to the destination y at the space location s_y within a given time limit t_2.

14.3 RESOURCE MANAGEMENT ARCHITECTURE: BASICS OF IEEE P1900.4

IEEE 1900 Standards Committee was sponsored jointly by the IEEE Communications Society and the IEEE Electromagnetic Compatibility Society in 2005. Its objective is to support future wireless protocols and advanced spectrum management standards. In March 22, 2007, IEEE Standards Board approved of reorganizing IEEE P1900 Work Group as Standards Coordinating Committee 41 (SCC41). The motive for establishing IEEE P1900.4 is to improve the overall wireless system capacity and the quality of service. They expect to design suitable system architecture and some entities which can help realize the exchange of information between network and terminals and optimize the usage of resources (Buljore, S. Merat, V. Harada, H., 2008).

Main entities definitions in IEEE 1900.4 are introduced here. The purpose of IEEE 1900.4 is to improve overall composite capacity and quality of service of wireless systems in a multiple Radio Access Technologies (RATs) existed environment. Specially, this is achieved by defining an appropriate system architecture and protocol that will facilitate the optimization of radio resource usage, and in particular, by exploiting information exchanged between network and mobile Terminals, whether or not they supporting multiple simultaneous links and dynamic spectrum access (IEEE 1900.4 WG.). Two main functional entities termed as the Network Reconfiguration Management (NRM) and Terminal Reconfiguration Management (TRM) is defined in IEEE P1900.4. There are also some other entities working at the terminal side and the RAN side respectively, as shown in Figure 3, which illustrates the proposed double cognition cycle (DCC) based on the utility bargaining process. For example, on the terminal side, it has the terminal measurement collector (TMC) and the terminal reconfiguration controller (TRC), while on the RAN side, it has the RAN measurement collector (RMC) and the RAN reconfiguration controller (RRC).

Another important definition is radio enabler (RE) in IEEE P1900.4 (Buljore, S., Harada, H. & Filin, S., 2009). RE can realize the exchange of information between NRM and TRM including the terminal context information to the NRM and the management policy information from the NRM to the terminal, as well as other entities. Certainly, there are also some other entities detail depicted in (IEEE 1900.4 WG.), and more detailed information and introduction can be found in the IEEE P1900.4 standard (IEEE 1900.4 WG.).

There are three typical use cases of original IEEE 1900.4 including dynamic spectrum assignment, dynamic spectrum sharing and distributed radio resource usage optimization. Specifically, in the dynamic spectrum assignment use case, frequency bands are dynamically assigned to the RANs among the participating networks in order to optimize spectrum usage. In other words, the assigned frequency bands are not fixed, but can

Figure 3. Interactive protocol between TRM and NRM based on utility function. A utility based RAN selection (UBRS) algorithm is designed for the terminal side. The functional blocks, e.g., TUE, TUP, are added into TRM.

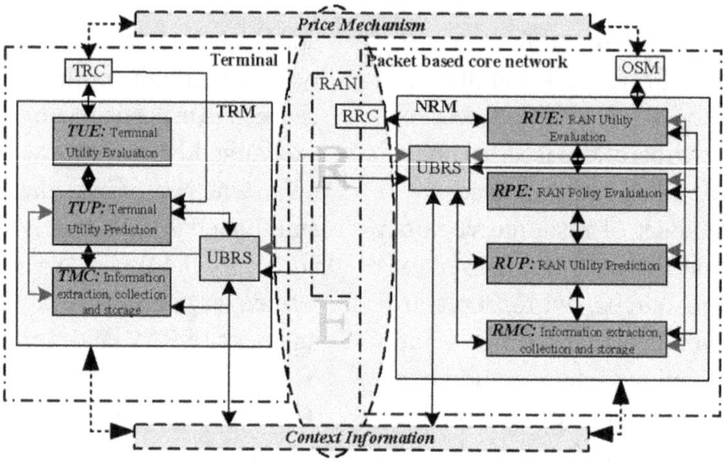

be dynamically changed. While in the dynamic spectrum sharing use case, frequency bands assigned to RANs are fixed. However, a particular frequency band can be shared by several RANs. In other words, the dynamic spectrum sharing use case describes how fixed frequency bands are shared and/or used dynamically by RANs and Terminals. As for the distributed radio resource usage optimization use case, frequency bands assigned to RANs are fixed. Reconfiguration of RANs is not involved in this use case. Distributed radio resource usage optimization use case considers Terminals with or without multi-homing capability. Decision on Terminal reconfiguration is made by its TRM and is supported by NRM.

The current version of IEEE P1900.4 standard only focuses on the architecture and functional description. However, it does not have the exact implementing algorithms and the concrete protocol, which remains to be done by the working group's next step task. In this chapter, we will investigate these functional entities based on the network utility maximization technology.

In Figure 3, it depicts Interactive protocol between TRM and NRM based on utility function The OSMs generate spectrum assignment policies expressing the regulatory framework and operators objectives for spectrum usage optimization. First, the OSMs provide these spectrum assignment policies to the corresponding NRMs. Then the NRMs analyze the spectrum assignment policies and available context information and dynamically make spectrum assignment decisions to improve spectrum usage and quality of service. After the new spectrum assignment decisions have been made, the NRMs request corresponding reconfiguration of their RANs. Following the RANs' reconfiguration, terminals need to reconfigure correspondingly. Meanwhile, NRMs analyze available context information and dynamically make spectrum access decisions to improve spectrum usage and quality of service. NRMs make these spectrum access decisions within the framework defined by spectrum assignment policies. Following these decisions, NRMs request corresponding reconfiguration of their RANs. On the other hand, NRMs dynamically generate radio resource selection policies and send them to their TRMs. These radio resource selection policies will guide these TRMs to their spectrum access decisions. TRMs analyze these radio resource selection policies and the available context information and dynamically make spectrum access decisions to improve spectrum usage and quality of service. These spectrum access decisions are made within the framework of the radio resource selection policies. Following these decisions, each TRM requests corresponding reconfiguration of its Terminal. These features are in consistent with the original standard related definition, but the operation objects are turned into the utility function. Resource management optimization architecture is introduced here: Double cognitive cycle design framework consisting of two main functional entities termed as the Network Reconfiguration Management (NRM) and Terminal Reconfiguration Management (TRM) are defined in IEEE P1900.4 (IEEE 1900.4 WG.).

Because of the factors, such as the complexity and the mobility of the existing traffic model, dynamic network topology and the unpredicted link property, the traditional mathematical tool will encounter some unprecedented difficulties. This will become even more serious in the context of the heterogonous network. Fortunately, the network utility maximization (NUM) technology, which is gaining wide attention of the academics and institute, can mathematically model the complex scenario and provide a powerful tool for addressing the above problems. Especially, we use the NUM model for the network resource management issues as well as cross-layer optimization. In addition, The decomposition technique as a novel mathematical method helps design the distributed algorithms to search the optimal strategy for the NUM model (Chiang, M., Low & S. H., 2007; Yang C. G. & Li J. D., 2009; Yang C. G., Li J. D. & Chen D., 2009; Yang C. G., Li J. D. & Tian Z., 2010; Yang C. G. & Li J. D., 2010).

Taking the end-to-end and overall network performance into account, we propose double-cycle learning structures based on the utility theory for the terminal and the RAN, respectively. With the aid of the radio enablers (REs) as the interface, the strategy and context information interaction protocol between TRM and NRM is investigated. Also, base on the double-cycle learning framework, the interactive protocol between the terminal and RAN side is roughly depicted as Figure 3. Meanwhile, we consider a simple heterogeneous scenario to verify the effectiveness of the proposed utility based double cycle learning architecture and the presented algorithms as well as the interactive protocol.

User-Dependent Utility Cycle as Figure 4 left shows: (1) Terminal Measurement Collector (TMC) is responsible for the terminal context information (e.g., user's preferences, QoS level requirements, terminal capabilities, we capture all these in a uniform utility function in this chapter) collection, extraction, storage, and also sending them to NRM entity through RE (below). (2) When each TRM acquires context information also including pricing (above) and network context information through REs, Terminal Utility Prediction (TUP) predicts the terminal utility, meanwhile, Terminal Utility Evaluation (TUE) further evaluates the utility requirement predicted by TUP. Step (2) is of crucial importance largely dependent on the terminal/pricing/network context information. (3) If last terminal utility is not a reasonable request, then the terminal resorts to terminal reconfigure controller (TRC) to reconsider it. Steps (1)(2)(3) will continue until all the terminals successfully obtain a final utility request.

Here we give more detailed illustration of step (2) and (3) as follows: If TRM regards this as a reasonable strategy after TUE acts, TRM will provide the terminal utility request to the radio access network (RAN) side through the RE. Otherwise, as the terminal is always rational to maximize their own expectations of the utility function, the terminal resorts to the reconfigure controller (TRC) to reconsider the terminal utility function, and the terminal may obtain a low-grade utility based on the same context information. The terminal utility cycle will not stop until all the terminals achieve a reasonable terminal utility request after the process of the TUP, TUE and TRC many times. In the RAN side, RMC is

Figure 4. Cognition cycles for the terminal side (as shown in the figure left) and the RAN side, respectively (as shown in the figure right): here, TRM is spitted into two functional blocks termed as "TUP" and "TUE," and the authors do not define a new entity but in totally line with original definition of IEEE P1900.4; There are two functional modules in NRM entity of the right figure, termed as "RUP" and "RUE." All these newly defined functional blocks are based on the utility function, and they are described in detail.

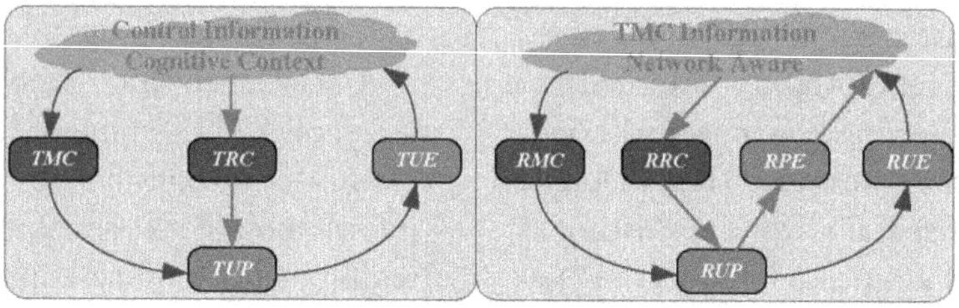

responsible for the network information collection, extraction and storage.

Network-dependent utility cycle as Figure 4 right shows: this is the cognition cycle on the RAN side. The RAN-side information contains goals of wireless resource usage optimization, RAN wireless transceiver capacity, the network context information observed by Network Measurement Collector (NMC), as well as the terminal context information received from REs. We map the policy decision process to the RAN-side utility prediction (RUP) process. Inside the NRM entity, the functional modules should contain the RAN utility prediction (RUP), the RAN utility evaluation (RUE) and the RRC entity controlling the RAN utility reconfiguration. These RAN-side utility based functional modules have the similar function as those of the TRM-side. After NRM achieves the entire terminal utility requirement obtained by the User-dependent utility cycle, it will start up these modules to build the RAN-side utility. Interaction protocol between TRM and NRM entities based on utility is referred to Figure 5.

From the practical perspective consideration, there ought to be interactively process between the entity NRM and the entity TRM, since the service requirement and the service provider relationship. This process is shown as Figure 5.

1. TRM informs NRM with terminal context/utility request information after User-dependent utility cycle, and all these become necessary for the decision-making process at the RAN-side.
2. Meanwhile, NRM can obtain the network context information as mentioned above. Also NRM employs a reverse engineering perspective. NRM builds its network utility function depending on the sum of the entire terminal utility requests, or it allocates a weight factor for the specific terminal.
3. NRM operates on the =1Network-dependent utility cycle functional modules. For malicious terminals, the OSM will relatively perform the pricing mechanism. In addition, during the interaction there are multiple RANs (referred to multiple service providers) and multiple terminals. As a result, serious competitions will emerge at both the TRM and the NRM side. To pursue more revenue, the OSM must carefully implement the pricing mechanism.
4. The interactive process won't stop until the terminals and the RANs both achieve the expectable utility function. The specific protocol will absolutely depend on the specific use case, and the interactive parameters will

Figure 5. Interactive protocol between TRM and NRM based on utility function. A utility based RAN selection (UBRS) algorithm is designed for the terminal side. The functional blocks, e.g., TUE, TUP, are added into TRM.

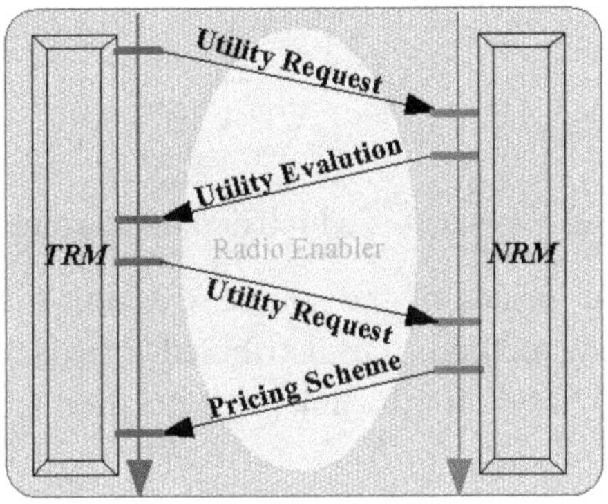

be verified in the next network selection use case under the above-proposed framework.

Here, we employ a novel resource management optimization approach, i.e., the Dual decomposition technique based distributed algorithm design. First, we introduce utility-based double cognition cycle and the interactive protocol between the terminal and the NRM through RE in the last section. Then under the case of dynamic spectrum sharing defined in IEEE P1900.4, we investigate the performance of the proposed framework in this section. We also design a distributed algorithm applying to the framework.

The basic idea of optimization decomposition is to decompose the original large problem into several sub-problems which are then coordinated by a master problem by means of some kind of signaling or message passing. If the decomposition is properly done, those sub-problems will be much easier to solve than the original problem. Most of the existing decomposition techniques can be classified into the primal decomposition and dual decomposition methods. A primal decomposition is achieved by decomposing the original problem by partitioning the variables into two sets, optimizing over one set of variables, and then over the remaining set. A dual decomposition is realized by decomposing the dual problem or, more exactly, forming the Lagrange expression by augmenting the original problem with respect to the coupling constraints (Chiang, M., Low & S. H., 2007). Why here we employ the dual decomposition technique:

1. The relationship between Network protocols and decomposition techniques: there can be many alternative decomposition techniques, leading to a choice of different layering architectures. What is surprising is that the elements of the decomposition process, e.g., NUM problem, sub-problems, decomposition methods and primal or dual variables, are well mapped as layers, layering methods, and the interface of the network protocols.
2. Distributed algorithm: the decomposition often leads to the corresponding distributed algorithms for each layer of the network protocol, respectively. Meanwhile, there are not weak conditions requirements for the convergence properties guarantee.
3. Better tradeoff: the overall optimal performance is always against the implementation complexity of the messaging exchange. The decomposing technique for the NUM problem always guarantees better tradeoff between them. Maybe the messages need to be transmit are the interface parameters which are reflected as the primal or the dual variables.

We consider a simple scenario, where three RANs co-exist to share the same multiple logic-links. The operator makes the spectrum policy directed by the OSM, and informs the NRM entity in the RAN side. Taking a reverse engineering point, dynamic spectrum sharing among multiple RANS can be formulated into the network utility maximization. Assume there are L links belong to three different RANs: 3G-TD-SCDMA, WLAN and WiMAX, and $L_{3G} + L_{WLAN} + L_{WiMAX} = L$, where L_{3G}, L_{WLAN} and L_{WiMAX} are the links owned by 3G-TD-SCDMA, WLAN and WiMAX, respectively. We further assume each link has the peak capacity limit as a result of specific property of the radio access network. Specifically, we assume there are S sources, the rate requirement is x_s, $s = 1, 2, \ldots S$. The traditional utility function in line with some type of fairness takes the form of

$$U_s(x_s, x_{-s}) = \left\{ \log x_s, \, for \; \alpha = 1; (1-\alpha)^{-1} x_s^{(1-\alpha)}, \; for \; \alpha \neq 1 \right\},$$

where x_s is the transmission rate requirement of the source s. The factor α is a fairness indicator detailed described in (Akyildiz, I. F., Lee, W.-Y., Vuran & M. C., Mohanty, S., 2006). We consider the use case of dynamic spectrum sharing in the utility-based double cognitive cycle IEEE P1900.4 framework, where the spectrum sharing issue is mapped into the dynamic link sharing problem in this chapter. The problem can be formulated as the typical sum-utility function maximization problem with several coupled constraint conditions. Here, we introduce the Lagrange multipliers λ_{l_i}, w_{l_i}, which will work as the interaction parameters between the primal and dual subproblems after we decompose the original problem. Meanwhile, we further have the terms $\lambda_{l_i}^s = \sum_{l_i} \lambda_{l_i}$, and it represents the sum of all the link prices of the user s. Based on the decomposition technique (Yang C. G. & Li J. D., 2009) (Reference therein), we can successfully obtain the dual and the primal problems, respectively.

That is to say, user-dependent utility cycle and network-dependent utility cycle depicted in Figure 6 will proceed for the available terminal utility with the form of the dual and the primal problems, respectively. Because of the convexity of the objective function termed as utility function in this chapter in dual sub-problems, which can be very simple to be proved, we can easily find the optimal transmission rate strategy for the terminal side. By mathematical derivation, we can obtain the optimal transmission requirement for each source as

$$x_s^\star(\lambda_l^s) = \left\{ 1/\lambda_l^s, \alpha = 1; \sqrt[\alpha]{1/\lambda_l^s}, \alpha \neq 1 \right\},$$

where $\lambda_{l_i}^s = \sum_{l_i} \lambda_{l_i}$, which depicts the sum of all the link prices of the source s. Using the subgradient method, the distributed approaches of the RAN-side for the link price iteration is (Akyildiz, I. F., Lee, W.-Y., Vuran & M. C., Mohanty, S., 2006),

$$\lambda_{l_i}(t+1) = \left[\lambda_{l_i}(t) - v_i \left(c_{l_i} - \left(\sum_{l_i} \lambda_{l_i}(t) \right) x_s^\star(\lambda_l^s) \right) \right]^+, \forall l_i.$$

At this point, the expressions of the terminal side and the network side become very clear shown as (1) and (2), respectively, where the Lagrange parameters λ_{l_i} work as the interaction variables between the "TRM" and the "NRM." Note: The use case of transmit rate strategy $x_s^\star$ and optimal link selection l_i at the terminal side (1) and the optimal link pricing scheme λ_{l_i} at the RAN side (2) under the proposed interaction protocol between TRM and NRM entities based on the utility well manifesto the proposed interactive protocol.

14.4 SIMULATION RESULTS AND ANALYSIS

We consider a heterogeneous wireless network scenario, where three radio access networks (RANs): 3G-TD-SCDMA, WLAN and WiMAX coexist in an identical area operated and managed by one operator. The operator works as the OSM

Figure 6. Convergence of the proposed interactive protocol between the RAN-centric and the user-centric and the proposed distributed PRICE based UBRS

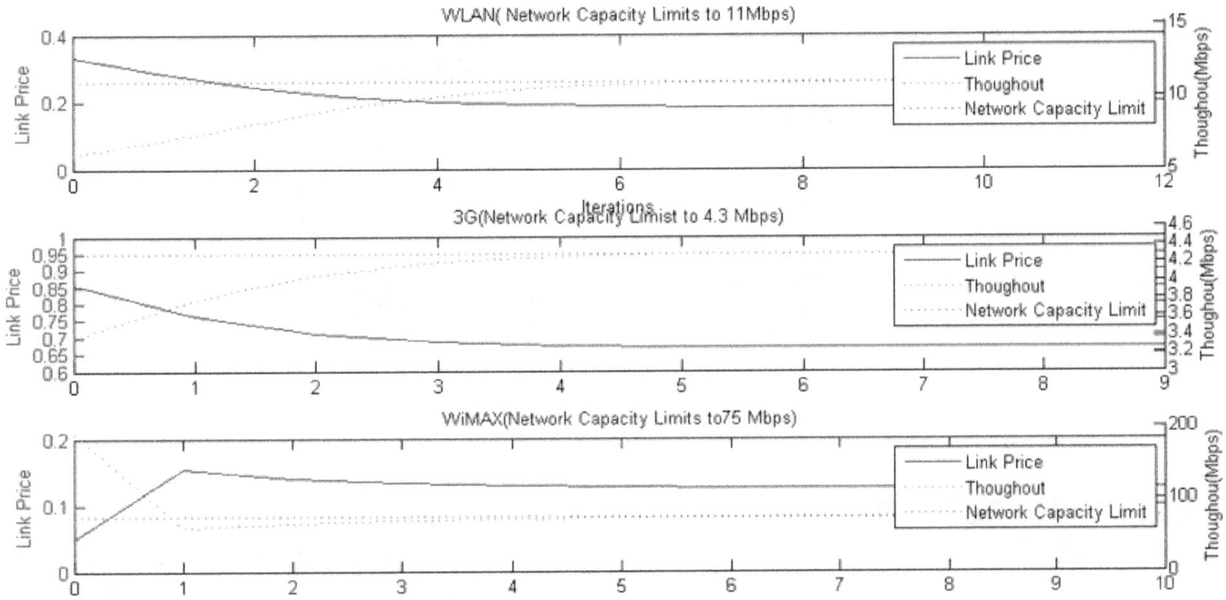

in the proposed protocol, and is responsible to design the frequency spectrum use policy. Each RAN is equipped with a NRM, which will interact with the potential TRM in the subscriber terminal. In addition, we further assume that each subscriber always has data transmission requirements. Meanwhile, every NRM hopes to have the largest number of subscribers to select and utilize his RAN for service request, until the highest overall network capacity is achieved. As a result, the spectrum revenue of each RAN operator is maximized. Moreover, the capacities for each type of 3G-TD-SCDMA, WLAN and WiMAX considered in the simulation is limit to 4.3(Mbps), 11(Mbps), 75(Mbps), respectively.

The initial RAN selection approach is based on the Received Signal Strength (RSS), and then the next step RAN churning behavior of the users is according to the network link price which is an experimental function of the NRM and the QoS requirement reflected by the successful terminal utility requirement in the TRM entity. So the joint resource management scheme for both network centric and user centric contexts is implemented in the designed double-cycle utility learning framework under the IEEE P1900.4 architecture.

Figure 6 is the convergence verification for the proposed interactive protocol. Each sub-figure in Figure 6 contains three lines, which represent the capacity limit, the link price and the achieving throughput. We can see that the presented protocol can well guarantee robustness after limited iterations. Meanwhile, efficient price scheme can help operator to always achieve the capacity-limit throughput. The feasibility of the presented protocol is verified.

Figure 7 depicts the subscriber density of our scheme and the RSS-based RAN selection approach. From Figure 7 we can find that our PRICE-based scheme can effectively adjust the churning behavior to efficiently balance the traffic load. For example, since the operator observes that WLAN may become congested, but at the same time the 3G and the WiMAX are with a relatively small service, our PRICE-based scheme can avoid over-loading by rational churning behavior. But the RSS-based approach can not do this; it may lead the WLAN network to be paralyzed.

Figure 7. Rational churning behavior, and PRICE-based proposed scheme can well prevent specific RAN form over-loading.

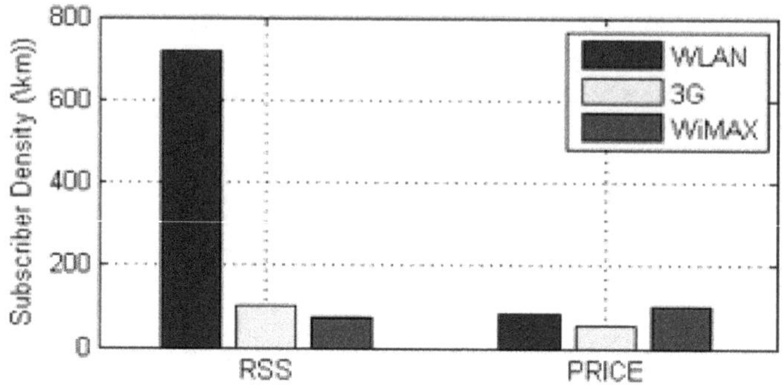

14.5 FUTURE RESEARCH DIRECTIONS

First, more attention should be paid to IEEE 1900.4 protocol, and further, more research interest and practical protocols should be raised in the academia and the industry. Specifically, extending the types of resource definitions from the heterogeneous networks perspective, and moreover, finding the possible uniformed resource management and scheme for every kinds of resource may be urgently needed. Second, we may use the game theory and pricing scheme to design more practical interaction protocol between the entity of TRM and NRM, and meanwhile, employ the utility function theory to investigate the optimal radio access network selection problem. Finally, we may also develop the local and global optimal gaming algorithm for TRM and NRM, respectively.

14.6 CONCLUSION

This chapter summarizes the current development of heterogeneous wireless access network, especially the unprecedented resource management and optimization problems during the development trend of the composite wireless network. Furthermore, we summarize the main blocks denoted in the architecture and functional framework of IEEE P1900.4. Then a novel double cognition cycle (DCC) is proposed in this chapter, which is based on the celebrated network utility maximization framework and the corresponding decomposition techniques for designing the distributed spectrum sharing approaches. Moreover, the two utility function based cognition cycles well capture the challenges of the resource optimization for the terminal (NRM) and the RAN (NRM), respectively. Meanwhile, a novel interactive protocol between the NRM and TRM is presented. Finally, using a simple composite wireless network scenario, we confirm the performance improvement of our proposed resource management approach and the feasibility of the presented protocol.

ACKNOWLEDGMENT

This work was supported in part by the National Science Fund for Distinguished Young Scholars under Grant no. 60725105, by the National Basic Research Program of China under Grant no. 2009CB320404, by the Program for Changjiang Scholars and Innovative Research Team in

University, by the 863 projects under Grant no. 2007AA01Z288, by the National Nature Science Foundation of China under Grant umbers 60572146, 60902032, 60902033, by the Research Fund for the Doctoral Program of Higher Education under Grant no. 20050701007, by the Key Project of Chinese Ministry of Education under Grant no. 107103, by ISN02080001, 2009ZX03007-004, 2010ZX03003-001, by the Natural Science Basic Research Plan in Shaanxi Province of China under Grant no. 2012JQ8012, by the Fundamental Research Funds for the Central Universities, and by the 111 Project under Grant no. B08038.

REFERENCES

Akyildiz, I. F., Lee, W.-Y., Vuran, M. C., & Mohanty, S. (2006). Next generation/dynamic spectrum access/cognitive radio wireless networks: A survey. *Computer Networks*, *50*(13), 2127–2159. doi:10.1016/j.comnet.2006.05.001

Buljore, S., Harada, H., & Filin, S. (2009). Architecture and enablers for optimized radio resource usage in heterogeneous wireless access networks: IEEE 1900.4 working group. *IEEE Communications Magazine*, *47*(1), 22–29. doi:10.1109/MCOM.2009.4752689

Buljore, S., Merat, V., & Harada, H. (2008). IEEE P1900.4 system overview on architecture and enablers for optimized radio and spectrum resource usage. *3rd IEEE Symposium on New Frontiers in Dynamic Spectrum Access Networks*, (pp. 253-260).

Chiang, M., & Low, S. H. (2007). Layering as optimization decomposition: a mathematical theory of network architectures. *Proceedings of the IEEE*, *95*(1), 255–312. doi:10.1109/JPROC.2006.887322

Filin, S., Ishizu, K., & Murakami, H. (2008). Dynamic spectrum assignment and access scenarios, system architecture, functional architecture and procedures for IEEE P1900.4 management system. *3rd International Conference on CrownCom 2008,* (pp. 1-7).

Haykin, S. (2005). Cognitive radio: Brain-empowered wireless communications. *IEEE Journal on Selected Areas in Communications*, *23*(2), 201–220. doi:10.1109/JSAC.2004.839380

IEEE. 1900.4 WG. (2011). *IEEE standard for architectural building blocks enabling network-device distributed decision making for optimized radio resource usage in heterogeneous wireless access networks.* Retrieved from http://ieeexplore.ieee.org.

Ishizu, K., Murakami, H., & Miyamoto, G. (2008). Design and implementation of cognitive wireless network based on IEEE P1900.4. *5th Annual IEEE Communications Society Conference on Sensor, Mesh and Ad Hoc Communications and Networks Workshops*, (pp. 1-8).

Li, J. D., & Yang, C. G. (2009). Pricing based power control game for cognitive radio networks. *International Conference on Telecommunications, ICT 2009*, (pp. 270-275).

MacKenzie, A. B., & Dasilva, L. A. (2005). *Game theory for wireless engineers* (pp. 1–86). Synthesis Lectures on Communications.

Marojevic, V., Salazar, J., Revés, X., & Gelonch, A. (2008). Resource modeling for a joint resource management in cognitive radio. *IEEE ICC 2008*, (pp. 4175-4180).

Niyato, D., & Hossain, E. (2009). Dynamics of network selection in heterogeneous wireless networks: An evolutionary game approach. *IEEE Transactions on Vehicular Technology*, *58*(4), 2008–2017. doi:10.1109/TVT.2008.2004588

Yang, C. G., & Li, J. D. (2009). *Joint resource management framework under IEEE P1900.4: A network utility maximization approach*. Poster 8, MobiCom 2009.

Yang, C. G., & Li, J. D. (2010). Dynamic resource management mechanism design for heterogeneous networks under IEEE1900.4 framework: A network utility maximization perspective. *2010 IEEE WCNIS*, (pp. 333–337).

Yang, C. G., Li, J. D., & Chen, D. (2009). *Optimal balancing between efficiency and fairness for resource management in cognitive radio networks: A dynamic game-theoretic approach*. CCNC 2009, Las Vegas, USA.

Yang, C. G., Li, J. D., & Tian, Z. (2010). Power control for cognitive radio networks with coupled constraints: A cooperative game-theoretic perspective. *IEEE Transactions on Vehicular Technology*, *59*(4), 1696–1706. doi:10.1109/TVT.2009.2039502

Chapter 15
Channel State Prediction in Cognitive Radio

Zhe Chen
Northeastern University, China

ABSTRACT

Spectrum sensing is the cornerstone of cognitive radio, which detects the availability of a spectrum band for the current time. In theory, the result of spectrum sensing reflects the current channel state, which is the ideal case. However, according to the author's measurements, hardware platforms can introduce a non-negligible time delay on the signal path, which undermines the accuracy of spectrum sensing. To reduce the negative impact of the hardware platform time delay, channel state prediction in cognitive radio is proposed and presented in this chapter. As examples, channel state prediction algorithms based on a modified hidden Markov model (HMM) are given and tested using recorded real-world data. Moreover, as a second stage, cooperative channel state prediction is also proposed and experimentally evaluated. The experimental results approve that channel state prediction in cognitive radio indeed helps improve the accuracy of spectrum sensing in practical cases.

15.1 RESPONSE DELAYS IN HARDWARE PLATFORMS

As we know, a number of spectrum sensing techniques for cognitive radio have been proposed. In addition to tremendous efforts on theoretical investigation, work on hardware implementation of spectrum sensing has been reported as well in (Chen, Guo, and Qiu, 2010; Mian, Zhou, Li, Hong, and Wu, 2009; Oh et al., 2009; Tkachenko, Cabric, and Brodersen, 2006). Experience gained in developing software defined radio (SDR) can benefit cognitive radio work, and existing SDR hardware platforms can be extended for developing cognitive radio transceivers. Implementing ef-

DOI: 10.4018/978-1-4666-2005-6.ch015

fective spectrum sensing schemes is a fundamental part of the development effort toward a cognitive radio network testbed (Qiu et al., 2009; Qiu et al., 2010). It is worth noting that measurement can be critical in guiding implementation work and verifying algorithm performance. In implementing the algorithms in hardware platforms, we have found, however, time delay introduced by hardware platforms becomes non-negligible, though in theoretical investigations such a time delay is usually ignored. Accurately quantifying this delay is necessary since it is wise to take into account the measured results in algorithm design and implementation.

There have been some hardware platforms that can be used for cognitive radio, such as universal software radio peripheral (USRP), universal software radio peripheral 2 (USRP2) (Ettus Research LLC, 2010), SFF SDR DP (Lyrtech Incorporated, 2010), wireless open-access research platform (WARP) (Rice University WARP Project, 2010), and Sora (Tan et al., 2009). There exist a few reports on discussions of the time delay of the USRP platform. In (Nychis, Hottelier, Yang, Seshan, and Steenkiste, 2009), the time delay between the GNU Radio (GNU Radio, 2010) and the field programmable gate array (FPGA) is reported. The value ranges from 289 μs to 9 ms. In (Valentin, von Malm, and Karl, 2006), an elapsed time is measured, where the elapsed time is the length from the start time of sending out a data link control (DLC) frame to the instance of completely receiving an acknowledgment DLC frame of the same size. The average of the time is 3.14 ms. In (Schmid, Sekkat, and Srivastava, 2007), the measured receive latency ranges from 1 ms to 30 ms, and the transmission latency ranges from 28.9 ms to 36.9 ms.

One of the reasons that the measured time delays of USRP differ greatly can be that different portions of the time delay are measured. We have not seen any reported measured time delays of USRP2 and SFF SDR DP.

In this section, the problem of the response delay is formulated and the minimum response delays of USRP2 and SFF SDR DP are measured.

15.1.1 Problem Formulation

A scenario of spectrum sensing is described as below. The time slot and the response delay are introduced.

15.1.1.1 Time Slot

Consider a scenario as shown in Figure 1, where a secondary user (SU) communicates with another SU or a secondary base station (SBS) through wireless channels, and both the uplink channel and the downlink channel are comprised of a sequence of time slots. Assume time slot level synchronization is performed and the length of the time slots is constant. Each time slot contains two phases: the spectrum sensing phase (the first phase) and the communication phase (the second phase). When in the spectrum sensing phase of a time slot, a potential SU sender or SBS sender senses the availability of the channel, it may start data transmission in the following communication phase if the sensed channel state is "idle."

In order to verify the sensed channel states, the actual channel states are required. The question is how SUs can get the actual channel states without the aid of primary users (PUs). Consider the following two cases: 1) The sensed channel state is "idle," so the SU sender can either (1a) send data in the next communication phase or (1b) hold on data transmission in the next communication phase. 2) The sensed channel state is "busy," so the SU sender holds on data transmission in the next communication phase. In (Hoang, Liang, Wong, Zhang, and Zeng, 2008), acknowledgment (ACK) and negative acknowledgment (NAK) messages are employed to indicate whether a transmission is successful or not. The same idea is borrowed here for case-1a verification. As shown in Figure 1, ACK or NAK mes-

Figure 1. Time slots

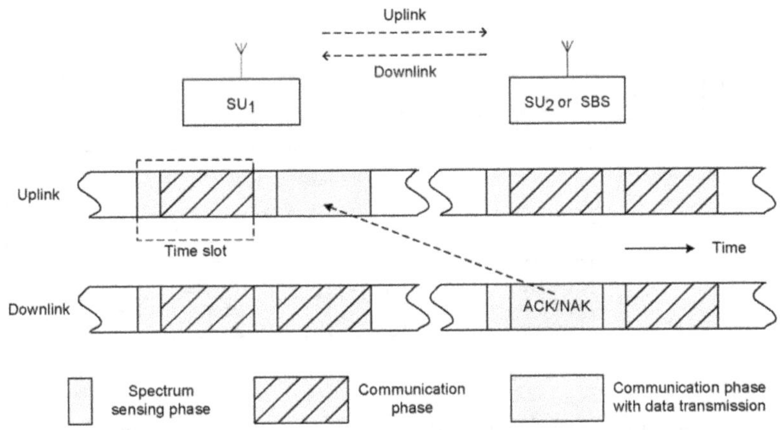

sages are sent from the targeted receiver to the sender together with other kinds of data in the communication phase. Take an uplink data transmission as an example, if an SU sender receives an ACK (or NAK) message from the targeted SU receiver, it means that the uplink data transmitted in the particular time slot have been received successfully (or unsuccessfully), which confirms that the uplink in that time slot is "idle" (or "busy"). Note that the verification process introduces a time delay caused by both propagation and signal processing. On the other hand, for case-1b and case-2 verification, it is assumed that the SU sender is able to verify the channel state by itself using the signals received in the following communication phase. In addition, it is assumed that there is a way for communications among SUs, so that spectrum sensing results from individual SUs may be shared and further cooperative decisions may be made.

15.1.1.2 Response Delay

To explain the response delay, a simplified data path of SU is shown in Figure 2. Received radio frequency (RF) signals are filtered, amplified, down-converted and digitized by an analog-to-digital converter (ADC). Then the digital signals are fed to a data processing module through a digital data interface, which ends the reception process. The transmission process starts with sending the digital data from the data processing module and ends at the transmitting antenna, through a data interface, a digital-to-analog converter (DAC) and an up-conversion module. Note that the down-conversion module and the up-conversion module are optional. For instance, in (Geirhofer, Tong, and Sadler, 2007), down-conversion is employed, whereas in (Qiu et al., 2009), wideband spectrum is measured directly without down-conversion. Also note that data interfaces and buffers on the data path may introduce delay.

The *response delay* is a delay as a cognitive radio device receives a signal from a channel over the air for spectrum sensing, then it transmits data to the air using the channel if the channel is believed to be available. This delay is contributed by three sections partitioned by test points A, B, C, and D (referring to Figure 2 and Figure 3). The corresponding delays are denoted by t_{rl}, t_{pl}, and t_{tl}, respectively. Then the total time delay t_l or response delay is:

$$t_l = t_{rl} + t_{pl} + t_{tl} \tag{1}$$

The response delay can vary depending on how intensive the data processing is. The *minimum*

Figure 2. Simplified data path of SU

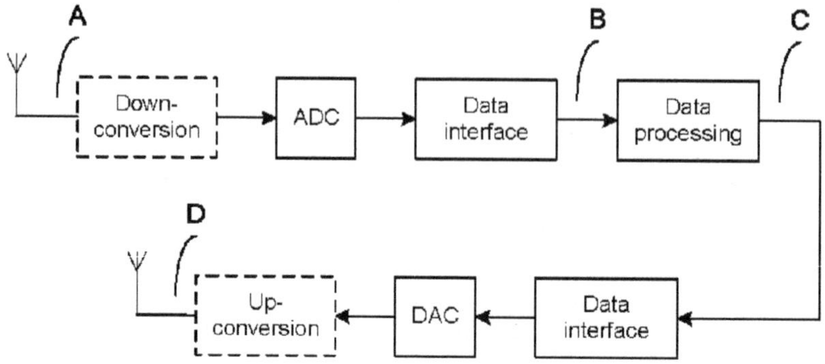

response delay refers to the response delay with minimum processing work, i.e., loopback (just passing through data without actual processing), in the data processing module.

If the response delay t_l is comparable with or greater than the length of a time slot, it cannot be ignored. Figure 3 shows a special case in which the response delay equals the length of a time slot. In this case, if an SU detects an "idle" channel state during the spectrum sensing phase of time slot 0, it may start a data transmission immediately. However, the actual data transmission over the air happens during the communication phase of time slot 1. Since the channel state may change in one time slot, the data transmission of the SU can interfere with PUs during time slot 1. In order to minimize such a negative impact caused by the

Figure 3. Response delay

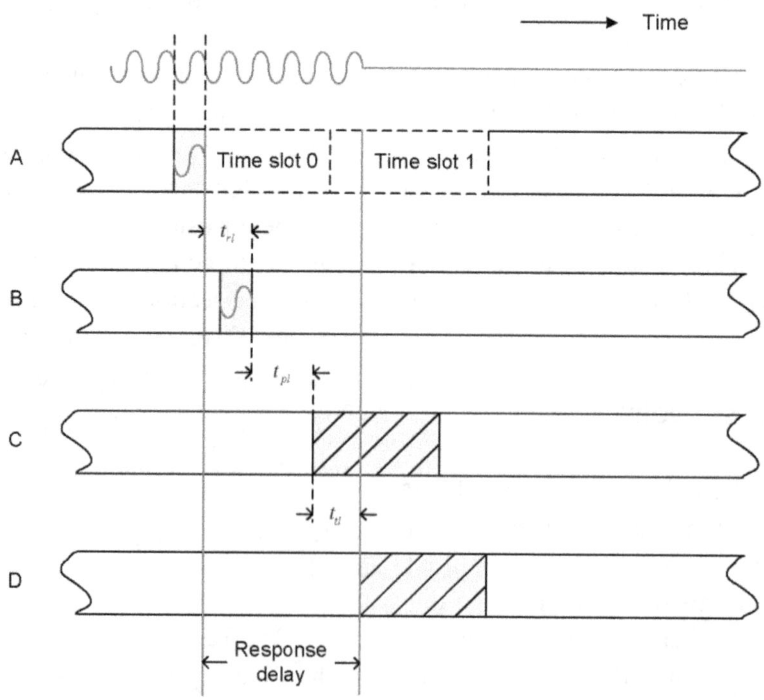

response delay, channel state prediction is proposed and presented in this chapter.

Response delay is inherent in many off-the-shelf hardware platforms for cognitive radio. Channel state prediction approaches can be employed in cognitive radio to help improve the accuracy of spectrum sensing and minimize the negative impact of the response delay caused by hardware platforms. As an example, Section 2 presents an approach for channel state prediction to show the benefits of channel state prediction in cognitive radio.

15.1.2 Measurement of Minimum Response Delays

As mentioned above, hardware platforms for cognitive radio introduce unwanted time delays. The questions are, how long they are and what the minimum response delays are. These can be answered using the measurements reported as below.

Two candidates are chosen from the existing commercial off-the-shelf hardware platforms that may be used for cognitive radio. One is USRP2, which is based on the architecture of host-based processing. In this architecture, RF frontend and data conversion modules are co-located on external boards, and the converted digital data are transferred to a host computer for further processing. The other one is SFF SDR DP, which is based on the standalone architecture and is able to perform intensive signal processing jobs on its boards.

Basically, the process consists of feeding a signal to the receiving antenna and checking the time delay at the transmitting antenna. An arbitrary waveform generator (AWG), Tektronix AWG7122B, generates a sequence of gated sinusoidal waveforms of 250 MHz. The sinusoidal burst lasts 50 μs (for the measurement of USRP2) or 500 μs (for the measurement of SFF SDR DP, since a sinusoidal burst of 50 μs does not work well in this case), and the burst appears once every 2 seconds. The duty cycle of 2 seconds is sufficient since none of the minimum response delays exceeds 2 seconds.

The output of the AWG is connected with both the receiving antenna port of the target hardware platform and a digital phosphor oscilloscope (DPO), Tektronix DPO72004, through a power divider. The transmitting antenna port of the target hardware platform is also connected with the DPO. The DPO is employed to display the signals at both the receiving antenna port and the transmitting antenna port of the target hardware platform on the same screen, so that the response delay can be read out. The DPO supports a maximum bandwidth of 20 GHz and a maximum sampling rate of 50 GS/s. It has four channels and has the ability of recording 250 megasamples per channel. The AWG supports a maximum sampling rate of 12 GS/s. It has two channels and is able to store 64 megasamples for each channel.

USRP2 consists of a motherboard and one or more selectable RF daughterboards (Ettus Research LLC, 2010). The major computing power on the motherboard comes from a Xilinx Spartan-3 XC3S2000 FPGA. The motherboard is also equipped with a 100 MS/s 14-bit dual-channel ADC, a 400 MS/s 16-bit dual-channel DAC, and a Gigabit Ethernet port for connecting to a host computer. Among the RF daughterboards available for USRP2, The one called WBX that covers a wide frequency range of 50 MHz to 2.2 GHz, with a nominal noise figure of 5-7 dB, is employed. When measuring the minimum response delay of USRP2, a USRP2 motherboard with the WBX RF daughterboard is connected directly to a host laptop computer via the Gigabit Ethernet. The GNU Radio companion (GRC) runs on the host computer with the Ubuntu Linux operating system using the default configuration. Using the GRC, a USRP2 source block and a USRP2 sink block are connected directly to form a loopback configuration.

Five hundred consecutive readouts of the minimum response delay of USRP2 are recorded. Figure 4 shows the distribution of the measured

Figure 4. Distribution of the measured minimum response delay of USRP2

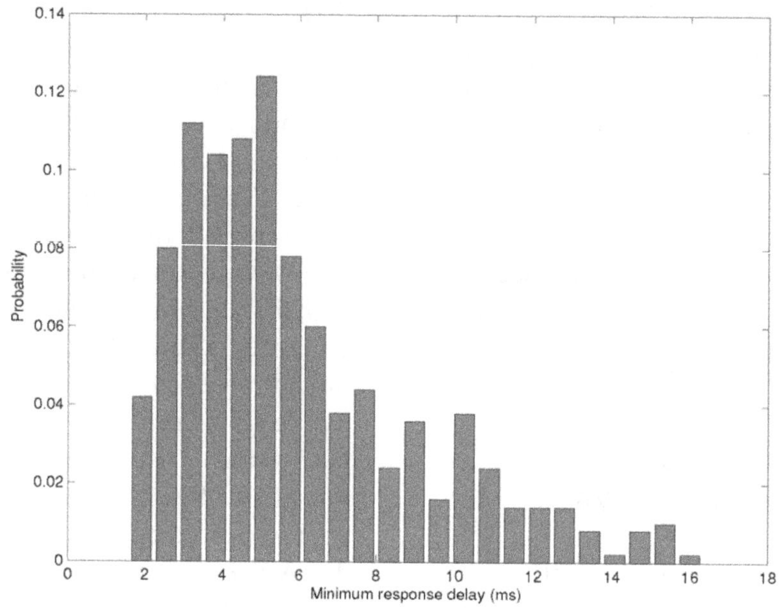

minimum response delays that spread from about 2 ms to about 16 ms. Random minimum response delays of USRP2 are observed, which is not surprising since both the Ethernet and the computer operating system can introduce randomness.

SFF SDR DP consists of three separate boards: a digital processing module, a data conversion module, and an RF module (Lyrtech Incorporated, 2010). The digital processing module is designed based on the TMS320DM6446 system-on-chip (SoC) and the Xilinx Virtex-4 SX35 FPGA. The TMS320DM6446 SoC has a C64x+ digital signal processor (DSP) core running at 594 MHz together with an advanced RISC machine 9 (ARM9) core running at 297 MHz. The digital processing module also comes with a 10/100 Mb/s Ethernet port. The data conversion module is equipped with a 125 MS/s 14-bit dual-channel ADC, a 500 MS/s 16-bit dual-channel DAC, as well as a Xilinx Virtex-4 LX25 FPGA. The low-band tunable RF module employed in this measurement can be configured to have either a 5 MHz or a 20 MHz bandwidth with working frequencies of 200-1050 MHz for the transmitter and 200-1000 MHz for the receiver. The nominal noise figure of this RF module is 5 dB. Other spectrum bands may be covered by several other RF modules. In the measurement of the minimum response delay for SFF SDR DP, an example project called *SFF_SDR_RF_Loopback_ADACIII* coming with the Lyrtech software package runs on both the FPGA and the DSP on the digital processing module. The function *TASK_Transmit* for the DSP in the example project is slightly modified to simply loopback all the received data.

One hundred consecutive readouts of the minimum response delay of SFF SDR DP are recorded. All the readout values are consistently around 48 ms. Although it is somewhat surprising to observe that the minimum response delay of SFF SDR DP is larger than that of USRP2, its constant minimum time delay is a desired feature for cognitive radio system design, since a constant time delay is easier to deal with. A larger minimum response delay may be caused by the interfaces and data buffers on the data path.

From these measurements one can see that the minimum response delays can be up to tens of milliseconds. The minimum response delays are measured without performing sophisticated baseband processing. In practice, an additional time delay will be added on the top of the minimum response delay. The total response delay and the uncertainty range have to be considered in cognitive radio system design. Note that the propagation delay can be ignored compared with the measured minimum response delay.

15.2 CHANNEL STATE PREDICTION USING MODIFIED HMM AND COOPERATIVE CHANNEL STATE PREDICTION

As measured in Section 1, the minimum response delays can be as large as tens of milliseconds. No matter how long the minimum response delay is, a time-slot-based channel state prediction approach can be applied to minimize the negative impact of response delay. For example, if the total response delay is 1 second and the length of the time slot is set to 500 ms (which means spectrum sensing is performed every 500 ms), then a 2-time-slot-ahead prediction can be applied. Or, if the total response delay is 40 μs and the length of the time slot is set to 20 μs, then the same 2-time-slot-ahead prediction can also be applied. In fact, channel state prediction approaches can be applied no matter how long the total response delay is.

In this section, traditional HMM is modified to take into account prediction, and a modified HMM based single-SU prediction is proposed and examined. When multiple SUs exist and their channel qualities are diverse, cooperative prediction can benefit the SUs as a whole. A prediction scheme with two stages is proposed and presented, where the first stage includes individual predictions conducted by all involved SUs, and the second stage further performs cooperative prediction using individual single-SU prediction results obtained at the first stage. In addition, a soft-combining decision rule for cooperative prediction is presented. In order to have convincing performance evaluation results, real-world Wi-Fi signals are used to test the presented approaches, where the Wi-Fi signals are simultaneously recorded at four different locations. Experimental results show that the presented single-SU prediction outperforms the 1-nearest neighbor (1-NN) prediction which uses a current detected state as an estimate of future states. Moreover, even with just a few SUs, cooperative prediction leads to overall performance improvement.

A spectrum sensing scheme uses received signals to detect channel states, and it predicts the channel states in the near future simply using previously detected channel states. Work on prediction for cognitive radio has been reported. In (Yarkan, and Arslan, 2007), channel occupancy status is converted into binary form, an autoregression (AR) model is used for predicting the binary channel status, and generated artificial GSM signals are used for testing the prediction. In (Su, and Wu, 2009), an autoregressive moving average (ARMA) model is employed to predict the power of television (TV) signals in time domain. In (Yu, He, and Quan, 2008), an algorithm based on support vector regression and empirical mode decomposition for RF spectrum prediction in frequency monitor system is introduced. In (Min, Qu, Cao, and Zhong, 2008), an interference time ratio that represents the fraction of a PU's burst interfered by secondary transmission is proposed to control the transmission probability for SUs, which is predicted using conditional probability. This scheme is tested only by simulation. The idea of predictive dynamic spectrum access is introduced in (Clancy, and Walker, 2006), which aims at the distribution of the time length that a channel is idle. The existence of the Markov chain for subband utilization by PU is validated in (Ghosh, Cordeiro, Agrawal, and Rao, 2009). HMM is used to predict the usage behavior of a spectrum band based on channel usage patterns in (Akbar, and Tranter, 2007), to decide whether or not to move to another spectrum band. And a channel status predictor using HMM based pattern recognition is proposed in (Park, Kim, Lim, and Song, 2007).

However, none of the previous work takes the time delay incurred by hardware platforms into consideration for prediction. Moreover, the ideas of using HMM for prediction in previous work are all based on pattern recognition. In fact, HMM can be exploited beyond pattern recognition. This section is summarized as follows. First, an approach for channel state prediction based on modified HMM is presented and examined. Second, a soft-combining decision rule for cooperative prediction using a group of individual prediction results is presented. Third, the performance of prediction is evaluated using the measured Wi-Fi signals both in the case of single-SU prediction and in the case of cooperative prediction among multiple SUs.

15.2.1 The Single-SU Prediction Approach

An HMM is defined by a tuple $\lambda = \{\boldsymbol{\pi}, \mathbf{A}, \mathbf{B}\}$ (Chen, and Qiu, 2010; Chen, Hu, and Qiu, 2009; Chen, 2003; Chen, Guo, Hu, and Qiu, 2011; Chen, 2011). $\boldsymbol{\pi}$ is the initial state probability vector,

$$\boldsymbol{\pi} = (\pi_1, \ldots, \pi_N) \tag{2}$$

$$\pi_i = \Pr(q_1 = \theta_i),\ i = 1, \ldots, N \tag{3}$$

where $\Pr(\bullet)$ denotes probability, N is the number of states of the Markov chain, $\{\theta_1, \ldots, \theta_N\}$ are the N states, q_t represents the state at time t, $q_t \in \{\theta_1, \ldots, \theta_N\}$. $\mathbf{A}$ is the state transition probability matrix,

$$\mathbf{A} = (a_{ij})_{N\times N} \tag{4}$$

$$a_{ij} = \Pr(q_{t+1} = \theta_j \mid q_t = \theta_i),\ i, j = 1, \ldots, N \tag{5}$$

and $\mathbf{B}$ is the emission probability matrix,

$$\mathbf{B} = (b_{jk})_{N\times M} \tag{6}$$

$$b_{jk} = \Pr(o_t = v_k \mid q_t = \theta_j) \equiv b_j(o_t),\ j = 1, \ldots, N,\ k = 1, \ldots, M \tag{7}$$

where M is the number of possible observation values in the observation space $\{v_1, \ldots, v_M\}$, o_t represents the observation value at time t, $o_t \in \{v_1, \ldots, v_M\}$.

Given a parameter tuple λ and a sequence of observation values $\mathbf{o} = \{o_1, \ldots, o_t, \ldots, o_T\}$, the state sequence that is most likely to have generated the input sequence $\mathbf{o}$ and the likelihood probability can be calculated using the Viterbi algorithm. Let $\delta_t(i)$ be the maximal probability of the state sequence of length t that ends in state i. A tailored Viterbi algorithm is shown as below.

Initialization.

$$\delta_1(i) = \pi_i b_i(o_1),\ i = 1, \ldots, N \tag{8}$$

Iteration.

$$\delta_t(j) = \max_{1\le i\le N}[\delta_{t-1}(i)a_{ij}]b_j(o_t),\ j = 1, \ldots, N,\ t = 2, \ldots, T \tag{9}$$

Termination.

$$P^* = \max_{1\le i\le N}[\delta_T(i)] \tag{10}$$

$$q_T^* = \arg\max_{1\le i\le N}[\delta_T(i)] \tag{11}$$

where P^* is the calculated likelihood probability and q_T^* is the estimated state at time T.

For spectrum sensing, the concern is the availability of some RF spectrum subbands of interest within a wider spectrum band. An architecture is given to predict such availabilities, as shown

Figure 5. The architecture for channel state prediction based on modified HMM

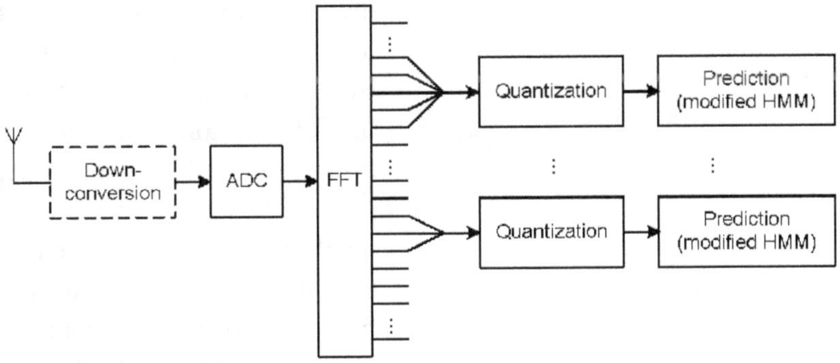

in Figure 5. In the spectrum sensing phase of every time slot, received time-domain signals are transformed into frequency-domain using fast Fourier transform (FFT). Then values from multiple frequency tones within a spectrum band of interest are quantified and fed into a modified HMM as a sequence of observation values. Denote the number of input frequency tones as Q. The quantization can be either scalar quantization or vector quantization. Multiple spectrum bands of interest can be processed in parallel using multiple modified HMMs. Thus, this approach can be applied to any wideband scenario.

For a spectrum band, at the end of the spectrum sensing phase of every time slot, an observation value o_t is obtained for a modified HMM. In each time slot, a spectrum band is associated with a certain channel state, i.e., "busy" or "idle." It is difficult to obtain the actual channel states in practice. As mentioned in Section 1.1, the state verification means can provide information about actual channel states. In this section, actual channel states are assumed to be determined by such verification means. The maximal verification delay is denoted by Y in time slots.

Channel state prediction is to use known observation values and actual channel states to estimate future channel states. However, in this approach, the definition of HMM is slightly different from the standard form. The modified HMM is defined by (2), (3), (4), (6), as well as the following two equations:

$$a_{ij} = \Pr(q_{t+(X+Y)} = \theta_j \mid q_t = \theta_i), \quad i, j = 1, \ldots, N \tag{12}$$

$$b_{jk} = \Pr(o_t = v_k \mid q_{t+X} = \theta_j), \; j = 1, \ldots, N, \quad k = 1, \ldots, M \tag{13}$$

where X is the span of prediction, in time slots, to take care of the maximal possible response delay.

Traditionally, the parameters of HMM are trained using a training algorithm like the popular Baum-Welch algorithm, given a sequence of observation values. However, in this section, a training algorithm for HMM is not employed. Instead, the parameters of HMM are obtained through a simple statistical process over training sequences.

Parameters of the modified HMM, $\{\pi, \mathbf{A}, \mathbf{B}\}$, are estimated statistically. The equations for extracting parameters of the modified HMM from a training sequence are listed as:

$$\pi_i = \frac{1}{L - X} \sum_{t=1}^{L-X} \mathrm{Eval}(q_t = \theta_i), \; i = 1, \ldots, N \tag{14}$$

$$a_{ij} = \frac{\sum_{t=1+Y}^{L-X} \mathrm{Eval}(q_{t+X} = \theta_j \mid q_{t-Y} = \theta_i)}{\sum_{t=1+Y}^{L-X} \mathrm{Eval}(q_{t-Y} = \theta_i)},$$
$$i, j = 1, \ldots, N \qquad (15)$$

$$b_{jk} = \frac{\sum_{t=1}^{L-X} \mathrm{Eval}(o_t = v_k \mid q_{t+X} = \theta_j)}{\sum_{t=1}^{L-X} \mathrm{Eval}(q_{t+X} = \theta_j)},$$
$$j = 1, \ldots, N,\ k = 1, \ldots, M \qquad (16)$$

where L is the length of the training sequence, in time slots, and

$$\mathrm{Eval}(\bullet) = \begin{cases} 1 & (\bullet)\text{established} \\ 0 & \text{otherwise} \end{cases} \qquad (17)$$

Then a one-step prediction for the channel state X-slot ahead can be performed based on the trained parameters $\{\pi, \mathbf{A}, \mathbf{B}\}$. There are two methods for the proposed one-step prediction.

One method named "π **B**" uses π and **B** for the prediction. The "π **B**" method is defined by (8), (10), and (11), with $T = 1$. The other method named "**AB**" uses **A** and **B** for the prediction, which is defined by (18), (10), and (11), with $T = 1$.

$$\delta_1(i) = a_{ji} b_i(o_1),\ q_{1-Y} = \theta_j,\ i = 1, \ldots, N \qquad (18)$$

Using either method, predicted channel state q_T^* in (11) and the corresponding likelihood probabilities P^* in (10) can be calculated.

15.2.2 Cooperative Channel State Prediction

Alternatively, cooperative prediction may be considered as an enhancement, if there are multiple SUs and they suffer different channel impairments, assuming that some sort of communication mechanism is available, so that all participant SUs can reliably talk with each other and share the information. Multiple SUs at different locations can combine their independent single-SU prediction results to achieve a joint prediction result. Cooperative prediction is expected to be effective especially when individual performance from each SU is significantly different. Of course, it imposes additional delays and causes increased complexity.

15.2.2.1 *m-out-of-n* Rule: Hard Combining

The $m - out - of - n$ rule ($m \le n$) is naturally suitable for the hard combining of multiple decisions, assuming there are n SUs predicting the channel states cooperatively. Each SU provides one-bit information d_i to reflect a future channel state, with $d_i = 1$ for "busy" state and $d_i = 0$ for "idle" state. Then the $m - out - of - n$ rule can be interpreted by:

$$\begin{cases} \sum_{i=1}^{n} d_i < m & \text{"idle"} \\ otherwise & \text{"busy"} \end{cases} \qquad (19)$$

Specially, when $m = 1$, the $m - out - of - n$ rule becomes the "OR" rule; and when $m = n$, the $m - out - of - n$ rule becomes the "AND" rule. In general, the prediction performance can be adjusted by changing the threshold m.

15.2.2.2 The Cooperative Prediction Rule: Soft Combining

An important feature of the presented prediction approach is that it outputs not only predicted channel states but likelihood probabilities for each possible channel state as well. This feature enables a soft combining of individual prediction outputs from multiple SUs. Associate with ith SU, denote the likelihood probability of "idle" state as P_{0i}, and the likelihood probability of "busy" state as P_{1i}, where the likelihood probabilities are from (10). Then the soft combining decision rule can be expressed as:

$$\begin{cases} \sum_{i=1}^{n} \frac{P_{0i} - P_{1i}}{P_{0i} + P_{1i}} > 0 & \text{"idle"} \\ otherwise & \text{"busy"} \end{cases} \tag{20}$$

This hypothesis compares two normalized likelihood probability summations $\sum_{i=1}^{n} \frac{P_{0i}}{P_{0i} + P_{1i}}$ and $\sum_{i=1}^{n} \frac{P_{1i}}{P_{0i} + P_{1i}}$, and these two summations can be viewed as approximations of accumulated probability of "idle" state and accumulated probability of "busy" state.

15.2.3 Performance Validation

In order to evaluate the performance of the proposed channel state prediction approaches using real-world data, Wi-Fi signals are measured and recorded. There are several reasons to consider Wi-Fi as PUs in evaluating channel state prediction approaches. First, the spectrum bands that Wi-Fi employs are unlicensed, which means experiments on these bands can be conducted without asking the regulators for permission. Second, the durations that Wi-Fi devices occupy the channel and the durations that the channel is kept idle are as small as microseconds. This fact enables recording plenty of Wi-Fi accesses in a short time. Third, the durations and intervals of Wi-Fi accesses are random, which poses an additional challenge for channel state prediction. It is hard to learn and predict Wi-Fi accesses. Thus, Wi-Fi signals are ideal for evaluating prediction approaches.

Wi-Fi signals have been measured using four antennas at four different locations. The experiment is conducted in an indoor environment. A laptop computer accesses the Internet through a wireless Wi-Fi router, downloading data at a data rate of 2.3 MB/s. In order to record the Wi-Fi signals at different locations simultaneously, an advanced DPO, Tektronix DPO72004, is connected with four antennas with a frequency range of 800 MHz to 2500 MHz distributed at four locations. Antennas 1, 2 and 3 are three, two, and two meters away from the router, respectively, but a metallic board is placed between antenna 3 and the router to emulate a non-line-of-sight (NLOS) propagation. Antenna 4 is placed closely to the router, so it can monitor the actual channel states. Since the DPO can record 250 megasamples per channel, by setting the sampling rate to 6.25 GS/s, the maximum duration of one measurement is 40 ms.

The sampling interval is 160 ps and 40 ms Wi-Fi signals are recorded at four different locations corresponding to channel 1, 2, 3, and 4. The measured Wi-Fi signals from channel 1, 2, and 3 are fed to three independent SUs for prediction, whereas the measured Wi-Fi signal from channel 4 is served as an indicator of channel states and fed to all the three SUs for reference.

Due to the limitation of measurement equipment, 40 ms is the maximum duration that Wi-Fi signals can be recorded. By setting the length of the time slot to 20 μs, there are 2,000 time slots available for performance evaluation. It may take one or two time slots for the channel state to change, which can reflect actual channel state changes. The duration of the spectrum sensing phase of a time slot is set to 4 μs, just one fifth of the length of a time slot. In the following, prediction spans of one to three time slots are considered in evaluating the prediction performances.

The presented prediction approach is configured as follows. Referring to the architecture shown in Figure 5, scalar quantization is used, quantified frequency-domain data from one frequency tone of 2.418 GHz are served as observation values, and they are fed into the modified HMM. Unless otherwise stated, the parameters *M*, *N*, and *Q* for the single-SU prediction are set to 288, 2, and 1, respectively. The $\{\pi, \mathbf{A}, \mathbf{B}\}$ parameters of the three modified HMMs for three single SUs are obtained beforehand using (14) (15) (16) and the measured Wi-Fi signals from channel 1, 2, and 3, respectively.

For comparison purpose, another single-SU predictor called 1-NN is employed as reference. 1-NN simply uses current detected or sensed channel state as an estimate of future channel states:

$$q_{t+X} = q_t \tag{21}$$

where q_t is the channel state for current time slot *t*, and q_{t+X} is the predicted future channel state X-slot ahead. q_t is determined by the following hypothesis:

$$q_t = \begin{cases} \text{"busy"} & \text{if } o_t > th \\ \text{"idle"} & otherwise \end{cases} \tag{22}$$

where o_t is the non-quantified observation value from the current time slot *t*, and *th* is a threshold. In our evaluation, all SUs use the same default value of *th*, which is predetermined using all the non-quantified observation values from channel 3, since the weakest measured Wi-Fi signal comes from channel 3. *th* simply takes the middle between two averages, one is the average of non-quantified observation values from all "idle" slots and the other is the average of non-quantified observation values from all "busy" slots. The 1-NN approach seems simple, but it is difficult to surpass it in terms of prediction performance.

The prediction performance is evaluated using two metrics: P_D and P_{FA}. Similar to their meanings in the case of detection, in the case of prediction, P_D means the rate that a prediction approach predicts the channel state correctly when the actual channel state is "busy," whereas P_{FA} means the rate that it fails to predict the channel state correctly when the actual channel state is "idle." Obviously, a combination of higher P_D and lower P_{FA} stands for a better prediction performance.

Considerations of choosing between the "π **B**" method and the "**AB**" method for the presented prediction approach are summarized as follows. When *Y*= 0, using the "**AB**" method can achieve a higher prediction performance with the measured Wi-Fi signals. However, in a typical case of $Y > 0$, the performance of the "**AB**" method would be degraded. On the other hand, the "π **B**" method is independent of *Y*, but its performance is slightly lower. Thus, it is recommended to use the "π **B**" method for the proposed prediction approach when *Y*> 0, and this method is employed in the following performance evaluation.

15.2.3.1 Performance of Single-SU Prediction

Three independent single-SU predictors of the same type run both the presented prediction approach and the 1-NN approach. Each single-SU predictor uses one channel of the measured Wi-Fi signals as its input. Figure 6 shows the performance of channel 3. Overall speaking, at the cost of slight increase in complexity, the proposed single-SU prediction is robust concerning channel conditions and outperforms the 1-NN predictor. It is also confirmed that the performance degrades fast as the prediction span increases, suggesting that the response delay is non-negligible.

15.2.3.2 Performance of Cooperative Prediction

The predicted states for the same time slot, output from the three independent single-SU predictors running the 1-NN approach or the proposed pre-

Figure 6. Performance of single-SU prediction (channel 3)

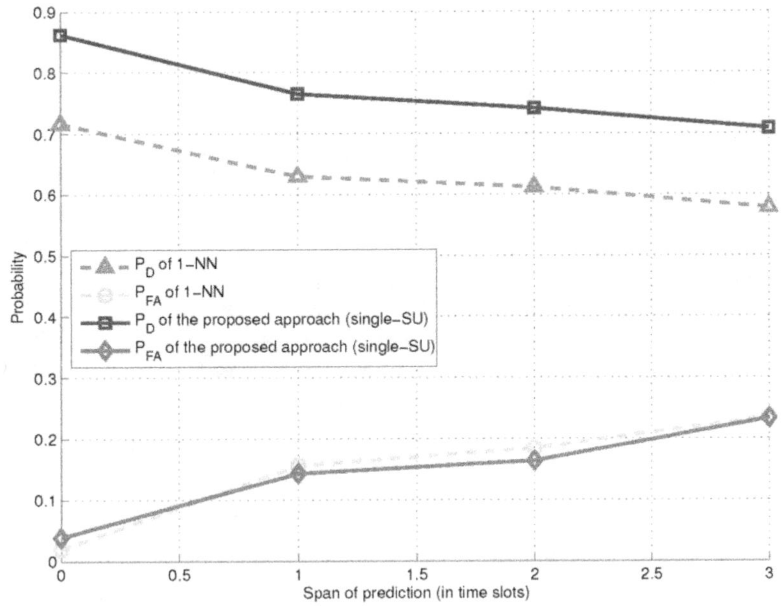

Figure 7. Performance of cooperative prediction using 1-NN prediction and 2-out-of-3 combining rule

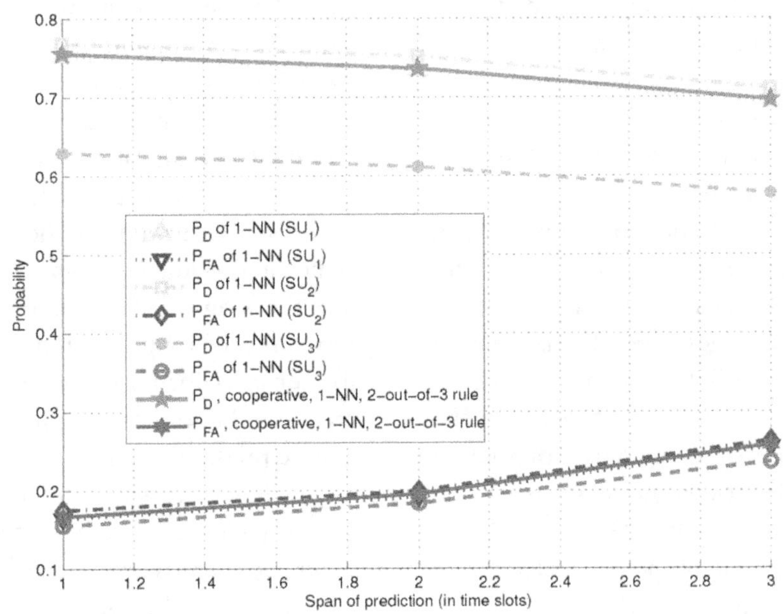

diction approach, can be combined to form a joint channel state prediction, using either the $m-out-of-n$ rule or the proposed soft-combining decision rule.

Figure 7 shows the performances of cooperative prediction using the $2-out-of-3$ rule. Since n=3 in this case, setting m to 2 yields a better overall performance. One can see that the

Figure 8. Performance of cooperative prediction using the proposed prediction approach and the proposed soft-combining decision rule

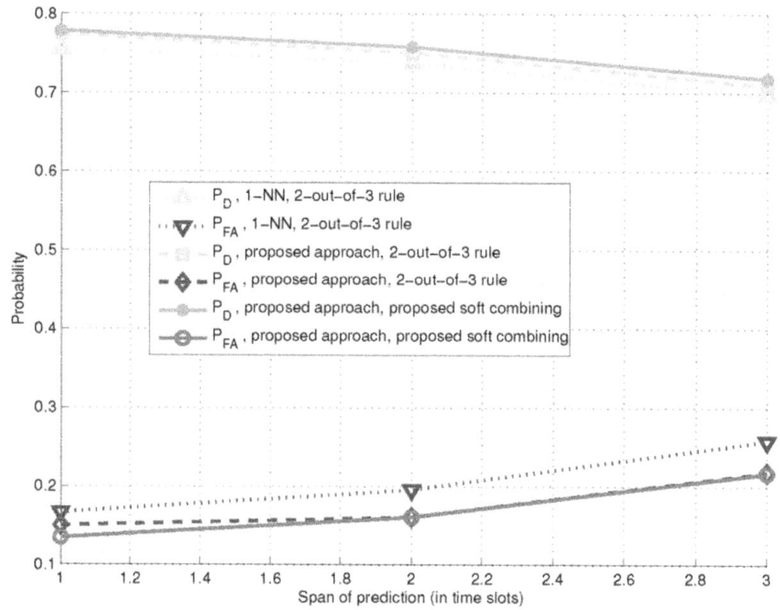

overall performance of cooperative prediction is better than that of most of the single-SU 1-NN predictors. The performance boost is not so significant, which may be due to limited number of SUs, and limited channel diversity in the SUs' propagation environments.

The performances of cooperative prediction using different combinations of approaches are depicted in Figure 8. It can be observed that the overall performance of the cooperative prediction using the proposed prediction approach is better than using the 1-NN approach. Moreover, the overall performance of cooperative prediction using the proposed soft-combining decision rule is slightly better than using the $2-out-of-3$ rule, which confirms the proposed decision rule is effective even for just a few SUs.

15.3 CONCLUSION

The minimum response delays of USRP2 and SFF SDR DP have been measured. The response delay has been taken into account in designing strategies for channel state prediction, and the strategies have been tested using Wi-Fi signals recorded at four locations simultaneously. An approach of single-SU prediction based on modified HMM and a soft-combining decision rule for cooperative prediction have been presented and evaluated. Evaluation results confirm that the single-SU prediction approach outperforms the 1-NN prediction approach, where the former asks for an insignificant complexity increase and does not need any thresholds. Cooperative prediction can be an alternative enhancement when there are a number of participant SUs and the channel information can be shared among them. Cooperative gain can be expected if there are a large number of SUs with diverse channel qualities. However, with just three SUs in the experiment, it can be seen that the cooperative prediction helps and the soft-combining decision rule outperforms the $m-out-of-n$ rule.

Many commercial systems have smaller response delays than those of the aforementioned two hardware platforms. What we would like to convey is that 1) the response delay exists and should not be ignored in cognitive radio, and 2) prediction has the potential to help reduce the negative impact of the response delay.

Although there are possibly many algorithms for prediction, the ones tested and reported are relatively easy to implement. The results based on measured data can guide our future work in developing the cognitive radio network testbed.

REFERENCES

Akbar, I. A., & Tranter, W. H. (2007). Dynamic spectrum allocation in cognitive radio using hidden Markov models: Poisson distributed case. *Proceedings of IEEE SoutheastCon*, (pp. 196-201).

Chen, Z. (2003). *A study and implementation of speech recognition system for mobile communication terminals*. Master's thesis, Hangzhou Dianzi University, China.

Chen, Z. (2011). *Cognitive radio network: Architecture, algorithms, and testbed.* Doctoral dissertation, Tennessee Technological University, USA.

Chen, Z., Guo, N., Hu, Z., & Qiu, R. C. (2011). Experimental validation of channel state prediction considering delays in practical cognitive radio. *IEEE Transactions on Vehicular Technology*, *60*(4), 1314–1325. doi:10.1109/TVT.2011.2116051

Chen, Z., Guo, N., & Qiu, R. C. (2010). Demonstration of real-time spectrum sensing for cognitive radio. *IEEE Communications Letters*, *14*(10), 915–917. doi:10.1109/LCOMM.2010.082610.100725

Chen, Z., Hu, Z., & Qiu, R. C. (2009). Quickest spectrum detection using hidden Markov model for cognitive radio. *Proceedings of IEEE Military Communications Conference*.

Chen, Z., & Qiu, R. C. (2010). Prediction of channel state for cognitive radio using higher-order hidden Markov model. *Proceedings of IEEE SoutheastCon*, (pp. 276-282).

Clancy, T., & Walker, B. (2006). Predictive dynamic spectrum access. *Proceedings of SDR Forum Conference*.

Ettus Research LLC. (2010). *Home page.* Retrieved October, 2010, from http://www.ettus.com

Geirhofer, S., Tong, L., & Sadler, B. M. (2007). Cognitive radios for dynamic spectrum access-dynamic spectrum access in the time domain: Modeling and exploiting white space. *IEEE Communications Magazine*, *45*(5), 66–72. doi:10.1109/MCOM.2007.358851

Ghosh, C., Cordeiro, C., Agrawal, D. P., & Rao, M. B. (2009). Markov chain existence and hidden Markov models in spectrum sensing. *Proceedings of IEEE International Conference on Pervasive Computing and Communications*.

GNU. (2010). *Radio*. Retrieved October, 2010, from http://www.gnuradio.org

Hoang, A. T., Liang, Y.-C., Wong, T. C., Zhang, R., & Zeng, Y. (2008). Opportunistic spectrum access for energy-constrained cognitive radios. *Proceedings of IEEE Vehicular Technology Conference*, (pp. 1559-1563).

Lyrtech Incorporated. (2010). *Home page.* Retrieved October, 2010, from http://www.lyrtech.com

Mian, O., Zhou, R., Li, X., Hong, S., & Wu, Z. (2009). A software-defined radio based cognitive radio demonstration over FM band. *Proceedings of International Conference on Wireless Communications and Mobile Computing*, (pp. 495-499).

Min, R., Qu, D., Cao, Y., & Zhong, G. (2008). Interference avoidance based on multi-step-ahead prediction for cognitive radio. *Proceedings of IEEE Singapore International Conference on Communication Systems*, (pp. 227-231).

Nychis, G., Hottelier, T., Yang, Z., Seshan, S., & Steenkiste, P. (2009). Enabling MAC protocol implementations on software-defined radios. *Proceedings of USENIX Symposium on Networked Systems Design and Implementation*, (pp. 91-105).

Oh, S. W., Le, T. P. C., Zhang, W., Ahmed, S. N. A., Zeng, Y., & Kua, K. J. M. (2009). TV white-space sensing prototype. *Wireless Communications and Mobile Computing*, *9*(11), 1543–1551. doi:10.1002/wcm.708

Park, C.-H., Kim, S.-W., Lim, S.-M., & Song, M.-S. (2007). HMM based channel status predictor for cognitive radio. *Proceedings of Asia-Pacific Microwave Conference*.

Qiu, R., Guo, N., Li, H., Wu, Z., Chakravarthy, V., Song, Y., & Chen, Z. (2009). A unified multi-functional dynamic spectrum access framework: Tutorial, theory and multi-GHz wideband testbed. *Sensors (Basel, Switzerland)*, *9*(8), 6530–6603. doi:10.3390/s90806530

Qiu, R. C., Chen, Z., Guo, N., Song, Y., Zhang, P., Li, H., & Lai, L. (2010). Towards a real-time cognitive radio network testbed: Architecture, hardware platform, and application to smart grid. *Proceedings of IEEE Workshop on Networking Technologies for Software-Defined Radio and White Space*.

Rice University. (2010). *WARP project*. Retrieved October, 2010, from http://warp.rice.edu

Schmid, T., Sekkat, O., & Srivastava, M. B. (2007). An experimental study of network performance impact of increased latency in software defined radios. *Proceedings of International Workshop on Wireless Network Testbeds, Experimental Evaluation and Characterization*, Montreal, (pp. 59-66).

Su, J., & Wu, W. (2009). Wireless spectrum prediction model based on time series analysis method. *Proceedings of ACM Workshop on Cognitive Radio Networks*, (pp. 61-66).

Tan, K., Zhang, J., Fang, J., Liu, H., Ye, Y., Wang, S., & Voelker, G. (2009). Sora: High performance software radio using general purpose multi-core processors. *Proceedings of USENIX Symposium on Networked Systems Design and Implementation*, (pp. 75-90).

Tkachenko, A., Cabric, D., & Brodersen, R. (2006). Cognitive radio experiments using reconfigurable BEE2. *Proceedings of Asilomar Conference on Signals, Systems, and Computers*, (pp. 2041-2045).

Valentin, S., von Malm, H., & Karl, H. (2006). *Evaluating the GNU software radio platform for wireless testbeds. Technical Report*. University of Paderborn.

Yarkan, S., & Arslan, H. (2007). Binary time series approach to spectrum prediction for cognitive radio. *Proceedings of IEEE 66th Vehicular Technology Conference*, (pp. 1563-1567).

Yu, C., He, Y., & Quan, T. (2008). Frequency spectrum prediction method based on EMD and SVR. *Proceedings of International Conference on Intelligent Systems Design and Applications*, (p. 3).

Chapter 16
Boosting Secondary-User Performance:
Challenges, Potential Solutions, and Expectations

Terry N. Guo
Tennessee Technological University, USA

ABSTRACT

This chapter addresses a few challenges and issues in developing Cognitive Radio Networks (CRNs), and provides unique solutions to enhance security at physical layer and to boost CRN computing power in a distributed manner. In this age of vast connectivity, network security becomes more and more prominent. In addition to the security means added at the upper layers, security can be further enhanced at physical layer. In particular, location based wideband channel characteristic as a unique signature can be utilized for security enhancement. Such a scheme is proposed and examined in different configurations. Lack of computing power is another critical issue, as CRN is expected to have more and more features. Instead of increasing onboard computing power, off-board computing resources can be connected to boost overall computing power. With increased computing power, the CRNs would be able to undertake computationally heavy tasks such as executing machine-learning algorithms and performing radio intrusion detection.

16.1 INTRODUCTION

Cognitive Radio Networks (CRNs) have been a hot topic for more than a decade since Joseph Mitola III described the concept in 1998 (Mitola III & Maguire Jr, 1999), and it has been catalyzed after enriching the concept by researchers (Mitola III & Maguire Jr, 1999; Haykin, 2005; Akyildiz, Lee, Vuran, & Mohanty, 2006; Ghasemi & Sousa, 2008; Steenkiste, Sicker, Minden, & Raychaudhuri, March 2009). The term CRN or simply CR is not just a concept any more. Experiments and

DOI: 10.4018/978-1-4666-2005-6.ch016

development work have been seen in the second half of the last decade (Pawelczak et al., 2011). In particular, the Federal Communications Commission (FCC) has adopted rules to allow unlicensed radio transmitters to operate in the broadcast television spectrum (TV white space) when that spectrum is not used by a licensed service, which sparks tremendous efforts in development of CRs in the TV white space (Nekovee, 2010). Despite explosive research and development as well as regulatory work throughout the world, commercial CR products have not appeared yet. Indeed, there is a long way to go before truly-cognitive radios are in front of average people. Advance in terms of performance, architecture and commercial development will continue. Seeing ten years of achievements, you may want to know what CR features people wish to see in the near future and what challenges are faced now by the researchers and developers.

This chapter focuses on just a few of many issues and concerns faced by the author in CRN related research and development. In next section some challenges and physical limitations are addressed, with emphasis on implementation aspect. Security concern and lack of computing power are among a list of critical issues. Following the next section, these two issues are addressed by proposing some approaches and showing evaluation results. Finally, the last section presents some expectations for certain features on future CRNs with enhanced computing power.

16.2 CHALLENGES AND PHYSICAL LIMITATIONS

Cognitive radios utilize and handle the spectrum in ways different from traditional radios do. Four basic functions, spectrum sensing, spectrum management, spectrum mobility and spectrum sharing, are required, which leads to many challenges in CRN research and development. In general, most of the challenges are related to frequency agility and computational power, and these two abilities are physically constrained.

Due to the nature of optimistic spectrum access, it is straightforward to consider multichannel option for the secondary users to maintain reliable connections. We have seen tremendous research effort regarding multichannel or multiband spectrum sensing and Media Access Control (MAC) protocols (Ganesan & Li, 2005; Zhang, Soong, & Xiao, 2007; Chong, Sung, & Sung, 2009). A frequency-agile physical layer is an essence to implement a multichannel version of dynamic spectrum access. It has to be pointed out that conventional transceiver design philosophy is no longer adequate in designing cognitive radio transceivers. Conventional radios are designed for fixed frequency band and known interferences, and typically preselect filters are used at the receiver input to notch unwanted frequency components. In contrast, without exactly knowing the signal frequency range, a cognitive radio receiver's Radio Frequency (RF) frontend has to have relatively wide bandwidth and be able to switch between frequency bands. The increased frequency bandwidth has to be coupled with increased linearity range in order to prevent the desired signal from being distorted. Programmable suppression of out-of-band interferences can be achieved at digital domain at the cost of additional resources for signal processing. At present processing bandwidth of 500 MHz to 1 GHz is very feasible, and the receiver can utilize large processing bandwidth in exchange for less frequency switching in analog circuits.

We are in the age of vast connectivity, and people especially enjoy freedom and mobility of wireless networks nowadays. Network security has become a prominent challenge. Compared to network security of wired networks, achieving wireless security is more difficult. This is because in radio communications signals are emitted in electromagnetic format, which is in favor of broadcasting fashion but harmful for security and privacy. In general confining a radio signal in a desired space is infeasible, thus wireless signal leakage is inevitable. Although signal leakage does not necessarily mean information leakage, reducing signal leakage can definitely enhance overall security. Encryption and authentication based security techniques have been proved effective (Xiao, Greenstein, Mandayam, & Trappe, 2008; Shiu, Chang, Wu, Huang, & Chen, 2011), but all of them need to consume additional radio resources. In addition to traditional security means implemented at the upper layers, there are some physical-layer techniques to strengthen security by reducing radio signal leakage. Joint Transmit-Receive (Tx-Rx) spatial-temporal processing is an ideal analytical framework that can explore wireless channel characteristics at physical layer. Techniques in joint Tx-Rx spatial-temporal processing include beamforming, multiple-input multiple-output (MIMO) precoding and Spatial Division Multiple Access (SDMA), etc. However, all of these mentioned techniques are at symbol level based on flat fading channel assumption. For frequency selective channels, Orthogonal Frequency-Division Multiplexing (OFDM) has been widely applied so each subchannel can be viewed as a flat fading channel. What has not been paid enough attention is true waveform level spatial-temporal processing by exploring wideband frequency selective channels.

There has been a trend that digital signal processing plays bigger and bigger role. Indeed, many functions can be effectively performed by using proper algorithms, provided sufficient computing resources are available. The intelligence of a true CRN relies heavily on the implementation of computational intelligence (Kulkarni, Forster, & Venayagamoorthy, 2011) and machine learning (Gao, Zhang, Liang, & Wang, 2010), etc. However, computation ability is very limited for most existing radio devices, and this is why we need to seek solutions to provide increased computing power.

Tasks for cognitive radios can be categorized into two groups: time critical tasks and delay-tolerable tasks. For example, spectrum sensing and collision avoidance must be performed timely, while updating knowledge database may be conducted in a less timely fashion. A parameter called response delay is defined as a total delay incurred in a secondary user transceiver (Chen, Guo, Hu, & Qiu, 2011). Let t_1 be the instant when the channel state alters from "busy" to "idle" (or vice versa), and t_2 be the instant when the secondary user alters its transmitter state (from "silent" to "transmit", or vice versa) as a response to the change of channel state. The response delay t_r is the interval between these two time instants: $t_r = t_2 - t_1$. The test points for t_1 and t_2 are at the secondary user's receiving antenna and transmitting antenna, respectively. This delay is a physical limit and has significant impact on spectrum sensing performance. Because of the response delay t_r, the spectrum sensing task is actually to predict the channel state at least t_r seconds ahead (Chen et al., 2011). Apparently, the larger response delay, the less prediction accuracy (or worse spectrum sensing performance). Shown in Figure 1 and Figure 2 are measurement results based on two commercial Software Defined Radio (SDR) products. The Small Form Factor Software Defined Radio (SFF SDR) development platform is a standalone platform and inserts a constant response delay, while the GNU radio platform needs a host computer for control and signal processing, thus it suffers from random response delay. Since a processing delay varies depending upon which algorithm is chosen, in these measurements the processing delays are intentionally not counted, so the response delays reflect the minimums.

Very likely, these physical limitations will not be directly reduced in a short term. Instead, system approaches should be adopted to design a CRN. For instance, some physical-layer techniques may be used to enhance CRN security, collaborative sensing can be considered to increase spectrum sensing performance, and off-board computing resources may join with on-board computing resources to handle computational burden.

Figure 1. Measured minimum response delay of small form factor software defined radio (SFF SDR) development platform

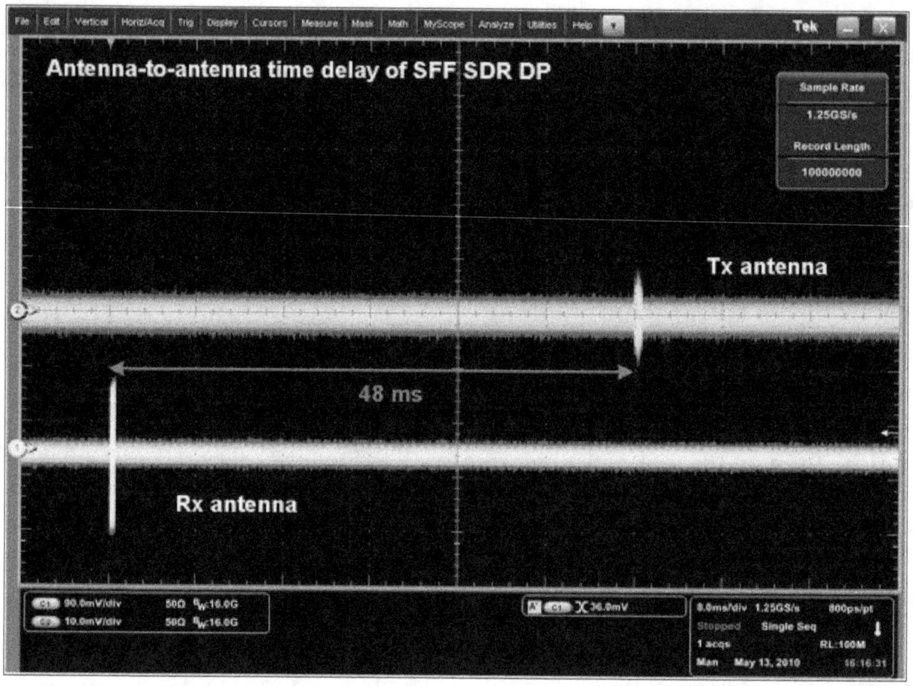

Figure 2. Distribution of measure minimum response delay of GNU radio USRP2 platform

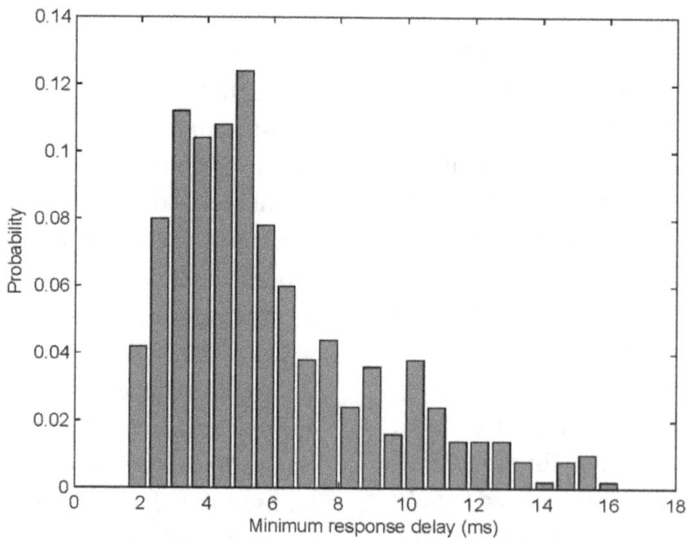

16.3 WIDEBAND TRANSMIT WAVEFROM DESIGN TO MINIMIZE INFORMATION LEAKAGE

Wideband communication systems can be benefited by spectral shaping via transmit waveform design, especially for wideband cognitive radios whose signal spectra have to be strictly controlled (Hu, Guo, & Qiu, 2010a, 2010b). With ability of programmable transmit waveforms and advanced spectrum sensing, the interferences generated by secondary users in the CRN can be well controlled. In addition to spectral shaping, pulse based (non-OFDM) wideband transmit waveform design as a physical-layer means may offer unique features like location based data delivery.

Physical-layer security techniques like those mentioned in previous section are complementary to the security means implemented at the upper layers. This section introduces a physical-layer security enhancement that is true waveform level spatial-temporal processing by exploring wideband frequency selective channels. In fact, the characteristic of a wideband frequency selective channel is location dependent, which can be utilized to offer certain security enhancement. One type of wideband channel characteristic is Channel Impulse Response (CIR) that contains rich multipath components due to electromagnetic scattering in the propagation space. The location dependent CIR is unique and can be used as a signature. Of course, the channel has to be quasi-static and known in prior by the transmitter, which is not an unrealistic assumption. The CIR can be estimated using many existing techniques.

To address wireless security problem, consider a simple scenario: a transmitter sending data to an intended receiver, and an unintended receiver or eavesdropper is around the intended receiver. If the channels between the transmitter and the two receivers are known, then theoretically there must be a best joint Tx-Rx strategy that is in favor of the intended receiver while minimizing the chance of information leakage to the eavesdropper. It is straightforward to put this problem in an information theory framework and use mutual information as a metric to jointly design a pair of transmit waveform and receive filter. Here we consider a transmitter-receiver pair with non-OFDM structure shown in Figure 3. In practice, controlling receiving signal strength at a given point might be easier than dealing with mutual information. The ultimate goal is to reduce signal leakage and increase the receiver bit error rate (BER) at the eavesdropper while maintaining an acceptable BER at the intended receiver. When BER reaches 20% or higher, it is difficult to recover the original transmitted data. Signal to Interference plus Noise Ratio (SINR) at receiver, and channel gain, are practical and easy-to-use metrics for this approach.

Among different transmit waveforms is the time reversal waveform that is simple yet effective (Qiu, Zhou, Guo, & Zhang, 2006; Guo, Zhang, Qiu, & Mo, 2007; Guo, Sadler, & Qiu, 2007; Zhou, Guo, & Qiu, 2009). Here reversal means a time reversed version of the CIR. Illustrated in Figure 4 are a CIR waveform (magnitude only) and its time reversed version. In (Guo, Qiu, & Sadler, 2005) a concept of location based security is proposed and evaluated using measured indoor Ultra-Wideband (UWB) channel data, where time reversal with binary differential modulation and autocorrelation receivers are utilized, and security enhancement is demonstrated in terms of a huge Bit Error Rate (BER) difference for the two receiver locations. Note that without perfect synchronization it is even more difficult for the eavesdropper to demodulate data correctly. The explanation is that, because the transmit waveform is designed based on the channel associated to the intended receiver, this receiver sees much higher SINR than the eavesdropper. This idea is further verified recently using a real-time test-bed with Quadrature Phase Shift Keying (QPSK) modulation, and the result is shown in Figure 5, where perfect synchronization has been assumed for the eavesdropper in

Figure 3. Analytical system model with joint Tx-Rx optimization

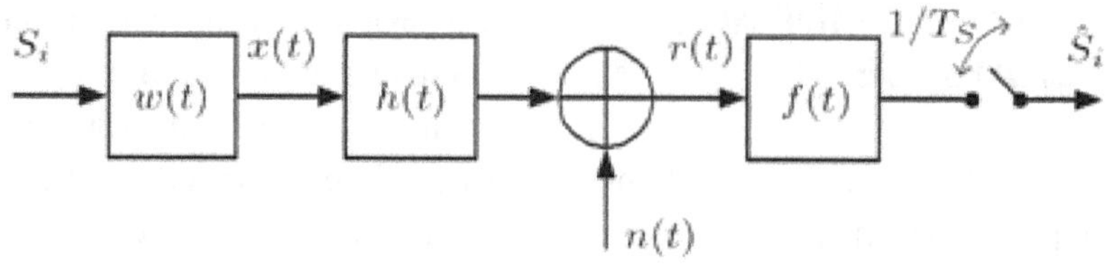

Figure 4. CIR and its time reversed version

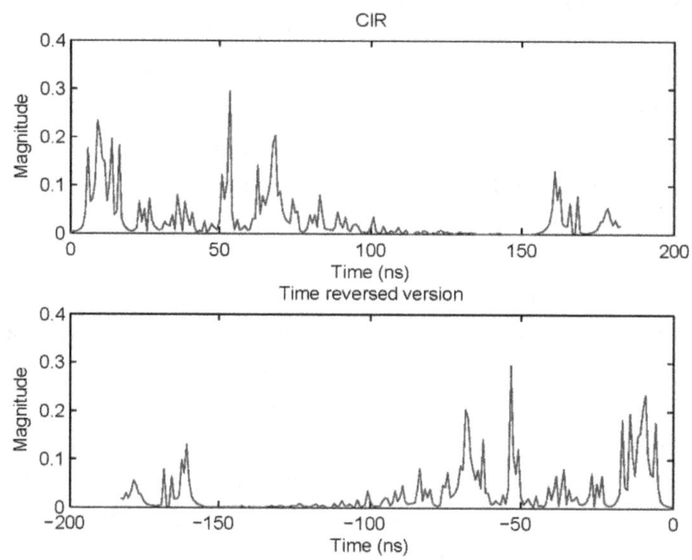

Figure 5. BER performances of the intended receiver and the eavesdropper

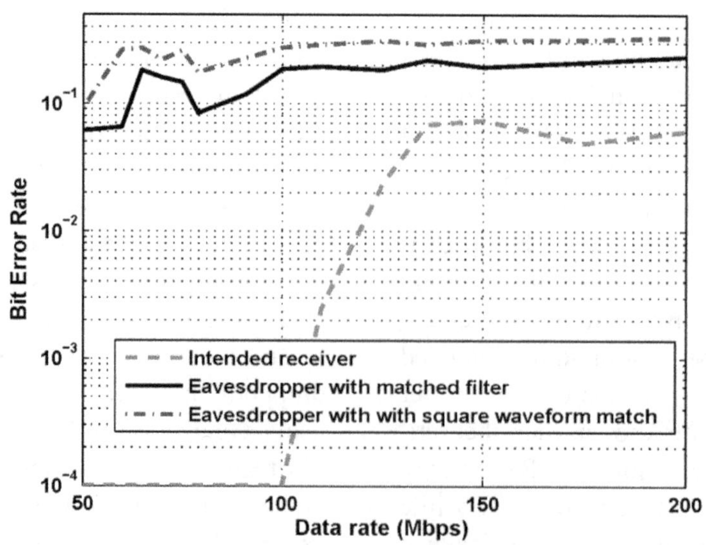

Figure 6. A wideband radio relay system with awareness of information leakage

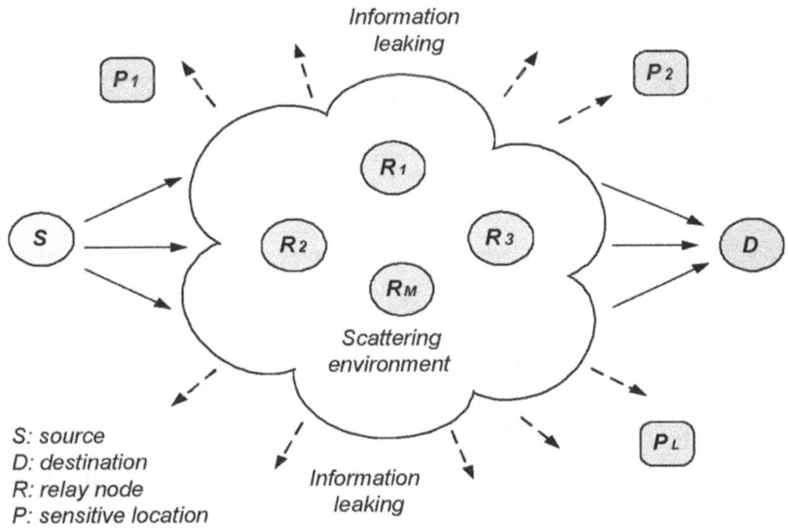

order to see a lower bound of BER corresponding to the worst case. From Figure 5 one can see that at certain data rate the eavesdropper is simply unable to recover the original data even under unrealistically favorable assumption.

The physical-layer security scheme discussed above actually takes advantage of rich multipath scattering, assuming both wide bandwidth and complex propagation environment. This scheme may not work well in an outdoor environment where a CIR usually does not contain a large number of resolvable multipath components. However, if a cluster of Amplify-and-Forward (AF) relay nodes are distributed in between a transmitter and receivers as illustrated in Figure 6, then each relay node acts like a radio mirror and virtually the number of multipath components in the effective CIR would increase. In such a distributed one-hop AF relay configuration, the proposed scheme could be applicable if the effective CIR can be practically estimated. To examine this idea, consider a data transmission scenario with low data rate and no Inter-Symbol-Interference (ISI). The best transmit waveform $w_{opt}(t)$ can be obtained via optimization technique such as Semidefinite Programming (SDP) (Xie&Guo, 2010), and the corresponding matched filter should be $f(t) = w_{opt}(t) \otimes h(t)$, where $h(t)$ is the CIR and $\otimes$ stands for convolution operation. Using IEEE 802.15.4a outdoor channel model (outdoor UWB) and SDP for waveform synthesis, it is demonstrated that the signal leakage can be reduced significantly in terms of channel gains measured at the matched filters' outputs.

Listed in Table 1 are simulation results based on the following setting:

- Two relay configurations, i.e., 4 and 25 relay nodes;
- Two waveform design cases, i.e., [Case 1] channel gain maximization for the intended receiver only, and [Case 2] maximizing the channel gain of the intended receiver while constraining the channel gain of the eavesdropper;
- Four levels of channel gains (Levels 1 to 4) are considered in Case 2.

It can be seen that huge channel gain advantage of the intended receiver over the eavesdropper can be achieved, even though the eavesdropper uses the best receiver filter. Case-1 waveform is

Table 1. Location based security enhancement via channel gain discrimination; channel gains are measured in dB; Gain0-4 and Gain0-25 are channel gains of the intended receiver for 4 and 25 relays; Gain1-4 and Gain1-25 are channel gains of the eavesdropper for 4 and 25 relays

	Case 1	Case 2 Level 1	Case 2 Level 2	Case 2 Level 3	Case 2 Level 4
Gain0-4	8.1627	6.5186	4.1976	3.2061	2.8653
Gain1-4	-8.1551	-20.0000	-40.0000	-60.0000	-80.0000
Gain0-25	28.3544	24.9332	23.9526	23.5496	23.4120
Gain1-25	13.3806	-20	-40	-60.0000	-80.0000

general and robust, but Case-2 waveform design takes into account the channel information for both types of receivers, leading to even larger channel gain differences with just a small penalty on the intended channel gain.

By using proper transmit waveforms, a wideband radio can enable reliable and secure data transmission as well as spectral shaping. With increased security and reliability, the CRNs can be employed for critical tasks like data transmission in smart grids (Raghuram et al., 2011).

16.4 ENHANCEMENT OF COMPUTING POWER— A PRACTICAL SOLUTION

Implementing added features requires additional signal processing and computing power, but the fact is that onboard computing power is very limited. Nowadays many electronic devices are highly networked, and it is likely a cognitive radio transceiver can take advantage of surrounding computing resources. In other words, time critical tasks like modulation/demodulation are handled by the onboard real-time processors, and delay-tolerable tasks like knowledge database updating are taken care of by the off-board networked computing engines in a collaborative fashion. A task management function is required to sort subtasks and assign them to different computing cores. Figure 7 shows a topology of such configuration with cognitive radio transceivers backed up by distributed computing engines.

With distributed computing, rich algorithm packages written in high level languages like MATLAB can be utilized, but this has to be under the condition that the connections between a cognitive radio transceiver and its supporting computing engines do not incur significant latency. Low-latency data connection and proper interfaces are essential to support multiple computing engines. Physical connection methods and their round trip transmission delays can be examined experimentally.

Among many protocols for data connection and interface is Ethernet which is most popular for a number of reasons. This type of connection method has been tested using a commercial software defined radio called Wireless Open Access Research Platform (WARP) (Rice University Wireless Open Access Research Platform (WARP), 2011). The test results show that 5 ms is required to transfer one full-length Datagram Protocol (UDP) packet of 1,480 bytes and get the response. With increasing number of user data packets, the round trip transmission delay increases roughly linearly. The throughput of data exchange is around 2 Mbps, without counting processing time that depends on specific algorithms. There is still some room to improve the interface performance by optimizing the implementation.

If Ethernet interfaces are not available, a serial port connection can be an optional choice but at the cost of higher transmission latency. Figure

Figure 7. Cognitive radio nodes connected with off-board computing resources

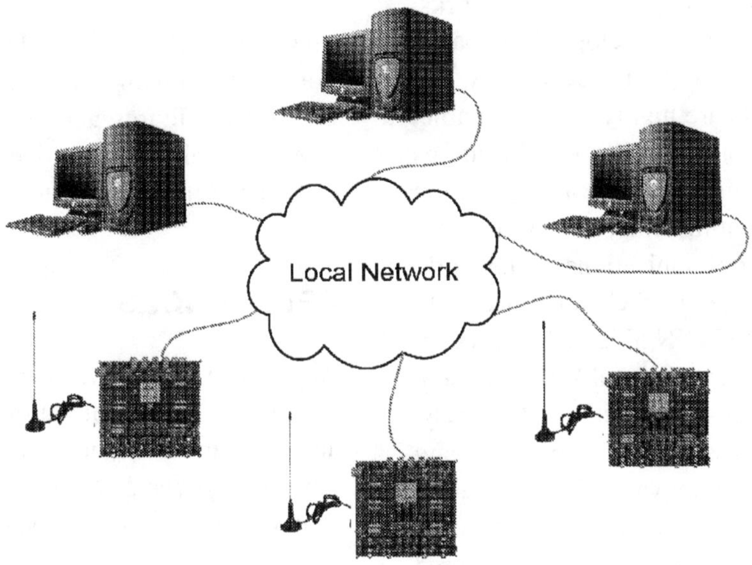

Figure 8. A serial connection setup

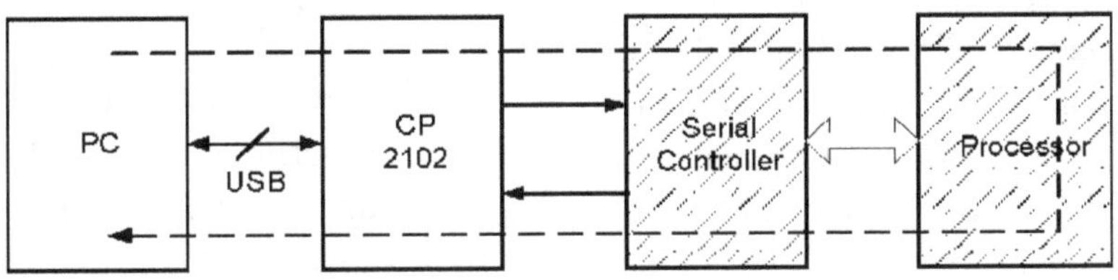

8 is a diagram of serial connection setup, where CP2102 is a USB-to-UART bridge controller to convert USB signal to RS-232 serial signal (USB = universal Serial Bus; UART = Universal Asynchronous Receiver/Transmitter). To transfer a matrix with 2 by 160 entries at baud rate 912.6 kbps, the round trip delay accumulated in transmission is a random number in the range of 2 to 4.5 ms, with an average about 2.8 ms.

These test results validate the feasibility of connecting the mainframe of a cognitive radio transceiver with one or multiple computing engines. Based on the experiments it is safe to say that the round trip transmission delay is in the order of milliseconds, and this figure can be considered in system design. With properly managed computing resources, advanced algorithms and features with heavy computation can be explored. These algorithms/features can be machine learning type of algorithms and RF/radio intrusion detection, etc.

16.5 MORE TO EXPECT

With spectrum sensing and certain intelligence added, a CRN can offer more than just communications function. One potential trend we may expect is convergence of CRNs and wireless sensor networks (WSNs). In a WSN the sensor nodes are severely constrained in terms of storage resources, computational capabilities, communication bandwidth and power supply (Kulkarni et al., 2011). In contrast, a CRN is a very advanced wireless communication network that is more intelligent than any currently deployed networks. There has been a trend to develop CRNs based on software defined radio (SDR) technology, and very likely the CRNs are highly programmable. The marriage of CRNs and WSNs will bring us some excitements. This marriage is a combination of intelligence, programmability, cooperation supported by enhanced communications and network, and massive deployment of cheap sensors. In (Mercier et al., 2008) WSN aided CRN concept is proposed, where the traditional WSN can help perform spectrum sensing function. Reversely, due to its powerfulness in communications and computing, a CRN can help a WSN play regular sensing function better (Yu et al., 2011). In general, the regular sensing function should be considered as an added feature in a CRN.

In addition to common security problems faced by most radio networks, there are some security threats specifically against CRNs (Bian & Park, 2008). One of promising techniques to secure the CRNs is RF anomaly detection which identifies RF intrusions, Sybil RF attacks and malicious spectrum occupation using detection and machine learning techniques. An emitted electromagnetic signal can form an RF fingerprint which is simply a recorded time series with time and location stamps. The cognitive radio transceivers can record and explore the RF fingerprint data to perform functions like Sybil attack detection (Xiao, Greenstein, Mandayam, & Trappe, 2009), object tracking and localization (Li, Jiang, & Guibas, 2010), as well as sparse attack cancellation (Lin et al., 2011). Existing detection, classification, and pattern recognition theories can be used to design new algorithms. These functions may be performed in a cooperative manner, and centralized data fusion is expected to deliver accurate outcomes. Of course, an RF fingerprint database has to be built in prior and maintained during network operation, which requires tremendous support of networked computing resources.

ACKNOWLEDGMENT

The author is grateful to Dr. Zhe Chen, Dr. Yu Song, Alex Zhang and Xia Li for providing a number of figures and some experiment data. Special thanks are also due to Professor Robert C. Qiu and his research team.

REFERENCES

Akyildiz, I., Lee, W., Vuran, M., & Mohanty, S. (2006). Next generation/dynamic spectrum access/cognitive radio wireless networks: A survey. *Computer Networks*, *50*(13), 2127–2159. doi:10.1016/j.comnet.2006.05.001

Bian, K., & Park, J. (2008). Security vulnerabilities in IEEE 802.22. *Proceedings of the 4th Annual International Conference on Wireless Internet* (p. 9).

Chen, Z., Guo, N., Hu, Z., & Qiu, R. (2011, May). Experimental validation of channel state prediction considering delays in practical cognitive radio. *IEEE Transactions on Vehicular Technology*, *60*(4), 1314–1325. doi:10.1109/TVT.2011.2116051

Chong, J., Sung, Y., & Sung, D. (2009). Rawpeach: Multiband CSMA/CA-based cognitive radio networks. *Journal of Communication Networks*, *11*(2).

Ganesan, G., & Li, Y. (2005). Cooperative spectrum sensing in cognitive radio networks. In *New Frontiers in Dynamic Spectrum Access Networks. First IEEE International Symposium on DySPAN*, (pp. 137–143).

Gao, F., Zhang, R., Liang, Y., & Wang, X. (2010). Design of learning-based MIMO cognitive radio systems. *IEEE Transactions on Vehicular Technology*, *59*(4), 1707–1720. doi:10.1109/TVT.2010.2042089

Ghasemi, A., & Sousa, E. (2008). Spectrum sensing in cognitive radio networks: Requirements, challenges and design trade-offs. *IEEE Communications Magazine*, *46*(4), 32–39. doi:10.1109/MCOM.2008.4481338

Guo, N., Qiu, R., & Sadler, B. (2005, Oct.). An ultra-wideband autocorrelation demodulation scheme with low-complexity time reversal enhancement. *IEEE Military Communications Conference (Milcom'05)*, (pp. 1–7).

Guo, N., Sadler, B., & Qiu, R. (2007, December). Reduced-complexity UWB time-reversal techniques and experimental results. *IEEE Transactions on Wireless Communications*, *6*(12), 4221–4226. doi:10.1109/TWC.2007.060251

Guo, N., Zhang, J., Qiu, R., & Mo, S. (2007, January 11-13). UWB MISO time reversal with energy detector receiver over ISI channels. *IEEE CCNC'07*, (pp. 629-633).

Haykin, S. (2005). Cognitive radio: Brain-empowered wireless communications. *IEEE Journal on Selected Areas in Communications*, *23*(2), 201–220. doi:10.1109/JSAC.2004.839380

Hu, Z., Guo, N., & Qiu, R. (2010a, November). *Wideband waveform design for relay cognitive network.* IEEE Military Communications Conference.

Hu, Z., Guo, N., & Qiu, R. (2010b, November). *Wideband waveform optimization for multiple input single output cognitive radio with practical considerations*. IEEE Military Communications Conference.

Kulkarni, V., Forster, A., & Venayagamoorthy, G. (2011). Computational intelligence in wireless sensor networks: A survey. *IEEE Communications Surveys &. Tutorials*, *99*, 1–29.

Li, M., Jiang, X., & Guibas, L. (2010). Fingerprinting mobile user positions in sensor networks. *2010 International Conference on Distributed Computing Systems*, (pp. 478–487).

Lin, F., Hu, Z., Hou, S., Yu, J., Zhang, C., Guo, N., et al. (2011). *Cognitive radio network as wireless sensor network (II): Security consideration*. National Aerospace & Electronics Conference.

Mercier, B., Fodor, V., Thobaben, R., Skoglund, M., Koivunen, V., & Lindfors, S. (2008). *Sensor networks for cognitive radio: Theory and system design*. ICT Mobile Summit.

Mitola, J. III, & Maguire, G. Jr. (1999). Cognitive radio: Making software radios more personal. *IEEE Personal Communications*, *6*(4), 13–18. doi:10.1109/98.788210

Nekovee, M. (2010). A survey of cognitive radio access to TV white spaces. *International Journal of Digital Multimedia Broadcasting*, 2010.

Pawelczak, P., Angeles, L., Nolan, K., Doyle, L., Oh, S. W., & Cabric, D. (2011). Cognitive radio: Ten years of experimentation and development. *IEEE Communications Magazine*, *49*(03), 90–100. doi:10.1109/MCOM.2011.5723805

Qiu, R. C., Zhou, C., Guo, N., & Zhang, J. Q. (2006). Time reversal with MISO for ultrawideband communications: Experimental results. *Antennas and Wireless Propagation Letters*, *5*, 269–273. doi:10.1109/LAWP.2006.875888

Raghuram, R., Robert, Q., Zhen, H., Shujie, H., Marbin, P., & Gang, Z. (2011). Cognitive radio for smart grid: Theory, algorithms, and security. *International Journal of Digital Multimedia Broadcasting*, 2011.

Rice University Wireless Open access Research Platform (WARP). (2011). Retrieved from http://warp.rice.edu/trac

Shiu, Y., Chang, S., Wu, H., Huang, S., & Chen, H. (2011). Physical layer security in wireless networks: A tutorial. *IEEE Wireless Communications*, *18*(2), 66–74. doi:10.1109/MWC.2011.5751298

Steenkiste, P., Sicker, D., Minden, G., & Raychaudhuri, D. (March 2009). *Future directions in cognitive radio network research.* NSF Workshop Report.

Xiao, L., Greenstein, L., Mandayam, N., & Trappe, W. (2008). Using the physical layer for wireless authentication in time variant channels. *IEEE Transactions on Wireless Communications, 7*(7), 2571–2579. doi:10.1109/TWC.2008.070194

Xiao, L., Greenstein, L., Mandayam, N., & Trappe, W. (2009). Channel-based detection of sybil attacks in wireless networks. *IEEE Transactions on Information Forensics and Security, 4*(3), 492–503. doi:10.1109/TIFS.2009.2026454

Xie, X., & Guo, N. (2010). UWB relay with physical layer security enhancement. *2010 IEEE International Conference on Ultra-wideband (ICUWB)*, Vol. 2, (pp. 1–4).

Yu, J., Zhang, C., Hu, Z., Lin, F., Guo, N., Wicks, M., et al. (2011). *Cognitive radio network as wireless sensor network (I): Architecture, testbed, and experiment.* National Aerospace & Electronics Conference.

Zhang, L., Soong, B., & Xiao, W. (2007). An integrated cluster based multi-channel MAC protocol for mobile ad hoc networks. *IEEE Transactions on Wireless Communications, 6*(11), 3964–3974. doi:10.1109/TWC.2007.06038

Zhou, C., Guo, N., & Qiu, R. (2009). Time-reversed ultra-wideband (UWB) multiple input multiple output (MIMO) based on measured spatial channels. *IEEE Transactions on Vehicular Technology, 58*(6), 2884–2898. doi:10.1109/TVT.2008.2012109

Compilation of References

3 GPP, TR 36.814. (n.d.). *Further advancements for E-UTRA physical layer aspects.*

3 GPP, TR 36.913. (n.d.). *Requirements for further advancements for evolved universal terrestrial radio access (E-UTRA).* Andrews, J. G., Choi, W., & Heath Jr., R. W. (2007). Overcoming interference in spatial multiplexing MIMO cellular networks. *IEEE Wireless Communications Magazine, 14*(6), 95-104.

3 rd Generation Partnership Project. (2008). *Evolved universal terrestrial radio access (E-UTRA) and evolved universal terrestrial radio access network (E-UTRAN).* Overall description. Tech. Rep. TS 36.300 V8.7.0.

3 rd Generation Partnership Project. (2009). *Technical specification group radio access network: Evolved universal terrestrial radio access (E-UTRA) and evolved universal terrestrial radio access network (E-UTRAN).* Overall description; Stage 2. Tech. Rep. TS 36.300 V9.0.0.

Aazhang, B., Lilleberg, J., & Middleton, G. (2004). Spectrum sharing in a cellular system. *IEEE Eighth International Symposium on Spread Spectrum Techniques and Applications*, (pp. 355 –359).

Agastya, M. (1999). Perturbed adaptive dynamics in coalition form games. *Journal of Economic Theory, 89*(2), 207–233. doi:10.1006/jeth.1999.2574

Aharon Ben-tal. L. E. (1998). Robust semidefinite programming. In H. Wolkowics, R. Saigal, & L. Vandenberghe (Eds.), *Handbook on semidefinite programming*, (pp. 139-162). Kluwer Academis Publishers.

Aharon Ben-Tal, A. N. (1998). Robust convex optimization. *Mathematics of Operations Research, 23*(4), 769–805. doi:10.1287/moor.23.4.769

Ahn, G. S., Campbell, A. T., Veres, A., & Sun, L. H. (2002). SWAN: Service differentiation in stateless wireless ad hoc networks. *IEEE Conference on Computer Communications (INFOCOM)*, (pp. 457–466).

Akbar, I. A., & Tranter, W. H. (2007). Dynamic spectrum allocation in cognitive radio using hidden Markov models: Poisson distributed case. *Proceedings of IEEE SoutheastCon*, (pp. 196-201).

Akyildiz, I. F., Lee, W.-Y., Vuran, M. C., & Mohanty, S. (2006). Next generation/dynamic spectrum access/cognitive radio wireless networks: A survey. *Computer Networks, 50*(13), 2127–2159. doi:10.1016/j.comnet.2006.05.001

Akyildiz, I. F., Vuran, M. C., & Mohanty, S. (2008). A survey on spectrum management in cognitive radio networks. *IEEE Communications Magazine, 46*(4), 40–48. doi:10.1109/MCOM.2008.4481339

Akyildiz, I., Lee, W. Y., & Chowdhury, K. (2009). Spectrum management in cognitive radio ad hoc networks. *IEEE Network, 23*(4), 6–12. doi:10.1109/MNET.2009.5191140

Akyildiz, I., Lee, W., Vuran, M., & Mohanty, S. (2006). Next generation/dynamic spectrum access/cognitive radio wireless networks: A survey. *Computer Networks, 50*(13), 2127–2159. doi:10.1016/j.comnet.2006.05.001

Al Hanbali, A., Altman, E., & Nain, P. (2005). A survey of TCP over ad hoc networks. *IEEE Communications Surveys & Tutorials, 7*(3), 22–36. doi:10.1109/COMST.2005.1610548

Almasaeid, H. M., & Kamal, A. E. (2010). Assisted-multicast scheduling in wireless cognitive mesh networks. *2010 IEEE International Conference on Communications (ICC)*, (pp. 1–5).

Al-Sbou, Y. A., Saatchi, R., Al-Khayatt, S., Strachan, R., Ayyash, M., & Saraireh, M. (2008). A novel quality of service assessment of multimedia traffic over wireless ad hoc networks. *The 1st IEEE International Workshop on Future Multimedia Networking (IEEE FMN'08) in The Second International Conference on Next Generation Mobile Applications, Services, and Technologies (IEEE NGMAST'08)*, (pp. 479-484).

Al-Sbou, Y. (2006). *Quality of service assessment and analysis of wireless multimedia networks*. Sheffield, UK: Sheffield Hallam University.

Al-Sbou, Y. A. (2010). Fuzzy logic estimation system of quality of service for multimedia transmission. *International Journal of Quality of Service Issues in Networking, 1*(1).

Anant Sahai, N. H. (2004). *Some fundamental limits on cognitive radio*. Forty-second Allerton Conference on Communication, Control, and Computing.

Apt, K. R., & Witzel, A. (2009). A generic approach to coalition formation. *International Game Theory Review, 11*(3), 347–367. doi:10.1142/S0219198909002352

Aragorn, F. P. (2008). *Adaptive reconfigurable access and generic interfaces for optimization in radio networks. Document Number D2.1: State-of-the-art*. Retrieved from Public Deliverable.

Arnold, T., & Schwalbe, U. (2002). Dynamic coalition formation and the core. *Journal of Economic Behavior & Organization, 49*(3), 363–380. doi:10.1016/S0167-2681(02)00015-X

Arshad, K., Imran, M. A., & Moessner, K. (2010). Collaborative spectrum sensing optimization algorithms for cognitive radio networks. *International Journal of Digital Multimedia Broadcasting*, 2010.

Attar, A., Krishnamurthy, V., & Namvar Gharehshiran, O. (2011). Interference management using cognitive base-stations for UMTS LTE. *IEEE Communications Magazine, 49*, 152–159. doi:10.1109/MCOM.2011.5978429

Aumann, R. J., & Hart, S. (1994). *Handbook of game theory with economic applications*. North Holland.

Aumann, R. J., & Peleg, B. (1960). Von Neumann-Morgenstern solutions to cooperative games without side payments. *Bulletin of the American Mathematical Society, 66*(3), 173–179. doi:10.1090/S0002-9904-1960-10418-1

Aune, F. (2004). *Cross layer design tutorial*. Norwegian University of Science and Technology: Published under Creative Commons License.

Ayyash, M. (2005). *A new robust quality of service routing protocol for wireless mobile ad hoc networks*. USA: Illinois Insitute of Technology.

Ayyash, M., Ucci, D., & Alzoubi, K. (2010). A proactively maintained quality of service infrastructure for wireless mobile ad hoc networks. [IJCNIS]. *International Journal of Communication Networks and Information Security, 2*(2).

Babaei, A., & Jabbari, B. (2010a). Throughput optimization in cognitive random wireless ad hoc networks. *Proceedings of IEEE Global Communications Conference*, (pp. 1-5).

Babaei, A., & Jabbari, B. (2010b). Distance distribution of bivariate Poisson network nodes. *IEEE Communications Letters, 14*(9), 848–850. doi:10.1109/LCOMM.2010.080610.100668

Baddeley, A. (2007). Spatial point processes and their applications. *Lecture Notes in Mathematics, 1892*, 1–75. doi:10.1007/978-3-540-38175-4_1

Baldo, N., & Zorzi, M. (2007). *Fuzzy logic for cross layer optimization in cognitive radio networks*. IEEE Consumer Communications and Networking Conference (CCNC'07).

Baldo, N., & Zorzi, M. (2008a). Learning and adaptation in cognitive radios using neural networks. *IEEE CNNC 2008 Proceedings*.

Baldo, N., & Zorzi, M. (2008). Fuzzy logic for cross-layer optimization in cognitive radio networks. *IEEE Communications Magazine, 46*(4), 64–71. doi:10.1109/MCOM.2008.4481342

Barkat, M., & Varshney, P. K. (1989). Decentralized CFAR signal detection. *IEEE Transactions on Aerospace and Electronic Systems, 25*(2), 141–149. doi:10.1109/7.18676

Barnhart, C., Wieselthier, J., & Ephremides, A. (1995). A neural network approach to solving the link activation problem in multihop radio networks. *IEEE Transactions on Communications*, *43*(234), 1277–1283. doi:10.1109/26.380169

Başar, T., & Jan Olsder, G. (1999). *Dynamic noncooperative game theory*. Philadelphia, PA: SIAM: Classics in Applied Mathematics.

Baynast, A. D., Mahonen, P., & Petrova, M. (2008). ARQ-based cross-Layer optimization for wireless multicarrier transmission on cognitive radio networks. *Computer Networks*, *52*(4), 778–794. doi:10.1016/j.comnet.2007.11.009

Ben Letaief, K., & Zhang, W. (2009). Cooperative communications for cognitive radio networks. *Proceedings of the IEEE*, *97*(5), 878–893. doi:10.1109/JPROC.2009.2015716

Bhagavatula, R., & Heath, R. W. Jr. (2011). Adaptive limited feedback for sum-rate maximizing beamforming in cooperative multicell systems. *IEEE Transactions on Signal Processing*, *59*(2), 800–811. doi:10.1109/TSP.2010.2090346

Bhagwat, P., Bhattacharya, P., Krishna, A., & Tripathi, S. K. (1996). *Enhancing throughput over wireless LANs using channel state dependent packet scheduling*. IEEE 15th Annual Joint Conference of the IEEE Computer Societies.

Bharghavan, V., Lu, S., & Nandagopal, T. (1999). Fair queuing in wireless networks: Issues and approaches. *IEEE Personal Communications*, *6*(1), 44–53. doi:10.1109/98.752787

Bhat, C. T. P. S. (2011). *Opportunistic scheduling, cooperative relaying and multicast in wireless networks.* Unpublished doctoral dissertation Arizona State University, Phoenix Metropolitan Area, Arizona, USA

Bian, K., & Park, J. (2008). Security vulnerabilities in IEEE 802.22. *Proceedings of the 4th Annual International Conference on Wireless Internet* (p. 9).

Bixio, L., Ciardelli, L., & Ottonello, M. Raffetto, Regazzoni, M., C., Alam, S. S., & Armani, C. (2011). *A transmit beamforming technique for MIMO cognitive radios*. Wireless Innovation Forum Conference on Communi- cations Technologies and Software Defined Radio, SDR'11.

Bixio, L., Ottonello, M., Raffetto, M., & Regazzoni, C.S. (2011). Comparison among cognitive radio architectures for spectrum sensing. *EURASIP Journal on Wireless Communications and Networking.*

Bixio, L., Cattoni, A. F., Regazzoni, C. S., & Varshney, P. K. (2008). *Embodied cognition-based distributed spectrum sensing for autonomic wireless systems. Autonomic Computing and Networking*. USA: Springer.

Bixio, L., Oliveri, G., Ottonello, M., Raffetto, M., & Regazzoni, C. S. (2010). Cognitive radios with multiple antennas exploiting spatial opportunities. *IEEE Transactions on Signal Processing*, *58*(8), 4453–4459. doi:10.1109/TSP.2010.2048322

Björnson, E., & Ottersten, B. (2009). On the principles of multicell precoding with centralized and distributed Cooperation. *Proceedings of IEEE Wireless Communication Signal Processing (WCSP)*, (pp. 1-5).

Björnson, E., Zakhour, R., Gesbert, D., & Ottersten, B. (2009). Distributed multicell and multiantenna precoding: characterization and performance evaluation. *Proceedings of the IEEE Global Communication Conference (GLOBECOM)*, (pp. 1-6).

Bocus, M. Z., Coon, J. P., Canagarajah, C. N., McGeehan, J. P., Armour, S. M. D., & Doufexi, A. (2011). Resource allocation for OFDMA-based cognitive radio networks with application to H.264 scalable video transmission. *EURASIP Journal on Wireless Communications and Networking*, *2011*, 1–10. doi:10.1155/2011/245673

Bogatinovski, M., & Gavrilovska, L. (2008). Overview of cross-layer optimization methodologies for cognitive radio. *TELFOR'08. ARAGORN (ICT-216856).*

Boudreau, G., Panicker, J., Guo, N., Chang, R., Wang, N., & Vrzic, S. (2009). Interference coordination and cancellation for 4G networks. *IEEE Communications Magazine*, *47*(4), 74–81. doi:10.1109/MCOM.2009.4907410

Brémaud, P. (1999). *Markov chains: Gibbs fields, Monte Carlo simulation, and queues*. Springer-Verlag.

Buckley, J., & Eslami, E. (2002). *An introduction to fuzzy logic and fuzzy sets*. Berlin, Germany: Springer-Verlag.

Buddhikot, M. M. (2010). Cognitive radio, DSA and self-X: Towards next transformation in cellular networks (Extended Abstract). *2010 IEEE Symposium on New Frontiers in Dynamic Spectrum*, (pp. 1-5).

Buljore, S., Merat, V., & Harada, H. (2008). IEEE P1900.4 system overview on architecture and enablers for optimized radio and spectrum resource usage. *3rd IEEE Symposium on New Frontiers in Dynamic Spectrum Access Networks*, (pp. 253-260).

Buljore, S., Harada, H., & Filin, S. (2009). Architecture and enablers for optimized radio resource usage in heterogeneous wireless access networks: IEEE 1900.4 working group. *IEEE Communications Magazine*, *47*(1), 22–29. doi:10.1109/MCOM.2009.4752689

Burbank, J. (2008). *Security in cognitive radio networks: The required evolution in approaches to wireless network security*. 3rd International Conference on Cognitive Radio Oriented Wireless Networks and Communications.

Cabric, D., Mishra, S. M., & Brodersen, R. W. (2004). Implementation issues in spectrum sensing for cognitive radios. *Proceedings of Asilomar Conference on Signals, Systems, and Computers*, (pp. 772–776).

Cabric, D., Tkachenko, A., & Brodersen, R. W. (2006). Experimental study of spectrum sensing based on energy detection and network cooperation. *Proceedings of the First International Workshop on Technology and Policy for Accessing Spectrum*, Article No. 12.

Cai, D. W. H., Quek, T. Q. S., & Tan, C. W. (2011). Coordinated max-min SIR optimization in multicell downlink -duality and algorithm. *Proceedings of the IEEE International Conference on Communication (ICC)*, (pp. 1-6).

Cai, D. W. H., Quek, T. Q. S., Tan, C. W., & Low, S. (2012). Coordinated max-min SINR optimization in multicell downlink - Duality and algorithms. *IEEE Transactions on Signal Processing*, 60.

Calin, D., Claussen, H., & Uzunalioglu, H. (2010). On femto deployment architectures and macrocell offloading benefits in joint macro-femto deployments. *IEEE Communications Magazine*, *48*(1), 26–32. doi:10.1109/MCOM.2010.5394026

Campbell, N. (1909). The study of discontinuous phenomena, mathematica. *Proceedings of the Cambridge Philosophical Society*, *15*, 117–136.

Cattoni, A. F., Minetti, I., Gandetto, M., Niu, R., Varshney, P. K., & Regazzoni, C. S. (2006). *A spectrum sensing algorithm based on distributed cognitive models*. SDR Forum Technical Conference, Orlando, Florida, USA.

Cesana, M., Cuomo, F., & Ekici, E. (2011). Routing in cognitive radio networks: Challenges and solutions. *Ad Hoc Networks*, *9*(3), 228–248. doi:10.1016/j.adhoc.2010.06.009

Chae, C.-B., Kim, S.-H., & Heath, R. W. Jr. (2009). Network coordinated beamforming for cell-boundary users: linear and nonlinear approaches. *IEEE Journal of Selected Topics in Signal Processing*, *3*(6), 1094–1105. doi:10.1109/JSTSP.2009.2035857

Chair, Z., & Varshney, P. K. (1986). Optimal data fusion in multiple sensor detection systems. *IEEE Transactions on Aerospace and Electronic Systems*, 98–101. doi:10.1109/TAES.1986.310699

Chakrabarti, S., & Mishra, A. (2001). QoS issues in ad hoc wireless networks. *IEEE Communications Magazine*, *39*(2), 142–148. doi:10.1109/35.900643

Chandrasekhar, V., & Andrews, J. G. (2009). Spectrum allocation in tiered cellular networks. *IEEE Transactions on Communications*, *57*(10), 3059–3068. doi:10.1109/TCOMM.2009.10.080529

Chang, Y.-L., & Hsu, C.-C. (2000). Routing in wireless/mobile ad-hoc networks via dynamic group construction. *Mobile Networks and Applications*, *5*(1), 27–37. doi:10.1023/A:1019175605040

Chen, R., & Park, J. M. (2006). Ensuring trustworthy spectrum sensing in cognitive radio networks. *1st IEEE Workshop on Networking Technologies for Software Defined Radio Networks*, (pp. 110-119).

Chen, S., & Wyglinski, A. M. (2009). Efficient spectrum utilization via cross-layer optimization in distributed cognitive radio networks. *Elsevier Computer Communications Journal, 32*(18).

Chen, Z. (2003). *A study and implementation of speech recognition system for mobile communication terminals*. Master's thesis, Hangzhou Dianzi University, China.

Chen, Z. (2011). *Cognitive radio network: Architecture, algorithms, and testbed.* Doctoral dissertation, Tennessee Technological University, USA.

Chen, Z., & Qiu, R. C. (2010). Prediction of channel state for cognitive radio using higher-order hidden Markov model. *Proceedings of IEEE SoutheastCon*, (pp. 276-282).

Chen, Z., Hu, Z., & Qiu, R. C. (2009). Quickest spectrum detection using hidden Markov model for cognitive radio. *Proceedings of IEEE Military Communications Conference*.

Chen, B., Jiang, R., Kasetkasem, T., & Varshney, P. K. (2004). Channel aware decision fusion in wireless sensor networks. *IEEE Transactions on Signal Processing*, *52*(12), 3454–3458. doi:10.1109/TSP.2004.837404

Chen, R., & Park, J. M. (2008). Defense against primary user emulation attacks in cognitive radio networks. *IEEE Journal on Selected Areas in Communications*, *26*(1), 25–37. doi:10.1109/JSAC.2008.080104

Chen, R., Park, J.-M., Hou, Y. T., & Reed, J. H. (2008). Toward secure distributed spectrum sensing in cognitive radio networks. *IEEE Communications Magazine*, *46*(4), 50–55. doi:10.1109/MCOM.2008.4481340

Chen, Y., Yu, G., Zhang, Z., Chen, H. H., & Qiu, P. (2008). On cognitive radio networks with opportunistic power control strategies in fading channels. *IEEE Transactions on Wireless Communications*, *7*(7), 2752–2761. doi:10.1109/TWC.2008.070145

Chen, Y., Zhao, Q., & Swami, A. (2008). Joint design and separation principle for opportunistic spectrum access in the presence of sensing errors. *IEEE Transactions on Information Theory*, *54*(5), 2053–2071. doi:10.1109/TIT.2008.920248

Chen, Z., Guo, N., Hu, Z., & Qiu, R. C. (2011). Experimental validation of channel state prediction considering delays in practical cognitive radio. *IEEE Transactions on Vehicular Technology*, *60*(4), 1314–1325. doi:10.1109/TVT.2011.2116051

Chen, Z., Guo, N., & Qiu, R. C. (2010). Demonstration of real-time spectrum sensing for cognitive radio. *IEEE Communications Letters*, *14*(10), 915–917. doi:10.1109/LCOMM.2010.082610.100725

Chiang, M., & Low, S. H. (2007). Layering as optimization decomposition: a mathematical theory of network architectures. *Proceedings of the IEEE*, *95*(1), 255–312. doi:10.1109/JPROC.2006.887322

Cho, H., & Andrews, J. G. (2009). Upper bound on the capacity of cognitive radio without cooperation. *IEEE Transactions on Wireless Communications*, *8*(9), 4380–4385. doi:10.1109/TWC.2009.071106

Chong, J., Sung, Y., & Sung, D. (2009). Rawpeach: Multiband CSMA/CA-based cognitive radio networks. *Journal of Communication Networks*, *11*(2).

Clancy, T. C., & Goergen, N. (2009). *Security in cognitive radio networks: threats and mitigation.* 3rd International Conference on Cognitive Radio Oriented Wireless Networks and Communications.

Clancy, T., & Walker, B. (2006). Predictive dynamic spectrum access. *Proceedings of SDR Forum Conference*.

Clark, D. D., Partridge, C., Ramming, J. C., & Wroclawski, J. T. (2003). *A knowledge plane for the internet.* ACM SIGCOMM.

Cordeiro, C., Challapali, K., Birru, D., & Shankar, S. (2006). IEEE 802.22: An introduction to the first worldwide wireless standard based on cognitive radios. *The Journal of Communication*, *1*(1), 38–47.

Coupechoux, M., et al. (2007). *Generic architecture of resource management and related protocols: spectrum occupancy: State of the art.* URC (Urban Planning for Radio Communication) Project, (Deliverable 1.3.1).

Couturier, S., & Scheers, B. (2009). *The state of the art of dynamic spectrum access.* Military Communication and Information Systems Conference.

Cox, D. (1970). *Renewal theory.* Methuen.

Delage, E. Y. Y. (2010). Distributionally robust optimization under moment uncertainty with application to data-driven problems. *Operations Research*, *58*(3), 595–612. doi:10.1287/opre.1090.0741

Diggle, P. J. (2003). *Statistical analysis of spatial point patterns* (2nd ed.). Arnold.

Digham, F. F., Alouini, M.-S., & Simon, M. K. (2003). On the energy detection of unknown signals over fading channels. *IEEE International Conference on Communications*, Vol. 5, (pp. 3575-3579).

Dilmaghani, R. B., & Rao, R. R. (2006). *Hybrid wireless mesh network deployment: A communication testbed for disaster scenarios.* The 1st International Workshop on Wireless Network Testbeds, Experimental Evaluation & Characterization.

Doyle, L. E., Nolan, K. E., Forde, T. K., et al. (2006). *A platform for dynamic spectrum experimentation.* 1st International Workshop on Technology and Policy for Accessing Spectrum.

Doyle, L. E. (2009). *Essentials of cognitive radio*. Cambridge University Press. doi:10.1017/CBO9780511576577

Dupcinov, M., & Krco, S. (2002). *Routing in ad hoc networks*. Ireland: Applied Research Lab, EEI, Ericsson.

Eastlake, D., & Jones, P. (2002). *US secure hash algorithm 1* (SHA1). RFC 3174, September 2001: 2002.

Elias-Fuste, A. R., Broquetas-Ibars, A., Antequera, J. P., & Yuste, J. C. M. (1992). CFAR data fusion center with inhomogeneous receivers. *IEEE Transactions on Aerospace and Electronic Systems*, *28*(1), 276–285. doi:10.1109/7.135453

Erez, U., & ten Brink, S. (2004). A close to capacity dirty paper coding scheme. *IEEE Transactions on Information Theory*, *51*(10).

Erlebach, T., & Grant, T. (2010). *Scheduling multicast transmissions under SINR constraints. algorithms for sensor systems* (pp. 47–61). Berlin, Germany: Springer.

Etkin, R., Parekh, A., & Tse, D. (2007). Spectrum sharing for unlicensed bands. *IEEE Journal on Selected Areas in Communications*, *25*(3), 517–528. doi:10.1109/JSAC.2007.070402

Ettus Research LLC. (2010). *Home page.* Retrieved October, 2010, from http://www.ettus.com

Farnham, T. Clemo, Haines, G., R., Seidel, E., Benamar, A., Billington, S., ... Mangold, P. (2000). IST-TRUST: A perspective on the reconfiguration of future mobile terminals using software download. *IEEE International Symposium on Personal, Indoor and Mobile Radio Communications*, London, UK, (pp. 1054–1059).

Fattah, H., & Leung, C. (2002). An overview of scheduling algorithms in wireless multimedia networks. *IEEE Wireless Communications*, *9*(5), 76–83. doi:10.1109/MWC.2002.1043857

FCC 05-57. (2005). *Facilitating opportunities for flexible, efficient, and reliable spectrum use employing cognitive radio technologies*. ET docket no. 03-108, Federal Communications Commission.

FCC. (2003). *Notice of proposed rule making and order*. ET Docket No. 03-322.

Felegyhazi, M., & Hubaux, J. (2007). *Game theory in wireless networks: A tutorial.* (EPFL Technical report: LCA-REPORT-2006-002).

Filin, S., Ishizu, K., & Murakami, H. (2008). Dynamic spectrum assignment and access scenarios, system architecture, functional architecture and procedures for IEEE P1900.4 management system. *3rd International Conference on CrownCom 2008,* (pp. 1-7).

Fitzek, F. H. P., & Katz, M. D. (Eds.). (2006). *Cooperation in wireless networks: Principles and applications: real egoistic behavior is to cooperate!*Dordrecht, The Netherlands: Springer. doi:10.1007/1-4020-4711-8

Fitzek, F. H. P., & Katz, M. D. (Eds.). (2007). *Cognitive wireless networks: Concepts, methodologies and visions inspiring the age of enlightenment of wireless communications*. Dordrecht, The Netherlands: Springer.

Foschini, G. J., Karakayali, K., & Valenzuela, R. A. (2006). Coordinating multiple antenna cellular networks to achieve enormous spectral efficiency. *IEE Proceedings. Communications*, *153*(4), 548–555. doi:10.1049/ip-com:20050423

Foukalas, F., Gazis, V., & Alonistioti, N. (2008). Cross-layer design proposals for wireless mobile networks: A survey and taxonomy. *IEEE Communications Surveys and Tutorials*, *10*(1), 70–85. doi:10.1109/COMST.2008.4483671

Fremlin, D. H. (2003). *Measure theory*. Torres Fremlin.

Fu, F., & Schaar, M. v. (2007). *A new theoretic foundation for cross layer optimization*. UCLA Technical Report.

Gandetto, M., & Regazzoni, C. S. (2007). Spectrum sensing: a distributed approach for cognitive terminals. *IEEE Journal on Selected Areas in Communications*, *25*(3). doi:10.1109/JSAC.2007.070405

Ganesan, G., & Li, Y. (2005). Cooperative spectrum sensing in cognitive radio networks. In *New Frontiers in Dynamic Spectrum Access Networks. First IEEE International Symposium on DySPAN*, (pp. 137–143).

Gao, F., Zhang, R., Liang, Y., & Wang, X. (2010). Design of learning-based MIMO cognitive radio systems. *IEEE Transactions on Vehicular Technology*, *59*(4), 1707–1720. doi:10.1109/TVT.2010.2042089

Garcia, I. P. (2010). *Interference management in cognitive radio systems*. Master's Thesis, Aalto University.

Gardner, W. A. (1988). Signal interception: A unifying theoretical framework for feature detection. *IEEE Transactions on Communications*, *36*(8), 897–906. doi:10.1109/26.3769

Gardner, W. A. (1991). Exploitation of spectral redundancy in cyclostationary signals. *IEEE Signal Processing Magazine*, *8*(2), 14–36. doi:10.1109/79.81007

Garg, S., & Kappes, M. (2003). *Admission control for VoIP traffic in IEEE 802.11 networks*. IEEE Global Telecommunications Conference.

Gavrilovska, L., & Prasad, R. (2006). *Ad hoc networking towards seamless communications*.

Gazis, V., Houssos, N., Alonistioti, N., & Merakos, L. (2003). *On the complexity of "always best connected" in 4G mobile networks*. IEEE 58th Vehicular Technology Conference.

Geirhofer, S., Tong, L., & Sadler, B. M. (2006). Dynamic spectrum access in WLAN channels: Empirical model and its stochastic analysis. *Proceedings of the First International Workshop on Technology and Policy for Accessing Spectrum*.

Geirhofer, S., Tong, L., & Sadler, B. (2008). Cognitive medium access: constraining interference based on experimental models. *IEEE Journal on Selected Areas in Communications*, *26*(1), 95–105. doi:10.1109/JSAC.2008.080109

Geirhofer, S., Tong, L., & Sadler, B. M. (2007). Cognitive radios for dynamic spectrum access-dynamic spectrum access in the time domain: Modeling and exploiting white space. *IEEE Communications Magazine*, *45*(5), 66–72. doi:10.1109/MCOM.2007.358851

Georgiadis, L., Neely, M. J., & Tassiulas, L. (2006). *Resource allocation and cross-layer control in wireless networks*. Now Publishers Inc.

Gesbert, D., Kiani, S. G., Gjendemsjø, A., & Øien, G. E. (2007). Adaptation, coordination, and distributed resource allocation in interference-limited wireless networks. *Proceedings of the IEEE*, *95*(12), 2393–2409. doi:10.1109/JPROC.2007.907125

Ghasemi, A., & Sousa, E. S. (2005). Collaborative spectrum sensing for opportunistic access in fadinenvironments. *2005 First IEEE International Symposium on New Frontiers in Dynamic Spectrum Access Networks*, (pp. 131-136).

Ghasemi, A., & Sousa, E. (2008). Spectrum sensing in cognitive radio networks: Requirements, challenges and design trade-offs. *IEEE Communications Magazine*, *46*(4), 32–39. doi:10.1109/MCOM.2008.4481338

Ghasemi, A., & Sousa, E. S. (2008, February). Interference aggregation in spectrum-sensing cognitive wireless networks. *IEEE Journal on Selected Areas in Communications*, *2*(1), 41–56.

Ghosh, C., Cordeiro, C., Agrawal, D. P., & Rao, M. B. (2009). Markov chain existence and hidden Markov models in spectrum sensing. *Proceedings of IEEE International Conference on Pervasive Computing and Communications*.

Giuliano, R., Monti, C., & Loreti, P. (2008). WiMAX fractional frequency reuse for rural environments. *IEEE Wireless Communication Magazine*, *15*(3), 60–65. doi:10.1109/MWC.2008.4547524

Giupponi, L., & Ibars, C. (2009). Bayesian potential games to model cooperation for cognitive radios with incomplete information. *Proceedings of 2009 IEEE ICC*, June 14-18, (pp. 1-6).

Giupponi, L., & Pérez-Neira, A. I. (2008). Fuzzy-based spectrum handoff in cognitive radio networks. *3rd International Conference on Cognitive Radio Oriented Wireless Networks and Communications, CrownCom 2008*, (pp. 1-6).

GNU. (2010). *Radio*. Retrieved October, 2010, from http://www.gnuradio.org

Goldsmith, A., & Wicker, S. (2002). Design challenges for energy-constrained ad hoc wireless networks. *IEEE Wireless Communications*, *9*(4), 8–27. doi:10.1109/MWC.2002.1028874

Gong, S. P. W. (2012). *Robust threshold design for cooperative sensing in cognitive radio networks.* The 31st Annual IEEE International Conference on Computer Communications (INFOCOM'2012 -- Mini-Conference), Orlando, Florida, USA.

Gong, S. P. W. (2012). *Spectrum sensing under distribution uncertainty in cognitive radio networks.* IEEE ICC 2012 - Cognitive Radio and Networks Symposium (ICC'12 CRN), Ottawa, Canada.

Gradshtegn, I. S., & Ryzhik, I. M. (1980). *Table of integrals, series, and products*. New York, NY: Academic Press.

Gulati, K., Evans, B. L., Andrews, J. G., & Tinsley, K. R. (2010, December). Statistics of co-channel interference in a field of Poisson and Poisson-Poisson clustered interferers. *IEEE Transactions on Signal Processing*, *58*(12). doi:10.1109/TSP.2010.2072922

Guo, N., Qiu, R., & Sadler, B. (2005, Oct.). An ultra-wideband autocorrelation demodulation scheme with low-complexity time reversal enhancement. *IEEE Military Communications Conference (Milcom'05)*, (pp. 1–7).

Guo, N., Zhang, J., Qiu, R., & Mo, S. (2007, January 11-13). UWB MISO time reversal with energy detector receiver over ISI channels. *IEEE CCNC'07*, (pp. 629-633).

Guo, N., Sadler, B., & Qiu, R. (2007, December). Reduced-complexity UWB time-reversal techniques and experimental results. *IEEE Transactions on Wireless Communications*, *6*(12), 4221–4226. doi:10.1109/TWC.2007.060251

Haas, Z. J., Deng, J., & Hall, F. R. (2002). Dual busy tone multiple accesses (DBTMA)-A multiple access control scheme for ad hoc networks. *IEEE Transactions on Communications*, *50*(6), 975–985. doi:10.1109/TCOMM.2002.1010617

Haenggi, M. (2005, October). On distances in uniformly random networks. *IEEE Transactions on Information Theory*, *51*(10), 3584–3586. doi:10.1109/TIT.2005.855610

Haenggi, M. (2011). Mean interference in hard-core wireless networks. *IEEE Communications Letters*, *15*(8). doi:10.1109/LCOMM.2011.061611.110960

Haenggi, M., Andrews, J. G., Baccelli, F., Dousse, O., & Franceschetti, M. (2009). Stochastic geometry and random graphs for the analysis and design of wireless networks. *IEEE Journal on Selected Areas in Communications*, *27*(7), 1029–1046. doi:10.1109/JSAC.2009.090902

Haenggi, M., & Ganti, R. K. (2008). Interference in large wireless networks, foundations and trends in networking. *NOW Publishers*, *3*(2), 127–248.

Harris, M. J., Hirst, J. L., & Mossinghoff, M. J. (2008). *Combinatorics and graph theory*. Springer Verlag.

Harrison, K., Mishra, S. M., & Sahai, A. (2010). How much white-space capacity is there? *IEEE Symposium on New Frontiers in Dynamic Spectrum Access Networks (DySPAN 2010)*, Singapore, (pp. 6–9).

Hassibi, B., & Hochwald, B. M. (2003). How much training is needed in multiple-antenna wireless links? *IEEE Transactions on Information Theory*, *49*(4), 951–963. doi:10.1109/TIT.2003.809594

Hatfield, D., & Weiser, P. (2005). *Property rights in spectrum: Taking the next step*. The first IEEE Symposium on New Frontiers in Dynamic Spectrum Access Networks.

Haverinen, H., & Salowey, J. (2004). (Manuscript submitted for publication). EAP SIM authentication. *Work (Reading, Mass.)*.

Haykin, S. (2005). Cognitive radio: Brain-empowered wireless communication. *IEEE Journal on Selected Areas in Communications, 23*(2), 201–220. doi:10.1109/JSAC.2004.839380

He, G., Debbah, M., & Altman, E. (2010, January). A Bayesian game-theoretic approach for distributed resource allocation in fading multiple access channels. *EURASIP Journal on Wireless Communications and Networking - Special Issue on Femtocell Networks, 8.*

Heusse, M., Rousseau, F., Berger-Sabbatel, G., & Duda, A. (2003). *Performance anomaly of 802.11b.* IEEE 22nd Annual Joint Conference of the IEEE Computer and Communications Societies.

Hincapie, R., Tang, J., Xue, G., & Bustamante, R. (2008). QoS routing in wireless mesh networks with cognitive radios. *IEEE Global Telecommunications Conference*, (pp. 1–5).

Ho, W. W. L., & Liang, Y.-C. (2006). Precoder design for MIMO systems with transmit antenna correlations. *Proceedings of the IEEE Vehicular Technology Conference (VTC)*, (pp. 1382-1386).

Ho, W. W. L., & Liang, Y.-C. (2008). User ordering and subchannel selection for power minimization in MIMO broadcast channels using BD-GMD. *Proceedings of the IEEE Vehicular Technology Conference (VTC)*, (pp. 1-5).

Ho, W. W. L., Quek, T. Q. S., & Sun, S. (2010). Decentralized base station processing for multiuser MIMO downlink CoMP. *Proceedings of the IEEE Vehicular Technology Conference (VTC)*, (pp. 1-5).

Ho, Z. K. M., & Gesbert, D. (2010). Balancing egoism and altruism on interference channel: The MIMO case. *Proceedings of the IEEE International Conference on Communication (ICC)*, (pp. 1-5).

Ho, Z. K. M., Kaynia, M., & Gesbert, D. (2010). Distributed power control and beamforming on MIMO interference channels. *Proceedings of IEEE European Wireless.*

Hoang, A. T., Liang, Y.-C., Wong, T. C., Zhang, R., & Zeng, Y. (2008). Opportunistic spectrum access for energy-constrained cognitive radios. *Proceedings of IEEE Vehicular Technology Conference*, (pp. 1559-1563).

Hoang, A. T., & Liang, Y. C. (2008). Downlink channel assignment and power control for cognitive radio network. *IEEE Transactions on Wireless Communications, 7*(8).

Holma, H., & Toskala, A. (Eds.). (2009). *LTE for UMTS: OFDMA and SC-FDMA based radio access.* West Sussex, UK: John Wiley & Sons, Ltd.

Hong, S. (2010). *Multi-resolution Bayesian compressive sensing for cognitive radio primary user detection.* IEEE Global Telecommunications Conference.

Hongyan, L., Yang, L., & Dexian, Z. (2010). Dynamic spectrum access for cognitive radio systems with repeated games. *Proceedings of IEEE Wireless Communications, Networking and Information Security (WCNIS)*, (pp. 59-62).

Horn, R. A., & Johnson, C. R. (1990). *Matrix analysis.* Cambridge University Press.

Hossain, E., & Bhargava, V. (2007). *Cognitive wireless communication networks.* Springer Publishers. doi:10.1007/978-0-387-68832-9

Hossain, E., Niyato, D., & Han, Z. (2009). *Dynamic spectrum access in cognitive radio network.* Cambridge University Press. doi:10.1017/CBO9780511609909

How, K. C., Ma, M., & Qin, Y. (2011). Routing and QoS provisioning in cognitive radio networks. *Computer Networks, 55*(1), 330–342. doi:10.1016/j.comnet.2010.09.008

Hsien-Po, S., & Schaar, M. V. D. (2009). Distributed resource management in multihop cognitive radio networks for delay-sensitive transmission. *IEEE Transactions on Vehicular Technology, 58*(2), 941–953. doi:10.1109/TVT.2008.925308

Hu, Z., Guo, N., & Qiu, R. (2010a, November). *Wideband waveform design for relay cognitive network.* IEEE Military Communications Conference.

Hu, Z., Guo, N., & Qiu, R. (2010b, November). *Wideband waveform optimization for multiple input single output cognitive radio with practical considerations.* IEEE Military Communications Conference.

Huang, J. W., & Krishnamurthy, V. (2009). Truth reveling opportunistic scheduling in cognitive radio systems. *IEEE 10th Workshop on Signal Processing Advances in Wireless Communications*, (pp. 91–95).

Huang, S., Liu, X., & Ding, Z. (2008). Opportunistic spectrum access in cognitive radio networks. *2008 IEEE INFOCOM- The 27th Conference on Computer Communications*, (pp. 1427-1435).

Huang, C.-W., Huang, S.-M., Wu, P.-H., Lin, S.-J., & Hwang, J.-N. (2011). OLM: Opportunistic layered multicasting for scalable IPTV over mobile WiMAX. *IEEE Transactions on Mobile Computing*, *11*(3), 1–9.

Huang, H., Trivellato, M., Hottinen, A., Shafi, M., Smith, P. J., & Valenzuela, R. (2009). Increasing downlink cellular throughput with limited network MIMO coordination. *IEEE Transactions on Wireless Communications*, *8*(6), 2983–2989. doi:10.1109/TWC.2009.080179

Huang, J., & Krishnamurthy, V. (2010). Transmission control in cognitive radio as a Markovian dynamic game: Structural result on randomized threshold policies. *IEEE Transactions on Communications*, *58*(1), 301–310. doi:10.1109/TCOMM.2010.01.080157

Huang, M., Zhou, S., & Wang, J. (2008). Analysis of Tomlinson-Harashima precoding in multiuser MIMO systems with imperfect channel state information. *IEEE Transactions on Vehicular Technology*, *57*(5), 2856–2867. doi:10.1109/TVT.2007.914051

Hu, D., & Mao, S. (2010). Streaming scalable videos over multi-hop cognitive radio networks. *IEEE Transactions on Wireless Communications*, *9*(11), 3501–3511. doi:10.1109/TWC.2010.092810.100098

Hu, D., Mao, S., Hou, Y. T., & Reed, J. H. (2010). Scalable video multicast in cognitive radio networks. *IEEE Journal on Selected Areas in Communications*, *28*(3), 334–344. doi:10.1109/JSAC.2010.100414

Hu, F., Dong, D.-C., & Xiao, Y. (2009). Attacks and countermeasures in multi-hop cognitive radio networks. *International Journal of Security and Networks*, *4*(4), 263–271. doi:10.1504/IJSN.2009.028672

Hulbert, A. P. (2005). Spectrum sharing through beacons. *Proceedings of IEEE International Symposium in Personal, Indoor and Mobile Radio Communications (PIMRC)*, Berlin, Germany, (pp. 989–993).

Hwang, J. (2009). *Multimedia networking: From theory to practice*. Leiden, The Netherlands: Cambridge University Press. Retrieved from http://cdsweb.cern.ch/record/1253368

IEEE Standard 802. (2009). *IEEE standard for local and metropolitan area networks part 16: air interface for broadband wireless access systems.*

IEEE Std 1900.1-2008. (2008). *IEEE standard definitions and concepts for dynamic spectrum access: Terminology relating to emerging wireless networks, system functionality, and spectrum management.*

IEEE Std 802.11-2007. (2007). *IEEE standard for local and metropolitan area networks part 11: Wireless LAN medium access control (MAC) and physical layer (PHY) specifications.*

IEEE Std 802.21-2008. (2009). *IEEE standard for local and metropolitan area networks part 21: Media independent handover services.*

IEEE. 1900.4 WG. (2011). *IEEE standard for architectural building blocks enabling network-device distributed decision making for optimized radio resource usage in heterogeneous wireless access networks.* Retrieved from http://ieeexplore.ieee.org.

Ilow, J., & Hatzinakos, D. (1998). Analytical alpha-stable noise modeling in a poisson field of interferers or scatterers. *IEEE Transactions on Signal Processing*, *46*(6), 1601–1611. doi:10.1109/78.678475

Ishizu, K., Murakami, H., & Miyamoto, G. (2008). Design and implementation of cognitive wireless network based on IEEE P1900.4. *5th Annual IEEE Communications Society Conference on Sensor, Mesh and Ad Hoc Communications and Networks Workshops*, (pp. 1-8).

Isii, K. (1962). On sharpness of tchebycheff-type inequalities. *Annals of the Institute of Statistical Mathematics*, *14*(1), 185–197. doi:10.1007/BF02868641

Iyengar, E. E. (2004). Ambiguous chance constrained problems and robust optimization. *Mathematical Programming*, *107*(1), 37–61.

Jackson, M. O., & Wolinsky, A. (1996). A strategic model of social and economic networks. *Journal of Economic Theory*, *71*(1), 44–74. doi:10.1006/jeth.1996.0108

Jakimoski, G., & Subbalakshmi, K. P. (2008). Denial-of-service attacks on dynamic spectrum access networks. *IEEE International Conference on Communications Workshops*, (pp. 524-528).

Jiang, Y., Li, J., & Hager, W. W. (200). Joint transceiver design for MIMO communications using geometric mean decomposition. *IEEE Transactions on Signal Processing*, *53*(10), 3791–3803. doi:10.1109/TSP.2005.855398

Jing, S., Tse, D. N. C., Soriaga, J. B., Hou, J., Smee, J. E., & Padovani, R. (2008). Multicell downlink capacity with coordinated processing. *EURASIP Journal on Wireless Communications and Networking*, *2008*, 1–19. doi:10.1155/2008/586878

Jing, X., Mau, S. C., Raychaudri, D., & Matyas, R. (2005). *Reactive cognitive radio algorithms for coexistence between 802.11b and 802.16a networks* (pp. 2465–2469). IEEE GLOBECOM.

Jin, J., Xu, H., & Li, B. (2010). Multicast scheduling with cooperation and network coding in cognitive radio networks. *2010. Proceedings - IEEE INFOCOM*, 1–9. doi:10.1109/INFCOM.2010.5461917

Ji, Z., & Liu, K. (2007). Dynamic spectrum sharing: A game theoretical overview. *IEEE Communications Magazine*, *45*(5), 88–94. doi:10.1109/MCOM.2007.358854

Jondral, F. K. (2005). Software-defined radio–Basics and evolution to cognitive radio. *EURASIP Journal on Wireless Communications and Networking*, 275–283.

Joshi, T., Mukherjee, A., Yoo, Y., & Agrawal, D. P. (2008). Airtime fairness for IEEE 802.11 multirate networks. *IEEE Transactions on Mobile Computing*, *7*(4), 513–527. doi:10.1109/TMC.2007.70740

Kahan, J. P., & Rapoport, A. (1984). *Theories of coalition formation*. L. Erlbaum Associates.

Kamal, H., Coupechoux, M., & Godlewski, P. (2009). Inter-operator spectrum sharing for cellular networks using game theory. *Proceedings of IEEE PIMRC*, September 13-16, (pp. 425-429).

Kamerman, A., & Monteban, L. (1997). WaveLAN-II: A high-performance wireless {LAN} for the unlicensed band. *Bell Labs Technical Journal*, *2*(3), 118–133. doi:10.1002/bltj.2069

Kandori, M., Mailath, G. J., & Rob, R. (1993). Learning, mutation, and long run equilibria in games. *Econometrica: Journal of the Econometric Society*, *61*(1), 29–56. doi:10.2307/2951777

Kang, X., Zhang, R., Liang, Y. C., & Garg, H. K. (2009). Optimal power allocation for cognitive radio under primary user's outage loss constraint. *IEEE International Conference on Communications, ICC'09,* (pp. 1-5).

Katta, G., & Murty, S. N. (1987). Some NP-complete problems in quadratic and nonlinear programming. *Mathematical Programming*, *39*(2), 117–129. doi:10.1007/BF02592948

Kaur, P., Khosla, A., & Uddin, M. (2010). Adaptive bandwidth allocation scheme for cognitive radios. *International Journal of Advancements in Computing Technology*, 35-41.

Kaur, P., Uddin, M., & Khosla, A. (2009). An efficient spectrum mobility management strategy in cognitive radio networks. *IEEE International Workshop on Cognitive Wireless Systems*, (pp. 1-6).

Kaur, P., Uddin, M., & Khosla, A. (2009). Cognitive radio as common information exchange reference model for interoperability between heterogeneous networks. *IEEE Workshop on Cognitive Wireless Systems*, (pp. 1-4).

Kaur, P., Khosla, A., & Uddin, M. (2010). *Fuzzy based adaptive bandwidth allocation scheme in cognitive radio networks* (pp. 41–45). IEEE ICT-KE.

Kaur, P., Khosla, A., & Uddin, M. (2011). Cognitive radios: Need, capabilities, standards, applications and research challenges. *International Journal of Computers and Applications*, *30*(1), 31–38.

Kaur, P., Khosla, A., & Uddin, M. (2011). Markovian queuing model for dynamic spectrum allocation in centralized architecture for cognitive radios. *International Journal of Engineering and Technology*, *3*(1), 96–101.

Kaviani, S., & Krzymien, W. A. (2008). Sum rate maximization of MIMO broadcast channels with coordination of base stations. *Proceedings of IEEE Wireless Communication and Networking Conference (WCNC)*, (pp. 1079-1084).

Kawadia, V., & Kumar, P. R. (2005). A cautionary perspective on cross layer design. *IEEE Journal on Wireless Communications, 12*.

Kelly, F. P., Maulloo, A., & Tan, D. (1998). Rate control in communication networks: Shadow prices, proportional fairness and stability. *The Journal of the Operational Research Society*, *49*(3).

Kemeny, J. G., & Snell, J. L. (1976). *Finite Markov chains*. New York, NY: Springer-Verlag.

Khan, S., Peng, Y., Steinbach, E., Sgroi, M., & Kellerer, W. (2006). Application-driven cross layer optimization for video streaming over wireless networks. *IEEE Communications Magazine*, *44*(1), 122–130. doi:10.1109/MCOM.2006.1580942

Kim, H., & Shin, K. (2008). In-band spectrum sensing in cognitive radio networks: Energy detection or feature detection? *Proceedings of the 14th ACM International Conference on Mobile Computing and Networking*, (pp. 14-25).

Kim, I. G., & Choi, J. Y. (2004). *Formal verification of PAP and EAP-MD5 protocols in wireless networks: FDR model checking.* 18th International Conference on Advanced Information Networking and Applications.

Kim, K. W., Park, J., Cho, J., Lim, K., Razzell, C. J., & Kim, K. … Laskar, L. (2007). Interference analysis and sensing threshold of detect and avoid (DAA) for UWB coexistence with WiMax. *Vehicular Technology Conference*, (pp. 1731-1735).

Kim, M. (2011). *Project.* Retrieved from https://engineering.purdue.edu/~ee695n/projects/Misook_Kim.pdf

Kingman, J. (1993). *Poisson processes*. Oxford University Press.

Kliazovich, D., Granelli, F., & Fonseca, N. L. (2009). *Handbook on sensor networks: architectures and cross-layer design for cognitive networks*. River Edge, NJ: World Scientific Publishing Co, Inc.

Klir, G., & Yuan, B. (1995). *Fuzzy sets and fuzzy logic: theory and applications*. Upper Saddle River, NJ: Prentice-Hall PTR.

Knisely, D. N., Yoshizawa, T., & Favichi, F. (2009). Standardization of femtocells in 3GPP. *IEEE Communications Magazine*, *47*(4), 68–75. doi:10.1109/MCOM.2009.5277458

Knuth, R. M. (1996). On the lambert w function. *Advances in Computational Mathematics*, *5*, 329–359. doi:10.1007/BF02124750

Kolodzy, P. (2002, November). *Spectrum policy task force report*. ET Docket No. 02-135.

Konishi, H., & Ray, D. (2003). Coalition formation as a dynamic process. *Journal of Economic Theory*, *110*(1), 1–41. doi:10.1016/S0022-0531(03)00004-8

Kostylev, V. I. (2002). Energy detection of a signal with random amplitude. *IEEE International Conference on Communications*, Vol. 3, (pp. 1606-1610).

Kulkarni, V., Forster, A., & Venayagamoorthy, G. (2011). Computational intelligence in wireless sensor networks: A survey. *IEEE Communications Surveys &. Tutorials*, *99*, 1–29.

Kumar, A., Wang, J., Challapali, K., & Shin, K. G. (2010). Analysis of QoS provisioning in cognitive radio networks: a case study. *Wireless Personal Communications*, *57*(1), 53–71. doi:10.1007/s11277-010-0006-8

Lansford, J. (2004). UWB coexistence and cognitive radio. *International Workshop on Ultra Wideband Systems joint with Conference on Ultrawideband Systems and Technologies*, (pp. 35-39).

Larsson, E. G., Jorswieck, E. A., Lindblom, J., & Mochaourab, R. (2009). Game theory and the flat-fading Gaussian interference channel. *IEEE Signal Processing Magazine*, *26*(5), 18–27. doi:10.1109/MSP.2009.933370

Le, H. S. T., & Ly, H. D. (2008). Opportunistic spectrum access using fuzzy logic for cognitive radio networks. *Second International Conference on Communications and Electronics*, (pp. 240–245).

Le, H.-S. T., & Liang, Q. (2007). An efficient power control scheme for cognitive radios. *IEEE Conference on Wireless Communications and Networking*, (pp. 2559-2563).

Le, H.-S. T., & Ly, H. D. (2008). Opportunistic spectrum access using fuzzy logic for cognitive radio networks. *2nd International Conference on Communications and Electronics, ICCE 2008*, (pp. 240-245).

Leaves, P., Moessner, K., Tafazolli, R., Grandblaise, D., Bourse, D., Tonjes, R., & Breveglieri, M. (2004). Dynamic spectrum allocation in composite reconfigurable wireless networks. *IEEE Communications Magazine*, *42*, 72–81. doi:10.1109/MCOM.2004.1299346

Lee, B. O., Je, H. W., Sohn, I., Shin, O.-S., & Lee, K. B. (2008). Interference-aware decentralized precoding for multicell MIMO TDD systems. *Proceedings of the IEEE Global Communication Conference (GLOBECOM)*, (pp. 1-5).

Lee, B. O., Je, H. W., Shin, O.-S., & Lee, K. B. (2009). A novel uplink MIMO transmission scheme in a multicell environment. *IEEE Transactions on Wireless Communications*, *8*(10), 4981–4987. doi:10.1109/TWC.2009.081690

Lee, M.-L., Chung, H.-Y., & Yu, F.-M. (2003). Modeling of hierarchical fuzzy systems. *Elsevier Fuzzy Sets and Systems*, *138*(2), 343–361. doi:10.1016/S0165-0114(02)00517-1

Lee, Y., Lee, J., & Song, J.-S. (2007). Design and implementation of wireless PKI technology suitable for mobile phone in mobile-commerce. *Computer Communications*, *30*(4), 893–903. doi:10.1016/j.comcom.2006.10.014

Lei, H., Zhang, L., Zhang, X., & Yang, D. (2007). A novel multi-cell OFDMA system structure using fractional frequency reuse. *Proceedings of the IEEE International Symposium on Personal, Indoor and Mobile Radio Communication (PIMRC)*, (pp. 1-5).

Leon, O., Hernandez-Serrano, J., & Soriano, M. (2010). Securing cognitive radio networks. *International Journal of Communication Systems*, *23*(5), 633–652.

Li, J. D., & Yang, C. G. (2009). Pricing based power control game for cognitive radio networks. *International Conference on Telecommunications, ICT 2009*, (pp. 270-275).

Li, M., Jiang, X., & Guibas, L. (2010). Fingerprinting mobile user positions in sensor networks. *2010 International Conference on Distributed Computing Systems*, (pp. 478–487).

Li, S. (2010). *Channel allocation for smooth video delivery over cognitive radio networks: Perspective*. University of Waterloo. Waterloo, Ontario, Canada. Retrieved from http://hdl.handle.net/10012/5416

Li, Y.-Y., Macuha, M., Sousa, E. S., Sato, T., & Nanri, M. (2009). Cognitive interference management in 3G femtocells. *IEEE 20th International Symposium on Personal, Indoor and Mobile Radio Communications*, (pp. 1118-1122).

Liang, Y.-C., & Chin, F. P. S. (2001). Downlink channel covariance matrix (DCCM) estimation and its applications in wireless DS-CDMA systems. *IEEE Journal on Selected Areas in Communications*, *19*(2), 222–232. doi:10.1109/49.914500

Lien, S.-Y., Tseng, C.-C., Chen, K.-C., & Su, C.-W. (2010). Cognitive radio resource management for QoS guarantees in autonomous femtocell networks. *IEEE International Conference on Communications (ICC)*, (pp. 1-6).

Lieven Vandenberghe, S. B. (2007). Generalized Chebyshev bounds via semidefinite programming. *SIAM Review*, *49*(1), 52–64. doi:10.1137/S0036144504440543

Li, J. M. (2008). Soft combination and detection for cooperative spectrum sensing in cognitive radio networks. *IEEE Transactions on Wireless Communications*, 4502–4507.

Like, E., Chakravarthy, V., & Wu, Z. (2007). *Reliable modulation classification at low SNR using spectral correlation.* IEEE Consumer Communication and Networking Conference.

Lin, F., Hu, Z., Hou, S., Yu, J., Zhang, C., Guo, N., et al. (2011). *Cognitive radio network as wireless sensor network (II): Security consideration.* National Aerospace & Electronics Conference.

Lin, S., Ho, W. W. L., & Liang, Y.-C. (2008). Block diagonal geometric mean decomposition (BD-GMD) for MIMO broadcast channels. *IEEE Transactions on Wireless Communications*, *7*(7), 2778–2789. doi:10.1109/TWC.2008.070171

Lin, Y., Chen, B., & Varshney, P. K. (2005). Decision fusion rules in multi-hop wireless sensor networks. *IEEE Transactions on Aerospace and Electronic Systems*, *41*(2), 475–488. doi:10.1109/TAES.2005.1468742

Liu, K. (2011, April). Cognitive radio and game theory. *IEEE Spectrum*.

Liu, Y., & Zhou, Q. (2009). State of the art in cross-layer design for cognitive radio wireless networks. *IEEE 2009 International Symposium on Intelligent Ubiquitous Computing and Education*, (pp. 366-369).

Liu, F., Jiang, L., & He, C. (2010). Advanced joint transceiver design for block-diagonal geometric-mean-decomposition-based multiuser MIMO systems. *IEEE Transactions on Vehicular Technology*, *59*(2), 692–703. doi:10.1109/TVT.2009.2032555

Liu, Q. W., Zhou, S. L., & Giannakis, G. B. (2005, May). Cross-layer scheduling with prescribed QoS guarantees for adaptive wireless networks. *IEEE Journal on Selected Areas in Communications*, *23*(5), 1056–1066. doi:10.1109/JSAC.2005.845430

Low, T. P., Pun, M. O., & Kuo, C. C. J. (2008). Optimized opportunistic multicast scheduling over cellular networks. *IEEE Global Telecommunications Conference*, (pp. 1–5).

Lowen, R. (1996). *Fuzzy set theory: Basic concepts, techniques and bibliography*. Dordrecht, The Netherlands: Kluwer Academic Publishers.

Low, T.-P., Pun, M.-O., Hong, Y.-W., & Kuo, C.-C. (2010). Optimized opportunistic multicast scheduling (OMS) over wireless cellular networks. *IEEE Transactions on Wireless Communications*, *9*(2), 791–801. doi:10.1109/TWC.2010.02.090387

Ludwig, J. A., & Reynolds, J. F. (1998). *Statistical ecology: A primer on methods and computing*. New York: John Wiley & Sons.

Lu, J., & Ma, M. (2011). Cognitive radio-based framework and self-optimizing temporal-spectrum block scheduling for QoS provisioning in WiMAX. *Computer Networks*, *55*(6), 1291–1309. doi:10.1016/j.comnet.2010.12.003

Lunden, J., Kassam, S. A., & Koivunen, V. (2010). Robust nonparametric cyclic correlation-based spectrum sensing for cognitive radio. *IEEE Transactions on Signal Processing*, *58*(1), 38–52. doi:10.1109/TSP.2009.2029790

Luo, H., Ci, S., Wu, D., & Tang, H. (2010). Cross-layer design for real-time video transmission in cognitive wireless networks. *2010 INFOCOM IEEE Conference on Computer Communications Workshops*, (pp. 1-6).

Lyrtech Incorporated. (2010). *Home page*. Retrieved October, 2010, from http://www.lyrtech.com

MacKenzie, A. B., & Dasilva, L. A. (2005). *Game theory for wireless engineers* (pp. 1–86). Synthesis Lectures on Communications.

Mamdani, E. H., & Assilian, S. (1973). An experiment in linguistic synthesis with a fuzzy logic controller. *International Journal of Man-Machine Studies*, *7*, 1–13. doi:10.1016/S0020-7373(75)80002-2

Market, M. (2006). *Clash of the titans — WiMAX and 4G: The battle for convergence is joined.*

Marojevic, V., Salazar, J., Revés, X., & Gelonch, A. (2008). Resource modeling for a joint resource management in cognitive radio. *IEEE ICC 2008*, (pp. 4175-4180).

Marsch, P., & Fettweis, G. (2008). On multicell cooperative transmission in backhaul-constrained cellular systems. *Annals in Telecommunication*, *63*(5), 253–269. doi:10.1007/s12243-008-0028-3

Matérn, B. (1986). *Spatial variation* (2nd ed.). Springer Lecture Notes in Statistics.

Mathur, C. N., & Subbalakshmi, K. P. (2007). Digital signature for centralized DSA networks. *4th IEEE conference on Consumer Communications and Networking Conference*, (pp. 1037-1041).

Mathur, S., Sankar, L., & Mandayam, N. B. (2008). Coalitions in cooperative wireless networks. *IEEE Journal on Selected Areas in Communications*, *26*(7), 1104–1115. doi:10.1109/JSAC.2008.080908

Mchenry, M. (2003). *Spectrum white space measurements*. New America Foundation Broadband Forum.

Mercier, B., Fodor, V., Thobaben, R., Skoglund, M., Koivunen, V., & Lindfors, S. (2008). *Sensor networks for cognitive radio: Theory and system design*. ICT Mobile Summit.

Mian, O., Zhou, R., Li, X., Hong, S., & Wu, Z. (2009). A software-defined radio based cognitive radio demonstration over FM band. *Proceedings of International Conference on Wireless Communications and Mobile Computing*, (pp. 495-499).

Min, R., Qu, D., Cao, Y., & Zhong, G. (2008). Interference avoidance based on multi-step-ahead prediction for cognitive radio. *Proceedings of IEEE Singapore International Conference on Communication Systems*, (pp. 227-231).

Mishra, S. M., Sahai, A., & Brodersen, R. W. (2006). Cooperative sensing among cognitiveradios. *IEEE International Conference on Communications*, Vol. 4, (pp. 1658-1663).

Mitola, J. (2000). *Cognitive radio: An integrated agent architecture for software defined radio*. Ph.D. dissertation, Sweden, Royal Institute of Technology (KTH).

Mitola, J. (2001). Cognitive radio for flexible mobile multimedia communications. *Springer Mobile Networks and Applications*, *6*(5), 435–441. doi:10.1023/A:1011426600077

Mitola, J. III, & Maguire, G. Q. Jr. (1999). Cognitive radio: Making software radios more personal. *IEEE Personal Communications*, *6*(4), 13–18. doi:10.1109/98.788210

Mohamadian, M., & Stonier, R. J. (2000). Hierarchical fuzzy logic control. *IPMU Conference*, (pp. 622-629).

Murthy, C. S. R., & Manoj, B. S. (2004). *Ad hoc wireless networks: Architectures and protocols*. Prentice Hall PTR.

Myerson, R. B. (1977). Graphs and cooperation in games. *Mathematics of Operations Research*, *2*(3), 225–229. doi:10.1287/moor.2.3.225

Myerson, R. B. (1997). *Game theory: Analysis of conflict*. Harvard University Press.

Namvar Gharehshiran, O., & Krishnamurthy, V. (2010). Coalition formation for bearings-only localization in sensor networks. *IEEE Transactions on Signal Processing*, *58*(8), 4322–4338. doi:10.1109/TSP.2010.2049201

Neel, J. O. (2006). *Analysis and design of cognitive radio networks and distributed radio resource management algorithms*. Ph.D dissertation, Virginia Polytechnic Institute and State University.

Nekovee, M. (2010). A survey of cognitive radio access to TV white spaces. *International Journal of Digital Multimedia Broadcasting*, 2010.

Nesterov, Y., & Nemirovskii, A. (1994). *Interior-point polynomial algorithms in convex programming*. SIAM Studies in Applied and Numerical Mathematics. doi:10.1137/1.9781611970791

Ng, B. L., Evans, J. S., Hanly, S. V., & Aktas, D. (2008). Distributed downlink beamforming with cooperative base stations. *IEEE Transactions on Information Theory*, *54*(12), 5491–5499. doi:10.1109/TIT.2008.2006426

Ngo, D. T., Tellambura, C., & Nguyen, H. H. (2009). Resource allocation for OFDM-based cognitive radio multicast networks. *IEEE Wireless Communications and Networking Conference*, (pp. 1–6).

Nie, N., & Comaniciu, C. (2005). Adaptive channel allocation spectrum etiquette for cognitive radio networks. *Proceedings of IEEE DySPAN*, (pp. 269-278).

Niyato, D., & Hossain, E. (2009). Dynamics of network selection in heterogeneous wireless networks: An evolutionary game approach. *IEEE Transactions on Vehicular Technology*, *58*(4), 2008–2017. doi:10.1109/TVT.2008.2004588

Niyato, D., Hossain, E., & Han, Z. (2009). Dynamic spectrum access in IEEE 802.22-based cognitive wireless networks: A game theoretic model for competitive spectrum bidding and pricing. *IEEE Wireless Communications*, *16*(2), 16–23. doi:10.1109/MWC.2009.4907555

Nolan, K. E., Rondeau, T. W., Sutton, P. D., Bostian, C. W., & Doyle, L. E. (2006). *A framework for implementing cognitive functionality*. 51st SDR Forum General Meeting and 2006 Technical Conference.

Nöldeke, G., & Samuelson, L. (1993). An evolutionary analysis of backward and forward induction. *Games and Economic Behavior*, *5*(3), 425–454. doi:10.1006/game.1993.1024

Nychis, G., Hottelier, T., Yang, Z., Seshan, S., & Steenkiste, P. (2009). Enabling MAC protocol implementations on software-defined radios. *Proceedings of USENIX Symposium on Networked Systems Design and Implementation*, (pp. 91-105).

OFCOM. (2002). *Trading guidance notes: A practical user guide for spectrum trading*. (OFW224, OFCOM, Technical Report).

O'Hara, B., & Petrick, A. (1999). *The IEEE 802.11 handbook: A designer's companion*. New York, NY: Standards Information Network, IEEE Press.

Oh, S. W., Le, T. P. C., Zhang, W., Ahmed, S. N. A., Zeng, Y., & Kua, K. J. M. (2009). TV white-space sensing prototype. *Wireless Communications and Mobile Computing*, *9*(11), 1543–1551. doi:10.1002/wcm.708

Oner, M., & Jondral, F. (2007). On the extraction of the channel allocation information in spectrum pooling systems. *IEEE Journal on Selected Areas in Communications*, *25*(3), 558–565. doi:10.1109/JSAC.2007.070406

Ong, E. H., & Khan, J. Y. (2009a). *On load adaptation for multirate multi-AP multimedia WLAN-based cognitive networks*. IFIP Wireless Days Conference.

Ong, E. H., & Khan, J. Y. (2009b). *A unified QoS-inspired load optimization framework for multiple access points based wireless LANs*. IEEE Wireless Communications and Networking Conference.

Ong, E. H., Khan, J. Y., & Mahata, K. (2010). *A simple model for non-homogeneous and non-saturated IEEE 802.11 DCF infrastructure BSS*. IEEE 12th International Conference on Communication Systems.

Ong, E. H., Khan, J. Y., & Mahata, K. (2011). *Radio resource management of composite wireless networks: Predicitve and reactive approaches*. IEEE Transactions on Mobile Computing.

Ong, E. H. (2010). *A generalized cooperative and cognitive radio resoure management architecture for future wireless networks. Ph.D.* NSW, Australia: The University of Newcastle.

Ong, E. H., & Khan, J. Y. (2010a). Cooperative radio resource management framework for future IP-based multiple radio access technologies environment. *Computer Networks*, *54*(7), 1083–1107. doi:10.1016/j.comnet.2009.11.001

Ong, E. H., & Khan, J. Y. (2010b). On optimal network selection in a dynamic multi-RAT environment. *IEEE Communications Letters*, *14*(3), 217–219. doi:10.1109/LCOMM.2010.03.092378

Osborne, M. J., & Rubinstein, A. (1994). *A course in game theory*. MIT Press.

Otal, B., & Alonso, L. (2004). A cross layer energy saving mechanism for an enhancement of 802.11 WLAN systems. *IEEE 59th Vehicular Technology Conference (VTC 2004 Spring)*, (p. 3).

Oteri, O., & Paulraj, A. (2008). Multicell optimization for diversity and interference mitigation. *IEEE Transactions on Signal Processing*, *56*(5), 2050–2061. doi:10.1109/TSP.2007.909338

Owen, G. (1995). *Game theory*. San Diego, CA: Academic Press.

Palomar, D. P., & Chiang, M. (2006). A tutorial on decomposition methods for network utility maximization. *IEEE Journal on Selected Areas in Communications*, *24*(8), 1439–1451. doi:10.1109/JSAC.2006.879350

Pang, Q., Leung, V. C. M., & Liew, S. C. (2008). An enhanced autorate algorithm for wireless local area networks employing loss differentiation. *IEEE Transactions on Vehicular Technology*, *57*(1), 521–531. doi:10.1109/TVT.2007.905418

Papoulis, A., & Pillai, S. U. (2002). *Probability, random variables and stochastic processes* (4th ed.). McGraw Hill.

Pareek, U., Naeem, M., & Lee, D. C. (2010). *An efficient relay assignment scheme for multiuser cognitive radio networks with discrete power control*. IEEE 6th International Conference on Wireless and Mobile Computing, Networking and Communications (WiMob).

Park, C.-H., Kim, S.-W., Lim, S.-M., & Song, M.-S. (2007). HMM based channel status predictor for cognitive radio. *Proceedings of Asia-Pacific Microwave Conference*.

Parvin, S., Han, S., et al. (2010). Trust-based authentication for secure communication in cognitive radio networks. *2010 IEEE/IFIP International Conference on Embedded and Ubiquitous Computing*, (pp. 589-596).

Parvin, S., Han, S., Tian, B., & Xie, M. (2010). Authenticated spectrum sharing for secondary users in cognitive radio networks. *2nd IEEE International Conference on Network Infrastructure and Digital Content*, (pp. 509-513).

Pawełczak, P. (2008). *Technical challenges of cognitive radio-related systems*. First Competition and Regulation in Network Industries Conference, Brussels, Belgium.

Pawełczak, P., Pollin, S., So, H. S. W., Motamedi, A., Bahai, A., Prasad, R. V., & Hekmat, R. (2008). *State of the art in opportunistic spectrum access medium access control design*. The 3rd International Conference on Cognitive Radio Oriented Wireless Networks and Communications.

Pawelczak, P., Angeles, L., Nolan, K., Doyle, L., Oh, S. W., & Cabric, D. (2011). Cognitive radio: Ten years of experimentation and development. *IEEE Communications Magazine*, *49*(03), 90–100. doi:10.1109/MCOM.2011.5723805

Pedrycz, W., & Gomide, F. (1998). *An introduction to fuzzy sets: Analysis and design*. MIT Press.

Peng, Q., Zeng, K., Wang, J., & Li, S. (2006). A distributed spectrum sensing scheme based on credibility and evidence theory in cognitive radio context. *IEEE 17th International Symposium on Personal, Indoor and Mobile Radio Communications*, (pp. 1-5).

Peng, Q., Cosman, P. C., & Milstein, L. B. (2011). Spoofing or jamming: Performance analysis of a tactical cognitive radio adversary. *IEEE Communications*, *29*(4), 903–911.

Pensavalle, C., & Pierri, G. (2000). On existence of Nash equilibria of games with constraints on multistrategies. *Journal of Optimization Theory and Applications*, *107*(3), 601–613. doi:10.1023/A:1026403417009

Petrova, M., & Mähönen, P. (2007). Cognitive resource manager: A crosslayer architecture for implementing cognitive radio networks. In Fittzek, F., & Katz, M. (Eds.), *Cognitive wireless networks*.

Polacek, P., Yang, T.-Y., & Huang, C.-W. (2011). *Joint opportunistic spectrum access and scheduling for layered multicasting over cognitive radio networks*. IEEE International Workshop on Multimedia Signal Processing, 2011.

Qin, H., & Su, J. (2009). Opportunistic scheduling with power control in cognitive radio networks. *International Symposium on Computer Network and Multimedia Technology*, (pp. 1-4). CNMT.

Qiu, R. C., Chen, Z., Guo, N., Song, Y., Zhang, P., Li, H., & Lai, L. (2010). Towards a real-time cognitive radio network testbed: Architecture, hardware platform, and application to smart grid. *Proceedings of IEEE Workshop on Networking Technologies for Software-Defined Radio and White Space*.

Qiu, R. C., Zhou, C., Guo, N., & Zhang, J. Q. (2006). Time reversal with MISO for ultrawideband communications: Experimental results. *Antennas and Wireless Propagation Letters*, *5*, 269–273. doi:10.1109/LAWP.2006.875888

Qiu, R., Guo, N., Li, H., Wu, Z., Chakravarthy, V., Song, Y., & Chen, Z. (2009). A unified multi-functional dynamic spectrum access framework: Tutorial, theory and multi-GHz wideband testbed. *Sensors (Basel, Switzerland)*, *9*(8), 6530–6603. doi:10.3390/s90806530

Quan, Z., Shuguang, C., & Sayed, A. H. (2008). Optimal linear cooperation for spectrum sensing in cognitive radio networks. *IEEE Journal of Selected Topics in Signal Processing*, *2*(1), 28–40. doi:10.1109/JSTSP.2007.914882

Raghuram, R., Robert, Q., Zhen, H., Shujie, H., Marbin, P., & Gang, Z. (2011). Cognitive radio for smart grid: Theory, algorithms, and security. *International Journal of Digital Multimedia Broadcasting*, 2011.

Raju, G. V. S., & Zhou, J. (1993). Adaptive hierarchical fuzzy controller. *IEEE Transactions on System and Cybernetics*, *23*(4), 973–980. doi:10.1109/21.247882

Rakus-Andersson, E., Zettervall, H., & Forssell, H. (2011). Fuzzy controllers in evaluation of survival length in cancer patients. In *Recent Advances in Fuzzy Sets, Intuitionistic Fuzzy Sets, Generalized Nets and Related Topics,* Vol. 2: Applications, Polish Academy of Sciences, System Research Institute, Warsaw, (pp. 203-222).

Rakus-Andersson, E. (2007). *Fuzzy and rough techniques. Medical diagnosis and medication.* Berlin, Germany: Springer-verlag.

Rakus-Andersson, E. (2011in press). The Mamdani controller with modeled families of constraints in evaluation of cancer patient survival length. In Ramanna, S., Jain, L. C., & Howlett, R. J. (Eds.), *Emerging paradigms in machine learning.*

Rappaport, T. (2001). *Wireless communications: Principles and practice* (2nd ed.). Upper Saddle River, NJ: Prentice Hall PTR.

Rawat, A. S., Anand, P., Chen, H., & Varshney, P. K. (2010). *Countering byzantine attacks in cognitive radio networks.* IEEE International Conference on Acoustics, Speech, and Signal Processing.

Ray, D. (2007). *A game-theoretic perspective on coalition formation*. Oxford University Press. doi:10.1093/acprof:oso/9780199207954.001.0001

Razavilar, J., Liu, K. J. R., & Marcus, S. I. (2002). Jointly optimized bit-rate/delay control policy for wireless packet networks with fading channels. *IEEE Transactions on Communications*, *50*(3), 484–494. doi:10.1109/26.990910

Reddy, Y. B., & Bullmaster, C. (2008). *Cross-layer design in wireless cognitive networks*. Ninth International Conference on Parallel and Distributed Computing, Applications and Technologies.

Rice University Wireless Open access Research Platform (WARP). (2011). Retrieved from http://warp.rice.edu/trac

Rice University. (2010). *WARP project*. Retrieved October, 2010, from http://warp.rice.edu

Richard, J. L., & Anantharam, V. (2002). Optimal routing control: Repeated game approach. *IEEE Transactions on Automatic Control*, *47*(3), 437–450. doi:10.1109/9.989076

Rieser, C. J., Rondeau, T. W., Bostian, C. W., & Gallagher, T. M. (2004). Cognitive radio testbed: Further details and testing of a distributed genetic algorithm based cognitive engine for programmable radios. *Military Communications Conference*, Vol. 3, (pp. 1437-1443).

Rieser, C. (2004). *Biologically inspired cognitive radio engine model utilizing distributed genetic algorithms for secure and robust wireless communications and networking*. Virginia Polytechnic Institute and State University.

Riihijärvi, J., Wellens, M., & Mähönen, P. (2006). Link-layer abstractions for utility-based optimization in cognitive wireless networks. *Proceedings of CrownCom 2006*.

Rivest, R. (2009). The MD4 message digest algorithm. *Advances in Cryptology, CRYPT0'90*, (pp. 303-311).

Rivest, R. (1992). *RFC1321: The MD5 message-digest algorithm*. RFC Editor United States.

Rondeau, T. W., & Rieser, C. J. B. L., & Bostian, C. W. (2004). Cognitive radios with genetic algorithms: Intelligent control of software defined radios. *Proceedings of SDR Forum Technical Conference and Product Exposition*, (pp. C-3–C-8). Phoenix, USA.

Rubinstein, M. G., Moraes, I. M., Campista, M. E., Costa, L. H., & Duarte, O. C. (2006). A survey on wireless ad hoc networks. *IEEE Conference on Mobile and Wireless Communication Networks*, (pp. 1-33).

Saad, W., Han, Z., Debbah, M., Hjørungnes, A., & Basar, T. (2009). A game-based self-organizing uplink tree for VoIP services in IEEE 802.16j networks. *IEEE International Conference on Communications*, (pp. 1-5).

Saad, W., Han, Z., Debbah, M., Hjørungnes, A., & Basar, T. (2009). Coalitional game theory for communication networks. *IEEE Signal Processing Magazine*, *26*(5), 77–97. doi:10.1109/MSP.2009.000000

Saad, W., Han, Z., Debbah, M., Hjørungnes, A., & Basar, T. (2009). *Coalitional games for distributed collaborative spectrum sensing in cognitive radio networks* (pp. 2114–2122). IEEE INFOCOM.

Sadek, M., Tarighat, A., & Sayed, A. H. (2007). A leakage-based precoding scheme for downlink multi-user MIMO channels. *IEEE Transactions on Wireless Communications*, *6*(5), 1711–1721. doi:10.1109/TWC.2007.360373

Sadler, M. B., & Zhao, Q. (2007). A survey of dynamic spectrum access. *IEEE Signal Processing Magazine*, *24*(3), 79–89. doi:10.1109/MSP.2007.361604

Sandholm, T., Larson, K., Andersson, M., Shehory, O., & Tohmé, F. (1999). Coalition structure generation with worst case guarantees. *Artificial Intelligence*, *111*, 209–238. doi:10.1016/S0004-3702(99)00036-3

Saraydar, C. U., Mandayam, N. B., & Goodman, D. J. (2002). Efficient power control via pricing in wireless data networks. *IEEE Transactions on Communications*, *50*(2), 291–303. doi:10.1109/26.983324

Scaperoth, D., Le, B., Rondeau, T., Maldonado, D., Bostian, C. W., & Harrison, S. (2006). Cognitive radio platform development for interoperability. *IEEE Military Communications Conference. MILCOM* 2006, (pp. 1-6).

Schaar, M. v., & N, S. S. (2005). Cross-layer wireless multimedia transmission: Challenges, principles and new paradigms. *IEEE Wireless Communications*, *12*(4), 50-58.

Schmid, T., Sekkat, O., & Srivastava, M. B. (2007). An experimental study of network performance impact of increased latency in software defined radios. *Proceedings of International Workshop on Wireless Network Testbeds, Experimental Evaluation and Characterization*, Montreal, (pp. 59-66).

Schwartz, M. (2005). *Mobile wireless communications*. Cambridge University Press.

Scutari, G., Barbarossa, S., & Palomar, D. (2006). Potential games: A framework for vector power control problems with coupled constraints. *Proceedings of IEEE International Conference on Acoustics, Speech, and Signal Processing (ICASSP)*.

Shakkottai, S., Rappaport, T., & Karlsson, P. (2003). Cross-layer design for wireless networks. *IEEE Communications Magazine*, *41*(10), 74–80. doi:10.1109/MCOM.2003.1235598

Shamai (Shitz). S., & Zaidel, B. M. (2001). Enhancing the cellular downlink capacity via co-processing at the transmitting end. *Proceedings of IEEE Vehicular Technology Conference (VTC)*, (pp. 1745-1749).

Shan, Y. (2005). Cross-layer techniques for adaptive video streaming over wireless networks. *EURASIP Journal on Advances in Signal Processing*, *2*, 220–228. doi:10.1155/ASP.2005.220

Sharma, S., & Dubey, R. (2010). *Distributed shared spectrum techniques for cognitive wireless radio networks*. The International Conference on Computational Intelligence and Communication Networks (CICN2010).

Shellhammer, S. J. (2009). A comparison of geo-location and spectrum sensing in cognitive radio. *18th International Conference on Computer Communications and Networks (ICCCN)*, (pp. 1-6).

Sherman, M., Mody, A. N., Martinez, R., Rodriguez, C., & Reddy, R. (2008). IEEE standards supporting cognitive radio and networks, dynamic spectrum access and coexistence. *IEEE Communications Magazine*, *46*(7), 72–79. doi:10.1109/MCOM.2008.4557045

Shi, Y., & Hou, Y. T. (2008). A distributed optimization algorithm for multi-hop cognitive radio networks. *The 27th Conference on IEEE Computer Communications*, (pp. 1292-1300). INFOCOM.

Shim, S., Kwak, J. S., Heath, R. W. Jr, & Andrews, J. G. (2008). Block diagonalization for multi-user MIMO with other-cell interference. *IEEE Transactions on Wireless Communications*, *7*(7), 2671–2681. doi:10.1109/TWC.2008.070093

Shiu, Y., Chang, S., Wu, H., Huang, S., & Chen, H. (2011). Physical layer security in wireless networks: A tutorial. *IEEE Wireless Communications*, *18*(2), 66–74. doi:10.1109/MWC.2011.5751298

Shi, Y., & Hou, Y. T. (2007). *Optimal power control for multi-hop software defined radio networks*. IEEE Infocom.

Siddique, T. (2010). *Spectrum optimization in cognitive radio networks using genetic algorithm*. MSc Thesis, Sweden, Blekinge Institute of Technology.

Simeone, O., Somekh, O., & Poor, H. V., & Shamai (Shitz), S. (2009). Local base station cooperation via finite-capacity links for the uplink of linear cellular networks. *IEEE Transactions on Information Theory*, *55*(1), 190–204. doi:10.1109/TIT.2008.2008151

Simon, F. F.-S. (2007). On the energy detection of unknown signals over fading channels. *IEEE Transactions on Communications*, *5*, 21–24.

Simpson, W. (1994). *PPP challenge handshake authentication protocol* (CHAP). RFC Editor United States, 1996.

Sivanantha, A., Ehsan, S., Dhamodaran, A. N., & Gatimu, K. (2010). *A survey on quality of service (QoS) provisioning for multimedia transmission in cognitive radio networks (CRNs)*. Oregon State University.

Sodagari, S., & Bil, S. G. (2010). *Denial of service attacks in cognitive radio networks through channel eviction triggering*. IEEE Global Telecommunications Conference.

Somekh, O., Simeone, O., Bar-Ness, Y., & Haimovich, A. M., & Shamai (Shitz), S. (2009). Cooperative multicell zero-forcing beamforming in cellular downlink channels. *IEEE Transactions on Information Theory*, *55*(7), 3206–3219. doi:10.1109/TIT.2009.2021371

Song, G., & Li, Y. (2005). Utility-based resource allocation and scheduling in OFDM-based wireless broadband networks. *IEEE Communications Magazine*, *43*(12), 127–134. doi:10.1109/MCOM.2005.1561930

Sousa, A. G. (2005). Collaborative spectrum sensing for opportunistic access in fading environments. *IEEE International Symposium on New Frontiers in Dynamic Spectrum Access Networks*, (pp. 131-136).

Sousa, E. S., & Silvester, J. A. (1990). Optimum transmission ranges in a direct-sequence spread-spectrum multihop packet radio network. *IEEE Journal on Selected Areas in Communications*, *8*(5), 762–771. doi:10.1109/49.56383

Spectrum Policy Task Force. (2002). *Spectrum policy task force report*. Federal Communications Commission ET Docket 02-135.

Srivastava, V., & Motani, M. (2005, December). Cross-layer design: A survey and the road ahead. *IEEE Communications Magazine*, *43*(12), 112–119. doi:10.1109/MCOM.2005.1561928

Srivastava, V., Neel, J., Mackenzie, A. B., Menon, R., Dasilva, L. A., & Hicks, J. E. (2005). Using game theory to analyze wireless ad hoc networks. *IEEE Communications Magazine*, *7*(4), 46–56.

Std, F. C. C. (2003). *Notice of proposed rulemaking and order*. ET Docket No 03-222, The Federal Communications Commission (FCC) Std., (2003). Retrieved from www.fcc.gov

Std, T. R. A. I. (2012). *Telecom regulatory authority of India* (TRAI). Retrieved from http://www.trai.gov.in

Std, W. P. C. (2011). *National frequency allocation plan*. Wireless Planning & Coordination (WPC) Std. Retrieved from www.wpc.dot.gov.in

Steenkiste, P., Sicker, D., Minden, G., & Raychaudhuri, D. (March 2009). *Future directions in cognitive radio network research*. NSF Workshop Report.

Steenkiste, P., Sicker, D., Minden, G., & Raychaudhuri, D. (2009). *Future directions in cognitive radio network research*. NSF.

Steenkiste, P., Sicker, D., Minden, G., & Raychaudhuri, D. (2009). Future directions in cognitive radio network research. *Network*, 1–40. Retrieved from www.cs.cmu.edu/~prs/NSF_ CRN_Report_Final.pdf

Stevenson, C., Chouinard, G., Shellhammer, S., & Caldwell, W. (2009). IEEE 802.22: The first cognitive radio wireless regional area network standard. *IEEE Communications Magazine*, *47*(1), 130–138. doi:10.1109/MCOM.2009.4752688

Stoyan, D., Kendall, W., & Mecke, J. (1996). *Stochastic geometry and its applications*. John Wiley.

Sturm, J. F. (1999). Using SeDuMi 1.02, a {MATLAB} toolbox for optimization over symmetric cones. *Optimization Methods and Software*, *11*(1), 625–653. doi:10.1080/10556789908805766

Su, J., & Wu, W. (2009). Wireless spectrum prediction model based on time series analysis method. *Proceedings of ACM Workshop on Cognitive Radio Networks*, (pp. 61-66).

Su, H., & Zhang, X. (2008). Cross-layer based opportunistic MAC protocols for QoS provisionings over cognitive radio wireless networks. *IEEE Journal on Selected Areas in Communications*, *26*(1), 118–129. doi:10.1109/JSAC.2008.080111

Sun, C. H., Zhang, W., & Letaief, K. B. (2007). Cluster-based cooperative spectrum sensing in cognitive radio systems. *IEEE International Conference on Communications 2007*, (pp. 2511-2515).

Sutton, P. D., Lotze, J., Nolan, K. E., & Doyle, L. E. (2007). Cyclostationary signature detection in multipath Rayleigh fading environments. *IEEE International Conference on Cognitive Radio Oriented Wireless Networks and Communications.*

Sutton, P., Nolan, K., & Doyle, L. (2008). Cyclostationary signatures in practical cognitive radio applications. *IEEE Journal on Selected Areas in Communications*, *26*(1), 13–24. doi:10.1109/JSAC.2008.080103

Takagi, T., & Sugeno, M. (1985). Fuzzy identification of systems and its applications to modelling and control. *IEEE Transactions on Systems, Man, and Cybernetics*, *15*, 116–132.

Tan, G., & Guttag, J. (2004). *Time-based fairness improves performance in multi-rate WLANs*. USENIX Annual Technical Conference.

Tan, K., Zhang, J., Fang, J., Liu, H., Ye, Y., Wang, S., & Voelker, G. (2009). Sora: High performance software radio using general purpose multi-core processors. *Proceedings of USENIX Symposium on Networked Systems Design and Implementation*, (pp. 75-90).

Tandra, R., Sahai, A., & Mishra, S. M. (2009). What is a spectrum hole and what does it take to recognize one? *Proceedings of the IEEE*, *97*(5), 824–848. doi:10.1109/JPROC.2009.2015710

Tan, L. T., & Kong, H. Y. (2010). *A novel and efficient mixed-signal compressed sensing for wide-band cognitive radio. International Forum on Strategic Technology*. IFOST.

Taricco, G. (2011). Optimization of linear cooperative spectrum sensing for cognitive radio networks. *IEEE Journal of Selected Topics in Signal Processing*, *5*(1), 77–86. doi:10.1109/JSTSP.2010.2055537

Teo, K. M. (2007). *Nonconvex robust optimization*. Massachusetts Institute of Technology.

Tepandi, J., Andahhirov, I. T. X. C., & Vassiljev, S. (2010). Wireless PKI security and mobile voting. *Computer*, *43*(6), 54–60. doi:10.1109/MC.2009.418

Thomas, R. W. (2007). *Cognitive networks. Ph.D.* Blacksburg, VA, USA: Virginia Polytechnic Institute and State University.

Thomas, R. W., Frlend, D. H., DaSilva, L. A., & MacKenzie, A. B. (2006). Cognitive networks: Adaptation and learning to achieve end-to-end performance objectives. *IEEE Communications Magazine*, 51–57. doi:10.1109/MCOM.2006.273099

Thrall, R. M., & Lucas, W. F. (1963). N-person games in partition function form. *Naval Research Logistics Quarterly*, *10*(1), 281–298. doi:10.1002/nav.3800100126

Tian, C., & Yuan, D. (2008). Cross layer opportunistic scheduling for multiclass users in cognitive radio networks. *2008 4th International Conference on Wireless Communications, Networking and Mobile Computing*, (pp. 1-4).

Tinnirello, I., & Choi, S. (2005). *Temporal fairness provisioning in multi-rate contention-based 802.11e WLANs*. IEEE 6th International Symposium on a World of Wireless Mobile and Multimedia Networks.

Tkachenko, A., Cabric, D., & Brodersen, R. (2006). Cognitive radio experiments using reconfigurable BEE2. *Proceedings of Asilomar Conference on Signals, Systems, and Computers*, (pp. 2041-2045).

Tkachenko, A., Cabric, D., & Brodersen, R. W. (2007). Cyclostationary feature detector experiments using reconfigurable BEE2. *IEEE International Symposium on New Frontiers in Dynamic Spectrum Access Networks*, (pp. 216–219).

Tölli, A., Pennanen, H., & Komulainen, P. (2009). SINR balancing with coordinated multi-cell transmission. *Proceedings of IEEE Wireless Communication and Networking Conference (WCNC)*, (pp. 1-6).

Topkis, D. (1998). *Supermodularity and complementarity*. Princeton University Press.

Triantafyllopoulou, D.-K., Passas, N., & Kaloxylos, A. (2007). *A cross-layer optimization mechanism for multimedia traffic over IEEE 802.16 networks*. Paris, France: European Wireless.

Tsai, J. T., & Cruz, R. L. (2011). Opportunistic multicast scheduling for information streaming in cellular networks. *IEEE Transactions on Wireless Communications*, *10*(6), 1776–1785. doi:10.1109/TWC.2011.040411.100517

Tsekourasa, G., Sarimveisb, H., & George, E. K. (2005). A hierarchical fuzzy-clustering approach to fuzzy modelling. *Fuzzy Sets and Systems*, *150*, 245–266. doi:10.1016/j.fss.2004.04.013

Valenta, V., Maršálek, R., Baudoin, G., Villegas, M., Suarez, M., & Robert, F. (2010). *Survey on spectrum utilization in Europe: measurements, analyses and observations. Cognitive Radio Oriented Wireless Networks & Communications* (pp. 1–5). CROWNCOM.

Valentin, S., von Malm, H., & Karl, H. (2006). *Evaluating the GNU software radio platform for wireless testbeds. Technical Report*. University of Paderborn.

Velayos, H., Aleo, V., & Karlsson, G. (2004). *Load balancing in overlapping wireless LAN cells*. IEEE International Conference on Communications.

Venkataraman, J., & Haenggi, M. (2004). Optimizing the throughput in random wireless Ad hoc networks. *Proceedings of 42st Annual Allerton Conference on Communication, Control, and Computing*.

Visotsky, E., Kuffner, S., & Peterson, R. (2005). On collaborative detection of TV transmissions in support of dynamic spectrum sharing. *2005 First IEEE International Symposium on New Frontiers in Dynamic Spectrum Access Networks*, (pp. 338-345).

Viswanathan, H., Venkatesan, S., & Huang, H. (2003). Downlink capacity evaluation of cellular networks with known-interference cancellation. *IEEE Journal on Selected Areas in Communications*, *21*(5), 802–811. doi:10.1109/JSAC.2003.810346

Viswanathan, R., & Varshney, P. K. (1997). Distributed detection with multiple sensors I. Fundamentals. *Proceedings of the IEEE*, *85*(1), 54–63. doi:10.1109/5.554208

Von Neumann, J., & Morgenstern, O. (1944). *Theory of games and economic behavior*. Princeton University Press.

Von Neumann, J., Morgenstern, O., Rubinstein, A., & Kuhn, H. W. (2007). *Theory of games and economic behavior*. Princeton University Press.

Vu, M., Devroye, N., & Tarokh, V. (2009a). On the primary exclusive region of cognitive networks. *IEEE Transactions on Wireless Communications*, *8*(7), 3380–3385. doi:10.1109/TWC.2009.080454

Vu, M., & Tarokh, V. (2009b). Scaling law of single-hop cognitive networks. *IEEE Transactions on Wireless Communications*, *8*(8), 4098–4097. doi:10.1109/TWC.2009.080453

Wang, F. (2009). *Adaptive weighted scheduling in cognitive radio networks*. Waterloo, Canada: University of Waterloo. Retrieved from http://hdl.handle.net/10012/4452

Wang, F., & Wang, W. (2010). *Robust beamforming and power control for multiuser cognitive radio network*. IEEE Global Telecommunications Conference.

Wang, W., & Sun, Y. (2010). *Cross-layer attack and defense in cognitive radio networks*. IEEE Global Telecommunications Conference.

Wang, W., Li, H., & Sun, Y. (2009). *Attack-proof collaborative spectrum sensing in cognitive radio networks*. IEEE 43rd Annual Conference on Information Sciences and Systems.

Wang, C., Jiang, C., Li, X.-Y., & Liu, Y. (2010). Multicast throughput for large scale cognitive networks. *Wireless Networks*, *16*(7), 1945–1960. doi:10.1007/s11276-010-0237-3

Wang, D., Zeng, X. J., & Keane, J. A. (2006). A survey of hierarchical fuzzy systems. *International Journal of Computational Cognition*, *4*(1), 18–29.

Wang, J., Ghosh, M., & Challapali, K. (2011). Emerging cognitive radio applications: A survey. *IEEE Communications Magazine*, 74–81. doi:10.1109/MCOM.2011.5723803

Wang, L.-X. (1999). Analysis and design of hierarchical fuzzy systems. *IEEE Transactions on Fuzzy Systems*, *7*(5), 617–624. doi:10.1109/91.797984

Wang, P., Jiang, H., & Zhuang, W. (2006). IEEE 802.11e enhancement for voice service. *IEEE Wireless Communications*, *13*(1), 30–35. doi:10.1109/MWC.2006.1593522

Wang, X., & Yu, H. (2005). How to break MD5 and other hash functions. *Advances in Cryptology–EUROCRYPT*, *2005*, 19–35.

Wei, J., & Zhang, X. (2010). *Energy-efficient distributed spectrum sensing for wireless cognitive radio networks*. IEEE Conference on Computer Communications Workshops.

Weingart, T., Sicker, D., & Grunwald, D. (2006). Evaluation of cross-layer interactions for reconfigurable radio platforms. *First International Workshop on Technology and Policy for Accessing Spectrum: WiCon '06*.

White, L. (2000). Propertyzing the electromagnetic spectrum: Why it is important, and how to begin. *Media Law and Policy*, *19*, 29–30.

Wieselthier, J., Barnhart, C., & Ephremides, A. (1994). Neural network approach to routing without interference in multihop radio networks. *IEEE Transactions on Communications*, *42*(1), 166–177. doi:10.1109/26.275311

Willkomm, D., & Wolisz, A. (2010). Efficient QoS support for secondary users in cognitive radio systems. *IEEE Wireless Communications*, *17*(4), 16–23. doi:10.1109/MWC.2010.5547917

Wong, S., & Wassell, I. (2002). Application of game theory for distributed dynamic channel allocation. *Proceedings of IEEE VTC*, (pp. 404-408).

Won, H., Cai, H., Eun, D., Guo, K., Netravali, A., Rhee, I., & Sabnani, K. (2009). Multicast scheduling in cellular data networks. *IEEE Transactions on Wireless Communications*, *8*(9), 4540–4549. doi:10.1109/TWC.2009.080330

Wyner, A. D. (1994). Shannon-theoretic approach to a Gaussian cellular multiple-access channel. *IEEE Transactions on Information Theory*, *40*(6), 1713–1727. doi:10.1109/18.340450

Xia, X., & Liang, Q. (2005). Bottom-up cross layer optimization for mobile ad hoc networks. *IEEE Military Communications Conference (MILCOM)*, (pp. 1-7).

Xiao, H., Seah, W. K., Lo, A., & Chua, K. C. (2000). *A flexible quality of service model for mobile ad hoc networks* (pp. 445–449). IEEE VTC-Spring.

Xiao, L., Greenstein, L., Mandayam, N., & Trappe, W. (2008). Using the physical layer for wireless authentication in time variant channels. *IEEE Transactions on Wireless Communications*, *7*(7), 2571–2579. doi:10.1109/TWC.2008.070194

Xiao, L., Greenstein, L., Mandayam, N., & Trappe, W. (2009). Channel-based detection of sybil attacks in wireless networks. *IEEE Transactions on Information Forensics and Security*, *4*(3), 492–503. doi:10.1109/TIFS.2009.2026454

Xie, X., & Guo, N. (2010). UWB relay with physical layer security enhancement. *2010 IEEE International Conference on Ultra-wideband (ICUWB)*, Vol. 2, (pp. 1–4).

Xu, L., Tonjes, R., Paila, T., Hansmann, W., Frank, M., & Albrecht, M. (2000). DRiVE-ing to the internet: Dynamic radio for IP services in vehicular environments. *The 25th Annual IEEE Conference on Local Computer Networks*, (pp. 281–289).

Yan, Z., Ma, Z., Cao, H., Li, G., & Wang, W. (2008). Spectrum sensing, access and coexistence testbed for cognitive radio using USRP. *IEEE International Conference on Circuits and Systems for Communications*, (pp. 270-274).

Yang, C. G., & Li, J. D. (2009). *Joint resource management framework under IEEE P1900.4: A network utility maximization approach*. Poster 8, MobiCom 2009.

Yang, C. G., & Li, J. D. (2010). Dynamic resource management mechanism design for heterogeneous networks under IEEE1900.4 framework: A network utility maximization perspective. *2010 IEEE WCNIS*, (pp. 333–337).

Yang, C. G., Li, J. D., & Chen, D. (2009). *Optimal balancing between efficiency and fairness for resource management in cognitive radio networks: A dynamic game-theoretic approach*. CCNC 2009, Las Vegas, USA.

Yang, K., Gondal, I., Qiu, B., & Dooley, L. S. (2007). *Combined SINR based vertical handoff algorithm for next generation heterogeneous wireless networks*. IEEE Global Telecommunications Conference.

Yang, W., Cai, Y., & Xu, Y. (2007). A fuzzy collaborative spectrum sensing scheme in cognitive radio. *International Symposium on Intelligent Signal Processing and Communication System*, (pp. 566-569).

Yang, C. G., Li, J. D., & Tian, Z. (2010). Power control for cognitive radio networks with coupled constraints: A cooperative game-theoretic perspective. *IEEE Transactions on Vehicular Technology*, *59*(4), 1696–1706. doi:10.1109/TVT.2009.2039502

Yarkan, S., & Arslan, H. (2007). Binary time series approach to spectrum prediction for cognitive radio. *Proceedings of IEEE 66th Vehicular Technology Conference*, (pp. 1563-1567).

Yau, K.-L. A., Komisarczuk, P., & Teal, P. D. (2008). *C2net: A cross-layer quality of service (QoS) architecture for cognitive wireless ad hoc network*. IEEE Australasian Telecommunication Networks and Applications Conference (ATNAC'08).

Yong, Y., Lifeng, W., & Quan, Y. (2009). *Cross-layer architecture in cognitive ad hoc networks*. WRI International Conference on Communications and Mobile Computing.

Yoo, T., & Goldsmith, A. (2006). Capacity and power allocation for fading MIMO channels with channel estimation error. *IEEE Transactions on Information Theory*, *52*(5), 2203–2214. doi:10.1109/TIT.2006.872984

Yu, C., He, Y., & Quan, T. (2008). Frequency spectrum prediction method based on EMD and SVR. *Proceedings of International Conference on Intelligent Systems Design and Applications*, (p. 3).

Yu, J., Zhang, C., Hu, Z., Lin, F., Guo, N., Wicks, M., et al. (2011). *Cognitive radio network as wireless sensor network (I): Architecture, testbed, and experiment*. National Aerospace & Electronics Conference.

Yücek, T., & Arslan, H. (2009). A survey of spectrum-sensing algorithms for cognitive radio applications. *IEEE Communications Surveys & Tutorials*, *11*(1), 116–130. doi:10.1109/SURV.2009.090109

Yu, F. R., Sun, B., Krishnamurthy, V., & Ali, S. (2010). Application layer QoS optimization for multimedia transmission over cognitive radio networks. *Wireless Networks*, *17*(2), 371–383. doi:10.1007/s11276-010-0285-8

Yu, H., Shen, Z., Miao, C., Leung, C., & Niyato, D. (2010). A survey of trust and reputation management systems in wireless communications. *Proceedings of the IEEE*, *98*(10), 1755–1772. doi:10.1109/JPROC.2010.2059690

Yu, J., & Choi, S. (2006). Comparison of modified dual queue and EDCA for VoIP over IEEE 802.11 WLAN. *European Transactions on Telecommunications*, *17*(3), 371–382. doi:10.1002/ett.1128

Yu, W., & Ciof□, J. M. (2006). Constant-power waterfilling: Performance bound and low-complexity implementation. *IEEE Transactions on Communications*, *54*(1), 23–28. doi:10.1109/TCOMM.2005.861678

Zadeh, L. A. (1965). Fuzzy sets. *Information and Control*, *8*(3), 338–353. doi:10.1016/S0019-9958(65)90241-X

Zakhour, R., & Gesbert, D. (2009). Coordination on the MISO interference channel using the virtual SINR framework. *Proceedings of the ITG Workshop on Smart Antennas (WSA)*, (pp. 75-81).

Zakhour, R., Ho, Z. K. M., & Gesbert, D. (2009). Distributed beamforming coordination in multicell MIMO channels. *Proceedings of the IEEE Vehicular Technology Conference (VTC)*, (pp. 1-5).

Zenon, C., & Ruckshan, W. (2010). Security threats in cognitive radio applications. *2010 14th International Conference on Intelligent Engineering Systems (INES)*, (pp. 209-215).

Zettervall, H. (2011). *Fuzzy and rough theory in the treatment of elderly gastric cancer patients*. Licentiate Dissertation, Karlskrona, Sweden.

Zettervall, H., Rakus-Andersson, E., & Forssell, H. (2011). The Mamdani controller in prediction of the survival length in elderly gastric patients. *Proceedings of Bioinformatics 2011*, Rome, (pp. 283–286).

Zhai, H., Chen, X., & Fang, Y. (2005). How well can the IEEE 802.11 wireless LAN support quality of service? *IEEE Transactions on Wireless Communications*, *4*(6), 3084–3094. doi:10.1109/TWC.2005.857994

Zhang, L., Zeng, K., & Mohapatra, P. (n.d.). Opportunistic spectrum scheduling for mobile cognitive radio networks in white space. *IEEE Wireless Communications and Networking Conference (WCNC)*, (pp. 844–849).

Zhang, X., & Li, C. (2009). The security in cognitive radio networks: A survey. *International Wireless Communications and Mobile Computing Conference*, (pp. 309-313).

Zhang, Y., Xu, G., & Geng, X. (2008). Security threats in cognitive radio networks. *10th IEEE International Conference on High Performance Computing and Communications*, (pp. 1036-1041).

Zhang, H., & Dai, H. (2004). Cochannel interference mitigation and cooperative processing in downlink multicell multiuser MIMO networks. *EURASIP Journal on Wireless Communications and Networking*, *2*, 222–235.

Zhang, H., Mehta, N. B., Molisch, A. F., Zhang, J., & Dai, H. (2008). Asynchronous interference mitigation in cooperative base station systems. *IEEE Transactions on Wireless Communications*, *7*(1), 155–165. doi:10.1109/TWC.2008.060193

Zhang, L., Soong, B., & Xiao, W. (2007). An integrated cluster based multi-channel MAC protocol for mobile ad hoc networks. *IEEE Transactions on Wireless Communications*, *6*(11), 3964–3974. doi:10.1109/TWC.2007.06038

Zhang, R., & Liang, Y.-C. (2008). Exploiting multi-antennas for opportunistic spectrum sharing in cognitive radio networks. *IEEE J. Sel. Topics Signal Process.*, *2*(1), 88–102. doi:10.1109/JSTSP.2007.914894

Zhang, X., & Li, C. (2010). Constructing secured cognitive wireless networks: Experiences and challenges. *Wireless Communications and Mobile Computing*, *10*(1), 50–69. doi:10.1002/wcm.878

Zhang, Y., Zhang, Q., & Wu, S. (2010). Entropy-based robust spectrum sensing in cognitive radio. *IET Communications*, *4*(4), 428–436. doi:10.1049/iet-com.2009.0389

Zhao, Q., & Swami, A. (2007). A survey of dynamic spectrum access: signal processing and networking perspectives. *IEEE International Conference on Acoustics, Speech, and Signal Processing (ICASSP): Special session on Signal Processing and Networking for Dynamic Spectrum Access.*

Zhao, Q., Geirhofer, S., Tong, L., & Sadler, B. M. (2008). Opportunistic spectrum access via periodic channel sensing. *IEEE Transactions on Signal Processing*, *56*(2), 785–796. doi:10.1109/TSP.2007.907867

Zhao, Q., & Sadler, B. (2007). *A survey of dynamic spectrum access: Signal processing, networking, and regulatory policy. IEEE Signal Processing Magazine: Special Issue on Resource-Constrained Signal Processing*. Communications, and Networking.

Zhou, X., Ma, J., Li, Y., Kwon, Y. H., Soong, A. C. K., & Zhao, G. (2008). *Probability based transmit power control for dynamic spectrum access.* IEEE 3rd Symposium on New Frontiers in Dynamic Spectrum Access Networks.

Zhou, C., Guo, N., & Qiu, R. (2009). Time-reversed ultra-wideband (UWB) multiple input multiple output (MIMO) based on measured spatial channels. *IEEE Transactions on Vehicular Technology*, *58*(6), 2884–2898. doi:10.1109/TVT.2008.2012109

Zhu, L., & Mao, H. (2010). *Research on authentication mechanism of cognitive radio networks based on certification authority*. International Conference on Computational Intelligence and Software Engineering.

Zhu, L., & Mao, H. (2010). *Research on security issues of cognitive radio networks*. The Conference on Web Based Business Management (WBM 2010).

Zhu, L., & Mao, H. (2011). *An efficient authentication mechanism for cognitive radio networks.* 2011 Asia-Pacific Power and Energy Engineering Conference.

Zhu, L., & Mao, H. (2011). *Unified layered security architecture for cognitive radio networks*. Conference on Power and Energy Engineering Conference.

Zhu, L., & Zhou, H. (2008). Two types of attacks against cognitive radio network MAC protocols. *International Conference on Computer Science and Software Engineering*, Vol. 4, (pp. 1110-1113).

Zimmermann, H. J. (2002). *Fuzzy set theory and its applications* (4th ed.). Boston, MA: Kluwer Academic Publishers.

About the Contributors

Meng-Lin Ku received the B.S., M.S. and Ph.D. degrees in Communication Engineering from National Chiao Tung University, Hsinchu, Taiwan, in 2002, 2003, and 2009, respectively. Between 2009 and 2010, he held post-doctoral positions in the Department of Electrical and Computer Engineering at National Chiao Tung University and in the School of Engineering and Applied Sciences at Harvard University (in chronological order). In August 2010, he became the faculty member of the Department of Communication Engineering with National Central University, currently holding a position as an Assistant Professor. He is a member of the IEEE Communications Society and served as technical program committee member for numerous international conferences such as VTC, ICCT, et cetera. He is the Secretary of the 2012 IEEE International Conference on Telecommunications for Intelligent Transport Systems. His current research interests are in the wide areas of next-generation mobile and wireless communications, cognitive radios, and optimization for radio access.

Jia-Chin Lin received a B.S. and a M.S. degree from the Department of Electronics Engineering at National Chiao Tung University (NCTU) in Taiwan in 1994 and 1995, respectively, and a Ph.D. degree from National Taiwan University (NTU) in Taiwan in 1998. Jia-Chin Lin was in the obligatory military service from July 1998 to March 2000 and then joined the Microelectronics and Information Systems Research Center at NCTU as a Research Assistant Professor. In February 2001, he joined the Department of Electrical Engineering at National Chi Nan University (NCNU) in Taiwan as an Assistant Professor. In August 2004, he was promoted to serve as an Associate Professor. From July to August of 2004 and from August 2005 to July 2006, he held the visiting Associate Professorship in the Department of Electrical Engineering at Stanford University. In August 2006, he joined the faculty in the Department of Communication Engineering at National Central University (NCU) in Taiwan as an Associate Professor. In August 2008, he was promoted to serve as a Full Professor. In January 2011, he became a Distinguished Professor. Since August 2011, he has been serving as the Chairperson in the Department of Communication Engineering at NCU. His research interests include wireless transmission technologies, signal processing for communications and synchronization techniques. Jia-Chin Lin has been serving as an Associate Editor for the *IEEE Transactions on Vehicular Technology* since 2008. He is serving as an Associate Editor for the *IEEE Signal Processing Letters* and as the Lead Guest Editor for the *IEEE Transactions on Vehicular Technology* (Special Issue on Telematics Advances for Vehicular Communication Networks) in 2011. He has served as a technical program committee member for numerous IEEE conferences and as a reviewer for IEEE and IET journals. Jia-Chin Lin has published more than 20 papers in first-class international journals, presented more than 20 papers at international (principal) conferences and has one patent. Almost all of the publications are single-authored and based on his independent work. He has also served as the Vice Chair of Meetings and Conferences Committee for the 2008-2009 IEEE AP ComSoc. Jia-Chin Lin has won many awards, including the Dr. Wu Da-You Research Award from the National Science Council (NSC), Executive Yuan, the 2003 Award for Outstanding Faculty from the Ministry of Education, the Young Scientist Award issued by URSI, the Outstanding Researcher Award from the IEEE ComSoc APB, the Investigative Research Award from the Pan Wen Yuan Foundation, and the Taiwan Merit Scholarships Program from Executive Yuan. Professor Lin also won the 2009 Ten Outstanding Young Persons Award of Taiwan. Jia-Chin Lin is a Fellow of the IET and a Senior Member of IEEE.

* * *

Prathima Agrawal is currently the Samuel Ginn distinguished professor of electrical and computer engineering and the director of the Wireless Engineering Research and Education Center, Auburn University, Auburn, Alabama. She is also the Auburn site director of Wireless Internet for Advanced Technology (WICAT). WICAT is a National Science Foundation (NSF) Center under the Industry University Cooperative Research Center (IUCRC) program. Prior to her present positions, she was an assistant vice president of the Network Systems Research Laboratory and an executive director of the Mobile Networking Research Department, Telcordia Technologies (formerly Bellcore), Morristown, New Jersey, where she worked from 1998 to 2003. From 1978 to 1998, she worked in various capacities at AT&T Lucent Bell Laboratories, including creating and serving as head of the Networked Computing Research Department, Murray Hill, New Jersey. Her research interests include computer networks, mobile and wireless computing, and communication systems. She has published more than 200 papers in IEEE and other journals and conference proceedings in areas related to networks and VLSI. She holds 51 US patents. She received the Distinguished Alumnus Award from the Indian Institute of Science in 2006, the Telcordia CEO Award in 2000, and the Distinguished Member of the Technical Staff Award from AT&T Bell Laboratories in 1985. She is the recipient of the IEEE Computer Society's Distinguished Service Award in 1990 and the IEEE Third Millennium Medal in 2000. She is a Fellow of the IEEE and the Institution of Electronics and Telecommunications Engineers (IETE), India, and a member of the ACM. She received the BE and ME degrees in Electrical Communication Engineering from the Indian Institute of Science, Bangalore, India, and the PhD degree in Electrical Engineering from the University of Southern California in 1977.

Yazeed Al-Sbou received his PhD in Computer Engineering in 2006 from Sheffield Hallam University, Sheffield, UK. He received his MSc (2001) and BSc (1997) in Communications Engineering from Jordan University and Mu'tah University, respectively. Dr. Alsbou has more than seven years of teaching experience in higher education. Dr. Al-Sbou is currently an Assistant Professor at Faculty of Engineering of Mu'tah University, Jordan since 2006. His research interests concern quality of service, computer networks measurement, performance analysis, routing in wireless ad hoc networks, artificial intelligent, ad hoc networks, cognitive radio, and simulations.

Sk. Shariful Alam received the B.Sc. in Electrical and Electronic Engineering from Chittagong University of Engineering and Technology, Bangladesh in 2003. In 2010, he received his M. Sc. in Telecommunication Engineering from the University of Trento, Italy. From 2011, he is a Ph.D. student under the Department of Biophysical and Electronic Engineering (DIBE) at the University of Genoa,

Italy. He received the international summer school grant for cognitive wireless communications at Florence, July, 2011. Since 2004, he has been a faculty member of the Department of Electronics and Communication Engineering with the Khulna University of Engineering and Technology, Bangladesh. His recent research interests include spectrum sensing of cognitive radio networks, dynamic spectrum management for the distributed cognitive radio networks, and MIMO interference channels.

Alireza Attar was a Post-Doctoral Fellow at the University of British Columbia, 2009-2011. He received his PhD at King's College London where he was awarded the Mobile VCE scholarship. He has been an active IEEE member, serving in various capacities including reviewer for various Communications and Signal Processing society journals, TPC member for major ComSoc conferences, as well as guest editor for one *IEEE JSAC* and one *IEEE TVT* special issues.

Moussa Ayyash received his Ph.D. degree (2005) in Electrical Engineering (Communications and Networking) from Illinois Institute of Technology (IIT), Chicago-USA. Dr. Ayyash has more than ten years of teaching experience in higher education. He is currently an Assistant Professor at Chicago State University/USA. He worked for Motorola, Inc. as a Software Engineer. His research interests include wireless sensor and ad hoc networks, quality of service, routing, graph theory, grid computing, ad hoc networks topology management, cognitive radio, and Wi-Fi interference studies. Dr. Ayyash is a member of the IEEE and holds several industrial certifications in the area of wired and wireless networking.

Alireza Babaei received his BS and MS degrees in Electrical Engineering from K. N. Toosi University of Technology and Iran University of Science and Technology, Tehran, Iran in 2003 and 2005, respectively and his Ph.D. degree in Electrical and Computer Engineering from George Mason University (GMU), Fairfax, Virginia, USA in 2009. He is the recipient of an outstanding graduate student award from GMU and is currently a Postdoctoral Fellow at the Electrical and Computer Engineering Department of Auburn University. He has served as a guest editor for *IEEE Wireless Communications* magazine. His areas of active research are modeling and performance evaluation of wireless networks, information theory, and cognitive radio networks.

Zhe Chen received the B.S. degree in Telecommunications Engineering from Northeastern University, Shenyang, China, in 2000, the M.S. degree in Signal and Information Processing from Hangzhou Dianzi University, Hangzhou, China, in 2003 and the Ph.D. degree in Electrical and Computer Engineering from Tennessee Technological University, Cookeville, Tennessee, USA, in 2011. He is currently an Associate Professor with the College of Information Science and Engineering, Northeastern University, Shenyang, China. From 2003 to 2008, he was an Algorithm Engineer with UTStarcom Inc, Shanghai, China, a Research Engineer with STMicroelectronics, Shanghai, and a Senior System Engineer with Huaya Microelectronics Inc, Shanghai. His research interests include signal processing and cognitive radio.

Lokesh Chouhan is full time Research Scholar in ABV-IIITM, Gwalior. He is Doing PhD on "Design and analysis of MAC protocols for Cognitive Radio Networks." He has completed her Master of Technology (Information Technology) from SOIT-RGPV Bhopal. He received her Bachelor of Engineering (Computer Science and Engineering) from SGSITS, Indore. He has also experience in industry and academics. He has also published several research papers in International journal and conferences. His research interest is wireless communication, cognitive radio, and mobile ad hoc networks.

Wei-Ho Chung was born in Kaohsiung, Taiwan, in 1978. He received the B.Sc. and M.Sc. degrees in Electrical Engineering from National Taiwan University, Taipei City, Taiwan, in 2000 and 2002 respectively, and the Ph.D. degree in Electrical Engineering Department at University of California, Los Angeles, USA, in 2009. From 2002 to 2005, he was a system engineer at ChungHwa TeleCommunications Company. In 2008, he was a research intern working on CDMA systems in Qualcomm Inc. Since January 2010, Dr. Chung has been a faculty member holding the position as an assistant research fellow in Research Center for Information Technology Innovation, Academia Sinica, Taiwan. His research interests include wireless communications, signal processing, statistical detection and estimation theory, and networks. Dr. Chung received the Taiwan Merit Scholarship, sponsored by the National Science Council of Taiwan, from 2005 to 2009.

Rajni Dubey is an Assistant Professor in SRCEM, Banmore, Gwalior, and Madhya Pradesh, India. She has completed her Master of Technology (Information Technology) from SOIT-RGPV Bhopal. She received her Bachelor of Engineering (Computer Science and Engineering) from ITM, Gwalior. She has also published some research paper in international journal and conferences. Her present research interest is cognitive radio, mobile ad hoc networks, and related security issues.

Omid Namvar Gharehshiran received his B.Sc. from Sharif University of Technology, Tehran, Iran, in 2007, and the M.A.Sc. from University of British Columbia, Vancouver, Canada, in 2010, both in Electrical Engineering. He is currently a member of the Statistical Signal Processing Laboratory, Department of Electrical and Computer Engineering, University of British Columbia, working towards his Ph.D. He has been awarded the UBC Four Year Fellowship in 2010. His research interests span stochastic optimization, game theory, and learning in games with applications in wireless communication and sensor networks.

Shimin Gong received his B.E. and M.E. degrees from the department of Electronics and Information Engineering, Huazhong University of Science and Technology, China in 2008 and 2011, respectively. He is currently pursuing his Ph.D. in Nanyang Technological University, Singapore. He was a visiting student for three month in 2011 at the Chinese University of Hong Kong. His research interests include cognitive radio network and its application, resource allocation in wireless networks, convex and robust optimization techniques.

Terry N. Guo received the M.S. degree in Telecommunications Engineering from the Beijing University of Posts and Telecommunications in 1990 and the Ph.D. degree in Communications and Electronic Systems from the University of Electronic Science and Technology of China in 1997. He has over 15 years of industrial and academic experience balanced in R&D, teaching, project management, and laboratory work. From January 1997 to December 1999, he was a Post-Doctoral Researcher in University of California, San Diego. During the period from December 1999 to February 2003, he was with Golden Bridge Technology, Inc., as a System Engineer, and then with Ansoft Corp. as a Research Engineer. In January 2004, he joined Center for Manufacturing Research, Tennessee Technological University, and he is now a senior R&D Engineer at the center. His research interests include advanced radio systems, statistic signal processing, optimization and its applications, and implementation impacts on system performance.

Robert W. Heath Jr. received the Ph.D. degree in Electrical Engineering from Stanford University in 2002. He is a Professor at The University of Texas at Austin, President and CEO of MIMO Wireless Inc., and Chief Innovation Officer at Kuma Signals LLC. He has been an Editor, Associate Editor, or lead guest editor for several special issues in journals. He was a member of the IEEE Signal Processing Society SPCOM Technical Committee and Chair of the IEEE COMSOC Communications Technical Theory Committee. He is a co-author of several best student paper awards at conferences and was co-recipient of the 2010 *EURASIP Journal on Wireless Communications and Networking* best paper award. He is the recipient of the David and Doris Lybarger Endowed Faculty Fellowship in Engineering. He is a licensed Amateur Radio Operator, a registered Professional Engineer in Texas, and is a Fellow of the IEEE.

Winston W. L. Ho had been working on space-time wireless and broadband communications in the Institute for Infocomm Research, under the Agency for Science, Technology and Research since 2008. From 2011, he has been researching on signal processing for compressed sensing in the Technion, Israel Institute of Technology, with Prof. Y. C. Eldar. He obtained his Bachelors and Ph.D. degrees, both in Electrical Engineering, and both from the National University of Singapore, in 2004 and 2008, respectively. His Ph.D. advisors were Prof. K. M. Lye and Prof. Y. C. Liang. He served in the Executive Committee of the IEEE Vehicular Technology Society Singapore Chapter in 2009 and 2010. During his term, he made significant contribution towards the successful organization of the IEEE International Conference on Communication Systems 2010.

Chih-Wei Huang is Assistant Professor of Communication Engineering heading the Information Processing and Communications (IPC) Laboratory of National Central University, Taiwan, since 2010. He received his Ph.D. from University of Washington, Seattle, in 2009 on cross-layer resource management for multimedia over wireless networks. From 2006 to 2009, Dr. Huang was intern researcher at Siemens Corporate Research and Microsoft Research. His current research interests include wireless networking, multimedia communications, and digital signal processing. He received an M.S. degree from Columbia University, and a B.S. from National Taiwan University all in Electrical Engineering.

Jianwei Huang is an Assistant Professor in the Department of Information Engineering at the Chinese University of Hong Kong. He received Ph.D. in Electrical and Computer Engineering from Northwestern University in 2005, and worked as a Postdoc Research Associate at Princeton University during 2005-2007. Dr. Huang currently leads the Network Communications and Economics Lab (ncel.ie.cuhk.edu.hk), with focus on nonlinear optimization and game theoretical analysis of communication networks. He is the recipient of the IEEE Marconi Prize Paper Award in Wireless Communications in 2011, the International Conference on Wireless Internet Best Paper Award 2011, the IEEE GLOBECOM Best Paper Award in 2010, the IEEE ComSoc Asia-Pacific Outstanding Young Researcher Award in 2009, and Asia-Pacific Conference on Communications Best Paper Award in 2009. He is an Editor of *IEEE Journal on Selected Areas in Communications - Cognitive Radio Series* and *IEEE Transactions on Wireless Communications*. He is a senior member of the IEEE.

Bijan Jabbaeri is a Professor of Electrical and Computer Engineering at George Mason University, Fairfax, VA, USA, and an affiliated faculty with ENST-Paris, France. Dr. Jabbari served as the Editor for Wireless Multiple Access for the *IEEE Transactions on Communications*, was the International Division Editor for Wireless Communications of the *Journal of Communications and Networks*, and was on the editorial board of *Proceedings of the IEEE*. He is a coeditor of recent books on Multiaccess, Mobility and Teletraffic (Kluwer Publishing, Volume I and IV-VI). He is the past chairman of the IEEE Communications Society technical committee on Communications Switching and Routing. He is a Fellow of IEEE and IEE, is a recipient of the IEEE Millennium Medal in 2000 and the Washington DC Metropolitan Area Engineer of the Year Award, in 2003. He founded innovative laboratories for Internet and wireless communications research and is helping the industry adoption of new technologies. He continues research on multi-access communications and high performance networking. He received the Ph.D. degree in Electrical Engineering from Stanford University, Stanford, CA.

Hany Kamal received his PhD degree on Dynamic Spectrum Access (DSA) techniques for cellular systems, from the "Computing and Networks" department at Telecom ParisTech, France, 2010. He has received his B.Sc. degree in Communication Engineering from Alexandria University, Egypt, on 1999. He received his M.Sc. from TELECOM ParisTech, France, in 2002, where he worked on channel estimation for wireless LAN systems. After his Master's, he worked on overload algorithms for 3G networks within the R&D team at Alcatel (Velizy). He worked as a radio design engineer in Alcatel (-Lucent) mainly on indoor design projects for 2G and 3G cellular systems. He also joined the WiMAX Advanced Radio Engineering group within the company late 2006. Hany Kamal has co-authored several methodology-documents within Alcatel-Lucent. He also co-authored several deliverables within the French research project Urban Planning for Radio Communication (URC), 2008. His current research interests are in wireless networks, dynamic spectrum access techniques, and in Intelligent Transportation Systems (ITS). He is currently a co-founder at Rakedet, Egypt, an innovative technology start-up with interests in wireless systems' architecture and design.

Prabhjot Kaur did her Masters of Engineering from Punjab University, Chandigarh and has submitted her Ph.D. thesis on MAC for Dynamic Spectrum Access in Cognitive Radio Networks at National Institute of Technology, Jalandhar, India. She is working as Assistant Professor in Electronics and Communication department with ITM University, Gurgaon, Haryana, India. Her research interests include dynamic spectrum allocation, ad-hoc networks, green networks, MIMO, software defined radios and cognitive radios. She is member, IEEE and life member of IETE and ISTE societies. She has also been awarded with a research grant from AICTE, Govt. of India under research promotion scheme in April 2008 and an international travel grant for attending IEEE conference by Department of Science and Technology, Govt. of India.

Jamil Y. Khan received his Ph.D. from the Department of Electronic and Electrical Engineering specializing in Communications Engineering from the University of Strathclyde, Glasgow, UK in 1991. Since his Ph.D. he has worked as a research assistant in the University of Strathclyde 1991-92 in the European research program RACE, then worked as a Lecturer/Senior in the Massey University, New Zealand 1992-1999. In 1999 he joined the University of Newcastle as a Senior Lecturer in Telecommunications Engineering. He established the Telecommunications Engineering teaching and research programs in the university. Currently he is an Associate Professor in the School of Electrical Engineering & Computer Science. His main research interest areas are cognitive and cooperative wireless networks, smart grid communications, wireless network architecture, wireless sensor and body area networks. Jamil Khan is a senior member of the IEEE Communications Society and a member of the ACM.

Arun Khosla received his PhD degree from Indraprastha University, Delhi in the field of Information Technology. He is associate professor in Electronics and Communication Engineering department and is presently heading this department at National Institute of Technology, Jalandhar. India. His areas of interest are fuzzy modeling, biologically inspired computing, and high performance computing. Dr. Khosla is member, IEEE and has been reviewer for various IEEE and other national and international conferences and journals. He also serves on the editorial board of *International Journal of Swarm Intelligence Research.*

Vikram Krishnamurthy was born in 1966. He received his Bachelor's degree from the University of Auckland, New Zealand in 1988, and Ph.D. from the Australian National University, Canberra, in 1992. He is currently a Professor and holds the Canada Research Chair at the Department of Electrical Engineering, University of British Columbia, Vancouver, Canada. Prior to 2002, he was a chaired Professor at the Department of Electrical and Electronic Engineering, University of Melbourne, Australia, where he also served as deputy head of department. His current research interests include computational game theory, stochastic dynamical systems for modeling biological ion channels and stochastic optimization and scheduling. He has served as associate editor for several journals including *IEEE Transactions Automatic Control, IEEE Transactions on Signal Processing, IEEE Transactions Aerospace and Electronic Systems, IEEE Transactions Nanobioscience,* and *Systems and Control Letters.* In 2009 and 2010 he served as Distinguished Lecturer for the IEEE Signal Processing Society. From 2010 he has served as Editor in Chief of *IEEE Journal Selected Topics in Signal Processing.*

Wlodek J. Kulesza has been appointed Professor in Metrology at University of Kalmar, and then Blekinge Institute of Technology since 2001. He is also Guest Professor at Southeast University in Nanjing and Soochow University in Suzhou. His research focuses on measurement methods and wireless sensor networks for industrial and public applications. Among others he is involved in an interdisciplinary project, Human-Measurement-Quality, where a use a fuzzy neural network facilitates a measurement of fuzzy quantities. Another of his projects, Multidimensional Measurement in Security and Health Care, deals with an environment where distributed multiple sensors and intelligence create an intelligent space. A wireless sensor network is a technology used for the intelligent space.

Jiandong Li was graduated from Xidian University with Bachelor's Degree, Master's Degree, and Ph. D in Communications and Electronic System in 1982, 1985, and 1991, respectively. He is with Xidian University from 1985, an Associate Professor from 1990 to 1994, Professor from 1994, PhD student supervisor from 1995, and Dean of School of Telecommunication Engineering, Xidian University from 1997, respectively. Now, he also serves as Executive Vice Dean Graduate School, Xidian University. Dr. Li is a senior member of IEEE and China Institute of Electronics (CIE) and the fellow of China Institute of Communication (CIC). He was the member of PCN specialist group for China 863 Communication high technology program between Jan. 1993 and Oct. 1994 and from 1999 to 2000. He is also the member of Communication specialist group for The Ministry of Information Industry. His current research interest and projects are funded by the 863 High Tech Project, NSFC, National Science Fund for Distinguished Young Scholars, TRAPOYT, MOE and MOI.

Lucio Marcenaro enjoys over 10 years' of experience in image and video sequence analysis and authored over 30 technical papers related to signal and video processing for computer vision. An Electronic Engineering graduate from Genoa University in 1999, he received his PhD in Computer Science and Electronic Engineering from University of Genoa, Italy in 2003. From 2003 to 2010 he was CEO and development manager at TechnoAware SRL, Genoa, Italy. From March 2011, he became Assistant Professor in Telecommunications for the Faculty of Engineering at the Department of Biophysical and Electronic Engineering at the University of Genoa, Italy. His main current research interests are: video processing for event recognition, detection and localization of objects in complex scenes, distributed heterogeneous sensors ambient awareness systems, ambient intelligence and bio-inspired cognitive systems.

Abbas Mohammed is a Professor of Telecommunications Theory at the School of Engineering, Blekinge Institute of Technology (BTH), Sweden. He was awarded the PhD degree from Liverpool University, UK, in 1992 and the Swedish Docent degree in Radio Communications and Navigation from BTH in 2001. He is the recipient of the Blekinge Research Foundation Award "Researcher of the Year Award and Prize" for 2006. From 1993 to 1996, he was a Research Fellow with the Radio Navigation Group, University of Wales (Bangor), UK. From 1996 to 1998, he was with the University of Newcastle, UK, working on a European collaborative project within the ACTS (Advance Communications Technologies and Services) FP4 Research Programme that investigated 3G Satellite-UMTS systems. He was also employed by Ericsson AB, where he consulted on power control standardization issues for 3G mobile communication systems. He has been a visiting lecturer to several universities. He is a Fellow of The Institution of Electrical Engineering since 2004. In 2006 he received Fellowship of the UK's Royal Institute of Navigation "in recognition of his significant contribution in navigation and in particular to advanced signal processing techniques that have enhanced the capability of Loran-C." He is a Senior Member of IEEE, and a Life Member of the International Loran Association, USA. He is the Editor-in-Chief of *ACEEE International Journal on Communication* (USA) and the *Journal of Selected Areas in Telecommunications* of Cyber journals (Canada). He is a Board Member of the IEEE Signal Processing Swedish Chapter since its inception in 2000, and the UK's IET Teknologia Journal to promote the research from the Middle East/North Africa region. He is an Associate Editor to the *International Journal of Navigation and Observation* and the *Mediterranean Journal of Electronics and Communications,* and a former Editorial Board Member of the *Radio Engineering Journal.* He has also been a Guest Lead Editor for several special issues of international journals. He is the author of over 200 publications, including many invited articles in peer reviewed Journals special issues and over 25 book chapters, in the fields of telecommunications, navigation systems and signal processing. He has also developed techniques for measuring skywave delays in Loran-C receivers and received a Best Paper Award from the International Loran Association, USA, in connection to this work. He is the Swedish representative and member of the management committee to the European Community COST 280, 296, 297, IC0802, IC0902, and IC1102 Projects. From 2002-2010, he was the BTH Director of the Graduate School of Telecommunications (a collaborative consortium in research and education between 4 Swedish universities), chaired by the Royal Institute of Technology (KTH). Professor Mohammed's research interests are in space-time signal processing and MIMO systems, channel modelling, antennas and propagation, satellite and high altitude platform communications, cognitive radio and dynamic spectrum access, and radio navigation systems.

Eng Hwee Ong received his Ph.D. in Electrical Engineering from The University of Newcastle, Australia in 2010. Since 2010, he has been with the Radio Systems Laboratory, Nokia Research Center Helsinki, Finland where he is currently a Senior Researcher. He has been actively involved in the standardization activities of upcoming IEEE 802.11 amendments, and 3GPP Release 12 and beyond study items. His research interests are in areas of IEEE 802.11xx WLAN, 3GPP LTE/LTE-A, composite wireless networks, network selection, quality of service, radio resource management, and architectural design for future wireless networks. He has (co)authored various publications in scientific journals and proceedings, and pending patent applications in the areas of wireless networking and mobile communications. He is a member of the IEEE and Wi-Fi Alliance.

Pavol Polacek is a member of the Information Processing and Communications Laboratory at the Department of Communication Engineering at the National Taiwan University in Zhongli, Taiwan, where he is currently pursuing a Ph.D. degree since 2010. He received his B.S. and M.S. degrees both in Communication Engineering in 2007 and 2009, respectively, from the Slovak University of Technology in Bratislava, Slovakia. He is currently involved in research of opportunistic multicasting in cognitive radio networks. His research interests include cognitive radio, multimedia communications, and video quality of service. He coauthored several conference papers and a book chapter.

Tony Q. S. Quek received the Ph.D. degree in Electrical Engineering and Computer Science from Massachusetts Institute of Technology in 2008. Currently, he is an Assistant Professor with the Information Systems Technology and Design Pillar at Singapore University of Technology and Design (SUTD). He is also a Principal Investigator and Scientist with the Institute for Infocomm Research. He is currently an Editor for the IEEE Transactions on Communications, the IEEE Wireless Communications Letters, the Wiley Journal on Security and Communication Networks, and the Transactions on Emerging Telecommunications Technologies. He received the Singapore Government Scholarship in 1993, Tokyu Foundation Fellowship in 1998, and the A*STAR National Science Scholarship in 2002. He was honored with the 2008 Philip Yeo Prize for Outstanding Achievement in Research, the IEEE Globecom 2010 Best Paper Award, the 2011 JSPS Invited Fellow for Research in Japan, and the 2012 CAS Fellowship for Young International Scientist.

Elisabeth Rakus-Andersson is employed at Blekinge Institute of Technology in Karlskrona, Sweden. Her research focuses on methods of computational intelligence, especially imprecise mathematical theories like fuzzy set theory, rough set theory, immunological systems and neural networks. Having a background in mathematics she develops her own mathematically formalized models and adapts already existing patterns to generate new applications in medicine and technical environments.

Carlo S. Regazzoni received the Laurea degree in Electronic Engineering and the Ph.D. in Telecommunications and Signal Processing from the University of Genoa (UniGE), in 1987 and 1992, respectively. Since 2005 he is a Professor of Telecommunications Systems. Since 1988, Dr. Regazzoni is involved in research on signal and video processing and data fusion in cognitive telecommunication systems. His main research interests are: bio-inspired signal and video processing and recognition, distributed data fusion, signal processing for wireless communications and localization, ambient intelligence, cognitive radio, multimodal intelligent interfaces, and pervasive adaptation in embodied cognitive systems. Since

1998, he is responsible of the Video and Signal Processing for Telecommunications (www.isip40.it) research group at Department of Biophysical and Electronic Engineering (DIBE), at the University of Genoa, Italy. Intelligent Distributed Video Surveillance and Wireless Mobile Communications/Interaction Systems are the main focus of applications studied in this group. Dr. Regazzoni is the coordinator of the Erasmus Mundus Joint Doctorate on Interactive and Cognitive Environments. He is Regional Director at Large for Region8 (Europe, Africa and Middle East) of IEEE Signal Processing Society (SPS) and member at large of IEEE SPS Conference Board in 2011-2012 as well as Chair of the IEEE SPS Italian Chapter. He is author or co-author of more than 70 papers on peer reviewed international journals.

Sanjeev Sharma is an Associate Professor and Head of School Information Technology in RGPV Bhopal, Madhya Pradesh, India. He is also Deputy Registrar (Academic) of RGPV University. He obtained his PhD degree from RGPV University. His research interest is in the field of MANETs. He has published in over 150 national and international journals and conferences papers across India and other countries.

Moin Uddin is serving as Pro-vice Chancellor, Delhi Technological University, Delhi. Prior to this he was Director, National Institute of Technology, Jalandhar, India, and Professor and Head in Electrical Department at Jamia Millia Islamia University, New Delhi. He has over thirty years of teaching and research experience. Fourteen research scholars have completed their PhD under his guidance and four more are pursuing the same. He has designed the computer engineering curriculums of many international and national universities and is among the expert panel of various universities and institutions. He is Senior member, IEEE and life member, ISTE national society and member, board of studies of many institutions. He has completed several projects under Ministry of Human Resource Development and All India Council for Technical Education, Govt. of India.

Ping Wang received the B.E. in 1994 and M.E. degrees in 1997, from Huazhong University of Science and Technology, China, and the Ph.D. degree in 2008 from the University of Waterloo, Canada, all in Electrical Engineering. She is currently an Assistant Professor at School of Computer Engineering, Nanyang Technological University, Singapore. Her current research interests include QoS provisioning and resource allocation in multimedia wireless communications. She was a co-recipient of a Best Paper Award from IEEE ICC 2007. She is an Editor of *IEEE Transactions on Wireless Communications, EURASIP Journal on Wireless Communications and Networking*, and *International Journal of Ultra Wideband Communications and Systems.*

Chungang Yang received the Bachelor and Doctoral Degree in the Department of Electrical and Information Engineering, Xidian University, XiAn, ShaanXi, China in 2006 and 2011, respectively. Now, he is with Xidian University as an Assistant Professor. Dr. Yang is a member of IEEE. His research interests include cognitive radio networks, resource management and network optimization, cognitive media access protocol and algorithm design, and game theory for wireless networks.

Index

S

T

U

V

W

www.ingramcontent.com/pod-product-compliance
Lightning Source LLC
LaVergne TN
LVHW062144200926
848484LV00007B/1853